PETROLEUM GEOSCIENCE

This book is dedicated to our families, Theresa, Josef, Jorunn, Jacob, Catriona (J.G.), Alison, Joanna, Caroline, and Emma (R.S.)

PETROLEUM GEOSCIENCE

Jon Gluyas and Richard Swarbrick

Blackwell
Publishing

BLACKWELL PUBLISHING
350 Main Street, Malden, MA 02148-5020, USA
9600 Garsington Road, Oxford OX4 2DQ, UK
550 Swanston Street, Carlton, Victoria 3053, Australia

The right of Jon Gluyas and Richard Swarbrick to be identified as the Authors of this
Work has been asserted in accordance with the UK Copyright, Designs, and Patents
Act 1988.

First published 2004 by Blackwell Science Ltd

5   2007

Library of Congress Cataloging-in-Publication Data

Gluyas, J. G. (Jon G.)
    Petroleum geoscience / by Jon Gluyas and Richard Swarbrick.
        p. cm.
Includes bibliographical references and index.
    ISBN 978-0-632-03767-4 (alk. paper)
    1. Petroleum—Geology. I. Swarbrick, Richard Edward. II. Title.
TN870.5 .G58 2003
553.2'8—dc21

                                                    2002015310

A catalogue record for this title is available from the British Library.

Set in 9/11½ pt Garamond
by SNP Best-set Typesetter Ltd, Hong Kong
Printed and bound in Singapore
by COS Printers Pte Ltd

The publisher's policy is to use permanent paper from mills that operate a sustainable
forestry policy, and which has been manufactured from pulp processed using acid-free
and elementary chlorine-free practices. Furthermore, the publisher ensures that the text
paper and cover board used have met acceptable environmental accreditation standards.

For further information on
Blackwell Publishing, visit our website:
www.blackwellpublishing.com

BRIEF CONTENTS

FULL CONTENTS

PREFACE

We met at AAPG London 1992 and, unbeknown to each other at the time, we were both facing similar problems with respect to teaching petroleum geoscience within the industry (J.G.) and academia (R.S.). The main problem was the paucity of published information on the basics of the applied science — how the geoscientist working in industry does his or her job, and with which other disciplines the geoscientist interacts. R.S. had already taken steps to remedy this with a proposal for a book on petroleum geoscience sent to Blackwell. The proposal was well received by reviewers and the then editor Simon Rallison. Simon sought an industry-based coauthor and found one in J.G. At that time, J.G. was teaching internal courses at BP to drillers, reservoir engineers, petroleum engineers, and budding geophysicists with a physics background. Simon's invitation was accepted and by early 1994 work had begun.

It seemed like a good idea at the time . . . but the petroleum industry was changing fast, nowhere more so than in the application of geophysics, stratigraphic geology, and reservoir modeling. The use of 3D seismic surveys was changing from being rare to commonplace, 4D time-lapse seismic was being introduced, and multi-component seismic data was also beginning to find common use. Derivative seismic data were also coming to the fore, with the use of acoustic impedance, amplitude versus offset, and the like. The application of sequence stratigraphic principles was becoming the norm. Reservoir models were increasing in complexity manifold, and they were beginning to incorporate much more geologic information than had hitherto been possible.

Alongside the technological changes, there were also changes in the business as a whole. Frontier exploration was becoming less dominant, and many geoscientists were finding themselves involved in the rehabilitation of old oilfields as new geographies opened in the former Soviet Union and South America.

It was tough to keep pace with these changes in respect of writing this book, but as the writing progressed it became even clearer that information on the above changes was not available in textbooks. We hope to have captured it for you! This book is written for final-year undergraduates, postgraduate M.Sc. and Ph.D. students, and nongeologic technical staff within the petroleum industry.

Jon Gluyas
Acorn Oil & Gas Ltd., London,
and University of Durham

Richard Swarbrick
University of Durham
and GeoPressure Technology
Limited, Durham

ACKNOWLEDGMENTS

Many people helped us to bring this book to publication. They supplied material, reviewed chapters, drew figures, donated photographs, searched databases, cajoled weary authors, and altogether insured that you now have opportunity to read this book. We thank Karen Atkinson, Joe Cartwright, Colin Grey, Alan Heward, Andrew Hogg, Mark Hopkins, Stuart Jones, Steve Larter, Rod Laver, Norman Oxtoby, and Dot Payne.

We would like to thank Mike Bowman of BP in particular. Mike encouraged us to continue during difficult times, supplied masses of material, and reviewed the whole manuscript. We also give particular thanks to Ian Vann of BP, who gave permission for the publication of much previously unpublished material from BP.

CHAPTER I

INTRODUCTION

1.1 THE AIM AND FORMAT OF THE BOOK

The aim of this book is to introduce petroleum geoscience to geologists, be they senior undergraduates or postgraduates, and to nongeologists (petrophysicists, reservoir engineers, petroleum engineers, drilling engineers, and environmental scientists) working in the petroleum industry. We define petroleum geoscience as the disciplines of geology and geophysics applied to understanding the origin and distribution and properties of petroleum and petroleum-bearing rocks. The book will deliver the fundamentals of petroleum geoscience and allow the reader to put such information into practice.

The format of the book follows the path known within the oil industry as the "value chain." This value chain leads the reader from frontier exploration through discovery to petroleum production. Such an approach is not only true to the way in which industry works; it also allows the science to evolve naturally from a start point of few data to an end point of many data. It also allows us to work from the larger basin scale to the smaller pore scale, and from the initial superficial analysis of a petroleum-bearing basin to the detailed reservoir description.

Case histories are used to support the concepts and methods described in the chapters. Each case history is a complete story in itself. However, the case histories also form part of the value chain theme. Specific emphasis is placed upon the problems presented by exploration for and production of petroleum. The importance and value of data are examined, as are the costs—both in time and money—of obtaining data.

Petroleum economics are clearly a critical part of any oil company's activity. Value will be calculated according to the interpretation and conclusions of the petroleum geoscien-tists. In consequence, some of the case histories include petroleum economics as an intrinsic part of the story.

1.2 BACKGROUND

Petroleum geoscience is intimately linked with making money, indeed profit. The role of the petroleum geoscientist, whether in a state oil company, a massive multinational company, or a small independent company, is to find petroleum (oil and hydrocarbon gas) and help produce it so that it can be sold.

In years past, geoscientists overwhelmingly dominated the bit of the industry that explores for petroleum—they have boldly gone to impenetrable jungles, to scorching deserts, and to hostile seas in the search for petroleum. Getting the oil and gas out of the ground—that is, production—was left largely to engineers.

Today, the situation is different. Geoscience is still a key part of the exploration process, but finding oil and gas is more difficult than it used to be. There are many fewer giant oil-fields to be discovered. The geoscientists now need to work with drilling engineers, reservoir engineers, petroleum engineers, commercial experts, and facilities engineers to determine whether the petroleum that might be discovered is likely to be economical to produce as crude oil for market.

Geoscientists now also play an important role in the production of petroleum. Oil- and gasfields are not simply tanks waiting to be emptied. They are complex three-dimensional (3D) shapes with internal structures that will make petroleum extraction anything but simple. The geoscientist will help describe the reservoir and the trapped fluids. Geoscientists will also help determine future drilling locations and use information from petroleum production to help the interpretation of reservoir architecture.

1.3 WHAT IS IN THIS BOOK

This book is aimed at satisfying the needs of undergraduates who wish to learn about the application of geoscience in the petroleum industry. It is also aimed at nongeoscientists (petrophysicists, reservoir engineers, and drilling engineers) who, on account of their role in the old industry, need to find out more about how geologists and geophysicists ply their trade.

The book is divided into six chapters. The first chapter introduces both the book and the role of petroleum geoscience and geoscientists in industry. It also includes a section on the chemistry of oil. Chapter 2 examines the tools used by petroleum geoscientists. It is brief, since we intend only to introduce the "tools of the trade." More detail will be given in the body of the later chapters and in the case histories as needed.

Chapters 3 through 6 comprise the main part of the book. Chapters 3 and 4 cover exploration. We have chosen to divide exploration into two chapters for two reasons. To include all of the petroleum geoscience associated with exploration in one chapter would have been to create a massive tome. Moreover, it is possible to divide the exploration geoscience activity into basin description and petroleum exploration. Chapter 3 contains the basin description, with the addition of material on acreage acquisition and a section on possible shortcuts to finding petroleum. The final section of Chapter 3 is about petroleum source rocks, where they occur, and why they occur. Chapter 4 opens with sections on the other key components of petroleum geoscience; that is, the petroleum seal, the petroleum reservoir, and the petroleum trap. The second half of Chapter 4 examines the spatial and temporal relationships between reservoir and seal geometries and migrating petroleum. Risk (the likelihood of a particular outcome) and uncertainty (the range of values for a particular outcome) are both intrinsic parts of any exploration, or indeed appraisal, program. These too are examined.

Once a discovery of petroleum has been made, it is necessary to find out how much petroleum has been found and how easily oil or gas will flow from the field. This is appraisal, which is treated in Chapter 5. During appraisal, a decision will be made on whether to develop and produce the field under investigation. The geoscience activity associated with development and production is covered in Chapter 6. The production of petroleum from a field commonly spans a much longer period than exploration and appraisal. The type of geoscience activity changes throughout the life of a field, and we have tried to capture this within the case histories.

1.4 WHAT IS NOT IN THIS BOOK

The book introduces petroleum geoscience, a discipline of geoscience that embraces many individual and specialist strands of earth science. We deal with aspects of all these strands, but these important topics — such as basin analysis, stratigraphy, sedimentology, diagenesis, petrophysics, reservoir simulation, and others — are not covered in great detail. Where appropriate, the reader is guided to the main texts in these sub-disciplines and there are also suggestions for further reading.

1.5 KEY TERMS AND CONCEPTS

The source, seal, trap, reservoir, and timing (of petroleum migration) are sometimes known as the "magic five ingredients" without which a basin cannot become a petroleum province (Fig. 1.1). Here, we introduce these and other essential properties, before examining each aspect in more detail in the chapters that follow.

1.5.1 Petroleum

Petroleum is a mixture of hydrocarbon molecules and lesser quantities of organic molecules containing sulfur, oxygen, nitrogen, and some metals. The term includes both oil and hydrocarbon gas. The density of liquid petroleum (oil) is commonly less than that of water and the oil is naturally buoyant. So-called heavy (high specific gravity) oils and tars may be denser than water. Some light (low specific gravity) oils are less viscous than water, while most oils are more viscous than water. The composition of petroleum and its properties are given in Section 1.6.

1.5.2 The source

A source rock is a sedimentary rock that contains sufficient organic matter such that when it is buried and heated it will produce petroleum (oil and gas). High concentrations of organic matter tend to occur in sediments that accumulate in areas of high organic matter productivity and stagnant water. Environments of high productivity can include nutrient rich coastal upwellings, swamps, shallow seas, and lakes. However, much of the dead organic matter generated in such systems is scavenged and recycled within the biological cycle. To preserve organic matter, the oxygen contents of the bottom waters and interstitial waters of the sediment need to be very low or zero. Such conditions can be created by overproduction

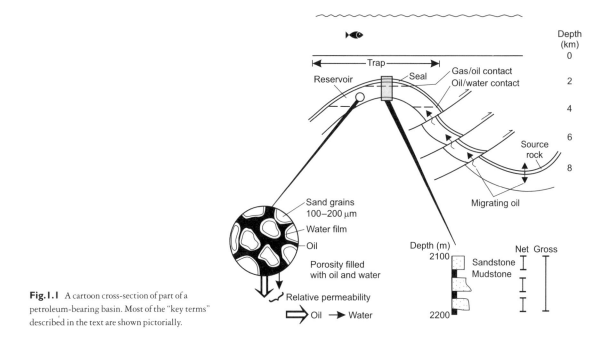

Fig.1.1 A cartoon cross-section of part of a petroleum-bearing basin. Most of the "key terms" described in the text are shown pictorially.

of organic matter, or in environments where poor water circulation leads to stagnation.

Different sorts of organic matter yield different sorts of petroleum. Organic matter rich in soft and waxy tissues, such as that found in algae, commonly yields oil with associated gas on maturation (heating), while gas alone tends to be derived from the maturation of woody tissues. Even oil-prone source rocks yield gas when elevated to high temperatures during burial. A detailed description of source rocks and source rock development can be found in Section 3.7.

1.5.3 The seal

Oil and gas are less dense than water and, as such, once they migrate from the source rock they tend to rise within the sedimentary rock column. The petroleum fluids will continue to rise under buoyancy until they reach a seal. Seals tend to be fine-grained or crystalline, low-permeability rocks. Typical examples include mudstone/shale, cemented limestones, cherts, anhydrite, and salt (halite). As such, many source rocks may also be high-quality seals. Seals to fluid flow can also develop along fault planes, faulted zones, and fractures.

The presence of a seal or seals is critical for the development of accumulations of petroleum in the subsurface. In the absence of seals, petroleum will continue to rise until it reaches

the Earth's surface. Here, surface chemical processes including bacterial activity will destroy the petroleum. Although seals are critical for the development of petroleum pools, none are perfect. All leak. This natural phenomenon of petroleum seepage through seals can provide a shortcut to discovering petroleum. Seals and seal mechanisms are described in Section 4.2.

1.5.4 The trap

The term "trap" is simply a description of the geometry of the sealed petroleum-bearing container (Section 4.5). Buoyant petroleum rising through a pile of sedimentary rocks will not be trapped even in the presence of seals if the seals are, in gross geometric terms, concave-up. Petroleum will simply flow along the base of the seal until the edge of the seal is reached, and then it will continue upwards toward the surface. A trivial analogy, albeit inverted, is to pour coffee onto an upturned cup. The coffee will flow over but not into the cup. However, if the seal is concave-down it will capture any petroleum that migrates into it.

The simplest trapping configurations are domes (four-way dip-closed anticlines) and fault blocks. However, if the distribution of seals is complex, it follows that the trap geometry will also be complex. The mapping and remapping of trap geometry is a fundamental part of petroleum geoscience at the exploration, appraisal, and even production phases of petroleum exploration.

1.5.5 The reservoir

A reservoir is the rock plus void space contained in a trap. Traps rarely inclose large voids filled with petroleum; oil-filled caves, for example, are uncommon. Instead, the trap contains a porous and permeable reservoir rock. The petroleum together with some water occurs in the pore spaces between the grains (or crystals) in the rock. Reservoir rocks are most commonly coarse-grained sandstones or carbonates. Viable reservoirs occur in many different shapes and sizes, and their internal properties (porosity and permeability) also vary enormously (Section 4.3).

1.5.6 The timing of petroleum migration

We have already introduced the concept of buoyant petroleum migrating upward from the source rock toward the Earth's surface. Seals in suitable trapping geometries will arrest migration of the petroleum. When exploring for petroleum it is important to consider the timing of petroleum migration relative to the time of deposition of the reservoir/seal combinations and the creation of structure within the basin. If migration of petroleum occurs before deposition of a suitable reservoir/seal combination, then the petroleum will not be trapped. If petroleum migrates before structuring in the basin creates suitable trap geometries, then the petroleum will not be trapped. In order to determine whether the reservoir, seal, and trap are available to arrest migrating petroleum, it is necessary to reconstruct the geologic history of the area under investigation. Petroleum migration is examined in Section 4.4.

1.5.7 Porous rock and porosity

A porous rock has the capacity to hold fluid. By definition, reservoirs must be porous. Porosity is the void space in the rock, reported either as a fraction of one or as a percentage. Most reservoirs contain >0% to <40% porosity.

1.5.8 Permeable rock and permeability

A permeable rock has the capacity to transmit fluid. A viable reservoir needs to be permeable or the petroleum will not be extracted. By definition, a seal needs to be largely impermeable to petroleum. Permeability is a measure of the degree to which fluid can be transmitted. The unit for permeability is the darcy (D), although the permeability of many reservoirs is measured in millidarcies (mD). Typically, the permeability of reservoirs is 10 D or less. At the lower end, gas may be produced from reservoirs of 0.1 mD, while oil reservoirs need to be 10× or 100× more permeable.

1.5.9 Relative permeability

Most reservoirs contain both oil and water in an intimate mixture. A consequence of there being more than one fluid in the pore system is that neither water nor oil will flow as readily as if there were only one phase. Such relative permeability varies as a function of fluid phase abundance.

1.5.10 Net to gross and net pay

A reservoir commonly contains a mixture of nonreservoir lithologies (rocks) such as mudstone or evaporite minerals interbedded with the reservoir lithology, commonly sandstone or limestone. The ratio of the porous and permeable interval to the nonporous and/or nonpermeable interval is called the "net to gross." Net pay is the portion of the net reservoir containing petroleum and from which petroleum will flow.

1.5.11 Water saturation

A petroleum-bearing reservoir always contains some water. The quantity of water is commonly expressed as a fraction or percentage of the pore space. There are, of course, comparable terms for oil and gas.

1.5.12 The formation volume factor

The pressure and temperature differ from trap to trap, and from the trap to the Earth's surface. Most oils shrink as they are lifted from the ground. This is partly a temperature effect, but it is largely due to exsolution of gas as the pressure is reduced. The ratio between the volumes of petroleum at reservoir conditions to that at standard conditions (also known as stock tank conditions) is called the "formation volume factor."

1.5.13 The gas to oil ratio

The gas that is exsolved when oil is raised from the trap to the surface is produced alongside the oil. The proportion of gas and oil in the produced fluid at stock tank conditions (see Table 1.1) is known as the gas to oil ratio (or "GOR").

Table 1.1 Conversion factors for British/US units to metric units.

Imperial/oilfield	Metric
1 foot	0.305 m
1 acre	4047 m^2
1 acre-foot	1233.5 m^3
1 cubic foot	0.02832 m^3
1 gallon (US)	0.003785 m^3
1 barrel = 42 gallons	0.159 m^3
1 cubic foot per barrel (GOR)	0.178 m^3 m^{-3}
	1 metric tonne of oil = 1280 m^3 gas (BTU basis)
	1 metric tonne of oil = 680 m^3 gas (chemical conversion basis)
1 barrel of oil = 6040 ft^3 (BTU basis)	
1 barrel of oil = 3200 ft^3 (chemical conversion basis)	
1 psi (pound-force per square inch)	0.06895 bar
1 psi (pound-force per square inch)	6.89 × 10^3 pascal
Degrees API (° API) 60/60°F = weight of oil volume (at 60°F divided by the weight of the same volume of water at 60°F)	$\dfrac{141.5}{\text{specific gravity } 60/60°F} - 131.5$
Degrees Fahrenheit (°F)	°C × (9/5) + 32
1 barrel of oil per day (bopd)	c. 50 t a^{-1}
1 barrel of water per day (bwpd)	0.1555 m^3 d^{-1}
1 standard cubic foot per day (scf d^{-1})	c. 10 m^3 a^{-1}

1.5.14 The gas expansion factor

Under pressure in the trap, gas is compressed. As pressure is released, the gas expands. The relationship between reservoir conditions and standard conditions is called the "gas expansion factor."

1.5.15 Timescales

In a book written by geoscientists about geoscience, but intended for at least some nongeoscientists, it is important to point out the difference between what is meant by time to a geoscientist and to a nongeoscientist. At around 4.5×10^9 years ($10^9 = 1$ US billion), the Earth is old. The oldest rocks that contain commercial quantities of oil are a mere half a billion years old or thereabouts. However, maturation migration and trapping of oil can occur quickly, at least in a geologic sense. Some of the oil in the southern Caspian Sea and some of the oil in eastern Venezuela are found in reservoir sands that are only a few million years old. Had *Australopithecus* wandered from Africa a little sooner, he would have been able to witness the deposition of sands that now host major oilfields in Venezuela and Azerbaijan. Thus, although a complete cycle of petroleum maturation, migration, and trapping can occur in a few million years — almost a geologic instant — that same instant covers all of recorded human history and most of that deduced from paleoanthropology.

Geologic time is divided, using a hierarchical scheme, into a variety of named units. The basic unit in this scheme is the period (individual periods lasted between 10 million and 100 million years). The sequence of periods with their attendant subdivisions and supra-divisions makes up the stratigraphic column (Fig. 1.2).

1.5.16 The units used in this book

For the most part, metric units are used in this book. Imperial terms appear where the original source data are in imperial measurements. However, the terms "barrels of oil" (bbl) and "standard cubic feet of gas" (scf) are generally employed in this book because they are used much more commonly than their metric equivalents. In these instances, the metric equivalents are bracketed (see the unit conversions in Table 1.1).

(a)

Era	Sub-era Period Sub-period			Epoch			Stage		Age (Ma)	Stage abbr.	Intervals (Ma)	
Paleozoic	Devonian		15	D_3 Late			Famennian		367.0	Fam	4.5	46
							Frasnian		377.4	Frs	10.4	
			9	D_2 Middle			Givetian		380.8	Giv	3.4	
							Eifelian		386.0	Eif	5.2	
			22	D_1 Early			Emsian		390.4	Ems	4.4	
							Pragian		396.3	Pra	5.9	
	46		D				Lochkovian		408.5	Lok	12.2	
	Silurian			S_4 Pridoll	Prd				410.7	Prd	2.2	31
			13	S_3 Ludlow	Lud		Ludfordian		415.1	Ldf	4.4	
							Gorstian		424.0	Gor	8.9	
			6.5	S_2 Wenlock	Wen		Gleedonian		425.4	Gle	1.4	
							Whitwellian		426.1	Whi	0.7	
							Sheinwoodian		430.4	She	4.3	
			8.5	S_1 Llandovery	Lly		Telychian		432.6	Tel	2.2	
							Aeronian		436.9	Aer	4.3	
	31		S				Rhuddanian		439.0	Rhu	2.1	
	Ordovician	Bala		Ashgill	Ash		Hirnantian		439.5	Hir	0.5	71
							Rawtheyan		440.1	Raw	0.6	
							Cautleyan		440.6	Cau	0.5	
			4				Pusgillian		443.1	Pus	2.5	
				Caradoc	Crd		Onnian		444.0	Onn	0.9	
							Actonian		444.5	Act	0.5	
							Marshbrookian		447.1	Mrb	2.6	
							Longvillian		449.7	Lon	2.6	
							Soudleyan		457.5	Sou	7.8	
							Hamagian		462.3	Har	4.8	
			25	Bal	21		Costonian		463.9	Cos	1.6	
		Dyfed		Llandeilo	Llo		Late		465.4	Llo 3	1.5	
			4.5				Mid		467.0	Llo 2	1.6	
							Early		468.6	Llo 1	1.6	
				Llanvirn	Lln		Late		472.7	Lln 2	4.1	
			12	Dfd	7.5		Early		476.1	Lln 1	3.4	
		Canadian		Arenig					493.0	Arg	17.0	
	71	0	34	Cnd		Tremadoc			510.0	Tre	17.0	
	Cambrian		7	Merioneth	Mer		Dolgellian		514.1	Dol	4.1	60
							Maentwrogian		517.2	Mnt	3.1	
			19	St. David's	StD		Menevian		530.2	Men	13.0	
							Solvan		536.0	Sol	5.8	
				Caerfai	Crf		Lenian		554	Len	18.0	
	325						Atdabanian		560	Atb	6.0	
	Pz	60	E	34			Tommotian		570	Tom	10.0	

Eon	Age	Era	Period			Stage						
Proterozoic	Pt_1	Sinian	Vendian	20	Ediacara	Edi	Poundian		580	Pou	10.0	40
							Wonokan		590	Won	10.0	
				40	V	20	Varanger	Mortensnes	Var	600	Mor	10.0	
		230	Z					Smalfjord		610	Sma	10.0	
			Sturtian							800	Stu	190	
	Pt_2	Riphean	Karatau							1050	Kar	250	
			Yurmatin							1350	Yur	300	
		850	Rlf	Burzyan						1650	Buz	300	
	Pt_3	Animikean								2200	Ani	550	
		Huronian								2450	Hur	250	
		Randian								2800	Ran	350	
Archean		Swazian								3500	Swz	700	
		Isuan								3800	Isu	300	
		Hadean	Early Imbrian							3850	Imb	50	
			Nectarian							3950	Nec	100	
		760	Basin Groups 1–9							4150	BG1-9	200	
Priscoan	4000	Hde	Cryptic							4560	Cry	410	

Fig. 1.2 A stratigraphic column. The primary unit of geologic time is the period. An alternative absolute timescale was published by Haq et al. (1988). (From Harland et al. 1990; reprinted with the permission of Cambridge University Press.)

(b)

Era	Sub-era Period Sub-period	Epoch	Stage	Age (Ma)	Stage abbr.	Intervals (Ma)
Cenozoic	Quaternary or Pleistogene 1.64	Holocene		0.01	Hol	0.01
		Pleistocene		1.64	Ple	1.63
	Neogene 22 Ng 18.1	Pliocene Pli 1 2	Placenzian	3.40	Pia	3.6
			Zanclian	5.2	Zan	
		Miocene Mio 3	Messinian	6.7	Mes	5.2
			Tortonian	10.4	Tor	
		2	Serravallian	14.2	Srv	5.9
			Langhian	16.3	Lan	
		1	Burdigalian	21.5	Bur	7.0
			Aquitanian	23.3	Aqt	
	Tertiary 22 Ng	Oligocene 12.1 Oli 2	Chattian	29.3	Cht	6.0
		1	Rupelian	35.4	Rup	6.1
	Paleogene 21.1 Eoc	Eocene 3	Priabonian	38.6	Prb	3.2
		2	Bartonian	42.1	Brt	11.4
			Lutetian	50.0	Lut	
		1	Ypresian	56.5	Ypr	6.5
65 Cz 63 TT 42 Pg 8.5		Paleocene Pal 2	Thanetian	60.5	Tha	4.0
		1	Danian	65.0	Dan	4.5
Mesozoic	Cretaceous 32 Gul	Gulf K₂ 23.5 Sen	Maastrichtian	74.0	Maa	9.0
		Senonian	Campanian	83.0	Cmp	9.0
			Santonian	86.6	San	3.6
			Coniacian	88.5	Con	1.9
			Turonian	90.4	Tur	1.9
		Gallic	Cenomanian	97.0	Cen	6.6
		43.3 Gal	Albian	112.0	Alb	15.0
	Early K₁		Aptian	124.5	Apt	12.5
			Barremian	131.8	Brm	7.3
		Neocomian	Hauterivian	135.0	Hau	3.2
	81 K 49 13.8 Neo		Valanginian	140.7	Vlg	5.7
			Berriasian	145.6	Ber	4.9
	Jurassic 20.9 J₂ Dog	Malm 11.5 J₃ Mal	Tithonian	152.1	Tth	6.5
			Kimmeridgian	154.7	Klm	2.6
			Oxfordian	157.1	Oxf	2.4
		Dogger	Callovian	161.3	Clv	4.2
			Bathonian	166.1	Bth	4.8
			Bajocian	173.5	Baj	7.4
			Aalenian	178.0	Aal	4.5
		Lias	Toarcian	187.0	Toa	9.0
			Pliensbachian	194.5	Plb	7.5
	62 J 30.0 J₁ Lia		Sinemurian	203.5	Sin	9.0
			Hettangian	208.0	Het	4.5
Mesozoic	Triassic	Tr₃ 27 Late	Rhaetian	209.5	Rht	1.5
			Norian	223.4	Nor	13.9
			Carnian	235.0	Crn	11.6
		Tr₂ 6 Mid	Ladinian	239.5	Lad	4.5
			Anisian	241.1	Ans	1.6
		Scythian	Spathian	241.9	Spa	0.8
			Nammalian	243.4	Nml	1.5
180 Mz 37 Tr 4		Tr₁ Scy	Griesbachian	245.0	Gri	1.6
Paleozoic	Permian 11 Zec	Zechstein	Changxingian	247.5	Chx	2.5
			Longtanian	250.0	Lgt	2.5
			Capitanian	252.5	Cap	2.5
			Wordian	255.0	Wor	2.5
			Ufimian	256.1	Ufi	1.1
		Rotliegend	Kungurian	259.7	Kun	3.6
			Artinskian	268.8	Art	9.1
			Sakmarian	281.5	Sak	12.7
	45 P 34 Rot		Asselian	290.0	Ass	8.5
	Carboniferous C₂ 33	Pennsylvanian	Gzelian 5 Gze	Noginskian	293.6	Nog	3.6
			Klazminskian	295.1	Kla	1.5
		Stephanian 13 Ste	Dorogomilovskian	298.3	Dor	3.2
		Kasimovian 8 Kas	Chamovnicheskian	299.9	Chv	1.6
			Krevyakinskian	303.0	Kre	3.1
		Moscovian 8 Mos	Westphalian 15 Wes	Myachkovskian	305.0	Mya	2.0
			Podolskian	307.1	Pod	2.1
			Kashirskian	309.2	Ksk	2.1
			Vereiskian	311.3	Vrk	2.1
		Bashkirian	Melekesskian	313.4	Mel	2.1
			Cheremshanskian	318.3	Che	4.9
		12 Bsh	Namurian 15 Nam	Yeadonian	320.6	Yea	2.3
			Marsdenian	321.5	Mrd	0.9
			Kinderscoutian	322.8	Kin	1.3
		Serpukhovian	Alportian	325.6	Alp	2.8
			Chokierian	328.3	Cho	2.7
			Arnsbergian	331.1	Arn	2.8
		10 Spk 15	Pendleian	332.9	Pnd	1.8
	Mississippian	Visean	Brigantian	336.0	Bri	3.1
			Asbian	339.4	Asb	3.4
			Holkerian	342.8	Hlk	3.4
			Arundian	345.0	Aru	2.2
		17 Vis	Chadian	349.5	Chd	4.5
		Tournaisian	Ivorian	353.8	Ivo	4.3
73 C 40 C₁		13 Tou	Hastarian	362.5	Has	8.7

The right-hand Intervals (Ma) column also shows grouped totals: 22, 42, 81, 62, 37, 45, 33, 40.

Fig. 1.2 *continued*

1.6 THE CHEMISTRY OF PETROLEUM

Petroleum is a mixture of hydrocarbons and other organic compounds that together dictate its chemical and physical properties. For example, typical oil from the Brent Field in the UK North Sea contains an average of 26,000 pure compounds. This oil will easily flow through the reservoir rock to reach a wellbore, and flow to the surface and along pipelines, up to 1000 km long, to the refinery. By contrast, oil found in parts of the Los Angeles basin (e.g., the Wilmington Field) is viscous and it needs to be heated by steam for it to flow. It would be too viscous to flow down a conventional pipeline, and it is refined at a surface installation above the oilfield. In a gross sense, it is possible to determine the physical properties of the petroleum if its chemical constituents are known. Thus we now review the main groups of hydrocarbon and associated compounds.

Hydrocarbons are molecules composed of hydrogen (H) and carbon (C) bonded together. Petroleum also contains lesser quantities of organic molecules that contain nitrogen (N), oxygen (O), and sulfur (S). Small but significant quantities of organometallic compounds (commonly with vanadium and nickel) are also present, as are a large array of elements in trace quantities. Examples of small and simple hydrocarbons include methane (CH_4), ethane (C_2H_6), and propane (C_3H_8). Each of these compounds contains only carbon and hydrogen. Since one carbon atom can bond with a maximum of four hydrogen atoms, methane is the simplest hydrocarbon molecule and, in the absence of oxygen, it is also chemically stable. It is said to be saturated; that is, there are no spare bonds and a change in chemistry involves replacement of one of the hydrogen atoms. Where spare bonds are available, the compound is unsaturated and less stable than the equivalent saturated compound. It will more readily react. The most common groups of hydrocarbons found in natural occurrences of petroleum (e.g., crude oil and natural gas) are saturated compounds found in the alkane or paraffins, and in the naphthene or cycloparaffins groups.

1.6.1 Alkanes (paraffins)

Methane, ethane, propane, and butane are all stable compounds with the maximum permitted number of hydrogen atoms. These saturated compounds form part of the group of hydrocarbons called alkanes, with a chemical formula C_nH_{2n+2}, where n is the whole number of carbon atoms. The chemical basis for the formula is illustrated in Fig. 1.3. Alkanes contain chains of carbon atoms but no closed loops. The smallest molecules up to the formula C_4H_{10} (butane) are

(a)

The alkane series	The alkene series
General formula: C_nH_{2n+2}	General formula: C_nH_{2n}
Methane CH_4	
Ethane C_2H_6	Ethene C_2H_4
Propane C_3H_8	Propene C_3H_6
Butane C_4H_{10}	Butene C_4H_8

(b)

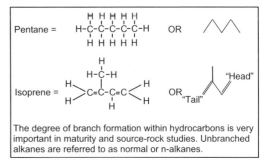

The degree of branch formation within hydrocarbons is very important in maturity and source-rock studies. Unbranched alkanes are referred to as normal or n-alkanes.

Fig. 1.3 (a) Chemical formulas for alkanes and alkenes, plus examples. (b) Standard molecular representation techniques. Straight-chain and branched alkanes share the same formula. n-Pentane and iso-pentane (C_5H_{12}) are used to illustrate the different molecular arrangements, which lead to different physical properties.

gases at standard conditions (i.e., at the Earth's surface) of temperature and pressure. Liquid compounds at room temperature range from C_5H_{12} (pentane) to $C_{16}H_{34}$ (hexadecane). Larger molecules have an increasing number of structural variations.

There are two important naturally occurring alkanes that are used in source rock analysis. Pristane and phytane are both branched alkanes of biological origin with an odd number of carbon atoms (C_{17} and C_{19}, respectively), whose relative abundance in crude oils is an indication of the depositional environment of their source rock.

1.6.2 Naphthenes (cycloalkenes)

In some hydrocarbon compounds the carbon atoms are linked in a ring, thereby reducing the number of sites for bonding with hydrogen atoms (Fig. 1.4). In natural petroleum, the most common group of compounds with a ring structure is the naphthenes. The most simple is cyclopropane (C_3H_6, Fig. 1.4), which is a gas. Cyclopentane (C_5H_{10}) and cyclohexane (C_6H_{12}) are liquids and are abundant in most crude oils. The

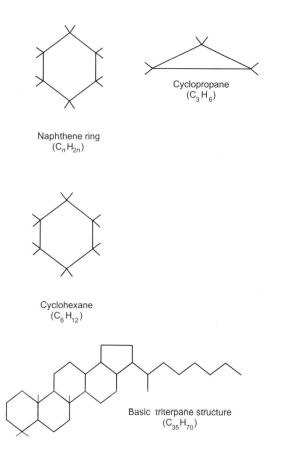

Naphthene ring
(C_nH_{2n})

Cyclopropane
(C_3H_6)

Cyclohexane
(C_6H_{12})

Basic triterpane structure
($C_{35}H_{70}$)

Fig.1.4 Naphthene, or ring, compounds have the chemical formula C_nH_{2n}. The smallest naphthene is cyclopropane. Compounds with both ring and branch-chain carbon atoms occur, and include the triterpanes, one of the most important groups of naphthenic hydrocarbon compounds.

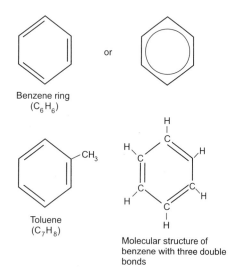

or

Benzene ring
(C_6H_6)

Toluene
(C_7H_8)

Molecular structure of benzene with three double bonds

Fig.1.5 Aromatic compounds are unsaturated ring compounds with the chemical formula C_nH_n. The simplest compound is benzene (C_6H_6). Side branches are common, and generate a variety of compounds with contrasting physical properties. Toluene is a common constituent of natural petroleum.

chemical formula for the naphthenes is C_nH_{2n}. Large naphthene molecules commonly have more than one ring. Some related compounds have a series of rings with straight-chain branches (Fig. 1.4). Thus the range of naphthene compounds is large. As a consequence, it is often possible to specify the exact source of a sample of crude oil from its unique chemical signature or to match the oil found in a reservoir to its source rock.

A second group of compounds exists with the chemical formula C_nH_{2n}. The alkenes form a group of unsaturated hydrocarbons in which some of the carbon atoms have more than one bond. Typical compounds include ethylene (strictly ethene, C_2H_4) and propylene (strictly propene, C_3H_6). These compounds occur only in small quantities in natural petroleum. They are also manufactured by the petrochemical in-

dustry because they are useful starting materials for the manufacture of plastics.

1.6.3 Aromatics

Unsaturated hydrocarbon ring compounds are called aromatics. Benzene (C_6H_6) is the most simple (Fig. 1.5). The simplest aromatics have a chemical formula C_nH_{2n-6}, but there are many complicated compounds that combine the benzene ring with naphthene and straight-chain branches to produce a wide range of compounds. For example, toluene ($C_6H_5CH_3$), in which one of the hydrogen atoms has been substituted by a methyl (CH_3) branch chain (Fig. 1.5), is a common constituent in aromatic crude oils. One characteristic of the purer aromatic compounds is their pleasant odor, in contrast to the naphtheno-aromatics, which, if they contain some sulfur, smell awful.

1.6.4 Asphaltenes

Complex hydrocarbon compounds that are relatively enriched in N, S, and O are known as asphaltenes or resins. They are characterized by high molecular weight and large size, and form some of the heaviest molecules in crude oils. These compounds are frequently found in immature oils and where the original oil has been altered due to biological activity, generally at low temperatures (below about 90°C).

Asphaltenes commonly present particular problems for reservoir, production string (the tubing from reservoir to surface), and facilities management. The production of oil from a reservoir requires a drop in pressure around the producing wellbore and drops in temperature in the production string and topsides facilities. Reductions in pressure and/or temperature can lead to asphaltene precipitation. This may occur in the reservoir near to the wellbore, blocking the pores in the rock and "killing" the well. It may also block any pipework. Cleaning, either mechanically or by solvent washing, is difficult and expensive in pipework and is not possible in the reservoir.

Naturally precipitated asphaltene may also occur in the reservoir. In such situations it is commonly present as discontinuous layers that may baffle cross-flow in the reservoir. A variety of processes can lead to natural precipitation. These include the mixing of two or more oils in the subsurface, uplift causing temperature and pressure drop, or precipitation of asphaltene on oil migration routes (Larter & Aplin 1995).

In summary, hydrocarbon compounds fall into a number of groupings that depend on the molecular arrangement of the carbon and hydrogen atoms. The molecular arrangement determines the physical and chemical characteristics of the group, but the size of the molecule dictates whether the compound is gas, liquid, or solid at standard conditions of temperature and pressure. The presence of minor amounts of N, S, and O will have a marked impact on the properties. Petroleum contains a range of compound mixtures, from relatively pure natural gases to highly complex mixtures found in some biodegraded crude oils.

1.7 GEOSCIENCE AND THE VALUE CHAIN

The components of the value chain are exploration (commonly divided into frontier exploration and basin exploitation), appraisal, development, production, reserves growth, and field reactivation or abandonment (Fig. 1.6). Geoscientists are employed in each part of this chain.

1.7.1 Exploration (Chapters 3 & 4)

Almost all of the world's petroleum occurs in sedimentary basins—areas that are now, or have at some time in the geologic past, been the sites of sediment accumulation and burial.

At the exploration stage the petroleum geoscientist needs to establish as a minimum that the four necessary ingredients are present in the basin: a petroleum source rock, a trap, a seal, and a reservoir. The relative timing of petroleum generation, trap formation, reservoir, and seal deposition also needs to be known. In some places around the world, the geoscientist may also need to "know" if he or she is drilling to find oil or gas, since the value of the two can be quite different.

In the earliest, frontier, phase of exploration in an unexplored basin, much effort is expended on determining if the basin has petroleum source rocks and whether such source rocks have yielded their petroleum (Chapter 3). Naturally, this activity runs in parallel with trying to find the largest structures, those that could contain the greatest quantities of oil and/or gas.

If a basin is proved to contain economically viable petroleum pools, then the nature of the work changes to one of basin exploitation (Chapter 4). It is during this phase that most of the largest and many of the medium-sized pools are discovered, as the companies involved in exploration begin to understand how, in that particular basin, the petroleum system works. Not only do the first companies to enter a new basin tend to find the biggest fields; they also get the chance to dominate the oil-transportation infrastructure. This enables those companies with early successes to levy charges upon other companies for the privilege of using their facilities.

1.7.2 Appraisal (see Chapter 5)

Once a petroleum discovery has been made, there can be much uncertainty about its size. The early stages of appraisal are concerned with defining the size and shape of the field and the position of the contacts between fluids (gas/water, oil/water, and possibly gas/oil). Appraisal of a field also starts to reveal the internal architecture of the reservoir and whether the field is divided into compartments, one isolated from another, how reservoir quality (porosity and permeability) varies spatially, and whether the reservoir is layered. At the same time, data are also generated on the variation in petroleum composition across the field.

These data and other commercial criteria are then applied to determine whether the field will return sufficient profit when brought into production.

Today, a typical field appraisal involves the acquisition of a dense grid of seismic data and the drilling of delineation wells. The quality and size of the seismic survey are determined by a number of factors, which include the structural and stratigraphic complexity, the estimated size of the

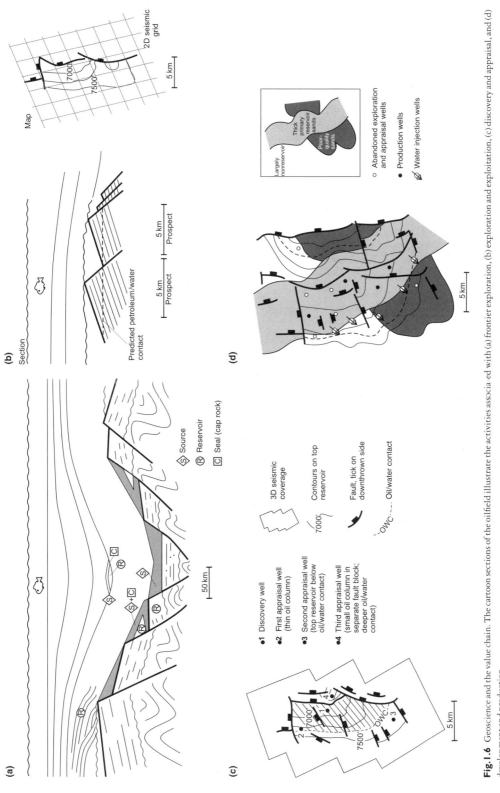

Fig.1.6 Geoscience and the value chain. The cartoon sections of the oilfield illustrate the activities associated with (a) frontier exploration, (b) exploration and exploitation, (c) discovery and appraisal, and (d) development and production.

discovery, the environment (marine surveys are much less expensive than land or land/sea transition surveys), the location, and the relative cost of wells versus seismic. Nowadays, a company is likely to shoot a 3D seismic survey to aid appraisal, although in some instances 3D seismic surveys are shot at the exploration stage.

The wells drilled during appraisal commonly carry large data acquisition programs. Often, core will be cut from the reservoir section. Some of the core may be preserved at reservoir conditions. Core is used by geoscientists, petrophysicists, and reservoir engineers to determine rock and fluid properties. Such data are used to build a reservoir model and calibrate wells to rock properties to seismic data. The reservoir intervals in wells are also subject to extensive wireline logging, to determine rock types, fluid types, fluid distribution, and seismic properties.

Oil, gas, and water samples will be collected to determine their composition and $P-V-T$ (pressure, volume, temperature) properties. Such data are used in the reservoir model and facilities design. For example, oil rich in dissolved H_2S will require stainless steel facilities, and apparatus to minimize corrosion and to stop the poisonous hydrogen sulfide from coming into contact with workers. An oilfield water rich in dissolved barium would behove the facility engineers to design production and water injection systems that would reduce the worst effects of scale (commonly barium sulfate) precipitation in pipework and in the reservoir adjacent to water injection wells.

1.7.3 Development (see Chapter 6)

Information gathered at the appraisal stage is used to construct a development plan — how many wells are needed, where they should be drilled, and what platform facilities are required — and the development plan is executed. The approach to development varies from offshore to onshore and from country to country. However, a very common plan nowadays is to drill the production wells while the production facilities are being built. This has the advantage that when the completed facilities are erected on site, the wells can be tied back very quickly and the field brought onto plateau production almost immediately.

The core of the development plan is the reservoir model. The model is used to simulate production of petroleum from the accumulation using computer modeling packages. Each model commonly comprises many tens of thousands of cells. Each cell represents a portion of the reservoir. It is populated with reservoir properties such as net to gross, porosity, perme-

ability, permeability anisotropy, pressure, and the like. Faults and other potentially impermeable barriers are woven through the array of cells. Cells may also have simulated wells, points at which petroleum is extracted from the model or water and/or gas introduced. When the reservoir simulation is running, petroleum is extracted from those cells that contain production wells. Changes in pressure and fluid content are transmitted between cells as the production continues. It is common to run models a number of times, and the results are used to optimize well location and offtake rate.

Reservoir models are commonly conditioned to static data, such as a geologic description of the trap shape, the reservoir architecture, the petrophysical properties, the gas to oil ratio, and the petroleum compressibility. They are also conditioned to any dynamic data that might be available, such as well flow test results. Geologic data continues to be collected and interpreted during the development process. It is used to help refine the reservoir model and thus the way in which petroleum is produced from the field.

1.7.4 Production (see Chapter 6)

Until the onset of production, the reservoir description has rested almost wholly on "static" data. As production begins, so "dynamic" data are generated. Such data include temporal changes in reservoir pressure and relative and absolute production rates of petroleum and water. These data yield a much greater insight into the internal architecture of the field and, used properly, they can help to improve the reservoir description and so improve the rate and quantity of petroleum production. During the production phase it is probable that the reservoir model will be rebuilt a number of times. First, the reservoir model will begin to fail to predict the performance of the reservoir. Secondly, as production continues a production history will be recorded. This history can be used to assess the success or otherwise of the early models by "history matching" the reservoir model to the real production data.

In most oilfields, man needs to help nature when it comes to getting the oil out. Most common amongst secondary recovery processes is the injection of water into the aquifer underlying the oil and/or injection of gas (usually methane) into the gas cap or top of the reservoir. For example, in the Beryl Field (UK North Sea) water and gas have been injected to maintain reservoir pressure and sweep oil to the producing wells (Karasek et al. 2003). A large gas cap was created. In a later stage of field development, the secondary gas cap will be produced. Here again, the reservoir description skills of the geoscientist are important. He or she will work with the reservoir engineer to design the best locations for water or gas injection wells.

1.7.5 Reserves additions and reserves growth (see Chapter 6)

As the rate of petroleum production from a field declines, expensive production equipment becomes underused. As a consequence, work will be undertaken to improve the percentage of petroleum that can be won from the field (reserves additions). Simultaneously, efforts will be made to discover, appraise, and develop additional pools of petroleum that lie near to the original development (reserves growth). At this stage in the value chain, small pools of petroleum can become economically attractive, because the incremental cost of developing them is low if the production facilities exist only a few kilometers away.

A geoscientist working on reserves growth can expect to see the whole range of activity from exploration through to production on a highly compressed timescale.

1.7.6 Field abandonment and reactivation

In most situations, a field will be abandoned when the production rate drops below a predetermined economic limit or, alternatively, the production facilities and wells require investment (repairs) that are deemed not economically viable. There is a tendency for the costs of operating to be reduced with the passage of time for most fields and hence abandonment postponed. However, sooner or later a field will be abandoned. But abandonment may not be the end of the story. Many oilfields are abandoned when the oil remaining in the ground is at least double that which has been produced. The technology available to describe, drill into, and produce the oil today (or in the future) may be that much better than was available during the original life of the field. In consequence, the abandoned field may be selected for reactivation (Chapter 6). The appraisal of an abandoned field for rehabilitation is commonly the job of a multidisciplinary team that includes geoscientists and engineers.

1.8 GEOSCIENCE ACTIVITY

The list of processes that make up the value chain will be familiar to those in the oil industry. They will be less familiar to the student geoscientist who is reading this book. However, the specific geoscience activities that are used to support exploration through to abandonment are similar to those used by all geologists and geophysicists, whether working in academia or in industry.

The history of geology, at least from the days of William Smith in the 18th century, can be characterized using a series of maps. The primary product from deciphering the stratigraphic column across the globe is the geologic map. Geoscientists working in the oil industry today also spend much time making maps. Such maps can be used to illustrate structure at a range of scales from basin geometry to the geometry of individual traps. Maps are also generated for specific properties, such as net reservoir thickness or porosity. The data used to generate maps comes from several sources. Outcrop mapping on the Earth's surface is not a common activity today, but it still occurs in some parts of the world as a cheap screening process, or where seismic acquisition is difficult. Seismic data provide the main components for map-making in most oil companies. Where seismic data are lacking, maps can be made from well data alone.

In order to create maps, there is a range of description and interpretation activities. These include core and outcrop description, structural and stratigraphic work on seismic and well data, paleontological analysis, geochemical analysis of rock and fluid, and a range of petrographic studies.

There is one activity that a petroleum geoscientist undertakes which does not have an adequate counterpart in academia. To find and produce commercial quantities of petroleum, wells must be drilled. In most instances it is a petroleum geoscientist, or a team of geoscientists, who will choose the well location. The choice of location will be supported by a prediction of what the well will encounter. This usually takes the form of a predicted stratigraphic column for the well, together with a more detailed description of the target horizon. The detailed description of the target is likely to include the expected rock type, porosity and permeability, whether the well is likely to encounter oil or gas, and whether there is an expectation of penetrating a fluid contact. These data will have been used to estimate the reserves likely to be encountered, and the reserves estimate will justify the expenditure and risk. The success of the well will be judged against the prediction. Although it is a given that petroleum geoscience is about finding oil and gas, a well result that is more positive than predicted is likely to be "forgiven" more quickly than a well that fails to find the petroleum anticipated.

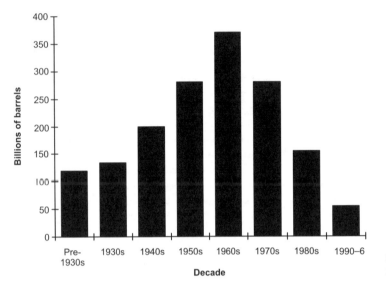

Fig. 1.7 The rate of oil discovery during the 20th century. (Data from Campbell 1997.)

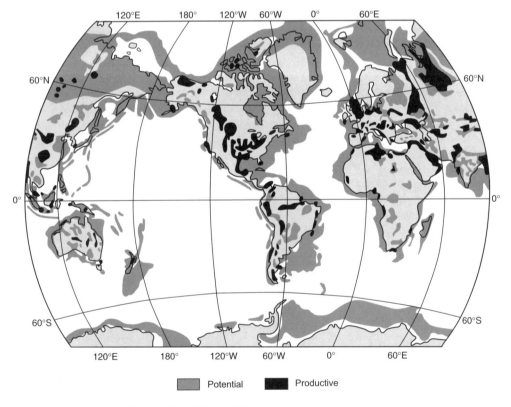

Fig. 1.8 The major petroleum basins of the world. (From Halbouty 1986.)

1.9 OIL, GAS, AND GEOSCIENTISTS — A GLOBAL RESOURCE!

In opening this book, we have attempted to describe the components of petroleum geoscience as applied to finding and producing oil and gas. Little has been said about the use of the resources — oil and gas — themselves. Two hundred years ago there was little use for, and thus little exploitation of, oil and gas. Some societies attributed medicinal properties to oil, while a little petroleum was used for heating and lighting. Small quantities of heavy oil and tar have also been used to caulk boats and pools since prehistoric times. Other people repelled invaders by pouring hot or burning oil down their fortress walls and onto the unfortunate soldiers below. One hundred years ago the demand was still small, with kerosene/paraffin being the key product from crude oil refining. Like crude itself, it was used for lighting, heating, and cooking. Natural bitumen deposits, together with the tarry residuum from oil refining, began to find use in road construction, but the internal combustion engine and thus the car was exceedingly rare on such roads.

Karl Benz sold his and the world's first gasolene-fuelled car in 1888. Fifteen years later, the Wright brothers used the same type of fuel to power their Flyer into aviation history. The demand for petroleum began to increase rapidly as greater use was made of petroleum products; particularly those used as engine fuels. World War I and then World War II amplified the demand for fuel. These wars also introduced large parts of the population in many countries to driving. Private ownership of vehicles grew at high rate throughout the second half of the 20th century.

During the whole of the period described above, the finding rate for oil greatly exceeded the rate of consumption. The known reserves were getting larger. The halcyon days for oil

exploration appear now to be over. Exploration successes, in terms of volume of oil discovered, peaked during the 1960s (Fig. 1.7). By the same time, most of the world's basins had been investigated and, it can be argued, most of the largest (giant) fields found. Indeed, most of the most prolific basins were initially discovered and exploited either late in the 19th century or early in the 20th century (southern Caspian Sea, eastern and western Venezuela, Texas and the Gulf of Mexico, and the Middle East — Yergin 1991; Fig. 1.8).

Geologists became involved in the search for oil in the latter part of the 19th century, and with the development of the so-called "anticline theory" they began to take the lead in exploration. Many of the earliest oil explorers relied solely upon seepage of petroleum as a means of identifying prospective drilling locations. Today, this might seem like an obvious and valid method, but since the composition of oil was not well understood, and its origins much less so, even the seeps method had its detractors. For example, the huge pitch (tar) lake of southwest Trinidad was categorically stated to have no value as an indicator of local oil potential by some of the first explorers on the island. Of course, they were wrong.

The anticline theory stated quite simply that oil, being less dense than water, would accumulate in parts of the Earth's crust where rocks were formed into convex-up folds. It was recognized that prospective folds would require a porous reservoir beneath an impermeable seal, or "cap rock." Thus geologic fieldwork — mapping — became the driving force behind exploration. Parts of the world with known seepage were mapped and anticlines drilled. As we shall see in the following chapter, the advent of seismic surveys allowed mapping to be carried offshore as the landlocked exploration of the first half of the 20th century gave way to near total exploration of the planet in the second half of the century.

Many would now argue that the exploration successes of the latter half of the 20th century delivered more than half

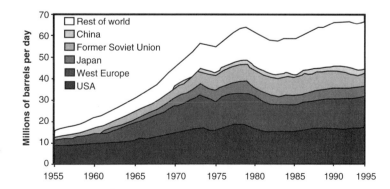

Fig. 1.9 The rate of oil consumption during the second half of the 20th century. (From Browne 1995.)

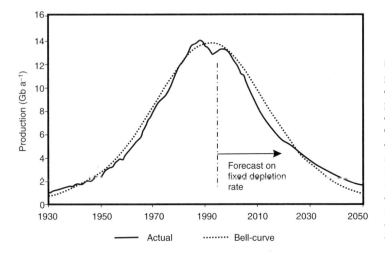

Fig.1.10 Historical oil production and projected production for non-OPEC countries for the coming 50 years. The solid line is the historical performance and predicted future trend. The dotted line is a "fitted" normal distribution. Campbell (1997) predicts that production will decline as a result of depletion of the global oil resource. The finding rate has fallen since the 1960s (Fig. 1.7) despite significant improvements in technology, while production and thus consumption have continued to rise (Fig. 1.9). The total volume for the non-OPEC countries is estimated to be about 850 giga-barrels (Gb). (From Campbell 1997.)

of the total global resource of conventional oil. By "conventional" we mean light- to medium-gravity oil that flows freely from the reservoir. The finding peak of the 1960s has become the production peak of today (Fig. 1.9). The search for oil is becoming more difficult, but pools are being — and will continue to be — found (Fig. 1.10). There is no doubt (to geoscientists at least) that this finite resource is running out. Much more controversial is the speculation surrounding the time when oil cannot be extracted in sufficient quantities to drive the world's economy. Campbell (1997) would argue that global peak production is occurring today, and that the major effects of declining production will appear within 50 years. The picture for gas is similar, but declining production will not occur until considerably further into the future. The effects of improved technology, which will allow more efficient finding of smaller pools and better recovery from pools, will undoubtedly extend the tail of the global production curves.

Quite clearly, as the oil resource declines the role of the geoscientist in the oil industry will change further, from new basin exploration to the search for the subtle trap, from exploration-dominated to production-dominated employment. As the conventional oil resource declines, greater use will be made of the massive accumulations of viscous and superviscous oils. Gas (methane) trapped in gas hydrates, gas dissolved in basinal brines, and that associated with coal will also be extracted. Indeed, coal-bed methane is already being exploited on a small scale in the USA and Europe. All these exploitation processes, be they for viscous oil or for gas, will require the skills of the geoscientist.

FURTHER READING

Campbell, C.J. (1997) *The Coming Oil Crisis.* Multiscience, Brentwood, UK.

Yergin, D. (1991) *The Prize: The Epic Quest for Oil, Money, and Power.* Simon & Schuster, London.

CHAPTER 2

TOOLS

2.1 INTRODUCTION

The petroleum geoscientist uses a wide range of tools to help explore for and produce petroleum. Some of these tools will be familiar to the student geoscientist, and some others will be familiar to the professional reservoir engineer. This "tools" chapter introduces the basic technology that is familiar to the petroleum geoscientist. Stated most simply, the petroleum geoscientist needs to describe the distribution, at basin to pore scale, of rock, fluid, and void in the Earth's subsurface. To do this, the geoscientist uses a large array of data types and methods (Table 2.1).

Data acquisition is examined briefly, as are methods of rock analysis. The products of the data, fluid, and rock analysis are described, along with a brief introduction to the application of the information. More detail on application resides in the individual chapters that follow.

2.2 SATELLITE IMAGES AND OTHER REMOTE SENSING DATA

2.2.1 Introduction

Satellite images, and gravity and magnetic surveys, are most commonly used during the early or mid-term phases of basin exploration. The methods tend to be used more often to evaluate a whole basin, or at least a substantial part of a basin, rather than an individual prospect or area. This is not a function of the resolution of the tools—particularly that of satellite images—but, rather, a function of the ease and low cost of obtaining information over a large area (Halbouty 1976). Thus application of such techniques tends to focus on evaluation of gross basin architecture: structural style and assessment of the likely basin fill (sediments or volcanic rock). Nevertheless, there are applications in which satellite image data have been used to identify petroleum seepage and thus individual traps (Section 2.8).

2.2.2 Satellite images

Image capture by satellite commonly takes one of two forms, spectral (compositional) data and photogeologic/photogeomorphological. Spectral data can be collected over a large range of wavelengths (0.2 μm to 50 cm) and includes ultraviolet, visible, near infrared, short-wavelength infrared, infrared, and microwave. However, within this range of wavelengths, parts of some bands are unusable due to atmospheric absorption. In photographic mode, satellite images are similar to those produced by aerial photography.

In using satellite data to try to image the solid subsurface or subsurface, we must consider that the image captured includes contributions from the atmosphere, hydrosphere, biosphere, soil, and subsoil. Nonetheless, satellite images can be used to identify large-scale basinal structures that would prove difficult and expensive to capture by other means (Fig. 2.1). Spectral data have yet to find common usage in the oil industry, and yet they have been successful in seep detection (Section 2.8).

2.2.3 Gravimetric data

Gravity data can be used to help define the regional tectonic regime, prioritize areas for seismic work, and identify the causes of seismic structure (e.g., reefs, salt, and basement uplift). They can also be obtained and used where seismic data are difficult to acquire, such as in inaccessible terrain. Moreover, gravimetric data can be obtained at much lower cost than seismic data. However, the resolution of gravimetric data is lower than that of seismic data, and interpretation of such data commonly yields ambiguous results.

Table 2.1 Data used in petroleum exploration and production.

Data type and source	Use
Satellite images and other remotely sensed data	Largely in frontier exploration (Chapter 3)
Seismic data, including:	
2D	Frontier exploration and exploitation (Chapter 4)
3D	Exploitation, appraisal (Chapter 5), development, and production (Chapter 6)
4D (repeat or time-lapse surveys)	Production
4C (component) includes shear wave signal	Used from exploration to production
Wireline log data	Exploration to production, but only once a well is drilled
Cuttings and cores from wells	Exploration to production, but only once a well is drilled
Fluid samples from wells	Exploration to production, but only once a well is drilled
Outcrop data	Frontier exploration, before wells are drilled, to production
Seepage of petroleum	Frontier exploration (Chapter 3) and exploitation (Chapter 4)

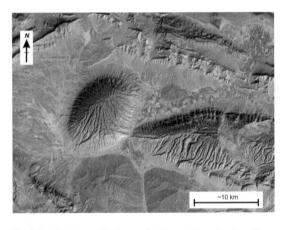

Fig.2.1 A Landsat satellite image of the Zagros Mountains, Iran. The center of the image shows a near-circular anticline above a salt dome. The radial erosion can be seen clearly.

Gravitational prospecting uses Newton's Law, which links the force of mutual attraction between particles in terms of their masses and separation. The law states that two particles of mass m_1 and m_2 respectively, and of small dimension compared with the distance r that separates their centers of mass, will be attracted to one another by a force F as follows:

$$F = G \frac{m_1 m_2}{r^2}$$

where G is the universal gravitational constant which, in the centimeter–gram–second (c.g.s.) system, equals

6.670×10^{-8}. This is a force, in dynes, exerted between two masses, each of 1 g, with centers 1 cm apart.

The acceleration (a) of a mass m_2 due to the attraction of mass m_1 at distance r can be calculated by dividing the attractive force by the mass m_2 (since force is mass × acceleration), thus:

$$a = \frac{F}{m_2} = G \frac{m_1}{r^2}$$

Acceleration is a force that acts upon a mass. It is the conventional quantity used to measure the gravitational field at any point. In the c.g.s. system, acceleration is measured in units of $cm\,s^{-2}$. This unit is referred to as a "Gal" (after Galileo). The gravitational attraction at the Earth's surface is about 980 Gals. However, in exploration geophysics it is likely that measurement differences in acceleration of about one millionth of the Earth's field will be required. In consequence, the commonly used unit is the milliGal (where 1000 milliGals = 1 Gal).

Gravity anomalies are caused by the heterogeneous distribution of rocks of different density (Table 2.2). However, because the measured gravity signal is a composite for all rocks underlying the area of study, there is often ambiguity as to the exact cause of a particular anomaly. Some of the ambiguity may be resolved by combining gravity and magnetic data, coupled with seismic data and wellbore density data if available.

Gravity data may be collected on land, at sea, or most commonly today by air. The land-based instruments can measure gravity anomalies with greater accuracy, precision, and resolution than other instruments. Airborne gravimetry trades

Table 2.2 Rock densities.

Rock type	Density range (g cm^{-3})	Average density (g cm^{-3})
Granite	2.50–2.81	2.64
Basalt	2.70–3.30	2.90
Acid igneous	2.30–3.11	2.61
Basic igneous	2.09–3.17	2.79
Metamorphics	2.40–3.10	2.74
Salt	2.10–2.50	2.22
Sandstone	1.61–2.76	2.35
Limestone	1.93–2.90	2.55
Dolomite	1.77–3.20	2.70
Shale/mudstone	2.58–3.42	2.40

accuracy, precision, and resolution against speed and areal coverage. Typically, airborne gravimeters are accurate to a few milliGals for gravity anomaly wavelengths of a few kilometers. Gravity data are subject to a variety of corrections before they may be used for interpretation, including the following:

- Isostatic corrections to minimize the influence of topography (for example, dense mountains adjacent to air-filled valleys will cause an anomaly because of the mountains themselves);
- tidal effects;
- temperature variations (that affect the instruments) and fatigue of equipment;
- latitude correction—gravity increases from the Equator to the poles due to the centrifugal force and flattening at the poles;
- free air correction—gravity diminishes with elevation above the Earth's surface, so gravity data are corrected to a sea-level datum;
- the Bouguer correction—this takes into account the mass between the data collection station and the chosen (sea-level) datum.

The corrected data are plotted as contoured isogravity maps. The maps combine the effects of all horizontal variations in density on both a regional and a local scale. Petroleum explorers are interested in anomalies of an areal extent large enough to be "economically interesting." The underlying principle is that the size, shape, and orientation of gravity anomalies should be similar to the shape, orientation, and areal extent of any petroleum traps, simply because both are largely controlled by structure.

2.2.4 Magnetic data

Magnetic data may be used to separate basin from nonbasin areas, define the shape and form of a basin, locate the major tectonic features within a basin, and identify the type of structure previously identified on seismic (e.g., salt pillow from shale lump). Magnetic data may be collected using an airborne tool: thus, like gravimetric data, they may be collected in areas that are not easily accessible for seismic acquisition.

The Earth's magnetic field may be divided into three components, only one of which is of interest to exploration geophysicists. The three components are the external field, the main field, and variations in the main field. The external field originates outside the Earth, fluctuating on diurnal and longer timescales. The main field is generated largely by the Earth's metallic core. It varies through time, but the rate of change is slow. Variations in the main field are commonly much smaller than the main field signal. They are produced by local magnetic anomalies in the near-surface crust and are relatively constant with time. It is these variations in the main field that are of interest to the oil explorer.

Magnetic field strength is measured in oersteds (Oe). The Earth's field varies between about 0.3 Oe at the Equator and 0.6 Oe at the poles. In exploration geophysics, anomalies are measured in gamma units—equivalent to the nanotesla (nT) in SI units—where 1 gamma = 0.00001 Oe.

Rock magnetism has two components, induced and remanent. The induced component is proportional to the Earth's magnetic field and the proportionality constant is called the "magnetic susceptibility." Thus a vector sum is used to describe the magnetization of a rock in terms of the remanent and induced components.

Magnetic susceptibility measures the degree to which an element or mineral can be magnetized. Rocks are commonly divided into three types. Diamagnetic substances (halite, anhydrite) exhibit a repulsive force to a magnetic field. Paramagnetic substances (including aluminum compounds) are attracted to magnetic fields. Ferromagnetic substances (magnetite, ilmenite) are paramagnetic materials with pronounced magnetic properties.

Remanent magnetism exists where there is no applied magnetic field. It is entirely independent of the induction effects of the present Earth's field. It is remanent magnetism that provides the evidence for our understanding of plate tectonics.

Magnetic data may be collected on land and with shipborne or airborne magnetometers. Once collected, data need to be corrected to remove the spatial variation in magnetic intensity caused by the Earth's main field and external fields.

The interpretation of magnetic data is more complex than that of gravimetric data, due to the inclined and varying properties of the magnetic field. However, the interpretation tools are much the same; that is, the generation of anomaly maps. Such maps are interpreted in terms of the effects of basement and cover structures.

2.3 SEISMIC DATA

2.3.1 Introduction

Seismic data are of critical importance for trap detection. They are the only widely used data that give a complete, albeit fuzzy, picture of the whole area of study, be it basin, play fairway, prospect, trap or reservoir. Few oil- or gasfields have been discovered in recent decades without the aid of seismic data. Indeed, it is difficult to conceive of petroleum exploration and production occurring today without seismic data. Such data are used extensively throughout the value chain. Seismic imaging of the Earth's shallow structure uses energy waves created at a sound source and collected some distance away (Section 2.3.2). A typical sequence of seismic acquisition from basin entry to oilfield production is as follows:

• Shoot regional two-dimensional (2D) lines across the basin. Individual lines are commonly tens of kilometers long, and may be hundreds of kilometers long. Such lines will be interpreted in terms of gross structural and stratigraphic geometries.

• Obtain 2D seismic across the area licensed to or owned by the exploration company. The density of coverage varies, but it needs to be sufficient to enable the interpreting geophysicists and geologists to define structural and stratigraphic configurations that might trap petroleum. At this stage there may be sufficient information on which to plan and execute an exploration well. However, it is possible that very closely spaced 2D or even 3D seismic will be obtained in order to define the trap and potential well locations more accurately. Once a drilling location has been chosen, a specific site survey is made. This will usually involve a sonar survey of the sea bed/lake bed in offshore locations, shallow coring both on land and possibly in water, and possibly a high-resolution seismic survey. Surveyors use all three techniques to check for hazards, such as shallow gas. However, if the target is near to the surface, the shallow seismic survey may also provide valuable information for the interpreting geophysicist.

• Following discovery of a petroleum pool, 3D or possibly further 2D seismic acquisition will be undertaken if the detail from the original seismic data is deemed insufficient. A discovery needs to be evaluated to assess the potential reserves and to determine the location of future appraisal and development wells.

• One or more repeat 3D surveys are now commonly obtained throughout the production life of large and giant fields. The purpose in taking a second or third 3D seismic survey has, up until the mid-1990s, largely been to take advantage of the rapidly improving quality of seismic surveys (Fig. 2.2). Recently, there has been a move to record repeat seismic surveys using, as near as possible, the same acquisition parameters, in an effort to detect differences in seismic response that can be linked directly to the response of the reservoir to production. The sorts of effects that might be detected include the movement of oil/water, gas/water, or gas/oil contacts; the generation of secondary gas caps; and compaction of overburden resulting from pressure decline in the reservoir. The terms "4D seismic" and "time-lapse seismic" have come into common parlance to describe such repeat 3D surveys that are used to examine temporal changes in the seismic response of an area.

• In well-developed basins, particularly those offshore, where seismic acquisition is inexpensive compared with drilling costs, 3D surveys may be acquired for exploration and reserves growth purposes (Section 1.7.5).

In recent years, increasing use has been made of shear wave data within the seismic signal. Shear wave data are not collected in conventional marine surveys, because shear waves are not transmitted through fluids (the water column). Only primary, compressional wave data are collected. In order to collect shear wave data, geophones are laid on the sea bed. Such surveys tend to be called ocean-bottom cable (OBC) surveys. In shear waves, the displacement along the waveform is orthogonal to the direction of propagation. As a consequence, it is possible to collect three sets of orthogonal shear wave data. These three sets of shear wave data, together with the primary compressional data, are known as four-component (4C) seismic.

The reason for collecting the shear wave data is that fluids do not transmit shear waves. This means that shear wave data are unaffected by gas plumes and gas "push-down" effects. Four-component seismic can be used to see through these gas effects. Moreover, because fluids do not transmit shear wave data, it is in some instances possible to subtract the shear wave signal from the compressional wave signal. In theory, this leaves only the seismic response due to the fluid. In this way, it has proved possible to image some petroleum accumulations directly.

Four-component seismic data can be collected in either 2D

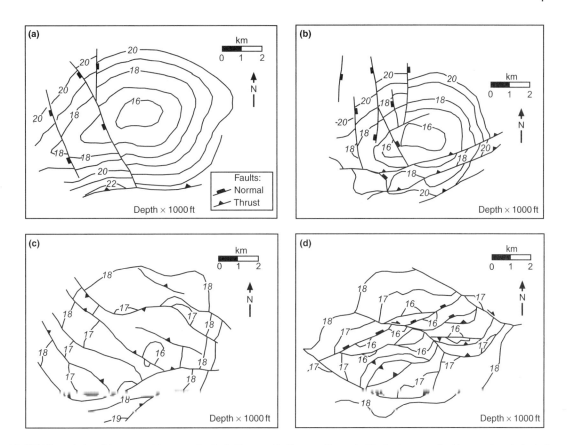

Fig.2.2 The evolution of the structural interpretation for the Boqueron Field, eastern Venezuela. The first interpretation (a) was made on the basis of 2D seismic only in 1988. Interpretations from 1990 (b) and 1992 (c) have no new seismic data, but incorporate the results from new wells. The 1993 interpretation (d) is based upon a combination of 2D seismic, well data, and a 3D seismic survey. The structural interpretation is radically different than the basically simple faulted dome interpretations based upon 2D seismic data. (From Lagazzi et al. 1994, reproduced courtesy of SOVG.)

or 3D arrays, and it can be—and is—used throughout the value chain.

2.3.2 The seismic method

The use of seismic data in petroleum exploration goes back to the second decade of the 20th century. However, its roots go further back, to the closing years of the 19th century and the first few years of the 20th century, when there was considerable interest in using seismic methods to record recoil (and thus location) from enemy heavy artillery.

The application of the seismic method to the search for petroleum began in the US Gulf Coast in 1924, when Orchard Salt Dome Field (Texas) was discovered (Telford et al. 1976). Here, a refraction seismic survey was used. It exploits the property whereby much seismic energy travels along the interface between lithologies that have different acoustic properties. Three years later (1927), a reflection seismic survey was employed to aid definition of the Maud Field (Oklahoma). Today, reflection seismic surveys, which are suited to relatively low-dipping strata, are used most commonly in the search for petroleum. Refraction surveys remain the favored method for those earth scientists who are studying the deep structure of the Earth. We will limit our discussion to reflection seismology.

The seismic method relies upon changes in acoustic properties of rock to alter the properties of sound waves transmitted through the rock. At its most basic, the requirements are an energy (sound) source, a receiver, and of course the layered Earth. A signal is transmitted from the source down into

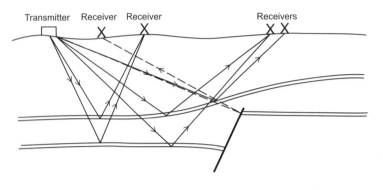

Fig.2.3 A schematic diagram showing ray-paths during seismic acquisition. In reality, many transmitters and receivers are operating simultaneously.

the Earth. Some of the transmitted energy is subsequently reflected at surfaces beneath the surface of the Earth, where there is acoustic contrast between different rock layers. The receiver (Fig. 2.3) collects the returned signals. In reality, many receivers are used. This enables the points of reflection to be located in a space defined by linear x and y coordinates as measured on the Earth's surface and z measured in time beneath the Earth's surface. The reflections generated from many sources delivering signals to many layers in the sub-surface, and collected at many receivers, are compiled to yield seismic cross-sections in 2D and seismic volumes in 3D.

The acoustic contrast is a combination of the contrasts in velocity and density, and is characterized by the reflection coefficient:

$$R = \frac{v_2\rho_2 - v_1\rho_1}{v_1\rho_1 + v_2\rho_2}$$

where R is the reflection coefficient, ρ is the density, v is the compression wave velocity, and the subscripts "1" and "2" refer to the upper and lower formations, respectively. The quality of reflectors and hence the ability to define successions of rocks and their characteristics depends initially on the natural variations in the rock. Strong lithological/mineralogical contacts may only have small values of R and vice versa. Quality also depends upon the type of seismic sources and receivers used.

Two-dimensional seismic cross-sections and 3D seismic volumes are most commonly displayed using true geographic coordinates and distances in the x and y directions but time, rather than linear depth, in the vertical (z) direction. In fact, two-way time—that is, the time taken for the seismic signal to propagate down to the reflector and back to the

surface—is plotted on the z-axis. There is a strong positive correlation between two-way time and depth, but in most instances it is nonlinear. As a consequence, seismic data plotted in the time domain can be interpreted as if it were a geologic cross-section. Transformation of two-way time to depth is obviously an important aspect of seismic interpretation if a prospect is to be drilled. It is covered later in this section.

For interpretation, seismic data are displayed as series of vertical, sinusoidal transverse waves arranged in time series, with zero two-way time at the top of the display. The amplitude of individual peaks or troughs on the waves (traces) is proportional to reflection strength. On black and white displays, either the peaks or the troughs are filled in. On color displays, the peaks may be filled with one color and the troughs with another. Individual traces are commonly spaced at a millimeter or less apart. In consequence, the peaks (or troughs) from one trace to the next commonly converge to yield a seismic reflector on the display (Fig. 2.4). It is the reflectors and their terminations that are mapped during seismic interpretation.

2.3.3 Seismic acquisition

Land

Dynamite and Vibroseis™ are the most common sources of energy for land-based seismic surveys. Dynamite has the longest history of use. Shots are placed in small-diameter (a few centimeters) hand- or machine-augured boreholes. The depth of the boreholes varies considerably, but is commonly between 6 m and about 30 m. Vibroseis™ comprises a heavy all-terrain vehicle that can lower a steel plate onto the ground surface (Fig. 2.5). The plate is then vibrated with a specific frequency distribution and amplitude. Vibroseis™ is the

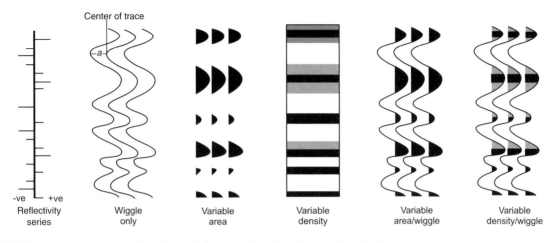

Fig.2.4 Examples of various types of seismic trace display: a = amplitude. (From Emery & Myers 1996.)

Fig.2.5 Land seismic acquisition using Vibroseis™ as an energy source. The photograph was taken in 1991 during a seismic survey of Europe's largest onshore oilfield, Wytch Farm, Dorset, England. (Copyright Sillson Photography.)

most common method in populated urban and rural areas. Dynamite is commonly used in areas where Vibroseis™* could not be employed, such as swamps, coastal flats, and other soft surfaces.

Other energy sources have been used from time to time. These include: the thumper or weight dropper; Dinoseis™*, which involves the explosion of a propane/air mixture in a chamber mounted below a truck; and Geoflex™*, which uses an explosive detonating cord, buried a few inches below the surface.

---

* Vibroseis™ is a trademark of Conoco. Dinoseis™ is a trademark of Arco. Geoflex™ is a trademark of ICI.

Water

The acquisition of seismic data from marine or, more rarely, from lacustrine surveys tends to be much less labor-intensive than that used on land or transitional (land/sea) surveys. Nowadays, the energy source for such surveys is almost exclusively the air gun. An air gun discharges a high-pressure pulse of air into the water (Giles 1968). The bubble of high-pressure gas vibrates within the water after release from the air gun chamber. The vibrational energy passes through the water column and into the underlying sediment. It is common to use a variety of air gun sizes so as to deliver energy with an array of frequencies to the subsurface. The air guns are towed behind boats.

Collection of seismic data: receivers

Seismic energy returning to the surface, on land or sea, is collected by an array of receivers. The arrangement of such "hydrophones" will depend upon the type of survey. Two-dimensional surveys, as their name indicates, use a linear array of receivers. Three-dimensional surveys are acquired in a variety of patterns, such as grids and brick patterns. The receivers are connected to a recording unit. Most often, cables are used, but in land areas with difficult access, such as swamps, mountains, or urban districts, radio telemetry may be an easier way of delivering data from receiver to recorder. The information delivered to the recording unit comprises a series of sinusoidal waves with differing phases, frequencies, and amplitudes. Since there are many energy sources and several types of generated waves and reflection surfaces, the seismic energy is delivered back to the receiver array over a time period and processing is required to correct for this.

2.3.4 Seismic processing

The aims of seismic processing are to enhance the interpretable (useful) seismic information relative to the noise in the signal and place the seismic reflectors in their correct x, y, z space. This is a complex subject, which is beyond the scope of this work: for a detailed description of seismic processing, refer to Telford et al. (1976). Here is a brief description of some of the processing steps:

- Fourier transformation, in which the signal is changed from the time to the frequency domain.
- Convolution/deconvolution processes, which are designed to allow determination of the effect of the Earth on the seismic signal.
- Frequency filtering, which allows management of the data output from the data input.
- The data are zero phased such that a consistent output results from the addition of waveforms. For example, the addition of equal signals that are completely out of phase will result in no signal.
- Common depth point stacking, which involves the arrangement of component data for a single depth point side by side.
- Muting of the first refraction breaks (signal arrivals), which reduces the masking of weak shallow reflections.
- Migration. Provided that the velocity only varies with depth (i.e., not laterally), the travel time for a diffraction is at a minimum directly above the point of diffraction. Clearly, the Earth rarely behaves with such simplicity. In consequence, migration is designed to restore seismic reflectors to their proper x–y position. Most seismic interpretations are made on sections or volumes in which the x and y dimensions are measured in space and the vertical, z, direction in time (strictly, two-way time). Depth-migrated data incorporates vertical and spatial variation in the velocity of the Earth's subsurface. Transformation of time to depth is of critical importance: many "exciting prospects" have turned out to be unreal — simply products of high-velocity rock causing underlying rock to be pulled up on a two-way time display.

2.3.5 Seismic interpretation

The object of seismic interpretation is to generate a coherent geologic story from an array of seismic reflections. At the most simple mechanical level, this involves tracing continuous reflectors across 2D grids of seismic lines or throughout 3D data volumes. The derived collection of lines or surfaces will then form the basis of the geologic interpretation. However, because a seismic survey delivers a fuzzy picture of the subsurface, it is rarely possible to generate a unique interpretation of a seismic dataset. The problem of deriving an interpretation that is close to reality is further exacerbated when data coverage is meager, such as in the early stages of exploration in a basin.

There is an alternative way to perform seismic interpretation. Ultimately, tracing of reflectors is needed in order to derive maps of stratal surfaces and faults. However, much of the seismic interpreter's job is intuitive, in that geologic experience is brought to bear on the interpretation of what is essentially just a collection of tiny black and white (or nowadays colored) waveforms. The approach differs slightly depending upon whether the interpreter is using paper copies of seismic lines or a computer workstation with a 2D or 3D dataset. The paper-based approach is to view the 2D paper seismic display as you would an impressionist painting — perhaps a Monet. Look at the seismic section from different angles, obliquely from the side, or upside down. The human brain is good at pattern recognition. Just as with an impressionist painting, the seismic section will probably deliver a suggestion of shapes that, to a geoscientist, will have a geologic analog. The mental analog may then be used to guide the detailed interpretation. If time permits, it is often useful to get a second or third interpretation of the same data by different geoscientists. Even if these second and third opinions are taken no further than the "impressionist" interpretation, the team will have delivered a powerful analysis of the uncertainties associated with interpretation of the seismic dataset. The alternative interpretations may prove extremely valuable if drilling results fail to confirm the initially preferred interpretation of the data. One of the alternative interpretations may better explain the drilling results and so allow rapid improvement in the understanding of the area drilled.

The approach using 3D seismic data on a workstation differs in that instead of taking a trip to the art gallery, the interpreter takes a trip to the movies. Three-dimensional seismic datasets are usually interpreted on a workstation. The computer files contain the whole seismic volume, which can be viewed or sliced in any direction. It is possible to make a movie of sequential slices through the data volume. The choice of orientation for the slices will depend on the perceived orientation of structural or stratigraphic elements within the subsurface. In addition to vertical slices, it is also possible to construct a movie of time slices (c.4 μs apart)

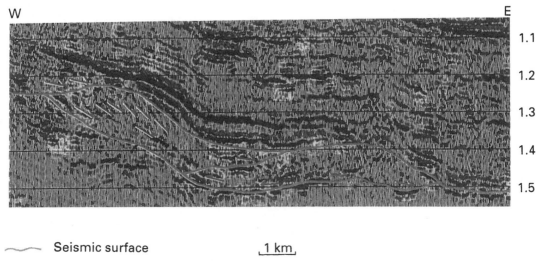

W   E

Two-way time (s)

1.1
1.2
1.3
1.4
1.5

⌒ Seismic surface   ⌐1 km⌐
↘ Reflection termination

Fig.2.6 Stratigraphic information derived from seismic: a lowstand systems tract on seismic data. The lowstand wedge is highlighted in grey on the left-hand half of the section. A lowstand fan unit occurs as the mounded feature on the right-hand side of the section, also outlined in grey. This example is from the Eocene of the North Sea (for an explanation of terms, see Section 3.6). (From Emery & Myers 1996.)

through the data. This could either be horizontal for strata with low dip or flattened to a marker horizon for high-dipping strata. In much the same fashion that the "Monet approach" to the paper displays of seismic will deliver an impression of the structure and stratigraphy, so too will the movie as the data volume is stepped through. Again, a mental analog will be created that can be used to help further detailed interpretation.

In an exploration setting, a structural description is commonly the primary objective of a seismic interpretation. The second level of interpretation is often stratigraphic (Fig. 2.6). More detailed interpretations of the geology from 2D and 3D seismic data are given within the appropriate chapters. Derivatives of seismic data, such as coherency, acoustic impedance, and amplitude versus offset, are also investigated in the chapters that follow.

The seismic method has revolutionized the search for, and development of, petroleum accumulations. Each new improvement in seismic data has also been accompanied by comparable improvements in exploration and production success. Nonetheless, it is not without its pitfalls and limitations, which should be borne in mind when interpretation is in progress. Some of the common limitations and pitfalls are as follows:

• *Resolution.* Even the most carefully gathered high-frequency 3D data are limited in their resolution to about 10 m. Most data have much poorer resolution.

• *Tuning effects.* Just as out-of-phase wave data will lead to signal destruction, so completely in-phase data from two reflectors can lead to amplification — the two reflectors appear as one strong reflector.

• *Multiples.* Bogus multiple reflections of a strong near-surface reflector (such as the bottom of the sea) can occur spread throughout a dataset. These can usually be identified because of their regular spacing.

• *Velocity "pull-ups" and "push-downs."* These are created by lateral variations in the seismic velocity of the overburden. They will disappear if the seismic is properly depth-migrated.

2.4 WIRELINE LOG DATA

2.4.1 Introduction

Wireline logs are tools that are attached to a "wireline," or steel cable, lowered to the bottom of a well after each major stage of drilling and then hauled back. As the tools pass by the various rock layers on the way back to the drilling rig, they record both intrinsic and induced properties of the rocks and their fluids. Over the past decade, conventional wireline logs

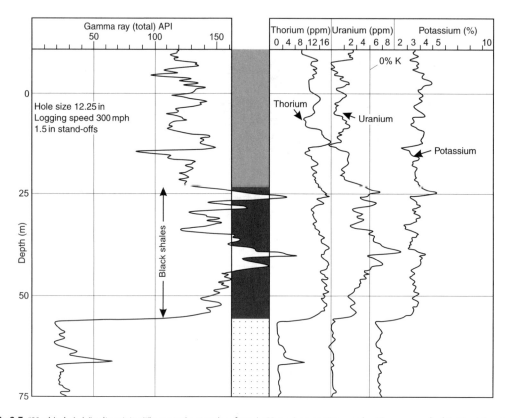

Fig.2.7 "Hot black shale" radioactivity. The spectral gamma log, from the Upper Jurassic, Kimmeridge Clay source rock of the North Sea, shows the significant contribution of uranium to the total radioactivity. (From Rider 1986, reproduced courtesy of Blackie.)

that are run after drilling have been supplemented with tools that can make much the same measurements but that are attached to the active drillstring. Such near-instantaneous information has added greatly to the ability to steer the well through the desired part of the stratigraphy.

Whether run during or after drilling, logs are used to give information about the rock, its pore space, and the fluids therein. Data on the condition of the wellbore can also be acquired. Pore fluid pressure can be measured using the Repeat Formation Tester (RFT^{TM}*) or a tool of similar type which, although not a wireline log in the strictest sense, is wireline-conveyed. The results from a suite of logs are integrated to yield a description of all three components. Unfortunately, no

single tool can give an unambiguous description of any of the three (rock, pore space, and fluid type) components. Combinations of tools are used, and the responses of these to the formations are calibrated with information gained from core.

2.4.2 Rock tools

Today, the prime tools for providing rock information are those that measure the natural radioactivity of the formations. Most common is the gamma ray tool. This tool provides a picture of the combined gamma ray output of all the radioactive elements in the rock formations. More sophisticated are the spectral gamma ray tools. These contain an array of receivers designed to measure radioactivity at different wavelengths. In so doing, the effects of potassium, uranium, and thorium can be differentiated. All of these tools record gamma ray activity in API units. This is a scale defined by the

---

* RFT^{TM} is a trademark of Schlumberger.

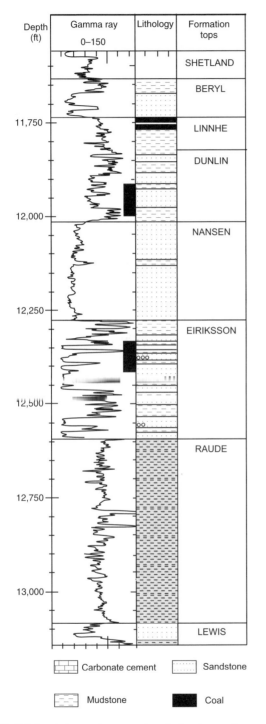

Depth (ft)	Gamma ray 0–150	Lithology	Formation tops
			SHETLAND
			BERYL
11,750			LINNHE
			DUNLIN
12,000			NANSEN
12,250			EIRIKSSON
12,500			
			RAUDE
12,750			
13,000			LEWIS

Carbonate cement Sandstone

Mudstone Coal

Fig.2.8 A typical gamma log response in a mixed sandstone and mudstone sequence, Beryl Field, North Sea. The gamma log has been used to "pick" sandstone and mudstone intervals in the well. The black bars in the gamma ray columns are the cored intervals. (From Robertson 1997, reproduced courtesy of the Geological Society of London.)

American Petroleum Institute, in which the natural radioactivities of rocks are compared with that of a known standard rock.

It is worthwhile considering the types of sedimentary rock in which potassium, uranium, and thorium are likely to influence log response. Potassium is the most common of these elements in sedimentary rocks. It occurs in some clay minerals that may be present as detritus in mudstone or as diagenetic cements in sandstones. Potassium also occurs in feldspars, a common mineral in many types of sandstone. However, the most concentrated sedimentary source of potassium is within sylvite (KCl), carnallite (KCl·MgCl$_2$(H$_2$O)$_6$), and polyhalite (K$_2$SO$_4$·MgSO$_4$). Naturally, rocks that contain an abundance of these evaporite minerals are much more radioactive than the average mudstone, which contains only few percent of potassium.

Uranium is a highly mobile element in oxidizing environments. It is soluble in the form of the uranyl ion (UO$_2$$^{2+}$), which is produced during weathering of igneous rocks. However, it is much less soluble in acidic and reducing environments, conditions that can lead to precipitation. It can also be adsorbed onto organic matter and precipitated in association with phosphates. Thus the conditions appropriate to fix uranium are much the same as those needed to generate source rocks (Chapter 3). In consequence, there is often an association between mudstones that are rich in organic matter and high in uranium radiation (Fig. 2.7). Mudstones that are rich in organic matter and have a high gamma ray response are often called "hot shales."

Thorium is weathered from igneous rocks, sharing its source with uranium. It is, however, extremely insoluble. It tends to adsorb onto terrestrial clays such as kaolinite and remain with those clays on its journey to the sedimentary basin. Thorium may become concentrated in the products of intense weathering, such as bauxites. It also occurs in heavy (dense) mineral lags.

In a geologic context, the gamma ray tools are commonly used to measure shaliness—shales are commonly more radioactive than sandstones. A "typical" clean sandstone may yield an API reading of 0–20, while a mudstone commonly gives a reading of >100 (Fig. 2.8). Such generalities are usually justified, but readers should be aware of situations in which highly radioactive sands or poorly radioactive mudstones may be present. For example, sandstones that contain abundant potassium-bearing feldspar (arkoses) will give high API readings. Similarly, sandstones that contain phosphate clasts or phosphate cements can have very high uranium contents (Cazier et al. 1995). Conversely, mudstones that accumulated in humid and hot conditions are likely

Well data

R_m	0.44 Ω	40.5°C
R_{mf}	0.68 Ω	25°C
R_{mc}	2.18 Ω	25°C
Borehole temp. 80°C		
R_w	0.27 Ω	25°C

Lithology

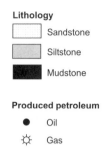

- [ ] Sandstone
- [ ] Siltstone
- [■] Mudstone

Produced petroleum

- ● Oil
- ☼ Gas

Fig.2.9 An example of a shale baseline and the SSP defined on an SP log. The shale (mudstone) baseline is the maximum positive deflection (in this example) and occurs opposite mudstones. The SSP is a maximum negative deflection and occurs opposite clean, porous, and permeable water-bearing sandstones. (From Rider 1986, reproduced courtesy of Blackie.)

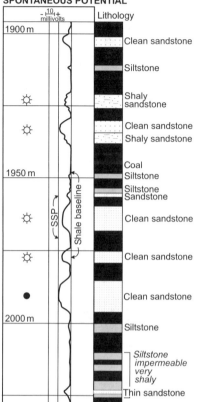

to be rich in kaolinite, potassium poor, and hence have low radioactivity.

A real example of where the gamma tools alone do not provide enough information for the lithologies to be differentiated is the Clyde Field of the Central North Sea (Turner 1993). It contains three main lithofacies; medium-grained arkosic sandstone, very fine-grained quartzose spiculite sandstone, and mudstones/siltstones. Only one of these lithofacies, the arkose, is an effective reservoir. It contains a relatively high abundance of potassium feldspar, and the gamma ray response of this arkose lies between that of the other two lithofacies in the field. The low-porosity, impermeable siltstones/mudstones have a very high gamma ray response. The highly porous but impermeable, very fine-grained sandstone, composed almost entirely of sponge spicules, has a very low radioactivity. Consequently, other wireline tools must be used in Clyde to identify reservoir sandstones. This is an unusual situation: in most instances, the data from any of the gamma tools provides sufficient information for sandstone and mudstone to be differentiated.

Limestones and dolomites commonly have low API responses such that the gamma tools can rarely be used to subdivide carbonate sequences with any degree of accuracy. Similarly, the gamma ray tools must be used in conjunction with other tools to identify other common rock types, such as coal and halite.

The spontaneous potential tool (SP) was once the most common tool for lithological identification; however, its use nowadays is less widespread than that of the gamma ray tool. The SP log records the difference in electrical potential between a fixed electrode at the surface and a movable electrode in the borehole. The hole must contain conductive mud in contact with formation, and so the tool cannot be used in oil-based muds, empty holes, or cased holes. Measurement is in millivolts. There is no absolute zero: only differences are recorded. The log essentially measures differences from a shale (mudstone) line. The theoretical maximum deflection of the SP opposite clean, permeable, and water-bearing sandstones is called the static SP (SSP, Fig. 2.9). At shallow depths where salinity is low, permeable sandstone beds may show a negative deviation from the shale line. In many basins, deeper sandstones with more saline formation waters will show a normal response. A comparison between a gamma ray log and SP curve is shown in Fig. 2.10. SP logs have limited value in carbonate reservoirs.

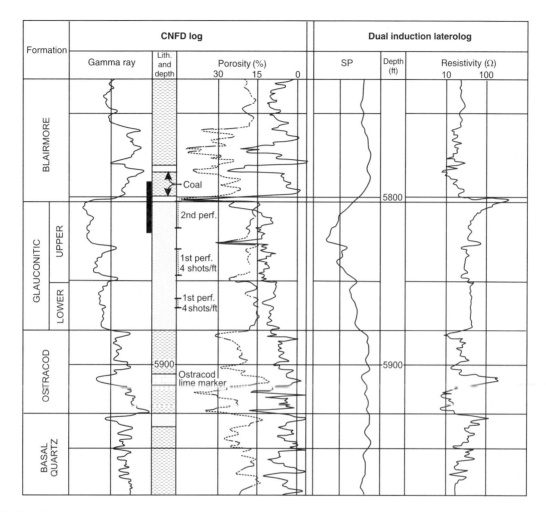

Fig.2.10 A comparison of gamma ray and spontaneous potential logs run at the same time in the same wellbore, Hoadley gasfield, Alberta, Canada. The gamma ray resolution is here much better than that of the SP resolution. (From Chiang 1990, reproduced courtesy of the American Association of Petroleum Geologists.)

A second class of lithological tools, which include dipmeters, Formation MicroImagers™*, and Formation MicroScanners™*, is used to provide structural and sedimentological information. Most of these tools use many small receivers to record the microresistivity of the formation. Features such as fracture systems and cross-bedding (Fig. 2.11) generate the microresistivity patterns.

Earlier in this chapter, we investigated the acquisition and use of seismic data. In order to tie and calibrate such seismic

* Formation MicroImager™ and Formation MicroScanner™ are trademarks of Schlumberger.

data to a well, it is usual to acquire seismic data from within a well. This may be either through a check shot survey or through a vertical seismic profile. Vertical seismic profiles (VSPs) are seismic surveys acquired with standard near-surface sources, but with receivers down a well. A simple VSP has data recorded as a function of time and receiver depth, rather than time and offset as in a conventional seismic survey. A variety of recording geometries are possible (Fig. 2.12). The simplest geometry is a zero-offset VSP, used on vertical wells where source offsets are typically 50–100 m. For deviated wells, the source must be moved to correspond with the wellbore orientation. The offset VSP can be either a fixed

Fig.2.11 (a) Dipmeter and core logs for the Balder Formation (Eocene) in the Gryphon Field, North Sea. The Gryphon reservoir was fluidized soon after deposition. This generated a very complex trap geometry, with reservoir sand injected into the overlying mudstone. The contrast between the coherent dipmeter response from the seal and the chaotic response from the reservoir is of value in definition of the Gryphon trap. (From Newman et al. 1993, reproduced courtesy of the Geological Society of London.) (b) Formation MicroImager* (FMI) and whole-core images. Planes that cross-cut the borehole (or core) occur as sinusoidal traces on both images. The planes may be bedding features, fractures, or cementation effects. Once calibrated to core, the FMI and the older FMS (Formation MicroScanner™) may be used in uncored wells to give sedimentological and fracture data at the same resolution as that expected from core studies. (Images reproduced courtesy of Schlumberger and Badley Ashton.)

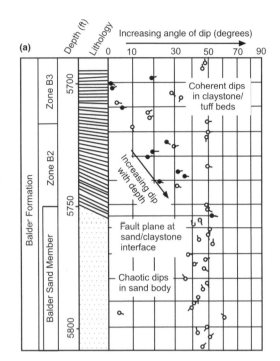

(a)

Coherent dips in claystone/tuff beds

Increasing dip with depth

Fault plane at sand/claystone interface

Chaotic dips in sand body

Oriented whole core

(b)

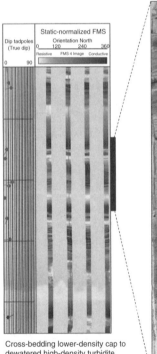

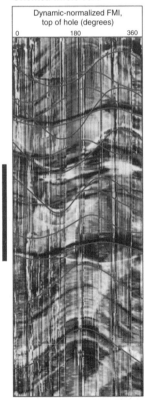

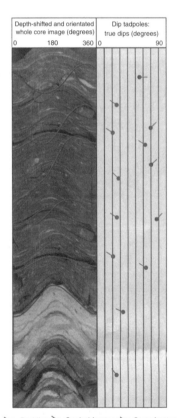

Cross-bedding lower-density cap to dewatered high-density turbidite provides paleocurrent data

Features in whole-core image matched to equivalent features resolved by FMI data in deviated well

Bedding   Sealed fractures   Open fractures

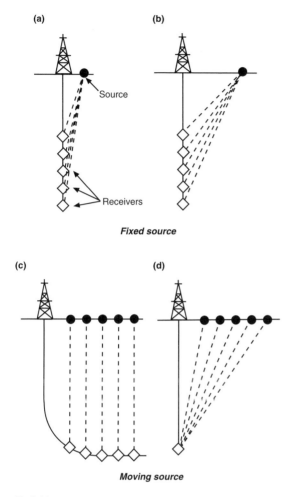

(a)

Source

Receivers

Fixed source

(b)

(c)

(d)

Moving source

Fig.2.12 Vertical seismic profile configurations. (a) Zero offset; (b) (long) offset, OVSP; (c) walk above, DVSP; (d) walkaway, WVSP.

offset, in which there is a single source and many receivers, or a walkaway VSP, whereby one receiver is used downhole with many sources on the surface. The term "offset" is generally applied to a VSP when the distance from well to seismic source is greater than about half the distance from wellhead to receiver location. The single most useful information from a VSP is that there is a firm tie between depth and time.

In structurally simple areas, VSPs can deliver an increased resolution relative to comparable 2D seismic surveys. This is because the seismic signal only passes one way, from surface source to downhole receiver. In practice, however, the improvement is minimal unless the near surface is highly attenuating, as can be the case when shallow gas is present.

Table 2.3 shows the sorts of situations in which VSPs can be used and the type of information that they might deliver. As with most of the tools described in this book, VSPs have applications in exploration, appraisal, development, and production. The only situation in which VSPs are inapplicable is in the very earliest stages of exploration, before a well has been drilled!

VSP data are commonly displayed as wiggle plots on an x–y graph, where the x-axis is the depth and the y-axis is the one-way travel time. Figure 2.13 contains some typical VSP data with signals from direct (down) arrival and its multiples, upwaves, and tube waves. Most of the information is dominated by the direct arrival and its multiples. The direct arrival times are used to estimate velocity and hence sonic log calibration. The amplitudes of the direct arrivals may also be used to estimate attenuation.

Tube waves are commonly regarded as noise. They travel slowly, at about the speed of sound, in the borehole fluid. However, they can be used to detect fractures. When a primary wave encounters a fracture, fluid is expelled from the fracture into the borehole (Beydoun et al. 1985). This process generates a tube wave. Tube waves might also be useful for detecting permeable zones in reservoirs.

Table 2.3 Uses of vertical seismic profile (VSP) data.

Exploration	Appraisal/development	Production	Description
✓	✓		To refine well ties and velocity model
✓	✓		To measure attenuation for survey design
✓	✓		To predict ahead of drill bit (well casing points)
	✓		Estimation of rock properties
	✓		Imaging around deviated wells and salt
	✓	✓	To constrain the stress direction and fractures
		✓	Reservoir monitoring

2850 ft   750 ft

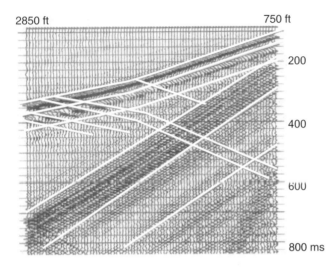

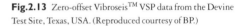

Fig.2.13 Zero-offset Vibroseis™ VSP data from the Devine Test Site, Texas, USA. (Reproduced courtesy of BP.)

2.4.3 Porosity and permeability tools

There are three common sets of tools that are used to measure how much of the formation is hole—in other words, pore space. These are the sonic, density, and the neutron tools.

The sonic tool works by sending a sound pulse into the formation and measuring the time taken for the sound wave to return to a receiver located further up (or down) the tool (Fig. 2.14). Typically, the distance between the transmitter and receiver is a few meters. The transit time in tight (nonporous) sandstone or limestone is short: in porous formations it is longer, and in mudstone longer still. Coal is "very slow." However, the responses of individual lithologies can overlap enormously. As such, the sonic tool cannot be used alone to identify lithology. Its main use is in the evaluation of porosity in liquid-filled holes. It may also be used to provide a calibration to a well for seismic data insofar as interval velocities and velocity profiles can be extracted from sonic log data. Used in conjunction with the density log, the acoustic impedance may be derived, from which a synthetic seismic trace may be constructed.

The formation density tool measures the electron density of a formation. It comprises a gamma ray source and detector. Gamma rays are emitted into the formation, where they interact with electrons and are scattered. The scattered gamma rays that reach the detector are counted as a record of forma-

tion (electron) density and the electron density can be related directly to bulk density (Fig. 2.15).

The plots from the tool rely on a straight-line, two-point interpolation between fluid (oil or water) with a density of $1\,\mathrm{g\,cm^{-1}}$ and rock with a mean density of $2.67\,\mathrm{g\,cm^{-1}}$. The porosity can thus be calculated from the measured value. A density calculation is the most reliable porosity indicator for both sandstones and limestones, because the density of the rock (solid) is well known. The same measurement is a less reliable indicator of porosity for mudstones, because of their variable density. The density tool is particularly useful for seeing gas-bearing sandstones. They yield artificially high porosities because gas is considerably less dense than oil or water.

Neutron logs determine the hydrogen atom concentration in a formation, and since almost all of the hydrogen occurs in the fluid phase (petroleum and/or water) such measurements can be transposed to porosity data. The neutron tools emit neutrons that lose energy as they collide with atoms in the host formation. Because of their similar size and mass, the most energy is lost when the neutrons collide with hydrogen atoms. Thus the maximum loss in energy equates to the highest volume of fluid-filled pores. In rocks that contain a large volume of "bound" water (water in the structure of minerals such as smectite or montmorillonite), the effective porosity is overestimated. Gas-filled pores also have a particular effect due to their low density. This leads to a low-

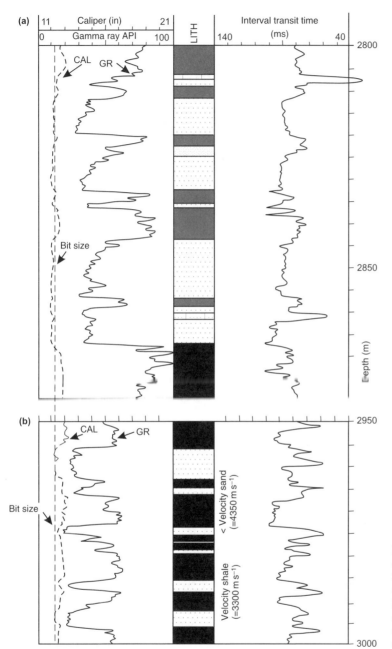

Fig.2.14 Sonic log responses in a sandstone and mudstone sequence. In (a) the sandstones have a lower sonic velocity than the shales (mudstones). In (b), the reverse is true. The large deflections at 2807 m and 2862 m are caused by pervasive carbonate cement in the sandstone. (From Rider 1986, reproduced courtesy of Blackie.)

ering of the value for neutron porosity and is called the "gas effect."

The search for a tool that is capable of directly measuring permeability has been the "Holy Grail" of petrophysicists for decades. The most common approach has been to calibrate resistivity logs to core or test permeability. However, there is a new generation of nuclear magnetic resonance tools (NMR) that can deliver permeability directly. The log works by in-

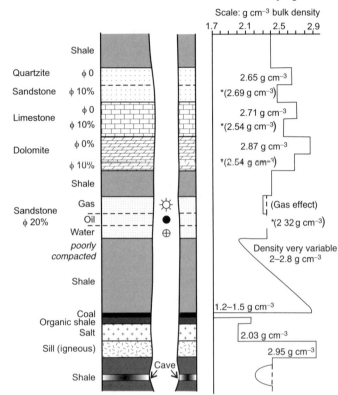

Fig.2.15 Typical responses of a density log. The density log records bulk density; hence porous rocks record a lower density than do rocks of similar mineralogy but lower density. *Density and porosity with fresh formation-water density 1.0 g cm⁻³. (From Rider 1986, reproduced courtesy of Blackie.)

ducing atom nuclei to align in a strong magnetic field and thus become magnetized. Specific radio frequencies (RF) are then used to disturb the alignment. This disturbance generates an induced RF field that can be measured. Only hydrogen delivers a signal that can be easily measured, and only the water and hydrocarbon in the rock contain appreciable hydrogen. Thus the NMR log sees only fluid. In order to measure permeability, the magnetism of the nuclei is rotated from parallel to round to perpendicular to the magnetic field. The time taken for the magnetic orientation to return to the parallel condition is called the "relaxation time." The relaxation rate of the NMR signal contains information about the surrounding fluids, such as pore size and pore throat size, factors that control permeability (van Ditzhuijzen & Sandor 1995; Fig. 2.16). NMR must be calibrated to core data.

Permeable-zone detection in fractured reservoirs might also be possible using tube waves generated during acquisition of a VSP.

2.4.4 Fluid tools

There is a large array of resistivity tools which, as their name implies, measure the electrical resistivity of the rock plus fluid (Fig. 2.17). The tools can be used to help identify rock type, but by far their most important application is for identification of petroleum. Petroleum, as either gas or oil, is much less conductive to electricity than water. By running a pair of tools in tandem, one that "sees" deep into the formation and one that sees only a little way, petroleum- and oil-bearing intervals in a well can be differentiated. The way in which the tools work is as follows. The deep tool records the resistivity deep in the formation (2–3 m from the borehole). This zone is unaffected by the activity of drilling the well. The other part of the tool reads close to the wellbore. This is in the zone that has been flushed of the original fluids by the water-based drilling mud. The drilling muds are solute-rich and therefore have low resistivity. Clearly, high-resistivity petroleum gives a very different signal. The separation between the two signals thus allows

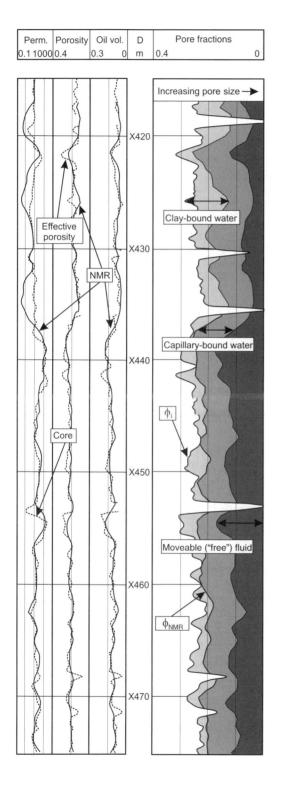

Perm.	Porosity	Oil vol.	D	Pore fractions
0.1 1000	0.4	0.3　0	m	0.4　　　　　　0

Increasing pore size →

Clay-bound water

Effective porosity

NMR

Capillary-bound water

ϕ_i

Core

Moveable ("free") fluid

ϕ_{NMR}

the identification of petroleum. For water-bearing formations, the signal from the deep part will be much like that from the shallow part of the tool. In situations where oil-based drilling muds are used, where formation water is fresh or reservoir beds are thin the methodology is not readily applicable.

2.4.5 Pressure tool

The most common pressure tool is the Repeat Formation Tester (RFT™), which is used to measure pressure at a number of points from a well. The pressure is recorded by (a) high-precision quartz gage(s) which is (are) sometimes backed up by a strain sage. The tool is also able to sample the fluid phase. The pressure data can be used to map pressure barriers and baffles, possible restrictions to fluid flow during production. RFT data from a well within a produced field can also yield critical information on reservoir depletion and cross-flow between reservoir intervals.

In addition to the logging tools outlined above, it is possible to measure temperature and wellbore geometry. Both are extremely important measurements for calibrating other logs. For example, if the caliper log shows that the wellbore is in poor condition, with washed out sections, then many of the wireline logs will record measurements that cannot be properly interpreted in terms of lithology, fluid content, and porosity. Borehole temperature measurements require correction for the cooling effect caused by the cold drilling mud in the borehole.

2.5 CORE AND CUTTINGS

2.5.1 Introduction

Core provides the basic geologic data from which to build a detailed description of a reservoir. It is also the only material on which high-quality (highly accurate) petrophysical properties can be measured (porosity, permeability, and relative permeability). Side-wall cores may also be taken from the wellbore. Cuttings will almost always be available, but are of limited use. The diameter of core commonly

Fig.2.16 An annotated NMR log. Curves are commonly provided for porosity, permeability, and oil volume (oil volume/porosity = oil saturation), together with fractional curves for clay-bound water, capillary-bound water, movable fluid, and NMR porosity. (From van Ditzhuijzen and Sandor 1995, reproduced courtesy of Shell.)

43/13a–1

TYPE LOG IN
BUNTER SANDSTONE RESERVOIR

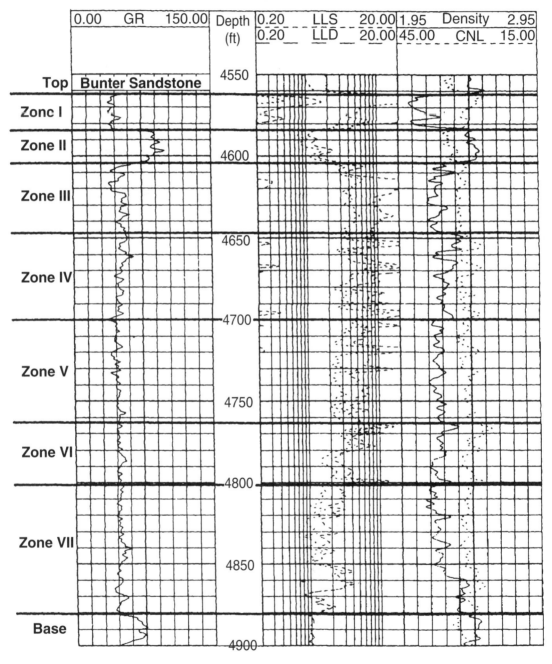

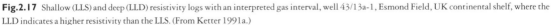

Fig.2.17 Shallow (LLS) and deep (LLD) resistivity logs with an interpreted gas interval, well 43/13a-1, Esmond Field, UK continental shelf, where the LLD indicates a higher resistivity than the LLS. (From Ketter 1991a.)

varies between about 5 cm (2 in) and 15 cm (6 in). It can be of any length, although the longest single core is normally about 27 m (90 ft). Much longer cores have been cut, and some slim hole-drilling operations use continuous coring. However, it is common to store core in slabbed 1 m or 1 yard lengths.

Core description can tell us about the depositional environments and may also give us the reservoir net to gross. Petrographic analysis, using either a conventional optical microscope or a scanning electron microscope, will yield information on reservoir quality and diagenesis. The same petrographic and mineralogical information will also be of use to drilling and production engineers when it comes to determining whether the formation will undergo chemical reactions with the drilling and well completion fluids. Such reactions can lead to formation damage in which the near-wellbore permeability is impaired (Chapter 5).

2.5.2 Conventional core analysis (porosity and permeability)

Core analysis provides measurements of key importance to the description of a reservoir. We will limit description here to porosity, permeability, and petroleum saturation. These properties are commonly determined in a process called "conventional core analysis." "Special core analysis" is a heterogeneous group of nonroutine analytic processes, such as relative permeability determination, core tomography, capillary pressure, and wettability measurement. Techniques in special core analysis tend to be used on an occasional basis rather than on every core cut, and most of the properties determined fall into the category of reservoir engineering or petrophysics rather than petroleum geoscience.

Sample acquisition

Normal coring involves the cutting of a long cylinder of rock, along the wellbore from the desired formation. A "core-barrel" is a tube with a special cutting bit attached to its lower edge and used on the end of a drillstring. Side-wall coring involves the collection of cylinders of rock from the outer edge of the wellbore. Most side-wall cores are obtained by percussion; firing a small steel tube into the wall of the borehole. The tube is then hauled back to the surface with the sample intact. It is less common to drill miniature core samples from the outside of the wellbore.

Porosity and permeability determinations are usually carried out on small right-cylinders cut from core. The measure-

ments can also be made on rock samples obtained by side-wall coring. Percussion side-wall coring often disrupts the fabric of the rock and therefore poroperm data obtained from side-wall core samples is rarely accurate.

Most laboratory measurements of porosity and permeability are carried out on cylinders of diameter 2.5–5.0 cm (1–2 in) and length 5–10 cm (2–4 in). The cylinders are most commonly cut perpendicular to the core and parallel to the core. However, where the formations are highly dipping, or where the wellbore is highly deviated and there is a strong fabric of sedimentary structures and/or fractures, the cylinders may be cut parallel and perpendicular to bedding and or fabric (Fig. 2.18). Care must be taken to obtain samples that adequately represent the property of interest.

Sample cleaning

Most of the analytic methods used for determination of porosity and permeability require clean core. The one exception is the summation of fluids method, which uses fresh core. There are a variety of cleaning techniques. The method used is dependent upon the degree to which the sample is indurated. Well-cemented, low-porosity reservoir rocks are commonly subject to high-energy cleaning (e.g., CO_2 toluene extraction, centrifugal extraction, and vacuum retorting), while low-energy techniques are used for poorly consolidated rocks (e.g., Soxhlet extraction, Dean–Stark extraction, vapor soaking, and flow-through cleaning).

The range of solvents used for cleaning also varies. Toluene is the most common, but if samples are known to be heat sensitive, a solvent with a lower boiling point is used. Methanol is commonly used to remove salt that has precipitated from included formation water. Flow-through cleaning uses a sequence of miscible solvents, each one designed to remove a component of the original fluid. The rationale is that by using a miscible sequence, no fluid interfaces will pass through the core samples. This can be important where delicate clays are present. The passage of a fluid interface through the core sample can cause the clays to flatten onto the pore walls, so increasing the permeability prior to permeability measurement.

Sample drying

Drying of the sample is as important as cleaning. For well-consolidated rocks with robust minerals, a simple convection oven can be used. The temperature range for drying is commonly between 82°C (180°F) and 115°C (240°F). A lower

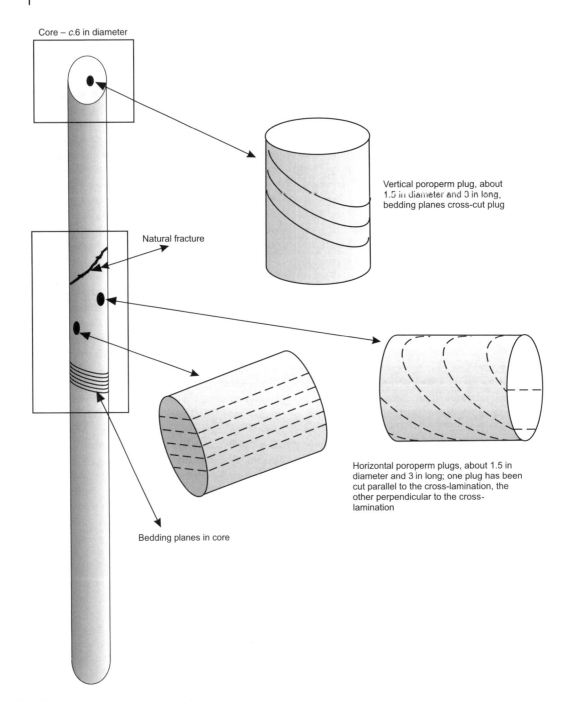

Core – *c*.6 in diameter

Vertical poroperm plug, about 1.5 in diameter and 3 in long, bedding planes cross-cut plug

Natural fracture

Horizontal poroperm plugs, about 1.5 in diameter and 3 in long; one plug has been cut parallel to the cross-lamination, the other perpendicular to the cross-lamination

Bedding planes in core

Fig.2.18 The preparation of core plug samples from core.

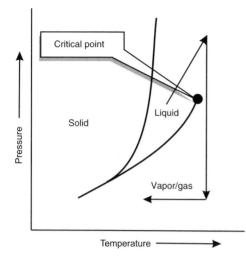

Fig.2.19 Critical point drying allows drying of the sample while avoiding a phase change in the cleaning fluid. The sample is heated under pressure until the critical point is exceeded and then, sequentially, the pressure and then the temperature are reduced such that the cleaning fluid is released from the sample as a gas.

temperature can be used if the samples are contained under partial vacuum.

Where delicate or hydrated phases are present in the mineral assemblage, different, gentler, drying techniques must be used. Such methods are designed to prevent either increasing the effective porosity of samples, by removing the water of hydration, or causing alterations to the pore fabric and hence permeability. For example, critical point drying (McHardy et al. 1982; Fig. 2.19) is commonly used to avoid damage to delicate clays (Heavyside et al. 1983).

Porosity measurements

In order to determine porosity, two pieces of data are required from three possible determinations: bulk volume (V_b), pore volume (V_p), and grain volume (V_g).

Bulk volume (V_b) is commonly measured using mercury displacement and Archimedes' principle. Mercury is nonwetting and will not enter most types of pore system through the sample surface. However, if the surface of the plug has large vuggy pores, as may be the case with carbonates, mercury displacement cannot be used. The mercury will partially fill surface vugs and the bulk porosity measurement will be too small. Here the core plug cylinders are simply measured: the

length and the diameter of the base being used to determine the bulk volume.

Two methods are in routine use to calculate pore volume (V_p). In liquid restoration, a dry, weighed sample is resaturated with a liquid of known density. The fully saturated sample is then reweighed and the volume of the fluid calculated from the increase in weight of the sample.

Pore volume from summation of fluids is calculated by adding together the gas, water, and oil volumes in a sample. Gas volume is determined by mercury injection, an extension of the bulk volume determination. The sample is then retorted to remove oil and water, and their respective volumes are measured.

Grain volume (V_g) measurement employs Boyle's Law. Helium is expanded from a reference cell (V_r) into a sample cell of known volume (V_c). The presence of a plug sample (V_s) in the cell reduces the absolute pore space by a volume that is the grain volume. The change in pressure from initial conditions (P_1) to final conditions (P_2) is also measured.

Thus grain volume is given by:

$$V_g = V_r + V_c - \frac{P_1 V_r}{P_2}$$

A grain density measurement can be obtained by dividing the weight of the plug by its grain volume. Grain density data are valuable for calibration of the wireline porosity log measurements.

Porosity can also be measured under a confining pressure designed to simulate conditions in the oilfield. A combination of the mercury displacement and Boyle's Law methods gives the most accurate results. Fluid summation porosity may be calculated on a few samples from a core as soon as it enters the core laboratory. Such data can be derived in less than 24 hours, and as such the well-site team may be able to make use of the information for further immediate operations.

Permeability measurements

Permeability is usually measured by flowing nitrogen through a core plug of known dimensions. Since the flow rate, core dimensions, and pressure differential are measured and the viscosity of the nitrogen is known, the permeability can be calculated using the Darcy equation:

$$k_{air} = \frac{2000 P_a \mu Q_a L}{(P_1^2 - P_2^2) A}$$

where k_{air} is the permeability (mD), μ is the viscosity of air/

nitrogen (cp), Q_a is the flow rate (cm^3 s^{-1} at atmospheric pressure), L is the length of the sample (cm), A is the cross-sectional area (cm^2), P_1 is the upstream pressure (atmospheres), P_2 is the downstream pressure (atmospheres), and P_a is the atmospheric pressure (atmospheres).

Even on the scale of a 2–3 in core plug, permeability is rarely uniform in all directions. It is therefore important to measure both horizontal and vertical permeability (Fig. 2.18). This is usually done on separate, albeit closely spaced, plugs. Similarly, in highly deviated wells and/or reservoirs with highly dipping strata it is desirable to measure permeability parallel to and perpendicular to bedding. Such a process may be taken one stage further in reservoirs with a strong fabric of sedimentary structures. For example, in aeolian dune reservoirs, permeability measurements may be made parallel to and perpendicular to the dune cross-bed sets.

Permeability to air in the laboratory may be a very poor indication of the permeability of the rock at reservoir conditions and with reservoir fluid in the pore spaces. It may be appropriate, therefore, to make permeability measurements while the sample is under confining pressure. The deterioration of permeability under pressure is commonly greater than that experienced by porosity. As an alternative to performing individual measurements at high pressure, simple conversion charts are available that allow the determination of reservoir permeability from laboratory measurements of air permeability (Fig. 2.20).

In rock that contains more than one fluid, it is important to measure the relative permeabilities of gas, water, and oil. The effects of relative permeability have already been mentioned in Chapter 1. Measurement of the effects is commonly by experimentation; that is, different two-phase (usually oil and water) mixtures are flowed through the rock and their relative permeabilites measured. Clearly, the fluids used must be either those found in the reservoir or simulated oils and brines. Moreover, both pressure and temperature should replicate that found in nature.

The matrix permeability of seals (mudstone, salt, etc.) also needs to be known, but measurement is difficult because such rocks have extremely low permeability, typically in the nanodarcy (nD = 10^{-6} mD or 10^{-21} m^2) range. It takes a long time to flow several pore volumes of fluid through these very low-permeability rocks, so a pressure decay method is used (Katsube & Coyner 1994). The measurement limit is currently about 0.1 nD and excludes direct measurement of some near-impermeable rocks such as halite. Relative permeability measurements in silt, mudstone, and other fine-grained rocks have not yet been performed successfully.

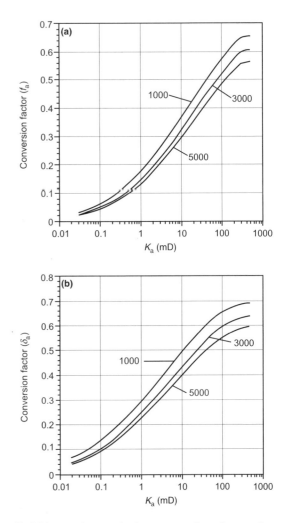

Fig.2.20 Permeability overburden corrections. Curves for conversion of (a) K_a and (b) K_L to K_{res} at various net overburden pressures (psi). (Reproduced courtesy of BP.)

2.5.3 Core logging

The interpretation of depositional environments from lithologies and sedimentary structures is the fundamental building block of a reservoir description. Much of the information available from core cannot be obtained from wireline logs, cuttings, or seismic data. A core description commonly comprises a graphic representation of the core with supplementary notes (Fig. 2.21).

The graphic core description includes a grain-size profile,

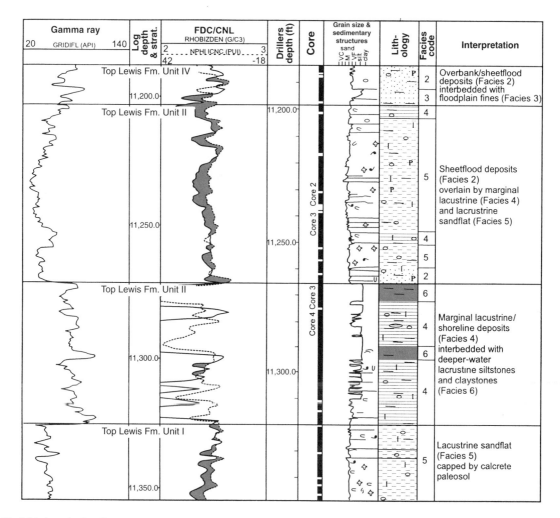

Fig.2.21 A core log from Triassic reservoirs within the Beryl Field, UK North Sea. The graphical depiction of the lithology and sedimentary structures has been plotted with the well logs and core coverage data. (From Bond 1997, reproduced courtesy of the Geological Society of London.)

lithology indicators, cartoon icons representing the sedimentary and biogenic structures in the rock, and a similar set of icons representing any visible mineral aggregations in the host rock (such as chert or carbonate nodules). These data are commonly displayed alongside a suite of wireline logs. Also displayed on the same diagram are the geologist's interpretations of the depositional environment for each of the different lithologies or lithofacies ("lithofacies" is a term used to describe a rock using both its lithology and sedimentary structures).

From the combined core description and wireline log data,

it is commonly possible to generate a series of (wireline) log facies. Such log facies may be used to describe the reservoir section in uncored, but logged, wells. The core description will also provide the basis for construction of a reservoir model (Section 5.8.4). The rock as described from core can be compared with analog information from outcrop to give an estimate of the size of sands or carbonate bodies. For example, the lithology and sedimentary structures produced on deposition of sands in a river channel are different from the lithology and sedimentary structures generated by an oolite shoal on a carbonate bank. The thickness, height, and width data obtained from the analog can be used in the reservoir model.

2.5.4 Petrography

Petrographic information often provides valuable components of the reservoir description. It can help us to determine provenance — where the sands came from — depositional environment, calibrations for wireline logs, and reservoir quality.

The petrographic techniques available to the petroleum geoscientist are much the same as those used by academics or by geologists in other industries. The basic tool remains the optical, petrological microscope; the basic sample is the petrographic thin section. The simple description of rocks seen beneath the microscope has changed little since Henry Clifton Sorby invented the method in the 19th century. The information provided by such an examination can yield much of the data that may be required to help calibrate wireline logs and determine the depositional environment. Other petrographic techniques that are commonly used include scanning electron microscopy (SEM) and "heavy" mineral analysis. In scanning electron microscopy, a small chip of rock is examined at very high magnification. This allows analysis of the surface texture and mineralogy of the pore space between grains. Heavy mineral analysis involves disaggregation of the rock and the separation of minerals that are more dense than the common rock-forming minerals (quartz and feldspar).

In the following subsections, we look at the way in which petrography can be used to help solve problems.

Provenance

Where did my reservoir sand come from? Such questions should be asked in the early stages of exploration. Knowledge of where the sediment came from and where it is now will contribute toward construction of a depositional environment. This may in turn allow sandstone distribution to be mapped. Simple petrography may yield sufficient information for the sediment source to be identified. For example, the Tertiary Vicksburg sandstones of the US Gulf Coast contain volcaniclastic grains that can be tied to local volcanic activity at the time of deposition.

Such unequivocal matches between reservoir sandstone and sediment source are rare. However, there are a number of other techniques — some petrographic and some geochemical — that can be used to supplement simple petrographic data. The most common of these other methods is heavy (dense) mineral analysis, in which minerals such as garnet, sphene,

apatite, rutile, tourmaline, zircon, and so on are used to identify the source of the sediment (Morton 1992). Both the simple petrography and the heavy mineral analysis can be taken a stage further by performing chemical analysis of the grains. This can help to further reduce uncertainty when it comes to identifying the provenance.

The depositional environment

The depositional environment for a reservoir is commonly determined using the sum total of the sedimentary structures combined with paleontological information. Petrography can supplement this information, and in carbonate rocks it can provide a large portion of the data required for determination of the depositional environment. For example, petrographic analysis of grainstone types (ooid and bioclastic) was used to help determine the depositional system and hence the detailed reservoir distribution for the Cretaceous reservoir of the Alabama Ferry Field in Texas (Bruno et al. 1991).

Reservoir quality

Scanning electron microscopy has joined conventional optical microscopy as a basic tool for reservoir quality evaluation. Optical microscopy is useful for looking at grains and bulky cements in a rock (Fig. 2.22). However, finely crystalline clays, microscopic pores, and some syntaxial or fringing cements are more readily observed by looking at a fracture surface of the rock beneath a scanning electron microscope (Fig. 2.23).

More sophisticated petrographic techniques, such as backscattered electron imagery and cathodoluminescence, particularly when combined with pore image analysis, can yield highly quantitative results for measurements such as porosity and mineral abundances, grain shapes, and perimeters (Emery & Robinson 1993).

2.5.5 Geochemistry

Geochemical analysis of petroleum and petroleum source rocks is of critical importance to our understanding of source-rock deposition, source-rock maturation, petroleum migration, and petroleum entrapment. Fluid samples may be extracted from core and the rock itself analyzed. Petroleum fluids are also sampled from wells and seeps, while potential source rock may be sampled from outcrop. Although the sampling interval in a core will be quite different to that achieved by a well test or outcrop study, the range of analytic

Fig.2.22 A thin-section photomicrograph of a fine-grained micaceous sandstone from the Marnock Field, UK continental shelf. The porosity in the sample has been infilled with a resin. Most of the grains are quartz. The fibrous grain on the left is biotite (mica). All of the grains are coated with chlorite (C). The photograph was taken using a combination of cross-polarized light and a quarter-wavelength plate. Field of view 650 μm × 450 μm. (Reproduced courtesy of BP.)

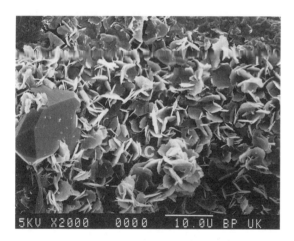

Fig.2.23 A scanning electron photomicrograph showing the same chlorite-covered grains as in Fig. 2.22 (Marnock Field, UK continental shelf). The chlorite plates are a few μm across and much less than 1 μm thick. The relatively large crystal on the left of the photograph is quartz. Field of view 55 μm × 45 μm. (Reproduced courtesy of BP.)

techniques applied to the petroleum or its precursor kerogen are much the same. In consequence, we discuss analysis of fluids in the following section, and analysis of rock samples later in this chapter.

Geochemical analysis of the mineral components of core and aqueous fluids from the wellbore or core may also be per-formed. The geochemical analysis of the rock can be used for much the same purposes as petrographic information: provenance, diagenetic history, and correlation. Emery and Robinson (1993) describe the large array of techniques and their applications. The techniques include stable isotope analysis (δ^{13}C, δ^{18}O, δ^{34}S, and ^{147}Nd/^{143}Sm), radiogenic isotope analysis (K/Ar dating), fluid inclusion analysis, and elemental analysis.

Geochemical analysis of the fluids yields information that is useful in describing compartmentalization in the reservoir or helping us to understand the diagenetic history of the reservoir rock. However, the main beneficiaries of fluid chemistry data are the drilling engineers, completions engineers, and production engineers. All three sets of engineers will be concerned to establish whether reactions between the natural fluids in the rock and those fluids introduced by drilling wells, completing wells, and producing wells will cause prob-lems such as wellbore instability, formation damage (perme-ability reduction), or precipitation of scale in the production equipment. Sampling methods and analytic techniques are examined in a little more detail below.

2.5.6 Biostratigraphy

Biostratigraphic analysis of fossil material derived from core and cuttings delivers information on the age of sedi-ments, the correlation surfaces between wells, and the deposi-tional environment. Such data will be used to establish the stratigraphy of the basin, enabling it to be tied to adjacent areas.

Almost all of the biostratigraphic analysis performed nowadays relies on the use of microfossils of either plant or an-imal origin (Section 3.6.3). Sample preparation depends on the rock and fossil type. For example, foraminifera can be hand-picked from a poorly indurated sand or mud, whereas dissolution of the rock matrix in hydrofluoric and other strong acids is used to release spores and pollen entombed in sedimentary rock. There are two sub-disciplines in bios-tratigraphy: palynology is the study of fossil pollen and spores, whilst micropaleontology is the study of animal fos-sils such as foraminifera and ostracoda. Not all rocks contain microfossils or palynomorphs. Some rocks—particularly those deposited in hostile environments such as deserts, or subject to oxidation after deposition—are commonly barren.

During the early phases of exploration, it is common prac-tice to try to apply a whole range of biostratigraphic methods to a new well, sampling both core and cuttings. It will often be revealed fairly quickly which methods are best applied to which parts of the stratigraphy in a well. For example, if the reservoir interval comprises largely fluvial sandstones and overbank mudstones, there is likely to be little point in trying to extract marine foraminifera from the rocks. Instead, palynology will be used to analyze pollen or spore data. The uses of biostratigraphy are given below, together with the issues associated with sampling.

Sampling

Biostratigraphic analysis can be performed on core, side-wall cores, and cuttings. The positions of core and side-wall cores in the wellbore are known with a high degree of accuracy, but the same is not true for cuttings which are cut by the drill bit and circulated in the mud system in the hole. The drilling bit generates cuttings. They are then entrained within the drilling mud and circulated to the surface. At this point, most of the cuttings are removed from the mud and the mud returned to the wellbore. The mud is a viscous fluid and its rate of circulation is known; hence the time taken for a cutting to reach the surface after being cut may be calculated. Quite clearly, even if the circulation rate for the mud is known precisely, there will be some smearing of the first and last cuttings returns from any particular horizon. This introduces uncertainty within the positional data for the cuttings. The establishment of where cuttings come from is of considerable importance if fossils extracted from them are to be useful for subdivision of the well.

A second, and commonly serious, problem is that sections of the borehole wall can break away and fall through the mud to the bottom of the well. Such out-of-place bits of rock are called "cavings" and, as can be imagined, the extraction of biostratigraphic data from cavings can lead to highly spurious results.

These mechanical issues over the position of cuttings in the wellbore can add to problems generated by the geology itself. Some fossils can be recycled; that is, fossils of one age can be eroded from their entombment, transported to a new site of deposition, and incorporated in younger rock.

Correlation

Biostratigraphic data are commonly used to correlate rocks of similar ages in different wells. The correlation surfaces provided by biostratigraphy are commonly much more reliable than those generated by lithostratigraphic correlation, a process that is apt to produce coherent reservoir bodies in the minds of the geologist when no such coherence occurs in reality.

Age dating

Age dating of sediments relies upon the fact of evolution, in that older fossils are different from newer ones. The ideal chronostratigraphically significant fossil would have evolved quickly, become extinct in a short time, been tolerant of a wide variety of environmental conditions, and spread quickly around the globe. Fossils, or at least the life forms from which they are derived, are rarely so obliging, and the micropaleontologist or palynologist needs to be able to estimate the degree to which a fossil assemblage in a series of wells is both diachronous and environmentally specific.

The depositional environment

To obtain information on the depositional environment requires fossils that are sensitive to the environment. Various foraminiferal species have been used as indicators of water depth or salinity during sedimentation, so aiding the sedimentological assessment of depositional environment. In a similar fashion, palynological assemblages are commonly used as precise indicators of specific depositional environments within near-shore and deltaic systems (Whitaker et al. 1992; Williams 1992).

2.6 FLUID SAMPLES FROM WELLS

2.6.1 Introduction

The composition and phases of oil, gas, and water within the petroleum pool carry important information as to how the accumulation formed, the degree to which the pool is segmented, and the presence of heterogeneity in the petroleum. Such data will also be used to calibrate the wireline logs during planning of the development scheme for the field and to help design the production facility. Offshore platforms and onshore production installations are expensive pieces of plant, with construction and installation costs commonly in excess of $100 million. A significant proportion of such costs will be spent tailoring the facility to the expected fluid production rate and composition. If the facilities specification is incorrect because the original fluid samples were extensively contaminated or unrepresentative, then the cost of remedial work can be prohibitive. Finally, the composition of the petroleum itself will determine its commercial value. In order to determine each of these factors, the fluids must be sampled and analyzed.

2.6.2 The sampling of fluids

The fluids may be sampled by flowing the well to the surface, capturing fluids down hole, or by extracting fluids from cores at the surface. Each method, whether it be for obtaining petroleum or water, suffers from problems of contamination and changes of phase and composition caused by sampling at a different pressure and temperature than those that exist in the petroleum pool.

The most reliable data come from fluids obtained during production tests and from those samples obtained at the ambient pressure and temperature of the reservoir. In most instances, the fluids are obtained as a secondary product to well flow rate testing or reservoir pressure analysis. In such instances the sample of petroleum may not be wholly representative of the fluid in the reservoir, due to fractionation of the fluid, gas exsolution, partial precipitation, or contamination with drilling fluids/solids. Similar problems exist for water samples insofar as they may be contaminated by solutes from drilling fluids, or exsolution of dissolved gases such as methane and CO_2, that will lead to changes in the Eh and pH values.

Samples of petroleum may also be taken for $P–V–T$ (pressure, volume, and temperature) purposes in order to produce equations of state to predict the behavior under different production schemes. These samples are taken downhole at the reservoir/wellbore interface. The petroleum or water is allowed to flow into and through a sealable chamber. The seals are then closed and the sample of a few liters brought to the surface. Once the physical properties have been determined, the samples are ideal for chemical analysis, because great care was taken when obtaining them.

Both oil and water may also be extracted from core. This can be done using organic solvents for the petroleum phase, washing with distilled water for the aqueous components, or centrifuging for either oil or water. Although core sampling cannot capture any of the physical properties of the fluids, and is clearly inappropriate for attempts to sample volatile components and gases, it does have some particular advantages. It is possible to sample the reservoir more finely when core is used than would be possible from a well test or any downhole sample. Moreover, some useful information may be extracted from old core. Long-chain and complex hydrocarbons will not evaporate from the core, although they may get a little oxidized; and although the water will evaporate from the core, residual salts may remain. Although, strictly speaking, we would no longer be sampling fluids, information from the residual hydrocarbons and salts is still information about the fluids within the reservoir. Residual salt analysis is used to help determine reservoir compartmentalization (Section 2.6.4).

2.6.3 Petroleum

Once extracted from well or core, chemical analysis of petroleum may be used to:
- Identify the source rock type for the oils;
- measure oil, and therefore source-rock, maturity;
- determine the origin of gases;
- correlate between fluids and from fluid to source rock.

If it is possible to sample two or more wells from a single accumulation, then information on reservoir unit correlation and field segmentation can be obtained.

The specific analytic tools that may be used to characterize the oil include the following:
- *Isotope measurement for oils*. It is possible to measure the ratio between ^{13}C and ^{12}C in whole oils and oil fractions. The ratio is often characteristic of a family of oils. In particular, it can be used to characterize oil produced from a high-maturity source rock, one in which the complex molecules that are normally used for characterization have been destroyed.
- *Gas classification*. Since gases are simple molecules, they too lack the complex biomarkers that are normally used for classification and identification of source rock. Instead, it is possible to use the ratio of methane to higher homologs such as ethane, propane, and butane, together with the isotope composition of each of the homologs, to characterize the gas and identify source.

Gas chromatography and gas chromatography (GC) fingerprinting

In gas chromatography, petroleum molecules are partitioned between a stationary, high molecular weight liquid phase and a moving gas phase. The petroleum under investigation is injected into a moving stream of an inert gas, such as helium or nitrogen. The gas with entrained petroleum flows through a capillary tube coated with the high molecular weight liquid. The tube itself may be from 0.3 m to 9 m (1 ft to 30 ft) long (Hunt 1979). Each hydrocarbon molecule within the petroleum moves in the liquid phase at a different rate. The short-chain paraffins are retained for less time than the longer and more complex molecules. The length of time between the point of injection and when the molecule reaches the detector is called the "retention time." Thus in the output of data from a gas chromatographic analysis abundance is plotted against

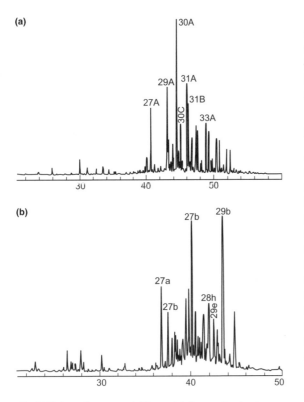

(a)

(b)

Fig.2.24 A gas chromatograph. The *x*-axis is the retention time in minutes and the peak height is a function of abundance. Compounds with more carbon atoms appear further to the right on the chromatograph, and there are relatively small differences in retention time for compounds with the same number of atoms but different structural configurations. The numerical codes on the peaks equal the number of carbon atoms in each different molecule. A detailed comparison of the chromatographic trace for peaks generated by compounds between nC-11 and nC-12 can be used as unique fingerprints for oils from different sources. (a) M/Z 191; (b) M/Z 217.

The output from a gas chromatograph can form the input to a mass spectrometer rather than a simple device that measures abundance. In gas chromatography–mass spectrometry (GC–MS), the mass spectrometer is used to specifically identify the petroleum molecules. The data can be used in the same way as for gas chromatography.

Molecular maturity

As oil forms within a source rock, the complex molecules created by the plants and animals change composition and structural chemistry. Different molecules undergo different reactions at different levels of thermal stress (temperature). Some of the reactions progress rapidly from the reactant to product stage. Other molecules react more slowly to increasing temperature. In consequence, analysis of an array of the changing molecules in a crude oil can be used as a measure of the degree of maturity of the host source rock. The types of molecules that can be used for these analyses include methyl phenanthrene, methyl biphenyl, and methyl dibenzothiophene. Reactions that involve racemic modification of biomarkers with chiral centers can also be used.

2.6.4 Water

The most common use for water composition data is in the calibration of resistivity logs. The salinity of the formation water, both in the water leg and as trapped water in the oil or gas leg of a field, will influence the response of the resistivity tool. Highly saline formation water is more conductive than fresh formation water. Since oil saturation is calculated using resistivity data, it is important to calibrate logs to the local formation water.

Bulk chemical analysis of formation water is commonly used to check for elements that will react with drilling and completions fluids and/or cause precipitation of what is known as "scale" within pipework. Scale-encrusted pipework may be more than a costly inconvenience; it may also be dangerous. For example, radium commonly coprecipitates with barium sulfate to produce radioactive scale.

Formation water characterization can, in a very similar fashion to that for petroleum, be used to help identify compartments in reservoirs. The premise is much the same: if the reservoir system is simple and homogeneous, the formation water composition should be uniform throughout the system (Warren & Smalley 1994).

The most common method using water to characterize reservoirs is called residual salt analysis (RSA). The tech-

retention time, different molecules occurring as peaks on the chromatogram. By reference to a standard and by using carefully controlled and repeatable experiments, it is possible to identify peaks on the chart as specific molecules (Fig. 2.24).

In GC fingerprinting, individual peaks are not specifically identified but, instead, the ratios between peak heights are calculated to generate a fingerprint for each oil sample. Moreover, it is common to select the small peaks that occur between the larger normal alkane peaks. It is such molecules that vary most between oils. The degree of difference between fingerprints from different oils may be used to identify barriers to fluid flow within fields or similar oils between fields.

nique involves measurement of the $^{87}Sr/^{86}Sr$ isotope ratio (Smalley et al. 1995) in formation water. The method is simple insofar as the strontium is extracted for measurement by simply washing core samples in distilled water. Care must be taken to calculate the extent of contamination of the isotope ratio from drilling mud. The ratio can be measured very accurately using a mass spectrometer and spatial variations in the ratio then used to interpret barriers to fluid flow in the reservoir.

For both the petroleum and water geochemistries, the product is a definition of the presence or absence of barriers or baffles to fluid flow in a reservoir. In neither case is it possible to identify the position of barriers between wells. The combination of these techniques with seismic analysis can, however, deliver both the presence and the position of barriers. Similar information on barriers can be derived from fluid flow tests (drill stem tests) in wells.

2.7 OUTCROP DATA

2.7.1 Introduction

Most rock is, of course, buried within the Earth. Where rocks break the surface, they are said to outcrop. The term "outcrop" may be slightly misleading to nongeoscientists,

because it carries no implication that the rock can be seen at the surface: it may be covered by soil and vegetation. Where bare rock is at the surface, it is said to be in an "exposure." For the most part, gathering data from outcrop really means gathering data only from exposures. Geologists of all kinds have always "gone into the field" to make maps from exposures of outcrops. In the earliest days of petroleum exploration, the main aims were to map structure and thus prospects for drilling, and to map the occurrence of seeping petroleum as proof positive of a working petroleum system. Such work would be the first step in an exploration program. In all but exceptional circumstances, such mapping would occur before leases were acquired, although the mapping operations may have been executed in a covert manner, so as not to alert competitors and landowners.

2.7.2 Maps

Most geoscientists reading this book will be familiar with the process of making geologic maps of areas on land. Other readers may not be so familiar. In the making of a land-based geologic map, the geologists survey the area, noting the location of exposed rock, its lithology, its dip and strike, and its internal characteristics, such as sedimentary structures, fossil content, and mineralogy. If the area has much rock

☐ Upper Morredge Sandstones and Mudstones
☐ Lower Morredge Shales and Sandstones
☐ Mixon Limestones
☐ Mixon Limestones and Shales
▨ Onecote Sandstones

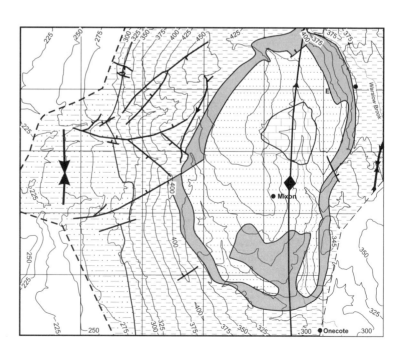

Fig.2.25 A field map of the Mixon Anticline (North Staffordshire, England). The structure is a four-way dip-closed anticline. An uneroded structure of similar size could hold a 100 mmbbl pool within Upper Carboniferous sandstones. Such structures exist beneath the younger Mezozoic cover to the west of the Mixon area.

exposed at surface, correlation of rock units across the area may be easy. If there are few exposures, the geologists must use their judgment to join together isolated exposures of what appear to be the same rock (lithology, fossil content, etc.). This may require that the geologist postulate the existence of faults to be able to make a consistent map. In this way a map will be produced showing the distribution of various rock units and the stratigraphic relationships between each of them. From such a map, cross-sections would be drawn to illustrate the distribution of the strata in the subsurface. A stratigraphic column would also be constructed.

Figure 2.25 is a much-simplified geologic map made by one of the authors (J.G.) many years ago as an undergraduate. It was not at the time conceived as a petroleum exploration project; it does, however, contain all the elements of the petroleum exploration process. The map is from the northernmost part of the county of Staffordshire in England. From the map and mapping process, it was possible to illustrate:

• A valid trapping geometry — the structure is an elongate dome (the Mixon Anticline);

• the presence of adequate shale seals of Namurian and possibly Viséan ages;

• potential reservoir horizons in Namurian sandstones and fractured Viséan limestones and subordinate dolomites;

• probable source rocks in Namurian marine mudstones and/or Viséan bituminous marls;

• a working source insofar as there are thin shaley limestones in the upper part of the Viséan with bitumen stain, and others which have a strong petroliferous odor when freshly broken.

The area also lies amid oil shows (Kent 1985) to the west (Keele), north (Gun Hill), and southeast (Ashby). Deep drilling for petroleum, into the center of the Mixon Anticline, would probably be a high-risk venture, since the search would be for stratigraphically deeper reservoirs and seals. However, the presence of the positive indications of trap, reservoir, seal, and source at Mixon may encourage the explorer to look further west beneath the Mesozoic cover, where Carboniferous structures may remain unbreached. On the basis of this Mixon analog, a similar structure with 100 m of net reservoir in the Morridge sandstones, a porosity of 15%, an area of closure of about 20 km^2, an oil saturation of 66%, and a formation volume factor of about 1.5 would contain approximately 100 mmbbl of in-place oil*. A similar line of reason-

ing, based on fieldwork in the central and eastern Pennines, delivered the failure at Edale in 1938 (Chapter 3) and led one year later to the discovery of the East Midlands (UK) oilfields (Kent 1985).

2.7.3 Reservoir analogs

Mapping of outcrops for the specific purpose of obtaining an exploration permit and drilling of wells is not common today, but its demise has not removed the need for fieldwork. Exposures of rock can be used to deliver analog information for reservoirs. Reservoir rocks, particularly those that are well exposed, can be studied in detail to reveal sand-body or carbonate-body geometries, internal reservoir architecture, porosity and permeability distribution, and heterogeneity and reservoir interconnectivity.

2.7.4 Rock sampling and analysis

Rock from outcrop may be sampled and analyzed in much the same fashion as core. Although drilling fluids do not contaminate such outcrop rock, it may be weathered. Weathering leads to oxidation of organic matter in source rocks, and to mineral dissolution and precipitation reactions in sandstone and carbonates. Hence data obtained from rocks at outcrop should be treated with the same care as would be afforded data from core. Earlier in this chapter, we examined the techniques applicable to analysis of reservoir in cores. The same techniques can be applied to outcrop data. As such, we do not repeat them.

Potential source rocks may also be sampled from outcrop as well as core. They may also be subject to some of the same types of analysis as performed on oils. These include stable isotope analysis and the determination of thermal and molecular maturity (Naeser & McCulloh 1989). Clearly, such analyses are used for the same purposes as their counterpart fluid analyses. There are, however, additional analytic methods that are applied to sediments but not to oils.

One of the more important of these additional methods is determination of S_1/S_2 (sometimes also referred to as P_1/P_2), a tool for measuring the petroleum potential of a source rock. S_1 and S_2 are measures of the petroleum yield from a source rock when heated (pyrolyzed) in the laboratory. Results are presented in kilograms of petroleum generated per tonne of source rock. S_1 is a measure of the free petroleum, while S_2 is a measure of all of the products generated on breakdown of the oil and gas precursor compounds (kerogen). Analysis of an immature high-quality source rock such as the Kimmeridge Clay (UK onshore and

---

* When completing the final draft of this book, it became known to the authors that a gas discovery had been made at Nooks Farm, to the west of Mixon, beneath the Mesozoic cover as predicted. Unfortunately, neither of the authors holds equity in this field!

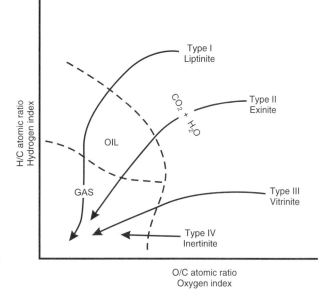

Fig.2.26 A van Krevelen plot, showing maturation pathways of the dominant kerogen types in terms of their atomic H : C and O : C ratios. Maturity increases toward the bottom left corner of the graph. Kerogen Types I and II (lipinite and exinite) are oil-prone, Type III (vitrinite) is gas-prone, and Type IV (inertinite) is largely inert.

North Sea) will commonly yield $15–20\,kg\,t^{-1}$. Rocks that yield less than about $5\,kg\,t^{-1}$ are unlikely to be considered as source rocks.

S_1/S_2 determination is commonly coupled with a measure of the hydrogen index, which is defined as the quantity of petroleum generated per unit weight of organic carbon:

$$HI = \frac{S_2}{TOC \times 100}$$

where S_2 is the pyrolysis yield (in $kg\,t^{-1}$), TOC is the total organic carbon, and HI is the hydrogen index.

The hydrogen index is an important measure of how oil-prone a source rock is, and of the ease with which petroleum will be expelled from it. An oil-prone source rock is likely to have a HI of >300, while that of a gas-prone source rock, such as a coal, is likely to range between 100 and 300. HI data are used with S_1/S_2 data to identify potential source rocks.

Source rocks may be classified according to their likely products, oil or gas. There are several ways in which potential source rocks may be analyzed to deliver such data, the most common of which is to cross-plot the H : C and O : C ratios (a van Krevelen plot). Data from potential source rocks tends to plot in one of four fields (I, II, III, and IV) on such a diagram (Fig. 2.26).

The temperature at which the S_2 peak occurs is known as T_{max} and is a measure of the thermal maturity of the source rock. If the source rock is mature or overmature, it will have a lower HI than its immature counterpart, because some of the original hydrogen will have been incorporated into the generated oil and gas. Other indicators of thermal maturity are vitrinite reflectance and the spore coloration index (Fig. 2.27). Vitrinite reflectance relies upon the fact that vitrinite (a component of organic matter) becomes progressively more reflective as the sediment is matured thermally. The spore coloration index is a semiquantitative measure whereby the spore color (from colorless through yellow, orange, brown, and black) is used to give a semiquantitative measure of thermal maturity.

2.8 SEEPAGE OF PETROLEUM

All petroleum traps leak and in consequence petroleum escapes naturally to the Earth's surface. Once at the surface, it can be detected by either remote sensing or direct observation. Satellites have been used to identify seepage either through secondary effects upon vegetation, from the oil's warm thermal signature, or offshore using the change in water reflectivity over slicks (Fig. 2.28). Slick or seep detection offshore can also be conducted using airborne laser fluorescence or a simple visual de Gama survey.

Airborne laser fluorescence uses the principle of absorption of ultraviolet (UV) light by petroleum, resulting in induced fluorescence. A UV laser, an imaging telescope, and a detector

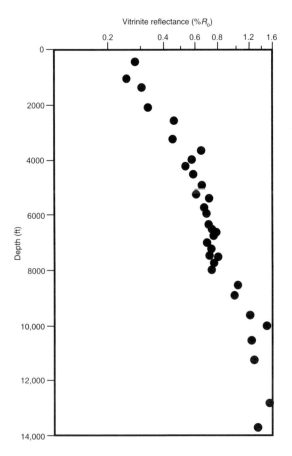

Fig.2.27 A plot of vitrinite reflectance against depth from well David River 1a, Bristol Bay, Alaska. The data clearly show an increase in thermal maturity with depth. Breaks in the $\%R_0$ versus depth are indicative of uplift and erosion. (Reproduced courtesy of BP.)

are mounted within an aircraft. The laser is used to illuminate an area of sea, and the detection and recording instruments capture and store the induced fluorescence signal.

A de Gama survey simply involves flying over the study area wearing a pair of polarizing sunglasses. Slicks will stand out as areas of particularly calm water and low reflectivity. Both the fluorescence and de Gama surveys require relatively calm seas. Moreover, repeat surveys should be conducted to guard against the possibility of wrongly identifying, as a natural seep, a slick caused by a ship cleaning its tanks. Oil leaking from shipwrecks can also lead to misidentification.

Follow-up work to identification of seeps either on land or at sea usually involves sampling the oil, and possibly gas, for laboratory analysis.

Fig.2.28 Oil that has leaked from sea-floor vents to the sea surface, usually via gas bubbles, can form micron-thin iridescent films, like this example from the Gulf of Mexico. (Reproduced courtesy of BP and the NPA Limited – TRIECo Limited joint venture partnership.)

FURTHER READING

Emery, D. & Robinson, A.G. (1993) *Inorganic Chemistry: Applications to Petroleum Geology*. Blackwell Scientific, Oxford.

McQuillin, R., Bacon, M., & Barclay, W. (1979) *An Introduction to Seismic Interpretation*. Graham & Trotman, London.

Naeser, N.D. & McCulloh, T.H. (eds.) (1989) *Thermal History of Sedimentary Basins: Methods and Case Histories*. Springer-Verlag, New York.

Oakman, C.D., Martin, J.H., & Corbett, P.W.M. (eds.) (1997) *Cores from the Northwest European Hydrocarbon Province: An Illustration of Geological Applications from Exploration to Development*. Geological Society, London.

Rider, M.H. (1986) *The Geological Interpretation of Well Logs*. Blackie/Halsted Press, Glasgow.

CHAPTER 3

FRONTIER EXPLORATION

3.1 INTRODUCTION

Frontier exploration is a term to describe the first episode of exploration in a basin. Nowadays, most newly explored basins will see closure of the frontier exploration phase in about five years, yet for some basins the same phase will last for many tens of years. The phase of frontier exploration is followed by an exploitation phase or, if the basin proves to be barren of petroleum, cessation of petroleum exploration activity.

The time that a basin spends in the frontier exploration category is controlled by both technical and political factors. The technical factors can be ones of access; that is to say, a hostile climate, difficult terrain, deep water, or impenetrable jungle may control the pace at which companies choose drilling locations and drill wells. The outcome of exploration drilling may also control the duration of the frontier exploration phase. A good example of this type of control is the Atlantic Margin area to the west of the Shetland Isles (north of the British mainland). The first few exploration wells were drilled in this area in the early 1970s. The giant Clair Field was discovered in 1977 (Fig. 3.1), so proving the area to be a major petroleum province, but due to a combination of factors it has not yet been possible to bring the Clair Field on stream. Exploration of the area continued throughout the 1980s, albeit at a low level compared with the adjacent North Sea. During this time further discoveries were made, at least one of which was a giant accumulation of highly viscous oil. Interest in the area remained low, despite the proof of these giant discoveries, until the early 1990s, when the Foinaven Field, which lies about 100 km southwest of the Clair Field, was discovered. Although the oil is moderately viscous, it flows well from high-permeability Tertiary reservoirs. Foinaven is a large field, with reserves of about 200 mmbbl. Further successes in the same play followed. An increase in activity (both licensing and drilling) led to more discoveries in a variety of plays.

Many oil explorers would probably now regard activity in the area to the west of Shetland to be at the end of the frontier exploration phase, 25 years after the first exploration wells. However, for about the first three-quarters of that period, the area was without a commercial discovery.

Political controls may also contrive to force a basin to remain in the frontier exploration phase. These controls may result from border disputes, financial constraints on state oil companies allowed to exploit the area, or from a strict licensing policy imposed by the licensing authority. Two examples of political changes that have opened up new basins come from China and the former Soviet Union.

The South China Sea was unexplored until the mid-1980s. Although China has a long tradition of onshore exploration, by the 1980s it had few ventures in offshore waters. In the early 1980s, the country opened up a little to foreign investment. The possibility of searching for oil within the South China Sea enticed many international oil exploration companies to compete for acreage in the first license round. Many of these companies had committed to drilling several wells. Most of the first wells drilled had oil shows; these were insufficient to yield an economic development but enough to ensure that exploration interest did not wane too quickly. However, in those first five, critical, exploration years only one, marginally economic, discovery was made. Moreover, the discovery was not where most of the companies had predicted, in Paleogene clastics, but much further offshore in a Miocene carbonate shelf limestone (Case history 3.9).

The traditional oil-producing areas in the former Soviet Union were, until the 1960s, the circum-Caspian states (Azerbaijan, Kazakhstan, Turkmenistan, and Russia). However, in the 1960s, the discovery of giant fields in Siberia led to a reduction of investment in the traditional circum-Caspian industry. In particular, exploration was limited to onshore, or in very shallow water offshore along the shallows of the Aspheron Ridge of the southern Caspian Sea. The breakup of

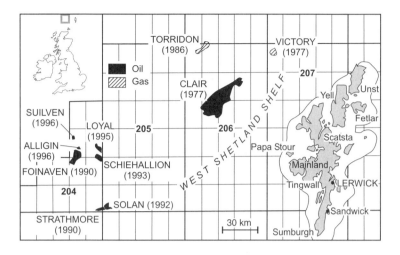

Fig.3.1 Oil- and gasfields on the Atlantic margin west of the Shetland Islands.

the Soviet Union led to renewed interest in the deep-water prospectivity of the Caspian Sea by the newly independent states and by international exploration companies.

The deep-water Gulf of Mexico (USA) is an area in which new technology has opened up a basin that was hitherto inaccessible. Improvements in drilling and production technology have allowed explorers to search for and produce oil in water depths of 1–2 km. Thus, in the late 1980s, the deep-water Gulf of Mexico became one of the most exciting areas for petroleum exploration. Several giant oilfields have been discovered and there have been many moderately sized, highly commercial discoveries. This same technology has now been applied in other deep-water basins around the world, and new, giant oil- and gasfields have been found offshore West Africa and offshore Indonesia.

The period during which a basin is categorized as having frontier exploration is also the period when many of the primary questions regarding its geology are asked and some are answered:

• What type of basin is being explored?
• Does the basin contain source rocks?
• Are the source rocks mature for petroleum generation?
• Will such source rocks, if proven, be oil- or gas-prone?
• What is the stratigraphy of the basin?

and hence:

• Where are the source rocks, reservoirs, and seals?
• What is the basin history? That is to say, did subsidence, sedimentation, and structural evolution combine to generate and trap petroleum?

The aim of this chapter is to examine the technical activities that accompany the entry of oil and gas exploration companies into a new basin, and to help answer the above questions. A description of how acreage is acquired by those companies who wish to drill oil and gas exploration wells is also given. The three case histories at the end of this chapter are all examples of frontier exploration, albeit from different times from the 1930s to the 1990s. For those readers who are unfamiliar with exploration, we suggest that the main text of Chapters 3 and 4 is read together before the case histories in these two chapters.

3.2 ACQUISITION OF ACREAGE

3.2.1 Early access to acreage

An oil company may have the best petroleum geoscientists and other technical staff in the world, but if that same company has no acreage in which it can apply those skills, then it will be doomed to failure. Obtaining the rights to explore and subsequently produce petroleum from an area is a primary prerequisite for any exploration program. The ways in which acreage may be acquired tend to change with time as the activity within a basin moves from frontier exploration to exploitation.

The importance of obtaining high-quality exploration acreage ahead of the competition cannot be over-stressed. This is particularly the case for offshore acreage. The first companies to obtain acreage are likely to be the first companies to make important discoveries. Having discovered petroleum, such companies will need to build production and transportation facilities to move the petroleum to market.

If these facilities are the first in a new area, it is highly likely that they will capture the petroleum produced from later discoveries. In other words, platforms and pipelines can still be used well after the field that caused them to be built has been depleted. Thus the company that owns the pipeline will profit from its own petroleum production and that of other companies who choose or are forced to use its export facilities.

Acreage can be acquired in two ways, through negotiation with the land or offshore rights holders or through a bid process. The owners may be private or state. A third method, that of sequestering prospective acreage, has been used by some state companies. However, such a method is unlikely to be readily available to an independent or multinational petroleum company!

In past decades, the most common method whereby acreage was acquired was through a negotiated deal with the landowners, be they private individuals or government ministries. Deals between oil exploration companies and private individuals are common on the onshore USA, but much less usual elsewhere in the world. They were, however, the norm for many oil exploration areas in the latter part of the 19th century and the early part of the 20th century. In one of the case histories presented at the end of this chapter (Edale No. 1, Case history 3.8), land for access and drilling was acquired from local landowners. The same was true during the early days of oil exploration in Trinidad (Higgins 1996). The effect in Trinidad, as in so many places around the world, was that some of their giant fields did not appear to be so. This is because an individual field may be discovered more than once by different companies drilling wells on different pieces of land, and thus a single field may have many names.

Governments tend to own larger tracts of land, or sea, than do individuals. In consequence, deals negotiated between oil companies and governments tend to deliver larger license areas than deals that involve private landowners. It is often possible for oil companies to carry out some sort of reconnaissance work on an area before negotiating with a government. Such work may not stretch to anything as intensive as a seismic survey, but would almost certainly include geologic fieldwork. Such fieldwork would aim to establish the presence of petroleum, or at least mature source rocks, potential reservoir intervals, and likely trapping geometries. Nowadays, it would be combined with an analysis of aerial photographs, satellite, and perhaps radar images, principally to identify petroleum seepage and structure.

3.2.2 The licensing process

The pure negotiated deal is relatively rare today. Much more common is the "bid" or "license" round. Such license rounds tend to be organized by ministries for energy or, alternatively, by the state oil companies. There are many possibilities for the way in which a license round can be organized. Some are based solely on a financial bid for a piece of acreage; others demand a technical bid in which a company will demonstrate its understanding of a piece of acreage by committing to collect data (usually seismic) and drill wells in accordance with its understanding. Whichever way, the bid round will cause exploration companies to spend money. The cost of an exploration license for a company can be divided into an access bonus and money that will be spent on exploration activity. In competitively tendered bid rounds, the company or consortium that bids highest usually wins. In areas that have a technical bid, offers of high spend may be ignored if the companies making them are judged not to be competent by the licensing authority.

The terms of a license will usually be defined before a bid round takes place. The terms will include the length of time for which the company may explore before the license either reverts, or partially reverts, to the government or other owner (a process termed relinquishment) or, if a discovery is made, the process by which it can be appraised and developed. In some parts of the world, appraisal and development will be accompanied by a diminution of the exploration company's (or companies') equity as the host government takes a greater share. The terms of the license will also include annual fees. Such fees commonly escalate with time, so encouraging active exploration early in the license period. Further royalties and taxes will then apply to any petroleum produced from fields within the license area.

Countries that host more than one exploration license round will commonly change the terms from round to round. The first round may have relatively benign fees and terms, to encourage prospective exploration companies to enter a new basin or area. This may not always be the case. For example, if the area is known to be an important petroleum province but has been closed to foreign investment, the first license round can be both highly competitive and costly.

Once exploration success is declared in a new area, the competition for new acreage will increase, encouraging the licensing authority to make the license terms tougher, which will have an impact on license fees and access costs. There may be a substantial increase in the tax and royalties payable on produced petroleum. Clearly, it is in the interest of the licensing authority to maximize its delivery of both energy and income, while maintaining appropriate exploitation of its petroleum resource. To this end, the terms and conditions are likely to be enough to "hurt" the participating companies without causing them to exit the country. It is clearly a fine balance, one that has on many occasions failed, with companies pulling out of some exploration areas, perhaps to return when fiscal regimes become less harsh. A common tool used by licensing authorities to encourage companies to explore is to allow tax relief on exploration from production. This means that a company will need to explore, discover, appraise, and develop a field before it can obtain tax concessions against new exploration.

3.2.3 License areas

License areas come in all shapes and sizes. Those on land may be bounded by geographic features, settlements, or old petroleum concessions, or tied to an arbitrary geographic grid (latitude and longitude or a national grid system). For example, the current license area for the Pedernales Field in eastern Venezuela is bounded to the south and east along old petroleum concessions. It is bounded to the northeast by a national boundary with Trinidad. The northwestern edge is defined by

the course of a creek running through Isla Cotorra—although the creek has changed course since the boundary was originally defined decades ago, the boundary itself has not. Finally, the coordinate system upon which the boundary is defined is centered upon Bolivar Square in the town of Maturin, some 50 km west of the field.

Offshore license areas tend to be much more simply defined using latitude and longitude. Deviations from such simplicity occur against land in near-shore areas and along national boundaries (Fig. 3.2). Many countries adopt a numerical or alpha-numerical and hierarchical coding system for license areas. For example, the North Sea and adjacent areas are divided into "quadrants" (or "quads"), which in turn are divided into "blocks." The quadrants occupy one degree of longitude and one degree of latitude. The quadrant system is common to Denmark, Germany, Norway, The Netherlands, and the UK, although each country uses a different coding system. The UK, Norway, and Denmark use numbers. Germany and The Netherlands use letter codes. Quadrants are divided into blocks, the size and system of numbering of which

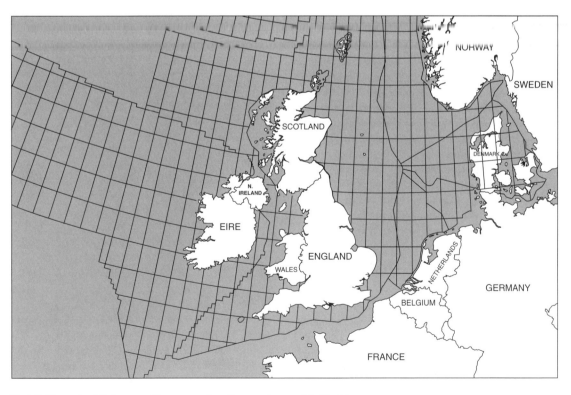

Fig.3.2 The quad and block system of license areas adopted by countries around the North Sea.

differs between the countries. For the most part, one block is one license. However, in some of the frontier areas (e.g., west of Shetland, UK, and Finnmark, Norway) a single license may cover several blocks. The same is true for the newly licensed areas around the Falklands Islands in the South Atlantic (Case history 3.10), where licenses were offered in large tranches, each of which contains several blocks. In the more mature areas such as the North Sea itself, relinquishment has yielded part blocks, denoted by a lower-case letter code. New license rounds commonly offer part blocks or amalgamations of part blocks into single licenses. That areas can be licensed, unsuccessfully explored, relinquished, relicensed, and reexplored demonstrates the technical complexity of oil exploration. The story of UK continental shelf Block 9/13 is a good example of this process. It was awarded to Mobil in 1972 (4th Licence Round). The Beryl Field was discovered and then 50% of the block (Block 9/13b) relinquished in 1978. Block 9/13b was then reawarded to Mobil in 1980 (7th Round) and in this part block, the Tay and Ness Fields were found. Fifty percent of the "b" part of the block (termed Block 9/13d) was relinquished in 1986. This Block 9/13d was then reawarded to Mobil in 1989 (11th Round). No further relinquishments have taken place to date (January 2000).

3.2.4 Farm-ins, farm-outs, and other deals

The company that discovers an oil- or gasfield may not be the company that abandons it. The ownership may change throughout the life of a field. The co-venturers may also change through time. There are many reasons why changes in ownership may occur, some of which have underlying technical causes that are discussed below.

In the frontier exploration phase, a common occurrence is for a company to obtain a license with perhaps no or few drilling commitments. During the course of its evaluation of the license, it may identify prospects for drilling that it might consider either too costly or too risky to execute. At this stage the company that owns the license may invite other companies to farm into its acreage. The form of the "farm-in" is often an exchange of equity for the cost of drilling a well (or wells), the so-called "promote." The initiative to exchange equities can of course come from outside the block. A third party may recognize exploration potential while the license holders are unaware of such potential. Throughout the life of a field partners and equities can continue to change, but many such changes will be driven by the different commercial considerations of the participating companies.

3.3 DIRECT PETROLEUM INDICATORS

3.3.1 Introduction

The first companies to make the major discoveries in frontier exploration areas will tend to end up controlling much of the infrastructure in a basin, and thus benefiting not only from their own exploration success but also from that of other companies; hence petroleum exploration is a race. In consequence, while much emphasis is placed upon "understanding the basin" during the first few years of petroleum exploration, considerable effort is also placed on determining whether there are any shortcuts to petroleum exploration success. There are two broad methods whereby the presence of petroleum in a basin may be proven ahead of acreage acquisition and drilling. Oil and gas may be found at the Earth's surface as seeps, products of leakage from underlying traps and or source rocks. In addition, petroleum may also be indicated on seismic data.

3.3.2 Petroleum leakage and seepage

All petroleum traps leak to a greater or lesser extent and consequently petroleum is able to seep toward the Earth's surface. In Chapter 2, the varied methods for seep detection were outlined. Here we examine more closely why seepage occurs, which fluids seep, their effects at the Earth's surface, and whether or not they can be detected. This last point is important, for while it is true to say that all traps leak, the rate may be exceptionally slow and the products of that leakage may not be readily detectable at the Earth's surface.

The seepage process can be divided into three stages: seal failure, tertiary migration, and dissipation in the near surface (Fig. 3.3). It is important to review all three stages, because the spatial patterns of seepage and phase (gas versus liquid) are influenced by all three stages in the process. To put it bluntly, petroleum seeps may not directly overlie petroleum accumulations and gas seeps may originate from oil pools.

Seal failure

Petroleum moves because a force acts upon it. Buoyancy (the density contrast between petroleum and subsurface brine) is the major force in most settings. Once expelled from the source, the tendency is for petroleum to move upward until a seal arrests migration. Seal development and characteristics are fully explored in Chapter 4. For the purposes of this

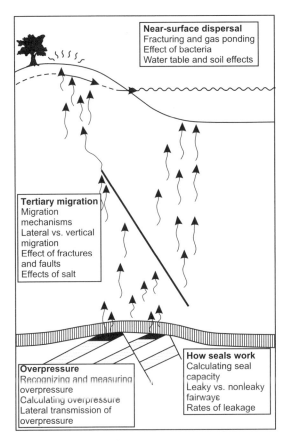

Fig.3.3 Seal failure, tertiary migration, and dissipation. (Reproduced courtesy of BP Amoco.)

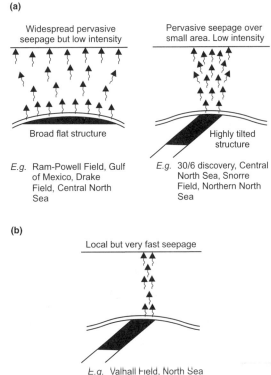

Fig.3.4 Capillary failure (a) and fracture failure (b) of a seal. Active faulting will focus flow; lateral migration will focus flow. (Reproduced courtesy of BP Amoco.)

section we need only consider how a seal counteracts the buoyancy of the petroleum and what happens during seal failure.

Capillary resistance opposes buoyancy. Petroleum must be forced through small pore throats for the seal to breach (Fig. 3.4). Thus smaller and narrower pore throats (capillaries) are more effective seals compared with larger and wider pore throats. Alternatively, a seal may fail if the driving force is sufficient to fracture the rock. Capillary failure is the normal mode of seal failure under hydrostatic and moderate-overpressure conditions, while fracture failure tends to occur in high-overpressure situations (Section 3.5.6).

Tertiary migration

Tertiary migration refers to the movement of petroleum from trap to surface. The physical processes are the same as those

that operate in secondary migration (movement of oil once out of the source rock to the trap), but the rate of supply is much higher when a seal fails compared with escape of petroleum from a source rock. Buoyancy is the prime mechanism and this may be assisted or retarded by overpressure gradients or hydrodynamics.

If a seal breaches through capillary failure then leakage of oil and gas will be pervasive over the whole of the trap. This will disperse the once-concentrated petroleum. However, the petroleum leaking from a trap can become concentrated in a high-permeability carrier bed, allowing it to migrate laterally many tens of kilometers. Fracture failure is likely to cause seepage to be concentrated in specific areas.

Attempts have been made to quantify the rates of petroleum leakage and subsequent tertiary migration. Capillary pressure calculations on mudstone samples indicate that vertical leakage rates in the order of tens of micrometers each year can occur through seals. Lateral migration in sandstones or other porous beds is 10^5 to 10^6 times faster than that in leaking seals (meters to tens of meters per year). The rate of migration of petroleum along faults is probably higher still, at least during periods when the faults are active and open.

The factors that act on migrating petroleum both near and at the surface are different from those in the subsurface. Both pressure and temperature are lower at the surface and at temperatures of <80°C bacteria may feed on the petroleum. This biodegradation of the oil can lead to partial or complete destruction of the petroleum, although alteration products may be deposited. Often, live petroleum is not the most obvious sign of a seep. Secondary effects such as pock marks, or topographic features such as mud volcanoes/diapirs (Fig. 3.5), may be more readily observed. Pock marks are shallow depressions (5–300 m diameter and 2–20 m deep) formed

Fig.3.5 A mud volcano with a fresh mud flow, Pedernales area, Venezuela. Seepage of both oil and gas occurs from mud volcanoes in eastern Venezuela and Trinidad. (From Bebbington 1996; reproduced courtesy of BP.)

by the catastrophic release of gas. Mud volcanoes erupt and diapirs rise when mud of reduced density moves vertically upward. The density reduction is associated with fluidization caused by earthquake activity or extreme sediment loading. While petroleum may not everywhere be associated with mud volcanism, some of the world's major petroleum provinces have abundant mud volcanoes that do release petroleum liquid and gases. Examples of mud volcanoes associated with petroleum seepage occur in California, Trinidad, Venezuela, Turkmenistan, Azerbaijan, Iran, the Gulf of Mexico, and the Niger Delta.

On land, important changes take place as a seep reaches the water table, while beneath water equally important effects come into play at the sediment/water interface. As seeping petroleum reaches the water table, gas will be exsolved and will separate from the oil. Beneath water, wave and tidal energy can both redistribute and disperse the petroleum. The released gas may also dissolve in the water.

3.3.3 The identification of petroleum on seismic data

Sometimes it is possible to identify the effects of both petroleum seepage and petroleum in place on seismic data. The presence of petroleum in the subsurface changes the acoustic properties of the sediment in which it occurs. In most instances the signal-to-noise ratio is such that the effects of the presence of petroleum go undetected. However, petroleum or petroleum products can be detected if the effect on the acoustic properties is large or if the architecture of the affected volume is distinctive. Finally, in the near subsurface, the presence of petroleum and seepage can create topographic effects that are visible on seismic data. Often, the products of seepage are more readily identified unequivocally than that of *in situ* petroleum; thus we shall deal with this first.

Pock marks can be seen on radar or by using underwater cameras, and are also picked up on high-resolution seismic surveys used for well-site investigation work. Both high-resolution seismic site survey data and conventional seismic data may be used for the identification of gas in shallow reservoirs, as plumes, and as gas hydrate mounds.

The effects of gas hydrates can be spectacular (Fig. 3.6). Gas hydrates—or "clathrates," as they are sometimes called —are crystalline compounds of gas and water. The water molecules form a hydrogen-bonded, 3D network within which gas molecules become trapped. They are stable at low temperature and moderate pressure. In consequence, they tend to form in polar regions and in the shallow subsurface beneath deep seas (e.g., offshore Finnmark, northern Norway, and the Gulf of Mexico). The hydrates may be sufficiently abundant

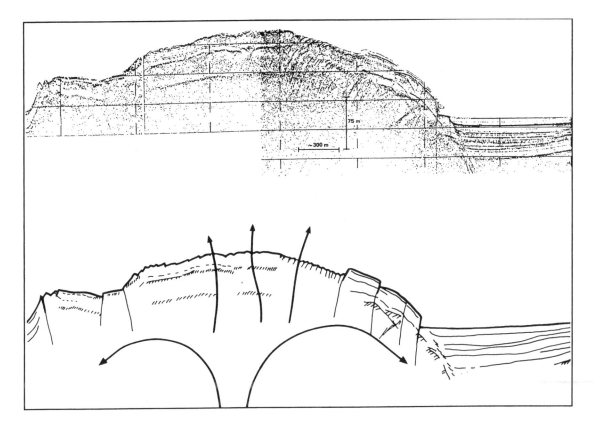

Fig.3.6 A gas hydrate mound. (From Hovland & Judd 1988.)

to form mounds, but even if there is no obvious topographic expression, because they are crystalline, they can affect the acoustic properties of the rock to produce a high-amplitude effect. In some places, this effect is so strong that resonance frequencies may also manifest themselves in the seismic data as multiples (artifacts) spaced at regular two-way time intervals throughout the seismic record. Such artifacts are known as bottom-simulating reflectors or seabed multiples (Fig. 3.7).

Petroleum, particularly gas, dispersed within a sediment column may have quite a different effect to that of hydrates. Dispersed gas tends to dissipate the seismic signal as it passes through the affected zone. Thus the affected volume often has an incoherent seismic signal. Moreover, because the effect of gas is to slow the seismic signal, the area beneath the gas plume will occur at a greater two-way time than the adjacent areas. This can lead to a missed opportunity. It is all too easy to forget when observing a seismic display that the z-axis is commonly two-way time, not depth. A valid, four-way dip-closed

structure in depth may disappear in the time domain if it underlies a gas plume (Fig. 3.8).

Gas, and on occasion oil, *in situ* in the reservoir may generate a "direct hydrocarbon indicator" (DHI) that is visible on seismic as a flat spot, bright spot or a dim spot. Just as sharp changes in the acoustic properties of rocks are responsible for generating seismic reflections, so too can abrupt changes in the acoustic properties of the formation fluids also create reflections. Where such reflections do occur, they are most commonly generated at gas/water or gas/oil contacts (Fig. 3.9). More rarely, DHIs can be generated at oil/water contacts. Such effects should be flat in the depth domain and can crosscut seismic reflectors generated from contrasting lithologies.

Finally, as far as seepage is concerned, care must be taken to determine the source of seeping gas. We have mentioned that gas may be detected above an oilfield or oil seep simply because of the near-surface separation effects, including the effects of gas on seismic. However, the presence of gas seepage or gas effects does not always presage the presence of oil. The

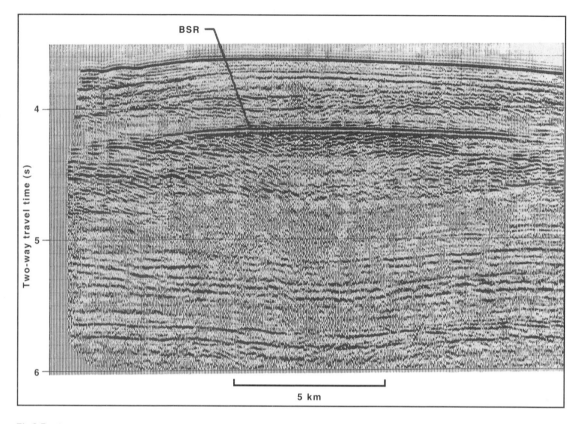

Fig.3.7 A bottom-simulating reflector (BSR), a seismic expression of shallow gas hydrates. (From Hovland & Judd 1988.)

source rock may only be gas-prone or may be overmature for oil (Section 3.7). Alternatively, the gas may be biogenic, produced from bacteria fermenting organic matter in the near subsurface. In order to differentiate between these sources, it may be necessary to sample the gas or its secondary products and carry out definitive chemical and stable isotopic analyses.

3.4 BASIN TYPES

3.4.1 Introduction

In Chapter 1 we said that almost all of the world's petroleum occurs in basins, and that basins are areas of sediment accumulation. In this section we examine the origin of basins and describe their external geometries. The following two sections contain information on the development history of basins and their internal architecture. Basins are generated by plate tectonics, the process responsible for continental drift.

The Earth's crust is made up of about 20 rigid plates (Fig. 3.10), formed from either thick continental crust (30–70 km thick) or thinner oceanic crust (4–20 km thick). These plates are in constant albeit slow motion, carried upon the viscous-liquid part of the upper mantle — the aesthenosphere (at the 1330°C isotherm; Fig. 3.11). The drive mechanism for plate tectonics appears to be convective turnover in the upper mantle. The relationship between the rigid crustal plates and the convecting mantle has been likened to the scum that forms atop boiling jam. Patches of the scum are torn apart or pushed together, and rotate past each other as the jam boils. These same relationships are seen in the Earth's crust.

Plates may be stretched and broken or pushed together, or may rotate past each other. Each of these processes — divergence (extension), convergence (compression and extension), and strike-slip (or wrench) — can lead to the formation of basins. The basins formed by the different processes have different overall geometries, different internal architectures, and different thermal and subsidence histories. Basins have

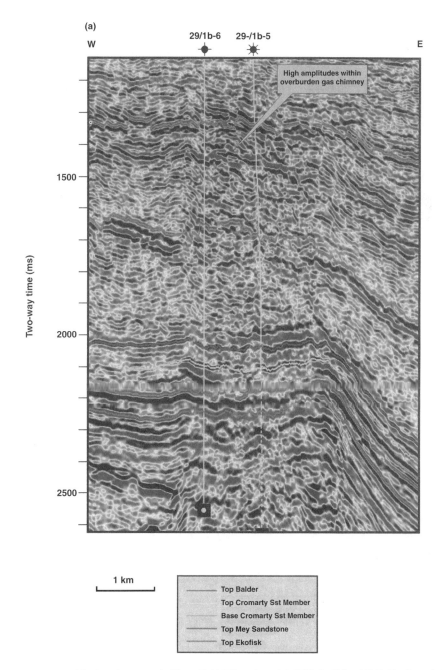

(a)

W   29/1b-6   29-/1b-5   E

High amplitudes within overburden gas chimney

Two-way time (ms)

1500

2000

2500

1 km

Top Balder
Top Cromarty Sst Member
Base Cromarty Sst Member
Top Mey Sandstone
Top Ekofisk

Fig.3.8 Two-way time seismic and depth sections across the Bittern Field, UK continental shelf (Blocks 29/1 and 29/6). The closure in depth is not seen on the time section (a) because gas overlying the field leads to a depression of the reflectors. Only when converted to depth (b) is the anticlinal structure obvious at the Cromarty Sandstone reservoir level. (Figures provided by Markus Levisham, reproduced courtesy of Shell UK.)

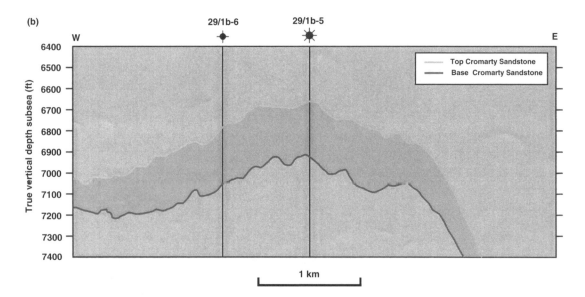

(b)

Fig. 3.8 *continued*

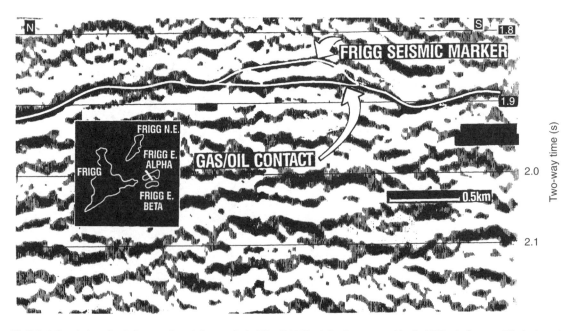

Fig. 3.9 A direct hydrocarbon indicator at the gas/oil contact in the Frigg Field, North Sea. In common with other DHIs, the flat spot will be horizontal on a depth-migrated section. (From Héritier et al. 1990.)

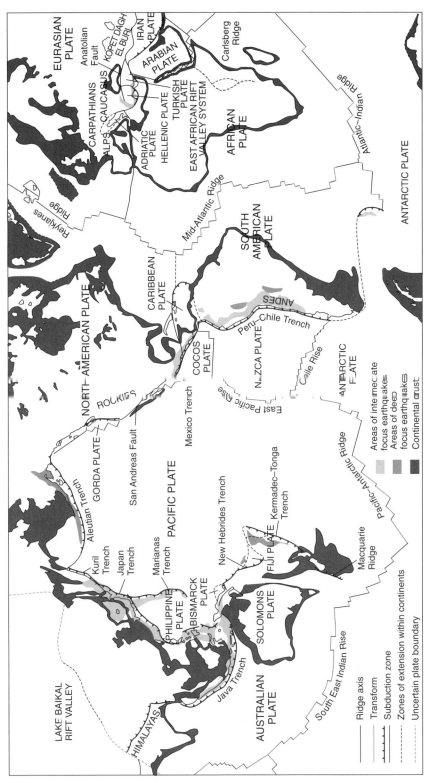

Fig. 3.10 The Earth's tectonic plates. (From Dewey 1972.)

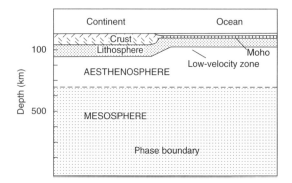

Fig.3.11 The main compositional and rheological boundaries of the Earth. The most important compositional boundary is between the crust and the mantle, although there are certain strong compositional variations within the continental crust. The main rheological boundary is between the lithosphere and the aesthenosphere. The lithosphere is rigid enough to act as a coherent plate. (From Allen & Allen 1990.)

been classified in a variety of ways. Here we adopt the genetic classification outlined above. This methodology is fully described by Allen and Allen (1990).

3.4.2 Extensional basins, generated by divergent plate motion

Two main mechanisms operate to create extensional basins. First, rifting can occur when a thermal plume or sheet impinges on the base of the lithosphere (active rifting). The lithosphere heats up, weakens, and can rift. An example is the East African Rift. The second mechanism (passive rifting) is continental stretching and thinning, which has happened during all major continental breakups.

Continued extension will eventually lead to the complete severance of the two halves of the affected plate and the introduction of oceanic crust along the line of parting. The two halves of the once single plate will then form passive continental margins. The most widely recognized "matching" pair of passive margins are South America and Africa, two parts of the massive Gondwana supercontinent that existed in the early part of the Mesozoic.

The South America and Africa example is, however, a final product of rifting. Extension may be attenuated at virtually any stage in the rift and drift process, and in consequence basins may form at any and all stages. The North Sea is an example of unsuccessful rifting, sometimes called a failed rift basin. The interplay between the fault-generating, tensional effects and the subsidence caused when the crust cools and becomes denser creates important differences between basins. Much work has been done on the relationship between these two processes, and on the magnitude and duration of the processes (McKenzie 1978; Royden & Keen 1980; Cochran 1983). Within the realm of petroleum geoscience, the interplay of these processes not only creates basins of differing geometry that may or may not evolve to yield petroleum; it also controls the thermal history and fluid movement within the basin. These in turn control the development and maturation of petroleum source rock (Section 3.7), together with reservoir diagenesis (Section 5.6).

Intracratonic basins: sags

For a long time, intracratonic (intracontinental) basins were regarded as somewhat enigmatic. Their alternative description, "sags," illustrates their form. Most tend to be broadly oval. The total sediment infill package increases from edge to center and major faults are conspicuous by their absence. Many are old (Paleozoic and older) and they occur on even older crust (Precambrian). It has been postulated that such basins could represent the products of extension where the degree of extension is almost zero, but where the area has suffered thermal subsidence following uplift due to a thermal plume or sheet.

A well-documented example of an intracratonic sag is the Michigan Basin in the USA (Burke & Dewey 1973). It contains a sediment pile that is only 4 km thick at its maximum, yet deposited over a period of 500 Ma. The basin is nearly circular and there is concentric thickening of the sediment pile from edge to center (Fig. 3.12). Although the Michigan Basin is a long-lived sag, the acquisition of gravity data (Hinze et al. 1975) and the drilling of a deep borehole in 1975 (Sleep & Sloss 1978) confirmed the presence of an ancestral Precambrian rift valley. A pertinent yet unresolved question is whether sag basins exist in the absence of ancestral rifts.

Rift basins

The specific locations of rifts may follow old lines of weakness (Wilson 1966). Alternatively, rifts may form above mantle hotspots. The hotspots cause thermal doming and the formation of triple junctions that then act as the rift lines (Burke & Dewey 1973).

Continental rifting is but the opening stage of a process that can lead to full breakup of a continental mass and the generation of oceanic crust between the fragmented parts.

(a)

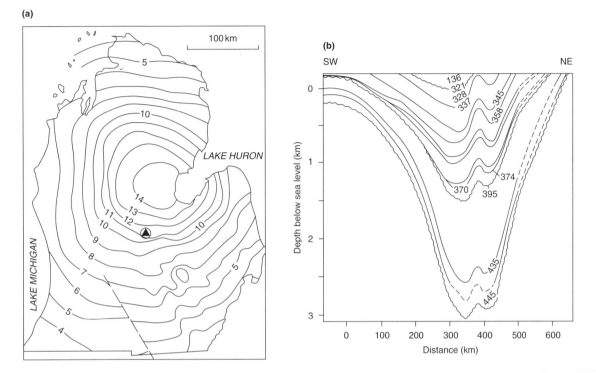

(b)

Fig.3.12 The geometry of the Michigan Basin, USA. (a) Structural contours in thousands of feet on the Precambrian basement surface in the Michigan Basin. The basement depth increases gradually toward the center of the basin, which is almost circular in plan view. (From Hinze & Merritt 1969; modified by Allen & Allen 1990.) (b) A cross-section of the Michigan Basin from Middle Ordovician to Jurassic. Younger units are found in the center of the basin. (From Sleep & Snell 1976; major unconformities are from Sloss 1963.)

Rifting is often associated with volcanism. The degree of uplift and the rate of rifting are controlled by the magnitude of the thermal disturbance and the stresses applied to the plate. For example, a plate that is largely under compression may experience rifting due to impingement of a mantle plume on the lithosphere. However, the confining forces will limit the rate and magnitude of extension across the rift. The duration of a period of rifting is also variable. Moreover, the intensity of movement may be episodic. For example, rifting in the North Sea lasted for about 100 Ma. There were periods of rifting in the Permian and Triassic, there was relative quiescence in the Lower Jurassic, and there was an acme of activity in the Upper Jurassic before rifting finally ceased in the Lower Cretaceous.

Passive margins

Continued rifting will eventually lead to the point at which extension can no longer be accommodated by fault block rotation alone. Oceanic crust will begin to form near the mid-point of the rift system, the so-called spreading center (Fig. 3.13). Naturally, in the earliest stages of ocean crust formation the configuration of fault blocks and sediment fill across the remainder of the rift is little different than before oceanic crust began to form. However, the proximity of the mantle plume will cause thermal gradients to be high or very high. For example, the Red Sea experiences thermal gradients of greater than 100°C km^{-1} close to the spreading center and about 60°C km^{-1} close to the margin of the rift.

The style of tectonics changes as the passive margin matures. Extensional faults typical of the rift phase become rare in the post-rift phase. Instead, gravity structures become more common. Large, listric (concave-up) gravity-driven faults are most common in areas with high sedimentation rates, such as in areas of delta progradation (Fig. 3.14). For example, along the eastern margin of the Atlantic Ocean, down to the ocean, major growth structures are developed within

(a)

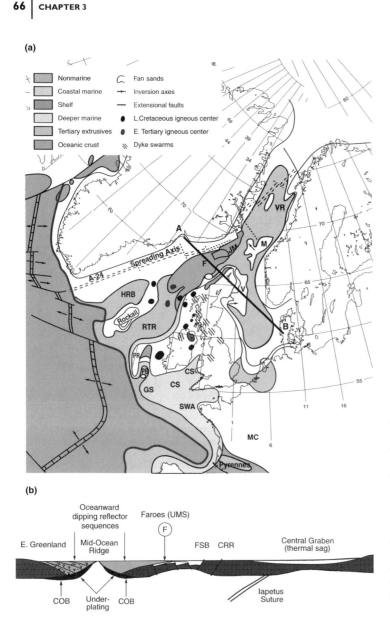

(b)

Fig.3.13 The reconstructed paleogeography and a cross-section of the North Atlantic during the Middle Thanetian (Tertiary). (a) Gross depositional environments in the North Atlantic: GS, Goban Spur; CS, Celtic Sea; SWA, South West Approaches; MC, Massif Central; PB, Porcupine Basin; PR, Porcupine Bank; RTR, Rockall Trough; HRB, Rockall–Halten Basin; F, Faroes; V, Viking Graben; M, Møre Basin; VR, Vøring Basin. (b) An NW–SE-trending schematic cross-section, the location for which is shown in (a): COB, continental/ocean boundary (×2 vertical exaggeration); F, Faroes–Shetland Escarpment; FSB, Faroes–Shetland Basin; CRR, Clair–Rona Ridge. (From Roberts et al. 1999.)

the sediment piles generated by the Niger and Congo deltas. Very large slide blocks may also be generated. Dingle (1980) traced ancient slide units off the coast of South Africa. These displaced "blocks" are several hundred meters thick, up to 700 km long, and were displaced up to 50 km down slope. In such situations, the glide surfaces often prove to be evaporite sediments.

In areas of low sediment input—that is, distant from major deltas—carbonate factories can form on top of the drowned rift blocks. In such situations, the intervening graben accumulate carbonate debris shed from shallow platform areas. This situation occurs today on the Bahamian Bank. It has also been described from the Mesozoic of Tethys (European Alps), the Exmouth Plateau on the Australian North West Shelf, and the Miocene of the South China Sea (Case history 3.9).

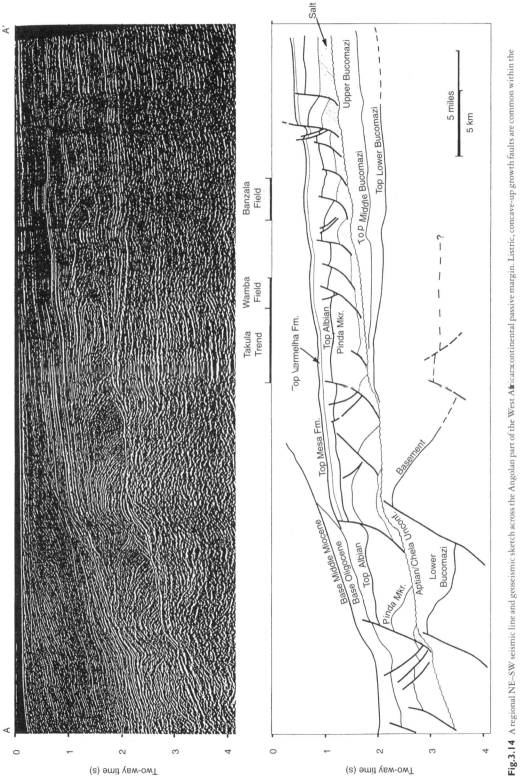

Fig.3.14 A regional NE–SW seismic line and geoseismic sketch across the Angolan part of the West African continental passive margin. Listric, concave-up growth faults are common within the Aptian–Albian interval, demonstrating substantial extension at this level. The faults sole out within Cretaceous evaporites. (From Dale et al. 1992.)

3.4.3 Basins generated during convergent plate motion

Convergent plate motion is of course compressive. However, the combination of both inhomogeneous stress distribution and thermal effects similar to those already described also produces areas of net extension. Thus both compressional and extensional basins can develop within the realm of convergent plate motion (Fig. 3.15).

The style of basin development associated with convergent plate motion is highly varied, and depends upon the interplay of several factors. These include the types of crust undergoing convergence: continental to continental, oceanic to oceanic, and oceanic to continental. The age of the oceanic crust undergoing subduction at the margin is also important. As oceanic crust ages and cools, so it thickens. Thick crust tends to subduct at a greater angle than thin crust. Particularly thick and old oceanic crust may, because of its density, subduct more quickly than the relative plate convergence demands. This situation, called roll-back, can lead to oceanward migration of basin formation. Given this complexity, there are many schemes that have been used to categorize basins that developed during convergent plate motion. The subdivision scheme used here is the same as that adopted by Allen and Allen (1990). Thus we discuss basins associated with arcs and those associated with continental collisions.

Arc systems

Arcs are characterized by six major components (Fig. 3.15; Dickinson & Seely 1979). From overridden oceanic plate to overriding plate, these components are as follows:

• *An outer rise on the oceanic plate.* This occurs as an arch on the abyssal plain, the flexural forebulge of the subducting oceanic plate.
• *A trench.* Commonly more than 10 km deep, the trench contains pelagic deposits and fine-grained turbidites. The trench is generated from subduction of the oceanic plate. Trenches are not considered to be prospective for petroleum exploration. Given current technology, their water depth alone precludes oil exploration.
• *A subduction complex.* This comprises stacked fragments of oceanic crust and its pelagic cover, together with material derived from the arc. Small basins may lie on top of the accretionary complex (Fig. 3.15).
• *A fore-arc basin.* This lies between the subduction complex and the volcanic arc. Petroleum provinces in fore-arc basins are rare. The Trujillo Basin on the Pacific coast of Peru is a late

Cretaceous to Recent fore-arc basin developed on continental crust: it contains petroleum.
• *The volcanic (magmatic) arc.* Magma is generated from the partial melting of the overriding and possibly subducting plates when the latter lies at a depth between about 100 km and 150 km.
• *The back-arc region.* This is floored by either oceanic or continental lithosphere. The back-arc region may or may not develop a basin. Extensional back-arc basins develop where the velocity of roll-back exceeds the velocity of the overriding plate. Moreover, as the basin migrates oceanward it can become starved of sediment derived from the continent.

Where there is a balance between the roll-back and overriding plate velocities, no basin develops. The same is true in a compressional situation, where roll-back is limited or absent. Such areas are dominated by thrust tectonics. Back-arc basins floored by oceanic lithosphere tend to have very high rates of subsidence and high heat flows. The Central Sumatra Basin is a well-described example (Eubank & Makki 1981). Where the back-arc region is floored by continental lithosphere, the area tends to undergo flexural subsidence coupled with major fold and thrust tectonics. Such areas are called foreland basins and they are commonly major petroleum provinces (e.g., the Western Canada Basin and the Andean basins). Given the importance of foreland basins in a petroleum context, they are discussed specifically in the following section.

Foreland basins

The importance of foreland basins as petroleum provinces outranks that of other basins generated by convergent plate motions. The basins are typically several thousands of kilometers long and parallel to the arc and thrust belt. Their widths range between 100 km and about 300 km, depending upon the rigidity of the plate. The basins are typically asymmetric, with maximum subsidence occurring toward the thrust complex and the advancing mountain front. Sediment accumulates both ahead of the frontal thrust area in the foredeep and within smaller basins that occur on top of the thrust complex (piggyback basins; Fig. 3.16).

3.4.4 Strike-slip basins

Strike-slip or wrench basins occur where sections of the crust move laterally with respect to each other. Although a wrench system taken as a whole can be of similar size to a rift, passive margin, or foreland complex, individual basins are much smaller than the other types of basin described above (Fig. 3.17). Strike-slip systems commonly involve some

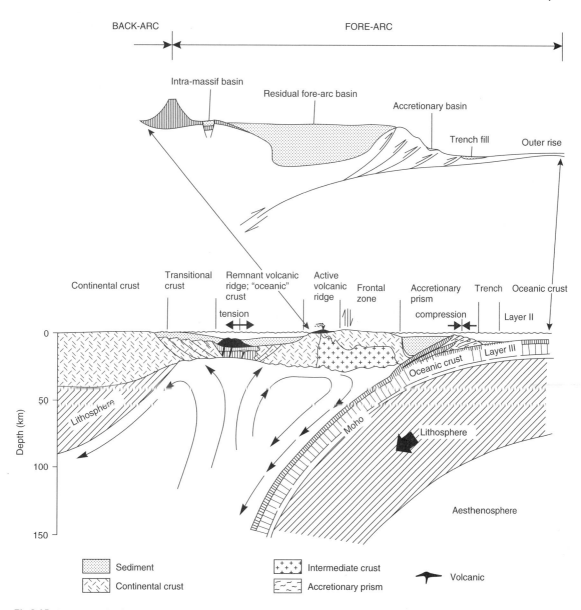

Fig.3.15 Compressional and tensional areas created during convergent plate motion and arc systems. (From Allen & Allen 1990.)

oblique relative movement of the plates to either side. In consequence, some parts of the system can be in tension and others in compression. The form of basins developed under transtension differs from those formed under transpression. Naturally, the tensional and compressional elements tend to produce morphologies similar to those already described.

3.5 BASIN HISTORIES

3.5.1 Introduction

As a basin develops, sediments become buried, as well as up-lifted, at different times and at different rates depending on a variety of controls, such as tectonic subsidence, compression and inversion tectonic forces, changes in the temperature of the base of the lithosphere, and lithospheric flexure. All of these factors shape the sequence of sediments and unconformities (breaks in the sedimentary record) that are preserved in the rock record.

The starting point for unraveling the history of a basin is the vertical sequence of rocks, either discovered by drilling a borehole, or present in exposure of rocks at outcrop. Once the rock types (lithologies) and their ages have been properly identified and the thicknesses established, a crude analysis of the burial history is possible. As further information is assem-

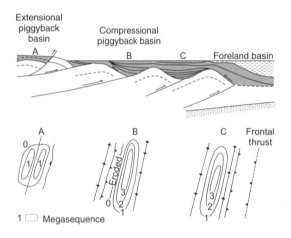

Fig.3.16 Piggyback basins (e.g., the Apennines) developed behind a foreland basin. (Reproduced courtesy of BP Amoco.)

bled, it is then possible to analyze the way in which the rocks have altered due to changes in temperature and stress, and the passage of fluids through the rocks. These factors will have helped to determine whether the reservoirs are of good or poor quality and the source rocks are mature for petroleum generation, and the relative timing of both petroleum maturation/migration and trap development.

3.5.2 Subsidence

There are two main controls on subsidence: tectonic, due to extensional and compressional forces acting on the lithosphere; and thermal, due to changes in lithospheric heat flow. A third coupled force is the response of the lithosphere due simply to the load of sediment acting downward on it. This causes the lithosphere to subside, creating more space for sediment to be deposited.

Sag basins created solely by the impact of a mantle plume at the base of the lithosphere have a rather different subsidence history. These basins have an early history of erosion at the base, with the maximum erosion close to the center of the plume. Conversely, the magnitude of subsidence is greatest in the center, decreasing to zero over a distance typically of 500–1000 km away. A good example is the mid-Jurassic plume that affected the central area of what is now the North Sea (Underhill & Partington 1993). This sag is superimposed upon the early part of the rift history for the basin. Underhill and Partington (1993) used a combination of detailed rock and age descriptions to show how a mantle plume caused uplift and created deep erosion in the center of the basin (Fig. 3.18). There was extrusion of lava associated with

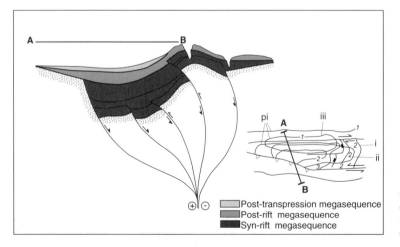

Fig.3.17 A strike-slip basin plan and cross-section, showing typical megasequence distributions for the syn-rift, post-rift, and transpression stages of the basin. (Reproduced courtesy of BP Amoco.)

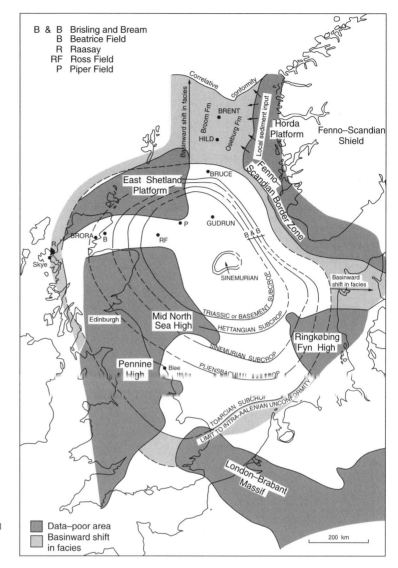

Fig.3.18 A basin developed on mantle plume. The map represents the concentric subcrop pattern beneath the Mid-Cimmerian Unconformity. It is best explained by invoking progressive erosion over an elliptical area of uplift. (From Underhill & Partington 1993.)

volcanism above the plume. When the plume ceased, thermal relaxation and contraction of the continental lithosphere induced subsidence that is documented in the progressive onlap of sediments that infill the basin.

In rift basins, which are created by stretching and thinning of the continental lithosphere, both tectonic and thermally induced forces will initially act in concert, but in opposite directions. McKenzie (1978) showed that the tectonic subsidence exceeds the thermal uplift (due to expansion of the lithosphere) when the crust is greater than about 18 km

thick. The main evidence for this stretching is found in large linear faults that cut deep into the upper crust, and fault blocks that have tilted in response to the stresses created. The subsidence creates space for early syn-rift sediments to be deposited. Sometimes, if they rise above base level—for example, during the Miocene history of extension of the Gulf of Suez (Steckler 1985)—the tilted fault blocks are eroded. Because the lithosphere has been thinned and the base of the lithosphere, the 1330°C isotherm, has been elevated, the heat flow through the crust and mantle are higher than they were

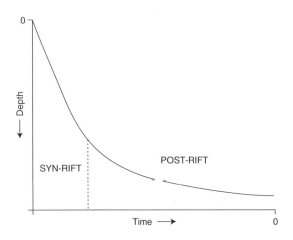

Fig.3.19 A schematic burial history for a rift basin. The syn-rift burial history is typically dominated by fault-controlled subsidence and coarse continental and shallow-marine sedimentation. The post-rift burial history reflects slowing subsidence (thermally controlled) and progressively deeper-water, fine-grained sedimentation.

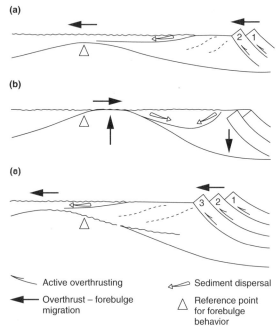

Fig.3.20 The effect of repeated deformation in a foreland basin. With each mountain-building episode, the load increases and there is an elastic response followed by a phase of relaxation. If the mountain belt then migrates into the foreland, the flexure also migrates. (a) Overthrust loading – flexural deformation; (b) relaxation phase – viscoelastic response; (c) renewed overthrust loading – flexural deformation.

prior to stretching. The response when the stretching stops is a reduction in heat flow as the lithosphere cools and returns eventually to its original thickness (about 125 km). The cooling induces contraction and further subsidence creates more space for the post-rift sediments to be deposited (Fig. 3.19).

In foreland basins, subsidence is created in response to lithospheric loading. The load generated by crustal shortening, itself a product of compressive tectonic forces, weighs down the lithosphere. The load acts like a weight on the end of a bending beam, creating both areas of uplift and areas of subsidence (Fig. 3.20). Consider a point on the Earth's surface as a large sedimentary wedge is pushed toward it. Uplift occurs first due to the effect of the peripheral bulge furthest away from the load. This bulge creates an unconformity if the underlying rocks are elevated and eroded. As the load approaches, subsidence takes over and the first sediments are deposited. As the load is transferred even nearer, the rate of subsidence increases, until eventually the tectonic forces cut beneath the basin and involve the sediments in the load itself.

Subsidence associated with strike-slip basins tends to be created by a hybrid of the forces described above. Tectonic forces are dominant though, and the amount of rapid subsidence created in pull-apart basins can be considerable. The Ridge Basin in California, for example, has an accumulated sediment thickness in excess of 12.0 km (40,000 ft), and the Dead Sea and Californian strike-slip systems each contain over 10 km of Neogene sediment (see the San Joaquin case

history in Chapter 4). Because such basins are small and sedimentation is rapid, facies belts are often attenuated. Coarse conglomerates and breccias adjacent to the bounding faults may pass with rapidity into alluvial fans and thence into lacustrine muds. Moreover, because the strong tectonic influence on individual basins is high, it is often difficult to correlate from one basin to its near neighbor. Indeed, the fills of adjacent basins can be quite different.

As wrenching continues, a basin and its fill can be detached from the sediment source areas. Such effects can be observed in modern wrench systems, where river drainage patterns are commonly displaced across a wrench system. A second consequence of the temporal and spatial fluctuation between transpression and transtension is that individual basins tend to be short-lived. Each may go through a complete cycle of formation, fill, uplift, and denudation in a few million years, or a few tens of millions of years. Just as the lateral movement forms sigmoidal fracture and fault systems, it also tends to form *en echelon* fold traces when transpression is operating.

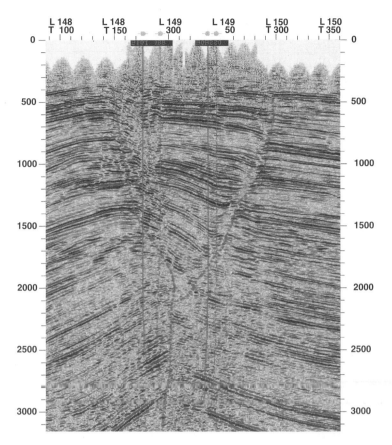

Fig.3.21 A flower structure, Burun Field, Turkmenistan. The flower structure is at depth, a single fault. At depths equivalent to less than 2 s TWT, the fault is divisible into several components bounding a crestal graben to the anticline. (Reproduced courtesy of Monument plc.)

Seen in cross-section, such systems tend to exhibit the highly characteristic flower structures (Fig. 3.21). The thermal consequences of strike-slip tectonics are restricted to these pull-apart basins where locally high heat flow can create uplift followed by subsidence.

3.5.3 Sediment supply

The rate at which the space created by tectonic and thermal forces is filled depends on the availability of sediment. Moreover, the type of sediment depends on the climate and the nature of the sediment supply. Some patterns of sediment accumulation can be used to characterize certain basin types. In rift basins, the fault-controlled and rapid early subsidence history creates coarse, nonmarine, alluvial and fluvial sediments. The sediments are locally derived and can be immature both texturally and mineralogically. Continued rifting and associated subsidence will eventually lead to inundation of the basin by the sea. However, the connection between the

new sea and its host ocean is commonly poor. This can have important consequences for the patterns of sedimentation. In areas where the connection between the new sea and host ocean is periodically severed and the climate is arid, evaporation can lead to the formation of thick sequences of anhydrite, halite, and other salts. Such a sequence of events is well described from the southern North Sea, where the Rotliegend sandstones and associated playa lake deposits are overlain by carbonates and evaporites of the Zechstein Series. The same situation existed during the Jurassic of the Gulf of Mexico and Texas (USA). Here, the Leuanne salt overlies the Norphlet aeolian sandstones and the Smackover, restricted-basin, carbonates. Another consequence of the formation of a restricted basin is that water turnover may be limited. This can occur in the arid, evaporating basins described above or ones in which a combination of humid climate and a connection with the host ocean maintains the sea level in the basin. Stagnation of the water column promotes the formation of sediments that are rich in organic matter. Such sediments can become petroleum source rocks (Section 3.7). Later stages of rift basins are characterized by a marine transgression and progressively finer-grained and deeper-water marine sediments, as the subsidence exceeds the sediment supply. Passive continental margins typify this sediment pattern, and many examples have been documented by drilling off the coasts of Africa and northwest Australia. The sediment thickness varies. Some parts of a continental margin may be starved of sediment, whilst other areas such as deltas can contain tens of kilometers of sediment. Sediment can also bypass the coastal area where it funnels down submarine slope canyons to find a final resting place at the base of the continental slope or out on the abyssal plain.

In foreland basins, the earliest deposits are laid down when the approaching load influences the emergent "foreland bulge." The sediment source is still a long way from this point, so the sediments are likely to be either limestones (in shallow water without a clastic input) or fine-grained clastics such as muds. As the foreland wedge of thrust sediments and other basement rocks moves nearer, the amount of sediment input increases and the sediments become progressively coarser. In an Alpine foreland basin this sequence is termed the "flysh" and it includes both deep-water muds and interbedded turbidite sandstones. There is then a transition from deep-water to shallow-water clastic rocks (shallow-marine and coastal sediments), followed by the emergence and rapid deposition of coarse clastic rocks in fluvial/alluvial conditions. In the Alpine foreland, this sequence is termed the "molasse." In summary, a foreland basin experiences an early basin history with a sedimentation rate that is slower than the subsidence rate (fine-grained sediments deposited in deeper water with time). In the later stage, the sediment supply exceeds the rate of subsidence (although the rate of subsidence is still increasing), which leads to a major upward-coarsening sequence of sediments (coarser-grained and shallower water with time).

3.5.4 Burial history

The burial history of a vertical sequence of rocks is reconstructed from age and depth information. The depth is recorded during drilling and logging of boreholes, or by measurement of sedimentary sections in outcrop. The age determination will vary in its detail depending on the type of material available (rock chippings, core, or outcrop) and any correlation with other rocks of known age. The technique of backstripping is used to draw the burial curve: the burial history of each unit is reconstructed by taking off the thickness of the overlying layers sequentially, with each new age assigned

by the known boundary ages (Fig. 3.22a). Two early assumptions are often made in the first attempt: (1) that there is no compaction and (2) that all sediments are deposited at sea level. In later reconstructions, compaction and paleowater depths need to be taken estimated. Compaction can only be accounted for when the lithology is known and an estimate of its compaction behavior is assumed. Paleowater depths are usually based on data such as benthic microfauna, interpreted to have inhabited water of specific depth ranges.

Mechanical compaction is the progressive loss of porosity due to imposed stresses and it leads to a reduction in the thickness of the original sedimentary unit as water is driven out of the bulk rock. As the rock loses porosity and becomes more rigid, the rate of loss of porosity slows down. When reconstructing a burial history, the rate of mechanical compaction is determined from typical porosity–depth or porosity–effective stress relationships, for which a starting porosity (or depositional porosity) is needed. Shales and carbonate muds typically have an initial porosity of about 60–80%, whilst sands have a starting porosity nearer to 40% (Section 5.6). The rate of decline of porosity is more rapid in shales than in sands, with mechanical compaction leading to a porosity of between 5% and 15% at 4.0 km. In sands, mechanical compaction is largely complete when the porosity has been reduced to about 25%.

3.5.5 Thermal history

The base of the lithosphere is defined by the 1330°C isotherm. Heat flows continuously from here to the Earth's surface. The thermal history of rocks in basins is determined by the boundary conditions of surface temperature (cold in polar and hot in the equatorial regions) and the basement heat flow. The temperatures at points in between are controlled by the thermal conductivities of the sediments. These parameters are linked through the heat flow equation:

$$Q = -K \cdot dT/dZ$$

where Q is the vertical component of heat flow, K is the thermal conductivity, and dT/dZ is the geothermal gradient (Blackwell & Steele 1989). Temperature gradients can be determined from petroleum boreholes by direct measurement. Detailed temperature profiles through sedimentary basins are rare. Boreholes drilled for oil and gas remain the prime source of direct temperature measurement in sedimentary basins. When boreholes are drilled, the effect of the drilling mud is to heat the shallow sediments (generally less than 500 m) and to cool the deeper rocks. During and immediate-

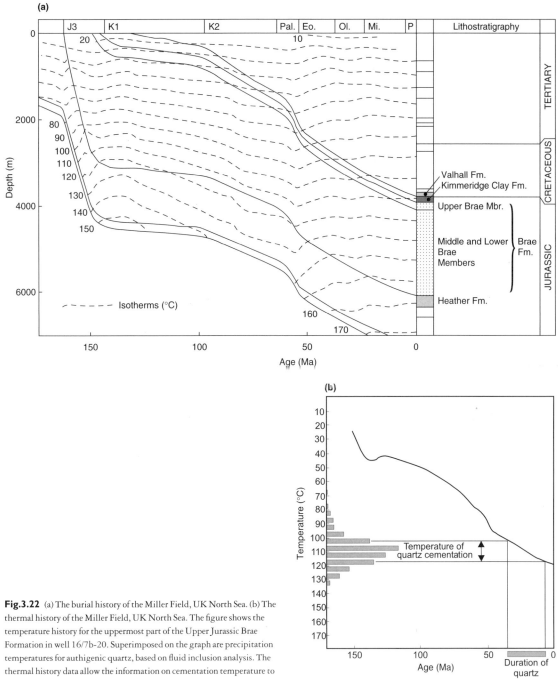

Fig.3.22 (a) The burial history of the Miller Field, UK North Sea. (b) The thermal history of the Miller Field, UK North Sea. The figure shows the temperature history for the uppermost part of the Upper Jurassic Brae Formation in well 16/7b-20. Superimposed on the graph are precipitation temperatures for authigenic quartz, based on fluid inclusion analysis. The thermal history data allow the information on cementation temperature to be converted directly to a time for cementation. (From Gluyas et al. 2000.)

ly after drilling, the rocks are out of thermal equilibrium and they may take a long time to reequilibrate to true formation temperatures. Conventionally, a maximum-recording thermometer is attached to wireline logging tools when they are lowered into the borehole. Corrections need to be made to any temperatures recorded. Such corrections take into account the time that has elapsed since circulation of the mud, a process that maintains the thermal disequilibrium between mud and host rock. In constructing a geothermal gradient from data collected during logging in the borehole, care must be taken to avoid excessive interpretation of temperatures from highly conductive lithologies such as salt, which may not be representative of the full rock succession. Finally, temperature estimates can be validated where the temperatures have been recorded during production of the formation fluids (e.g., during drill stem tests), corrected for the influence of the cooler mud during the drilling process. Once the corrected temperatures are available at two or more depth points of known stratigraphic age, a geothermal gradient, and hence a thermal history (Fig. 3.22b), can be determined. If the thermal conductivity of the sediments is known, the heat flow can be calculated using the heat flow equation. In practice, the thermal conductivity is often estimated from a look-up table of common values. Heat transport is governed by the thermal conductivity of the rocks, which varies for each lithology and in porous sediments due to the fluid contained therein. High-conductivity rocks include salt, quartzite, and some ultrabasic rocks. Low-conductivity rocks include coal and many mudrocks and shales. Water has a low conductivity relative to many rocks, but it is high in comparison with oil, and particularly with gas. Further complications in assigning correct values exist because matrix thermal conductivity is temperature dependent and most sediments are porous, so the bulk thermal conductivity is required.

The boundary conditions of surface temperature and basement heat flow vary through time during the evolution of a sedimentary basin. In the North Sea area, the basement heat flow is linked to rifting events about 220 and 150 Ma. The basement heat flow was higher then than today. There has been progressive decay of the thermal pulse associated with continental rifting, especially over the past 100 Ma. During the same period, the surface temperature has fluctuated as the climate has alternated between subtropical and ice-house. During the burial of the sediments, their thermal conductivity has altered as the rocks have compacted. The boundary conditions have also changed. A computer program is usually used to model these basin conditions and to estimate the thermal history of a sedimentary basin. Two- and three-dimensional modeling also takes into consideration the effects of fluid movements, especially flow through regional aquifers. Aquifers of meteoric water will introduce cooler waters than expected, whilst upward- and out-flowing waters will elevate the regional geotherm in the vicinity of the aquifer. The natural radioactivity in the sediments will also influence the temperature of the sediments through time.

Temperature controls the maturation of source rocks, although time is also important (see Section 3.6.4). Using the Arrhenius equation (the reaction rate increases exponentially with temperature), the cumulative effect of changing temperature through time on maturation can be assessed. Basin modeling is routinely used to estimate timing of maturation of source rocks and availability of petroleum fluids to fill reservoirs and traps. The parameters that are used to estimate the cumulative effects of the thermal history of sediments include vitrinite reflectance ($\%R_o$), illite crystallinity, clay dehydration states, various biological markers (Curiale et al. 1989), apatite fission track analysis, and T_{max}. Vitrinite reflectance is the most common method, as it is a relatively easy optical technique and therefore cheap. A common component of sedimentary organic matter, the reflectivity of vitrinite increases in a consistent manner as a function of the temperature to which it has been subjected. Plots of vitrinite reflectance at regular spaced sample intervals down a borehole section yield powerful insights into geothermal history. Such plots also provide a way to estimate the maturation level of any source-rock intervals (Fig. 2.27).

3.5.6 Uplift

Compressional and lithospheric rebound forces can lead to uplift, whereby the sedimentary section is raised above base level, leading to erosion. Erosional truncation is an indication of sediments above base level. Widespread uplift follows continental collision, leading to deep erosion. Uplift may be followed by further burial. A history of uplift is recognized in the sedimentary record by erosion and missing time in the rock record, and in temperatures which are lower than those associated with maximum burial (Fig. 3.23). Several thermal parameters measured in sediments reveal a history of uplift, including vitrinite reflectance, illite to smectite ratios, fluid inclusions, and fission tracks. Zircon and apatite fission track analysis provides a powerful tool to identify not only how much uplift there has been, but also the timing of the uplift. A study from Upper Cretaceous sandstones in the Green River Basin, Wyoming (Naeser & McCulloh 1989), reveals higher temperatures in the past. The fission track analysis supports rapid uplift about 4–2 million years ago (in the Pliocene).

When uplift and erosion take place, the overburden is gradually reduced, decreasing the vertical stress imposed on the sediments. Compaction is a largely irreversible process, so

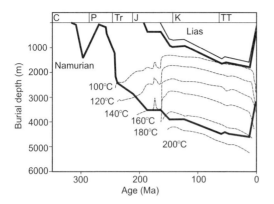

Fig.3.23 The uplift history of the Ravenspurn Field, UK North Sea. There were phases of uplift in the area in the late Carboniferous and again in the Tertiary. (From Gluyas et al. 1997b.)

uplift only leads to very small increase in porosity as the elastic component is recovered. Joints commonly develop as a response to this elastic rebound during unloading.

3.5.7 Pressure history

Porous sediments buried below the vadose zone or under the sea contain water (and more rarely oil and gas). The pressure of the pore fluids is controlled by the depth, the fluid gradient (a function of the density of the fluid), and stresses. Matrix stress (the vertical weight of the overburden and complementary horizontal stresses) will only affect the pore pressure if there is incomplete dewatering to balance the strength of the bulk matrix. According to Terzaghi's principle, the total stress (S) is equal to the pore pressure (P_p) plus the mean effective stress (σ). On a plot of pressure versus depth, the hydrostatic pressure gradient is a reference line for the normal water gradient pressure and the lithostatic stress gradient is used for the overburden. Direct pressure measurements (Repeat Formation Tester, RFT™, and drill stem test) provide a profile of pressures, but can only be collected in permeable units such as reservoirs. Pore pressures in nonreservoir rocks can sometimes be inferred from a comparison of the expected porosity at normal pressures (as if they were fully compacted) and the actual porosity (measured by wireline logging). Where pore pressures are between hydrostatic and lithostatic the sediments have "overpressure," whereas "underpressure" is used to describe sediments with pore pressures below hydrostatic. Natural underpressure is unusual (Swarbrick & Osborne 1998) in relation to uplift, and is often associated with gas-prone, low-permeability rocks. Overpressure, on the other hand, is typical of most basins, especially below about 3.0 km (Hunt 1990).

Several processes (Osborne & Swarbrick 1997) create overpressure (pore pressure in excess of hydrostatic). The most frequent cause is rapid loading of sediments, in which the fine-grained rocks cannot dewater fast enough to remain in equilibrium with the vertical stress of the overburden. This mechanism is termed "disequilibrium compaction" and is associated with anomalously high porosity relative to the depth of burial. If burial slows or stops, the excess pressure will dissipate and the rate will be dependent on the permeability structure of the overlying and adjacent sediments. Lateral stresses may also cause overpressure. Fluid expansion is an alternative mechanism for overpressure, created by changes in the volume of the solid and/or fluid phases. Examples include aquathermal pressuring (thermal expansion), clay dehydration reactions (such as smectite or anhydrite), mineral transformation (such as smectite to illite, or kerogen to oil and gas plus residual kerogen), and oil to gas cracking. Each example involves either volume change and/or alteration of the compressibility of the rock framework, leading to additional loading of the pore fluids. Each of the fluid expansion mechanisms is likely to be secondary under the normal range of basin conditions (20–40°C km⁻¹ geothermal gradient; up to 2.0 km Ma⁻¹ sedimentation rate) and realistic low-permeability seals (100 m of sediment with a permeability of 1.0 nD or less; Swarbrick et al. 2002). It is also possible that overpressure is created by changes in the solid-to-liquid ratio of the sediments; for example, when solid kerogen transforms to liquid oil and residual kerogen plus some gases. If the kerogen is load bearing, then the load could be passed onto the fluid, increasing its pore pressure.

Pressure gradients in excess of hydrostatic are termed "pressure transition zones" (Swarbrick & Osborne 1996). They indicate low-permeability rocks; that is, those sediments that prevent equilibration of pressures between the successive permeable units. Hunt (1990) called these rocks "pressure seals," but their history as seals is ephemeral and changing conditions can allow the pressures to equalize due to slow dissipation. The control on dissipation is principally permeability, but also the storage volume of the higher-pressured unit. Young, rapidly deposited, fine-grained sediments exhibit a continuous pressure gradient above hydrostatic, leading to increasing overpressure below the point at which fluid retention commences (Fig. 3.24). In this case there is a continuous pressure transition zone from the fluid retention depth. In this situation, heavier and heavier mudweight is needed to drill a borehole to the objective reservoir horizon. In older rock sequences, the fluids organize themselves into intervals that are dominated by fluid gradient, separated by discrete pressure transition zones. Over time, and depending on burial conditions, pressures will

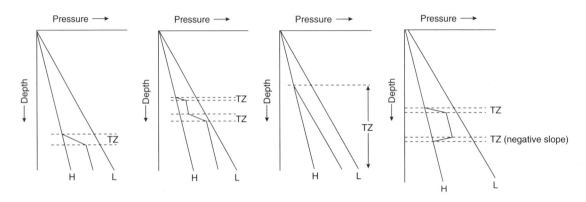

Fig.3.24 Pressure transition zones reflect fluid retention due to low sediment permeability (commonly mudstones, carbonates, and evaporates). Pressure transition zones are recognized where pressure gradients on pressure–depth plots exceed the range of naturally occurring subsurface fluids, including brines. (From Swarbrick & Osborne 1996.)

dissipate upward as well as equilibrate locally to create fluid-dominated pressure gradients.

Pressure histories are estimated using fluid flow simulation in basin modeling software (Waples 1998). Calibration for the results includes direct pressure measurement from reservoir units, as well as matching to any porosity and/or permeability data that are available for the rocks today. Paleopressure can be estimated using fluid inclusions. Aplin et al. (1999) describe a technique using petroleum and aqueous fluid inclusions trapped simultaneously in mineral cements. Fluid inclusions are time capsules that provide information about both the temperatures and the pressures of the pore fluids at the time of trapping. Using a confocal laser scanning microscope to measure the liquid-to-vapor ratio accurately, it is possible to determine the effective composition (expressed as the gas to oil ratio, GOR) of the petroleum. Reconstruction of the P–V–T relationships leads to an estimation of the paleopressures. A case study from the North Sea (Swarbrick et al. 2002) illustrates an increasing amount of overpressure during cementation over a time interval of about 2 Ma during rapid burial of the sediments in Pleistocene–Quaternary times. The maximum overpressure is found today in the reservoir, which is 22 MPa (3200 psi) above hydrostatic pressure at a depth of 3400 m (11,150 ft).

3.5.8 Integrated basin modeling

Powerful computer simulations of rock and fluid behavior during both burial and uplift are available today in commercial software packages. The oil industry has led the way in building these packages, first to model the timing of maturation of source rocks and later to couple this geochemical modeling with fluid flow modeling. Input for the models includes sediment parameters for each of the main lithologies, such as initial porosity, bulk rock and fluid compressibility, thermal capacity and conductivity, and organic richness. Algorithms in the modeling software attempt to describe the behavior of the rocks and fluids during progressive burial, as temperatures and stresses increase and sediments compact and try to expel fluids. The choice of a suitable algorithm and associated parameters to describe shale compaction, for example, is not trivial, but would be improved with multiple measurements of actual rock data for calibration. Wireline data may give insights into porosity behavior during compaction, for example, but direct measurement of both permeability and thermal conductivity of shales is rare.

Although the temperature profile in the sediment sequence can be known from borehole measurements today, the paleogeothermal gradients will not be well known and will need to be calibrated using the full range of geothermal history indicators. At present, it is not possible to capture fully the effects of fluid-rock chemistry on sediment parameters, particularly porosity and permeability, in basin models. Young sediments, in which there has been limited or no time at elevated temperatures (>80°C or thereabouts), are the most readily modeled using these simulation packages, since the effects of chemistry can be largely ignored.

3.6 STRATIGRAPHY

3.6.1 Introduction

Stratigraphy is the study of temporal and spatial relationships between bodies of sedimentary rocks. The establishment of the geologic column, from Precambrian to Holocene, is a product of the stratigraphic analysis of many basin-fill sequences, in many parts of the world, performed by generations of geologists over the past two centuries. Thus the goal of any stratigraphic analysis is to establish the temporal sequence of sedimentary rocks in the area under investigation. Petroleum exploration without stratigraphic analysis degrades to "drilling the big bumps"—that is, simply drilling the largest structures seen on seismic—without attempting to determine whether the source, reservoir, and seal exist.

In frontier exploration, stratigraphic analysis of seismic and well data is used to determine the disposition and age of the main subdivisions of a basin fill. Acquisition of such knowledge will help the geoscientist determine the most likely locations of the source, reservoir, and seal. This in turn will drive geochemical and structural modeling to determine the likely timing and products of source-rock maturation (Section 3.7) and the quality of reservoirs (Chapter 5).

As exploration, appraisal, and development progress, so the emphasis of stratigraphic analysis changes to a smaller scale. At this stage in the value chain, accurate correlation of reservoir intervals becomes important. In this section we examine the main methods of stratigraphic analysis used in petroleum exploration and production.

Given that there is but one main goal in stratigraphic analysis, that of establishing what rocks were deposited where and when, it may appear strange to the nongeoscientist that we describe five methods of stratigraphic analysis. The truth is that it is very difficult to establish a stratigraphic column that is an accurate representation of the past. The cause of this is manifold. To begin with, there is no correct answer with which to compare our attempts at a stratigraphic analysis. Secondly, very little of past time is represented by the rocks, because as well as deposition of sediments, geologic history also contains periods in which no deposition occurred, and it also contains periods of erosion. Finally, each of the methods described below has shortcomings. Used together, they commonly provide a powerful tool for understanding basin development, reservoir distribution, and source-rock accumulation.

Stratigraphy carries with it a large array of terminology. We will do our best to minimize unnecessary exposure to too many different words. However, the use of some terms is unavoidable. Moreover, some words that have been used in the past as general, nonspecific descriptors now have precise technical meanings. It is easy to mix usage of the general and the specific. We will try to avoid such confusion. Above all, one word used in stratigraphy—unconformity—demands description before all others. An unconformity is a surface that separates a body of older rock from a body of younger rock (Fig. 3.25). The surface is equivalent to a period of time in which erosion and/or nondeposition occurred. The length of time represented by the unconformity may be large or small (billions of years to a fraction of a million years). The geometric relationship between the overlying and underlying rocks may be a dramatic contrast (as in flat rocks overlying folded and tectonized strata) or may have no apparent contrast (parallel unconformity). Unconformities have been used to subdivide strata for at least 200 years (Hutton 1795). They retain their importance today.

3.6.2 Chronostratigraphy

Unlike all of the other stratigraphic methods that are described in this section, chronostratigraphy is better described as a product rather than a tool. The chronostratigraphic analysis of a basin, or of an area within a basin, is commonly derived from some combination of seismic stratigraphy, sequence stratigraphy, and biostratigraphy, possibly supplemented by other stratigraphic methods. Direct dating of sediments using radiogenic (radioactive decay) methods is rarely possible, since the radiogenic elements in sediments have commonly undergone several cycles of deposition and erosion. Volcanogenic deposits can usually be dated, but while this can give an absolute age it is the widespread nature of the ash fall or flow that provides the stratigraphic marker in the area under investigation.

The deduced or perceived chronostratigraphy for an area is commonly displayed on a chronostratigraphic chart (Fig. 3.26). The chart is a 2D cross-plot of geologic time versus distance. Time is displayed on the y-axis, with the youngest at the top. Distance is displayed on the x-axis, which usually follows a line of section across the area under study. The line of section need not be straight. A chronostratigraphic section differs from a conventional geologic cross-section in the use of (geologic) time, rather than depth, for the y-axis. In so doing, erosive and nondepositional events can be displayed easily. Similarly, the temporal fluctuations in deposition can be portrayed. Chronostratigraphic charts are a powerful tool when it comes to capturing and describing the geologic history of an area. However, they provide no information on the thickness of sediments deposited.

3.6.3 Biostratigraphy

Biostratigraphy is an old and well-established tool. Its origin is based on observations made by geologists in the early part of the 19th century, that the fossil assemblages in similar rocks

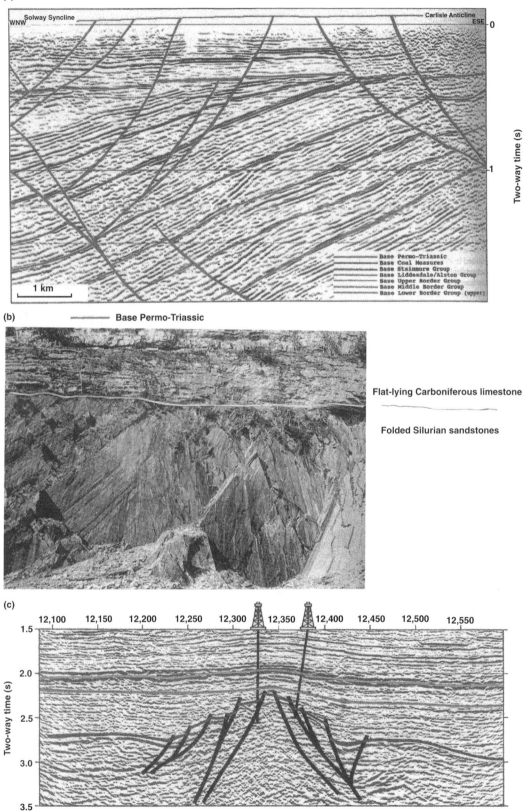

(a)

Solway Syncline   Carlisle Anticline
WNW   ESE 0

Two-way time (s)

1

———— Base Permo-Triassic
———— Base Coal Measures
———— Base Stainmore Group
———— Base Liddesdale/Alston Group
———— Base Upper Border Group
———— Base Middle Border Group
———— Base Lower Border Group (upper)

1 km

(b)   ———— Base Permo-Triassic

Flat-lying Carboniferous limestone

Folded Silurian sandstones

(c)

12,100   12,150   12,200   12,250   12,300   12,350   12,400   12,450   12,500   12,550

1.5

Two-way time (s)

2.0

2.5

3.0

3.5

———— Base Lower Cretaceous

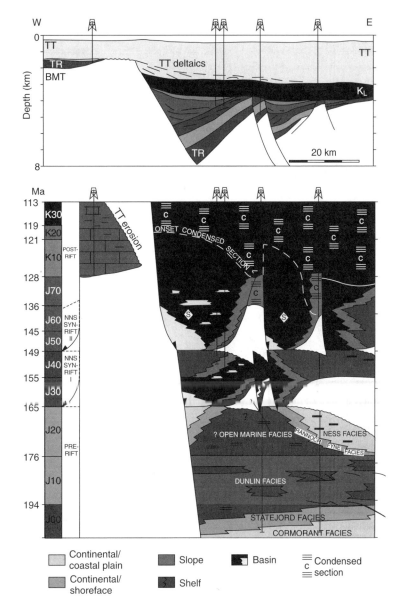

Fig.3.26 A local-scale chronostratigraphic diagram and cross-section, Frigg Area and Terrace Province, northern North Sea. (From Rattey & Hayward 1993.)

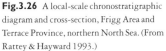

Continental/ coastal plain
Continental/ shoreface
Slope
Shelf
Basin
Condensed section

Fig.3.25 Unconformities. (a) An angular unconformity beneath the base Permo-Triassic on the flank of the Carlisle Anticline, Northumberland–Solway Basin, UK. (From Chadwick et al. 1993). (b) Flat-lying Carboniferous limestone unconformable on folded Silurian sandstones, Horton-in-Ribblesdale, Yorkshire, UK. (From Leeder 1992). (c) A topographic unconformity, Buchan Field, UK continental shelf. The figure shows a horst formed in Devonian sandstone and buried beneath Cretaceous and younger sediments. The base Cretaceous is the major unconformity surface across the horst. (From Edwards 1991.)

of different ages are dissimilar. Moreover, it was recognized that in progressively older rocks the suite of fossils contained therein have fewer and fewer similarities with extant species. The antithesis of these observations is the basis for biostratigraphy; rocks containing similar assemblages of fossils are of the same age. The biostratigraphic method was well established before Darwin offered an explanation as to why it worked — evolution. Thus biostratigraphy can be used to correlate rocks over areas in which exposure is poor, across land or sea.

The quality and resolution of a biostratigraphic analysis depends upon a number of factors. The ideal animal or plant for biostratigraphic purposes would have evolved quickly and become extinct shortly thereafter. During its short time on Earth it would have spread around the globe instantaneously, permeated all ecological systems (desert, river, ocean, etc.), flourished abundantly, and retained the same but distinct form from inception to extinction. Naturally, such a species has not existed, and does not exist at the present time. However, even to the casual observer of nature today, it will be obvious that some species of animals and plants are much more restricted in their range than others. For example, free-swimming animals, plankton, and wind-blown spores and pollen are much more widely distributed than sessile animals or those that are locked into a specific environmental niche. Abundance is another important factor. You have to find the fossil in the rock! In consequence, although the tympanic (ear) bones of whales have been used for biostratigraphic purposes in the Pliocene of southern Europe, they are not nearly so abundant and thus useful as microfossil foraminifera.

Biostratigraphy in petroleum geoscience encounters problems in addition to those of general biostratigraphy. The first is size. For example, ammonites and their cousins, goniatites, make wonderful biostratigraphic zone fossils. They swam freely, migrated across large parts of the globe, evolved quickly, and individual species are commonly highly distinctive. The Jurassic is subdivided in detail using a progression of ammonites. Indeed, ammonites are used to subdivide much of the Mesozoic, and goniatites, parts of the Paleozoic. However, they are large. A typical zonal ammonite is 5–

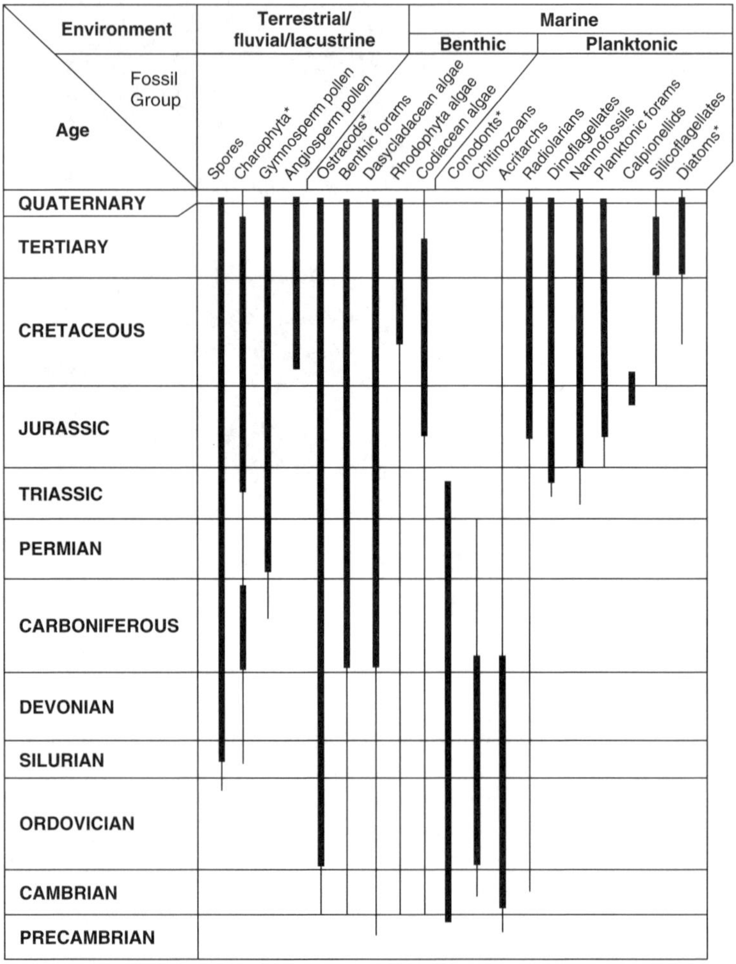
Environment
Terrestrial/
fluvial/lacustrine
Marine
Benthic
Planktonic
Fossil
Group
Age
Spores
Charophyta*
Gymnosperm pollen
Angiosperm pollen
Ostracods*
Benthic forams
Dasycladacean algae
Rhodophyta algae
Codiacean algae
Conodonts*
Chitinozoans
Acritarchs
Radiolarians
Dinoflagellates
Nannofossils
Planktonic forams
Calpionellids
Silicoflagellates
Diatoms*
QUATERNARY
TERTIARY
CRETACEOUS
JURASSIC
TRIASSIC
PERMIAN
CARBONIFEROUS
DEVONIAN
SILURIAN
ORDOVICIAN
CAMBRIAN
PRECAMBRIAN

Fig.3.27 A biostratigraphic range diagram. (From Emery & Myers 1996.)

Table 3.1 Examples of the resolution of fossil groups by age and by geography. (From Sturrock 1996.)

Fossil group	Age range	Geography	Average resolution (million years)
Planktonic foraminifera	Neogene	Tropical	1.2
Planktonic foraminifera	Neogene	Subtropical	1.4
Planktonic foraminifera	Paleogene	Tropical	1.7
Planktonic foraminifera	Paleogene	Southern temperate	3.0
Nanofossils	Neogene	Undifferentiated	1.0–1.3
Nanofossils	Paleogene	Undifferentiated	1.3–1.6
Radiolaria	Neogene and Paleogene	Undifferentiated	1.9–2.0
Diatoms	Neogene and Paleogene	Undifferentiated	1.4–2.4
Dinoflagellates	Neogene and Paleogene	Undifferentiated	5.7
Dinoflagellates	Neogene	North Sea	3.3
Dinoflagellates	Paleogene	North Sea	1.1
Planktonic foraminifera	Cretaceous	Tropical	2.5
Planktonic foraminifera	Cretaceous	Temperate	4.0
Nannofossils	Cretaceous	Undifferentiated	3.0
Radiolaria	Cretaceous	Undifferentiated	10.0
Palynomorphs	Cretaceous	Undifferentiated	6.5
Palynomorphs	Late Jurassic	North Sea	1.0
Palynomorphs	Early–Middle Jurassic	North Sea	2.0–2.5

10 cm in diameter. Cores cut in wells are commonly from 5 cm to 15 cm diameter and cuttings are millimeter sized. Ammonites have been found in core, as have fish, logs, dinosaur bones, and other types of large fossil. However, recognizable-size chunks are rare and, as such, large fossils are seldom of use for stratigraphic purposes in the oil industry.

A second problem encountered in the petroleum industry is that of "caving." Most fossils are extracted from the cuttings produced when a well is drilled. The cuttings become entrained within the drilling mud and eventually circulated to the surface, where they are extracted. Although the muds are viscous, there is still some smearing of the cuttings from any one depth. Moreover, material can fall from higher up the walls in a well. Such fallen material is referred to as cavings. Fossils extracted from cavings will seem to come from a lower part of the stratigraphy than they really do. This means that the species inception points that are most commonly used to erect biozones where samples are removed from outcrop cannot easily be applied to wells. Instead, biostratigraphy in wells tends to rely on extinctions—tops (Fig. 3.27).

Instead of using macrofossils for stratigraphic purposes, the oil industry uses microfossils, nanofossils, and palynomorphs (Table 3.1). Palynomorphs include pollen and spores. Microfossils are the tests (shells) of tiny animals and nanofossils are also the skeletal parts of biota, albeit smaller. All of these fossil forms are small, ranging in size from mi-

crometers (nanofossils) to millimeters. All can be extracted from cuttings as well as core. Those with tests made from organic matter tend to be extracted from rocks using an acid mix that contains hydrofluoric acid. Those with carbonate skeletons or shells formed from sand grains are commonly picked by hand from disaggregated rock. Microfossils in well-indurated limestones may be viewed in petrographic thin sections using a microscope.

Stratigraphic schemes using microfossils, nanofossils, and palynomorphs have been developed for most parts of the Phanerozoic (Paleozoic, Mesozoic, and Cenozoic; Fig. 3.28). Such schemes have been cross-linked with the older schemes based on macrofossils. At their best, the stratigraphic resolution of such schemes is precise. For example, the nanofossil biozonation used for the Late Miocene to Pleistocene of the Gulf of Mexico has an average resolution of 0.375 million years. However, the resolution can be improved by using combined nanofossil and foraminiferal ranges. In the Gulf of Mexico example, a combined scheme delivers a resolution of 0.2 million years (Sturrock 1996).

Until the widespread availability of seismic data, biostratigraphy was the main stratigraphic correlation tool used in the petroleum industry. It remains a key component of many stratigraphic studies on a range of scales from basin analysis to detailed field evaluation. The only situation in which biostratigraphic studies yield little information is in

Age	Planktonic foraminiferal zones		Formations in south and southeast Trinidad	
			Rich in planktonic foraminifera	Predominantly benthonic foraminifera
HOL	Globorotalia truncatulinoides truncatulinoides	• Globorotalia fimbriata	Planktonic foraminifera are poorly represented in this interval	
PLEIST.		• Globigerina bermudezi		Cedros
		• Globigerina colida colida		
		• Globoratalia hessi		
		• Globoratalia crassaformis viola		Erin / Palmiste
PLIOCENE U	• Globorotalia truncatulinides cf. tasgensis.			
PLIOCENE M	• Globorotalia morgaritica	• Globorotalia exilis		Marne l'Enfer
		• Globigerinoides trilobus fistulosus		Forest / Mayaro
PLIOCENE I	Globorotalia morgaritaé	Globorotalio margaritae evoluta		
		Globorotalia margaritae margaritae		Crose / Gros Morne, Lower Cruse
MIOCENE U	Globoratalia dutertrei			
	Globoratalia acostaensis			
MIOCENE M	Globoratalia menardii		Lengua	Karamat
	Globoratalia mayeri			
	Globoratalia ruber		Hiatus	
	Globoratalia fohsi robusta			Herrera Mbr.
	Globoratalia fohsi lobarta			
	Globoratalia fohsi fohsi			
	Globoratalia fohsi peripheroranda			
MIOCENE L	Praeorbulina glomerosa		Cipero	
	Globigerinatella insueta			
	Globigerinita stainforthi			Nariva
	Globigerinita dissimilis			
	Globigerinoides primordius			
OLIGOCENE U	Globorotalia kugleri			
	Globigerina ciperoensis ciperoensis			
OLIGOCENE M	Globorotalia opima opima			
	Globerina ampliapertura			
OLIGOCENE L	Cassigerinella chipolensis/Hastigerina micra		? Hiatus	
EOCENE U	Globoratalia cerroazulensis s.l.		San Femando	
	Globigerinatheka semiinvoluta			
EOCENE M	Truncorotoloides rohri		Navet	
	Orbulinoides beckmanni			
	Globorotalia lehneri			
	Globigerinatheka subconglobata subconglobata			
	Hantkenina aragonensis			Pointe-à-Pierre
EOCENE L	Globorotalia palmerae		Upper Lizard Springs	
	Globorotalia aragonensis			
	Globorotalia formosa formosa			
	Globorotalia subbotinae			
	Globorotalia edgari		Hiatus	
PALEOCENE U	Globorotalia velascoensis		Lower Lizard Springs	
	Globorotalia pseudomenardii			
PALEOCENE M	Globorotalia pusilla pusilla			Chaudière
	Globorotalia angulato			
	Globorotalia uncinata			
	Globorotalia trinidadensis			
PALEOCENE L	Globorotalia pseudobulloides		Hiatus	? Hiatus
	Globigerina eugubina			

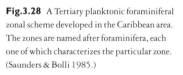

Fig.3.28 A Tertiary planktonic foraminiferal zonal scheme developed in the Caribbean area. The zones are named after foraminifera, each one of which characterizes the particular zone. (Saunders & Bolli 1985.)

fluvial and desert sediments, which often tend to be barren of fossils or contain few that have any stratigraphic value.

In describing biostratigraphy and its value to stratigraphic correlation, we have omitted to report a second useful aspect of biostratigraphic analysis — environmental evaluation. Fossils of animals and plants that had restricted environmental ranges may not be of much use for stratigraphic purposes. However, they may give critical environmental information,

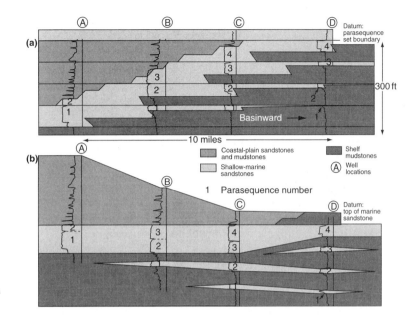

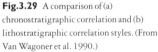

Fig.3.29 A comparison of (a) chronostratigraphic correlation and (b) lithostratigraphic correlation styles. (From Van Wagoner et al. 1990.)

such as water depth, salinity range, dissolved oxygen levels, and depositional environment. Such information contributes to the detective story when establishing a geologic history for an area.

3.6.4 Lithostratigraphy

Lithostratigraphy is the oldest weapon in the stratigraphic armory. It is also now the most maligned. Lithostratigraphy relies on the correlation of like lithologies—rocks. Sedgwick practiced it most carefully during his establishment of the Cambrian system in Wales (1831–52; Secord 1986). It was also the cornerstone of de la Beche's defense of the ancient age of the coal-bearing Culm in Cornwall (England). De la Beche claimed (de la Beche 1839) that the contorted graywacke of the Culm were older than the Old Red Sandstone, and therefore that coal-bearing strata could be found by mining beneath the Old Red Sandstone in other parts of Britain. As the first head of the Geological Survey in England and Wales, and in effect the world's first professional geologist, his job was at risk. He was wrong, a fact pursued relentlessly—and often with acrimony—by Murchison, who preferred to use biostratigraphic methods. This episode culminated in the establishment of the Devonian system (Rudwick 1985). The Culm was demonstrated to be Carboniferous and the Old Red Sandstone Devonian. In Britain at least, there is no coal beneath the Old Red Sandstone. The Culm looks more

"ancient" than the Old Red Sandstone because the area in which it now outcrops was tectonized during the Hercinian orogeny, a fate escaped by the areas in which the Old Red Sandstone is found.

There is a message in the above text. Clearly, like lithologies can be deposited in different places at different times. Correlation of a Devonian arkose in the Midland Valley of Scotland with a Triassic arkose in the North Sea simply on the basis of their composition and sedimentary structures would be erroneous, although at first glance, and in isolation, the error would not necessarily be obvious. With help from biostratigraphic, magnetostratigraphic, or chemostratigraphic analysis, large errors such as the ones outlined above may be identified. However, lithostratigraphic correlation can be just as misleading when trying to correlate reservoir sandstones/limestones between wells within an individual oilfield (Fig. 3.29).

Lithostratigraphic correlation should only be applied with great care, and only within a well-defined biostratigraphic and/or sequence stratigraphic framework (Case history 5.10).

3.6.5 Seismic stratigraphy

The sediments that fill a basin accumulate by virtue of the tectonic processes that generate both the basin and the sediment source. Major changes in the tectonic regime are reflected by the gross architecture of the sediment packages, as can be major changes in sea level. The latter effects may be global, rather than linked to individual basins. The major tectonic/stratigraphic units in a basin fill are called "megasequences." Regional-length 2D seismic lines (from tens to hundreds of kilometers) are often ideal data on which to recognize and map megasequences. Indeed, as the title of this subsection suggests, seismic stratigraphy is based on interpretation of seismic data. Coarse stratigraphic analysis of seismic data comprises two steps. The first step is the mapping of the major unconformity surfaces in the data, while the second is the description of the interval between each major unconformity (megasequence). The latter will be described in terms of lateral thickness changes, seismic facies variations, and relationships between the major structural elements and the form of the megasequence.

The seismic stratigraphic methodology is based on the premise that almost all seismic reflectors are time surfaces (Payton 1977; some that are not, including direct hydrocarbon indicators, have already been described earlier in this chapter). Although it uses seismic data as its source, seismic stratigraphy is no different in terms of its analysis of the subsurface to traditional outcrop mapping. The major unconformities and gross morphology of the intervening interval may be difficult to recognize and map at outcrop, but the products are similar; a geologic understanding of how the basin originated and filled with sediment. The major tectonic/stratigraphic units, megasequences, have characteristic forms in compressional and tensional basins. They are described below.

Rift basins

In rift basins, packages of sediment deposited before, during, and after the active rift phase can often be clearly identified on regional seismic lines (Fig. 3.30). The pre-rift megasequence commonly shows uniformity of thickness over large areas. The syn-rift megasequence commonly comprises a multitude of sediment wedges in fault hanging walls (Fig. 3.31), while the post-rift megasequence passively infills the residual topography of the rift and also that created during the phase of thermal subsidence. The post-rift sequence commonly lies with pronounced unconformity on the pre-rift at the crests of fault blocks, while it may be conformable on top of the syn-rift close to the main axes of deposition.

Passive margins

The passive margin is, of course, a development of a successful rift. The post-rift megasequence of the rift basin is therefore the passive margin megasequence. However, the seaward end of this megasequence in the passive margin setting may lie directly upon new oceanic crust, rather than upon the pre- or syn-rift megasequences as it does in a rift basin. Moreover, since a passive margin can develop over a substantial period of time, the megasequence can be enormously thick. The low level of tectonic activity experienced by a passive margin also means that the influence of sea-level fluctuations on the stratal patterns is much more readily observed than in regions of high tectonic activity. Passive margins were the areas upon which the sequence stratigraphic methodology was developed and tested (see the following subsection).

Foreland basins

A foreland basin megasequence forms contemporaneously with the flexuring of the lithosphere caused by thrust loading. The shape is commonly elongate parallel to the thrust belt. Widths are typically in the region of 100 km to 300 km. The exact amount depends upon the rigidity of the lithosphere. The megasequence is commonly triangular in cross-section. The hypotenuse forms the upper surface, while the maximum height (depth!) of the triangle lies closer to the thrust

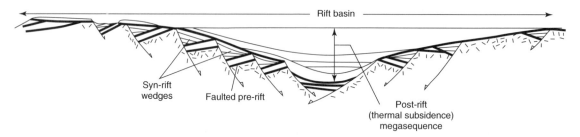

Fig.3.30 A cross-section through a rift basin. The geometry of such rift basins has often been likened to a steer's head – a product of the thermal subsidence.

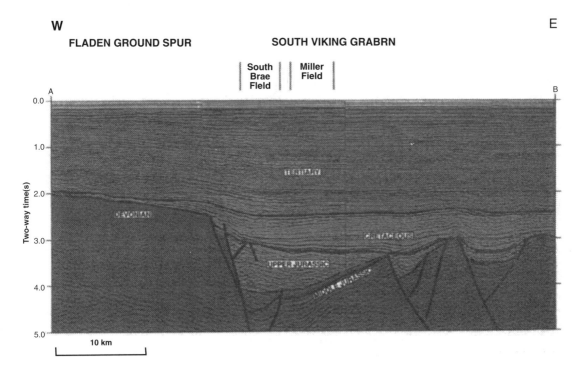

Fig.3.31 A syn-rift wedge of Upper Jurassic deep-water sandstones and mudstones in the South Viking Graben. (From McClure & Brown 1992.)

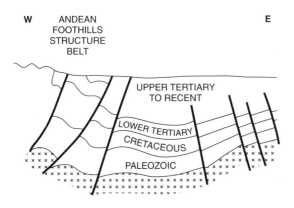

Fig.3.32 A geologic section across the Llanos Basin, Colombian Andes. The internal architecture of the basin fill comprises a series of wedges that taper from the fold belt in the west to the shield area in the east. (From McCollough & Carver 1992.)

belt than the foreland (Fig. 3.32). The wedge thins toward the flexural bulge. Continued development of the thrust belt commonly results in uplift and erosion of the early proximal parts of the megasequence by the growing mountain front.

Wrench systems

Megasequences developed during strike-slip are commonly of small aerial extent. The sequences may be symmetric, as in transtensional systems, or asymmetric, as in transpressional systems. The overall geometry of the megasequences is commonly like that of their syn-rift and foreland counterparts in extensional and compressional basins. Typically, the megasequences have length-to-width ratios of about 3:1 and lengths of a few tens of kilometers. The thickness of the megasequence can be deceptive. Although thick, 5 km not being uncommon, apparent thicknesses in excess of 20 km have been recorded. In the Hornelan Basin (Norway), Steel and Aasheim (1978) showed that an apparent thickness of 25 km of sediment pile was accommodated by a shift of the basin depocenter during continued wrenching.

3.6.6 Sequence stratigraphy

Sequence stratigraphy was developed from seismic stratigraphy in the 1970s, by workers in the Exxon research facility (Mitchum et al. 1977). It was founded on the same principle as used in seismic stratigraphy, that seismic reflectors are time surfaces and that unconformites are bounding surfaces that separate strata into time-coherent packages. Moreover, it was recognized from the circum-Atlantic passive margins that the stratal pattern in one area could be correlated with others that were far distant. The stratal patterns were as distinctive as the biostratigraphic correlations. The two also matched. It was clear that the stratal patterns in these areas of low tectonic activity were the signatures of sea-level rise and fall (Fig. 3.33).

The implications of sequence stratigraphy are profound. An explanation of strata in terms of relative sea-level fluctuations and a combination of eustatic sea-level change and tectonic subsidence allows an understanding of why sediment packages develop where they do (Emery & Myers 1996). It can therefore provide a predictive tool for determining the likely presence of source rocks, and the distribution of reservoirs and seals.

The basic unit in sequence stratigraphy is, of course, the sequence (Fig. 3.34). This has been defined as "a relatively conformable, genetically related succession of strata bounded by unconformities or their correlative conformities" (Mitchum et al. 1977). Sequence boundaries form when the water depth decreases. A sequence is made up of parasequences and parasequence sets. A parasequence is defined as a relatively conformable, genetically related succession of beds or bed sets, bounded by marine-flooding surfaces or their correlative unconformities (Van Wagoner et al. 1990). The parasequence set is a group of genetically linked parasequences that form a distinctive stacking pattern. Such parasequence sets are typically bounded by major marine-flooding surfaces or their correlative surfaces (Van Wagoner et al. 1990). Smaller units of subdivision are beds and laminae, both sometimes grouped into sets. The key property of the bounding surfaces to sequences, parasequence sets, parasequences, and beds is that they are chronostratigraphically significant surfaces. Thus each surface is a physical boundary that separates all the rocks above from those below.

It is important to point out that the terminology adopted above is derived from the "Vail" school of thought (Mitchum et al. 1977). An alternative sequence scheme has been developed by Galloway (1989), in which the maximum flooding surfaces are used to bound what he terms "genetic sequences." In the Galloway genetic sequences, the unconformity is in the middle. This has led to widespread confusion. There are merits to both schemes. On seismic data the unconformities tend to be most easily identified, while on wireline log data maximum flooding surfaces are commonly more obvious. Given the more widespread usage of the Vail terminology, we will use it exclusively hereafter.

We mentioned earlier the linkage between sequence development and changes in sea level. Sediments accumulate where accommodation space exists and where a sediment supply exists. Accommodation space is created and destroyed by the interplay of tectonic subsidence or uplift and sea level (in the marine environment). Changes in absolute sea level (eustasy) and basement can be considered as changes in relative sea level. For example, basement subsidence equates to relative sea-level rise, and basement uplift to relative sea-level fall. The effect of relative sea-level rise will be a landward shift of facies belts, while the opposite will occur during sea-level fall. Because the angle of slope from land to shelf to continental slope to basin is not constant, the effects of sea-level fall and rise change dramatically depending upon the magnitude of the sea-level change. The critical points on the shelf–slope–basin profile are the shelf/slope break and the point of coastal onlap (Fig. 3.35). The shelf/slope break is the boundary between the less steeply dipping shelf (0.1°) and the more steeply dipping slope (2–7°). The point at which sedimentation begins on the coastal plain is the point of coastal onlap. Above this point is an area of nondeposition and/or erosion.

During cycles of relative sea-level fall, two types of sequence can develop depending on whether or not the fall in the sea level is sufficient to expose the shelf/slope break to erosion. When the sea level falls sufficiently to expose the shelf/slope break, the sequence is termed Type 1. The large drop in sea level will cause erosion of the shelf area and the coastal plain beyond. Incised valleys and canyons may form up slope from the shelf/slope break, while deposition of coarse clastics occurs in the basin (the basin-floor fan). As sedimentation responds to the sea-level drop, the slope fan and slope wedge succeed the basin-floor fan. Collectively, these three elements are termed the "lowstand systems tract" (LST). Moreover, much of the LST could contain reservoir lithologies dominated by sediment gravity flows, including turbidites. The LST will in turn be buried beneath the "transgressive systems tract" (TST), which is formed as the sea level rises. This will culminate in a "maximum flooding surface" (MFS). Typically, during periods of maximum flooding, large areas of shelf lie beneath shallow water. In consequence, circulation of the water column can be poor. Such conditions are often ideal environments for source-rock development (Section 3.7). A high sea level also reduces physical erosion, resulting in deposition of fine-grained lithologies, which are potential seals. The "highstand systems tract" (HST) succeeds the transgressive systems tract. The HST builds over and downlaps onto the TST. Initially, the sediment pile aggrades until accommodation space is exhausted and then progradation

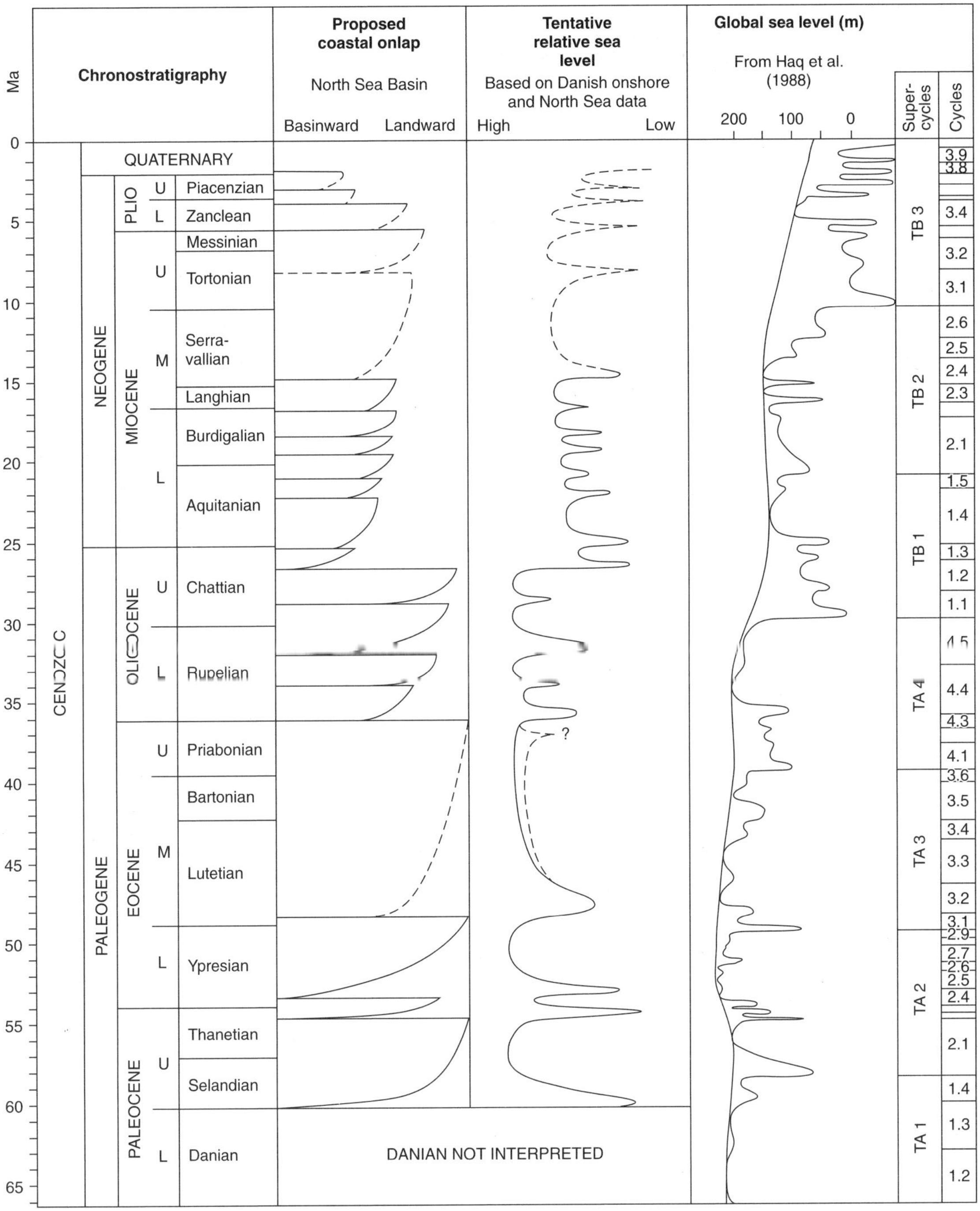

Fig.3.33 A sea-level curve for the Cenozoic of the eastern North Sea. The local (North Sea) curve is flanked by the current global sea-level curve and a chronostratigraphic panel. (From Michelsen et al. 1998.)

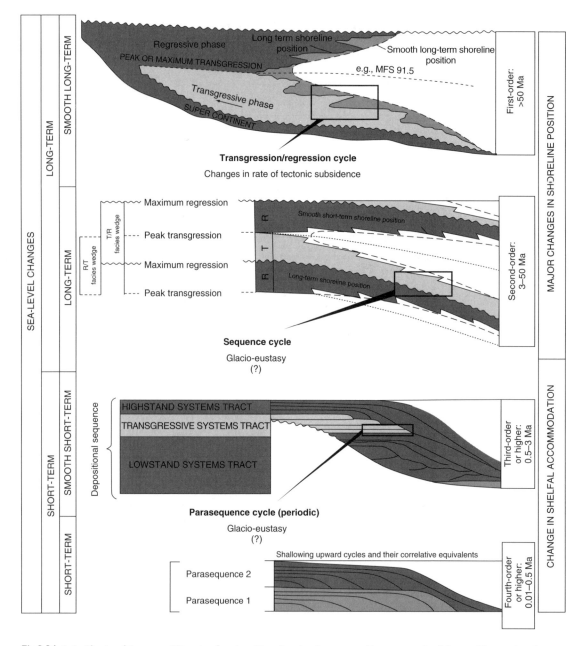

Fig.3.34 A classification of the types and hierarchy of stratigraphic cycles using short-term and long-term sea-level changes. The eustatic cycles – smooth long-term, long-term, smooth short-term, and short-term – have specific time durations as shown on the figure. (From Duval et al. 1998.)

takes over (Fig. 3.36). As the sea level falls again, the next sequence boundary is created.

In situations where the sea level does not fall below the shelf/slope break, extensive coarse-grade sedimentation does not occur in the basin but, instead, occurs at the shelf margin,

the so-called "shelf margin systems tract" (SMST). The complete shelf margin systems tract, with the transgressive systems tract and the highstand systems tract, collectively form a Type 2 sequence (Fig. 3.37).

The situation in carbonate systems is quite different from

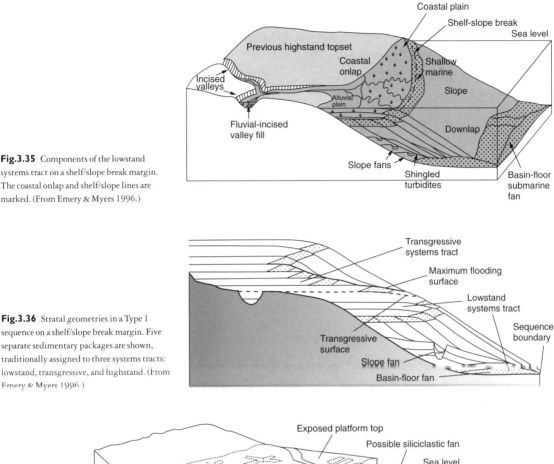

Fig.3.35 Components of the lowstand systems tract on a shelf/slope break margin. The coastal onlap and shelf/slope lines are marked. (From Emery & Myers 1996.)

Fig.3.36 Stratal geometries in a Type 1 sequence on a shelf/slope break margin. Five separate sedimentary packages are shown, traditionally assigned to three systems tracts: lowstand, transgressive, and highstand. (From Emery & Myers 1996.)

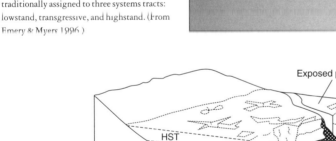

Fig.3.37 A carbonate lowstand systems tract (LST) on a humid escarpment-margin system. Here the sea level has fallen significantly below the margin, and the exposed platform top is karstified and may become incised by fluvial channels. Siliciclastic sediment may be deposited in the basin, onlapping the carbonate slope, and failure of the margin may also result in the deposition of talus cones at base of slope. *In situ* carbonate production is likely to be minor on very steep margins. HST = highstand systems tract; TST = transgressive systems tract; SB = sequence boundary. (From Emery & Myers 1996.)

that in clastics. Because carbonates predominantly develop *in situ* in shallow shelf settings, large sea-level falls can shut down the carbonate factory (Fig. 3.38). Exposure of the shelf often promotes chemical weathering, the formation of karst above the water table, and meteoric diagenesis (cementation) below the water table. Thus sea-level fall may generate no sediment into the basin. At times of high sea level, the oppo-

site can be true. Productivity is high when the shelf is flooded. Large areas of shelf may overproduce carbonate material that may be shed from the shelf into deep water during sea-level highstand (Fig. 3.38).

Sequence stratigraphic analysis of seismic data is similar to seismic stratigraphy, albeit at a higher resolution. Attempts are made to map both the large- and small-scale unconformi-

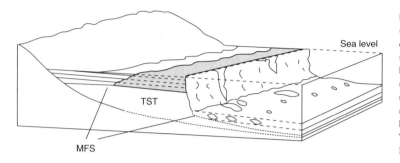

Fig.3.38 A carbonate highstand systems tract (HST) of an escarpment margin, showing continued aggradation but a progressive thinning of topsets. In the basin and at the basin slope, carbonate shed from the platform top during highstand onlaps the talus apron of the transgressive systems tract. Some talus also may be shed from the carbonate margin during highstand. MFS = maximum flooding surface; TST = transgressive systems tract. (From Emery & Myers 1996.)

ties on the seismic data, using the presence of reflector terminations to guide placement of the unconformities. Ideally, wireline log data will be available from wells. This would be tied to the seismic data and the positions of the unconformities identified. Maximum flooding surfaces can commonly be identified as peak readings on logs that measure the rocks' natural radioactivity (gamma and natural gamma ray tools, NGT logs; Chapter 2). The intervals between the maximum flooding surfaces and the unconformities (sequence boundaries) can be characterized in terms of the sediment stacking patterns as identified from the logs. For example, upward cleaning cycles in shallow-marine sequences are typical of aggradational sediment accumulation in the highstand sequence tract.

3.6.7 Chemostratigraphy and magnetostratigraphy

Neither chemostratigraphy nor magnetostratigraphy are widely used in the oil industry, although both methods are well established and both have potential for more extensive application.

Chemostratigraphy relies on producing correlatable chemical fingerprints for two or more stratigraphic sections by analyzing a suite of elements. The best known chemostratigraphic marker is the iridium anomaly (peak), which is coincident with the KT event (at the Cretaceous/Tertiary boundary). The abundance of iridium at this point has been linked to the impact of a comet with the Earth. This event, with an extraterrestrial origin, has of course been implicated as the cause of the mass extinctions at the end of the Cretaceous, including that of the dinosaurs.

In most circumstances, a piece of chemostratigraphic analysis would not rely on a single element or on an extraterrestrial cause. Events that affect single basins, such as a change in sediment supply or fluctuations between oxia and anoxia, will impart a chemical signature on the sediments. These can be either direct, as will happen with a change

in sediment supply, or indirect, as fluctuations in bottom-water Eh and pH values control mineral precipitation and dissolution.

Magnetostratigraphy relies on two phenomena. The first is that the magnetic polarity of the Earth's field switches from normal (as at present) to reversed, in which the magnetic north pole occupies a position near the geographic South Pole. The second is that of "polar wander," a term coined before full recognition of continental drift or plate tectonics. Polar wander measures the apparent movement of the poles across the Earth's surface through time. It is really a measure of the migration of lithospheric plates across the Earth's surface. "Polar wander" is a much slower process than pole reversal. Discontinuities in the polar wander path in a stratigraphic section may help to identify unconformities. Both the normal/reverse polarity patterns and the "polar wander" information may be used as correlation tools for well-to-well correlations. Although magnetostratigraphy is used less widely than other forms of stratigraphy, it can have particular value in nonmarine sediments and red-bed sequences, which are commonly barren of fossils.

3.6.8 Stratigraphic tests

Stratigraphic tests are wells drilled in a basin, or part of a basin, not with the aim of finding petroleum but to help define the local stratigraphy, and to provide information from which the presence of reservoir, seal, and source rocks might be identified. Such wells are commonly drilled under license from government ministries, although in many instances the cost of drilling is borne by the companies that wish to explore the area. Stratigraphic tests are commonly drilled in the basin center, where the stratigraphy is expected to be most complete.

For example, five stratigraphic test wells (Zhu 1–5) were drilled in the Pearl River Mouth Basin (China) in the early 1980s, ahead of any exploration wells. Cores were cut in some of the wells and both core and cuttings were used for analysis of the stratigraphy of the basin fill. The presence of traces of live oil in one of the wells undoubtedly stimulated the high expectations held by the companies involved in oil exploration at the time (Case history 3.9).

A second example of stratigraphic tests is that performed by the Norwegian Continental Shelf Unit (IKU). IKU has drilled many shallow boreholes along the Norwegian coast ahead of exploration wells proper. The Baltic Shield, including Norway, was uplifted during the Late Tertiary. In consequence, potential reservoir, seal, and source horizons in the offshore basins subcrop the sea bed and/or a thin cover of recent sediments in the shallow waters near to the shore. A series of short boreholes drilled progressively further out to sea provides a complete coverage of the stratigraphy of the basin.

3.7 SOURCE ROCK

3.7.1 Introduction

Oil and gas form from organic matter—dead plants and animals. The composition of oil and gas (petroleum) has already been examined in Chapter 1. Here, we look at what organic matter is, how and where it accumulates to form source rocks, and the process of maturation of organic matter into petroleum.

3.7.2 The origin of petroleum from living organisms

Organic matter

The chemical composition of organic matter is diverse because the organisms from which it is derived are complex. The principal biological components of living organisms are proteins, carbohydrates, lipids, and lignin. Animal tissue and enzymes are partly composed of proteins, built from amino acids. Carbohydrates are also found in animal tissue, being a principal source of energy for living organisms. Lipids are fatty organic compounds, insoluble in water, and found in most abundance in algae, pollen, and spores. Lipids are rich in hydrogen, and hence yield high volumes of hydrocarbon molecules on maturation (Section 3.7.4). The lipid group contains a special group of compounds called isoprenoids, which are found in chlorophyll and include pristane and phytane. These molecules are preserved during petroleum formation. Their abundance and composition in petroleum can be indicative of the depositional environment in which the organic matter accumulated.

Under normal conditions, organic matter is very dilute in sediments. In global average terms, claystones and shales (excluding oceanic sediments) contain only 0.99 wt% (weight percent) of organic carbon, compared with 0.33 wt% in carbonate rocks and 0.28 wt% in average sandstones (Hunt 1979). In comparison, most source rocks contain in excess of 1.0 wt% of organic carbon, rich source rocks contain >5.0 wt%, and the value can reach as high as 20 wt%.

Preservation of organic matter

The two basic requirements for the generation and preservation of organic matter in sediments are (1) high productivity and (2) oxygen deficiency of the water column and the sea bed. The supply of organic matter to any depositional site is controlled by primary productivity (commonly within the top 50 m of the water column) and the depth of water through which the material must settle. Preservation beneath the sediment/water interface is a function of the rate of burial and oxygenation of the bottom waters. Both productivity and oxygen deficiency at the site of deposition can combine to produce excellent source rock, although some source rocks may result from a dominance of only one control. Environments of high organic productivity include (1) continental margins, (2) lagoons and restricted seas, (3) deltas in warm latitudes, and (4) lakes (Demaison & Moore 1980; Fig. 3.39).

Oceanic waters tend to be stratified (Fig. 3.39), although water circulation from the surface to sea bed also occurs. The stratification results from high organic productivity in the photic zone (i.e., in the top c. 50 m of the water column, influenced by sunlight). Periodical overturning of the oxygenated waters from the surface supplies oxygen to organisms whose habitat is the deep ocean, including the ocean floor. Between the periods of overturning, the high biological activity at the surface causes oxygen deficiency in the layer immediately beneath. Where this anaerobic layer intersects the sea bed on the shelf and upper slope of the continental margin, organic debris is preserved, since there is a relative scarcity of organisms here to scavenge the debris. Along some parts of the continental margins, upwelling of nutrient-rich waters creates a favorable niche for even higher levels of organic productivity; for example, on the western sides of the African and North and South American continents. Periodic algal blooms, which are most frequent in conditions of calm and warmth, can also act to poison the microplankton, leading to high deposition and preservation rates in these settings.

Lagoons and restricted seas are favorable for high preservation of organic matter. A lack of circulation of waters from the oxygenated surface layer to the bottom waters induces anoxic (oxygen-deficient) conditions (Fig. 3.39). Dead organisms

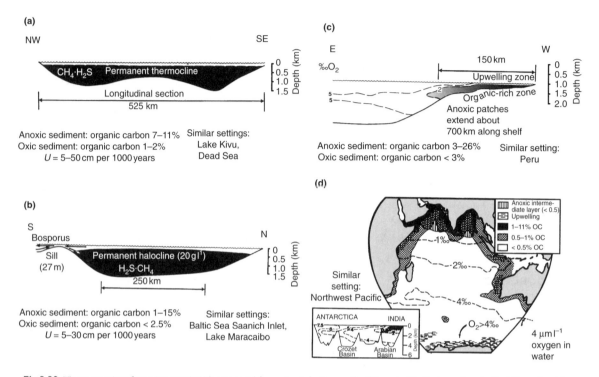

Fig.3.39 The preservation of organic matter in (a) large anoxic lakes such as Lake Tanganyika, (b) anoxic silled basins such as the Black Sea – a "positive water balance" basin, (c) anoxic layers caused by upwelling, such as the South West African shelf, and (d) open ocean anoxic layers, such as in the Indian Ocean. (From Demaison & Moore 1980.)

that sink to the sea bed are not scavenged there. Present-day examples include the Gulf of California and Lake Maracaibo, where the amount of organic matter in the seabed sediments is as high as $10.0\,wt\%$ (Peridon 1983). Landlocked seas such as the Black Sea and the Caspian Sea have similarly high preservation rates of organic matter.

Extensive shelf seas created during sea-level highstands are also important sites of organic matter production and preservation. The Liassic (Toarcian) source rocks of the Wessex (UK) and Paris (France) Basins were created under such conditions.

Deltas are typified by some of the highest sedimentation rates of any depositional environment. It is this factor that results in some deltaic sediments being source rocks. Rapid deposition leads to quick burial below the zone of organic scavenging near the sea bed. The same rapid deposition and subsidence generate massively thick sediment piles which, in total, contain a great deal of terrestrially derived, organic matter. Today and in the Neogene, the Mississippi and Niger Deltas are, and were, sites of source-rock accumulation.

Freshwater lakes on continents are sites for high productiv-

ity and preservation in the anoxic bottom waters that characterize the lake bed. The dominant organisms that create lacustrine oil shales are algae and fungi/bacteria. Some lakes have a low clastic sediment input; producing thick accumulations on the lake bed of slowly deposited, but very organic-rich, mud. Ancient examples of lake sediments form the richest petroleum source rocks in the world, including the Green River Shale Formation in the central USA. Lacustrine source rocks are also responsible for generating much of the oil onshore China, Thailand, and Burma (Case history 3.9). Water depth is not a critical factor. In fact, very shallow waters are often indicated in ancient lacustrine oil shales by the presence of desiccation cracks and an association with evaporite minerals, which are indicative of periods of drying during low lake levels.

3.7.3 Kerogen

Organic matter can be usefully divided into two components: bitumen, which is composed of compounds that are soluble in organic solvents; and kerogen, the insoluble components

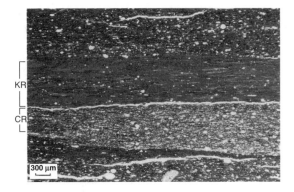

Fig.3.40 Microfacies of laminated sapropelic limestone, Lower Cretaceous carbonate source rock, France. The figure shows laminations in which organic matter is concentrated (kerogen-rich laminae, KR), and light carbonate-rich laminae (CR) containing small amounts of kerogen. In the remaining laminae, the organic matter is composed of unicellular filamentous marine algae that occur parallel to stratification. (From Machour et al. 1998.)

(Fig. 3.40). Bitumen or extractable organic matter includes the indigenous aliphatic, aromatic, and N–S–O compounds, including migrated hydrocarbons generated from elsewhere within the sedimentary sequence. The proportion of the original organic matter that is kerogen is commonly high, about 85–90% in shales.

Kerogen is the most abundant organic component on Earth (Brooks et al. 1987) and is the term used for a set of complex organic compounds, the composition of which depends on the original organic source. Kerogen is composed of varying proportions of C, H, and O. Such elemental data for kerogen can be displayed on a diagram first developed by Van Krevelen (1961; Fig. 2.26).

Kerogen type

Conventionally, kerogen is subdivided into four main "types" on the basis of maceral content; that is, original organic source material. The main maceral groups are liptinite, exinite, vitrinite, and inertinite. It is useful in petroleum geology to be able to identify the depositional environment of the source rock for a number of reasons. First, the kerogen type is dependent on the types of organic material preserved in each sedimentary environment. Secondly, each kerogen type matures under different burial conditions, controlling the timing of petroleum generation and expulsion from the source rock. Finally, each kerogen produces contrasting petroleum products, and in differing yields. The kerogen type or types present in a source rock can be recognized on the basis of optical properties such as color, fluorescence, and reflectance.

Liptinite (Type I) has a high hydrogen to carbon ratio but a low oxygen to carbon ratio. It is oil-prone, with a high yield (up to 80%). It is derived mainly from an algal source, rich in lipids, which formed in lacustrine and/or lagoonal environments. Liptinite fluoresces under UV light. Examples of liptinite-rich kerogen are found in the Devonian Orcadian shales of northeast Scotland, the Carboniferous oil shales of the Lothians, southern Scotland, the Eocene Green River Shale Formation of the central USA, and the Upper Permian lacustrine shale of northwest China. Type I kerogen is relatively rare.

Exinite (Type II) has intermediate hydrogen to carbon and oxygen to carbon ratios. It is oil- and gas-prone, with yields of 40–60%. The source is mainly membranous plant debris (spores, pollen, and cuticle), and phytoplankton and bacterial microorganisms in marine sediments. Exinite fluoresces under UV light. Examples of exinite-dominated source rocks include the Toarcian black shale of the Paris Basin and the Kimmeridge Clay Formation of the North Sea basin. Some exinite-rich kerogen contains a high proportion of sulfur and is termed Type II-S kerogen. The presence of sulfur influences the timing and rate of kerogen maturation. An example of a Type II-S kerogen is found in the coastal Monterey Formation of California. Type II kerogens are the most abundant.

Vitrinite (Type III) has a low ratio of hydrogen and high ratio of oxygen relative to carbon, and therefore forms a low-yield kerogen, principally generating gas. The primary source is higher plant debris found in coals and/or coaly sediments. Vitrinite does not fluoresce under UV light; however, it is increasingly reflective at higher levels of maturity and therefore can be used as an indicator of source-rock maturity. Examples of vitrinite-dominated source rocks include the Carboniferous Coal Measures of the southern North Sea basin and the Triassic source coals of the Australian North West Shelf.

Inertinite (Type IV) is the nonfluorescing product of any of the above kerogens. It is high in carbon and very low in hydrogen, and is often termed "dead-carbon," having no effective potential to yield oil and gas (Brooks et al. 1987).

The quantity and quality of kerogen

The environment at the site of deposition of the organic matter controls the quantity and quality of the kerogen found in a source rock. These include the rate of deposition and burial, the ratio of terrestrial to marine plant input, the oxidation state of the depositional environment, and the amount of reworking of the sediment prior to burial.

The quantity of kerogen in a rock defines its richness as a source rock, which in turn relates to its petroleum potential in two ways. First, the richer the source rock, the larger is the volume of hydrocarbons that can be generated. Secondly, the higher the proportion of the rock that is organic material, the greater is the efficiency of migration of hydrocarbons out of the source rock. The quantity of kerogen in a source rock is determined from the total organic carbon (TOC) and reported as a weight percentage of the rock (Chapter 1).

The quality of kerogen in a rock determines the hydrocarbon yield — that is, the volume of hydrocarbon generated for each volume of source rock — which is usually expressed in kilograms of hydrocarbon per ton of rock (kg HC t^{-1}). A range of techniques is used to evaluate potential source rock samples, including visual inspection of the kerogen type, elemental analysis, and pyrolysis (Chapter 1).

3.7.4 Maturation of source rocks: kerogen to oil to gas

Kerogen is composed of large hydrocarbon molecules that are stable at low temperatures, but will break down into smaller molecules of liquid and gaseous hydrocarbon compounds with progressive exposure to higher temperatures. In addition, nonhydrocarbon gases such as CO_2 and H_2O are produced, and a nonreactive residue remains. The transformation to smaller and lighter compounds is controlled by the reaction kinetics; that is, the strength of the bonds between the atoms and thus the energy required to break those bonds. Many studies have shown that the most important control is temperature (Tissot et al. 1987). Lesser controls are the nature and abundance of the kerogen in the source rock and pressure.

Temperature

In sedimentary basins, temperature increases with depth. Average temperature gradients for sedimentary basins fall mainly between 20°C km^{-1} and 40°C km^{-1}. There are exceptions: "cool" basins, with gradients of less than 20°C km^{-1}, include the Caspian Sea (as low as 10°C km^{-1}). The Caspian Sea is experiencing very high active sedimentation rates. "Hot" basins include the Gulf of Thailand (48°C km^{-1}), the Rhine Graben (50–80°C km^{-1}), and central Sumatra (60–120°C km^{-1}). All of these hot basins are underlain by thinned lithosphere. The source of the heat in the sediments comes primarily from the basement (i.e., it originates from the center of the Earth), coupled with a local contribution from the decay of radionuclides (commonly in clays). Basement heat flow varies according to the thickness and nature of the lithosphere, and proximity to thermal anomalies in the mantle.

In the laboratory, the temperature needed to generate oil from a source rock is around 430–460°C, whereas in a typical sedimentary basin time is expressed in millions of years and temperatures are in the range 80–150°C. In other words, there is a combination of time and temperature that must be integrated if we are to make an assessment of the thermal maturity of a source rock, and determine its petroleum generation history and any remaining potential. There are a number of ways in which sediments reflect their thermal exposure during basin evolution, which are known as thermal maturity indicators (Section 2.6.3).

The kinetics of hydrocarbon generation

The rate at which a kinetically controlled reaction proceeds (k_i) is a function of the absolute temperature (T), the activation energy (E_i), the gas constant (R), and a constant (A) given by the Arrhenius equation:

$$k_i = A \exp(-E/RT)$$

In practice, a whole series of parallel reactions is under way simultaneously during source-rock maturation, including secondary cracking of oil to gas at higher temperatures. Determination of the activation energy for each reaction is achieved in the laboratory, and there are now a number of published and widely used kinetic models, including all three main kerogen types as well as a Type II-S kerogen. Profiles of activation energies against product yield are shown in Fig. 3.41. The narrow spread of activation energies for a Type I kerogen (Green River Shale), coupled with a very high petroleum yield, is compared with the wider profile of activation energies required to generate a lower petroleum yield from a Type II kerogen (Toarcian Black Shale). In practical terms, this difference results in the high-yielding Type I kerogen generating its petroleum over a relatively narrow range of temperatures (and therefore depths) compared with a Type II kerogen.

Reaction products

Early alteration of kerogen in the shallow subsurface (generally less than 1.0 km) results in the production of CO_2 and H_2O. These products are derived from short –COOH side chains that combine with any available oxygen in the sediments. Thereafter, the products are dominated by a mixture of oils and gases at depths of 2.0 km and greater. There are a number of schemes that depict the main phases of product generation in terms of temperatures and depths of burial. We illustrate one such scheme in Fig. 3.42, in which there is a

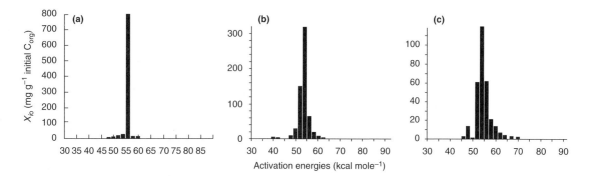

Fig.3.41 Activation energies and product yield. (a) Green River shales (Type I); (b) Paris Basin (Type II); (c) Mahakam Delta (Type III).

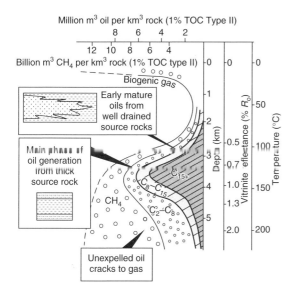

Fig.3.42 The maturation of oil from source rock as a function of depth and temperature. The curves show product generation from a source rock with 1% total organic carbon (TOC) as Type II kerogen. The curve is based on pyrolysis – gas chromatography. (From Cornford 1990.)

peak oil-generation phase followed by a phase of wet gas, and finally peak gas generation from a typical Type II source rock. The product mix will depend on a number of factors. These include the distribution and richness of the kerogen in the original source rock, the rate of temperature increase, the primary migration route for products out of the source rock once maturation commences, and its efficiency, and the distribution of pressure within and outside the source rock (Chapter 4).

With increasing temperature, the product mix moves increasingly toward dry gas. Gas is derived as a result of direct

alteration of kerogen (as the primary product in Type III kerogen, and as a secondary reaction in Types I and II), and by cracking of the remaining oil. Peak production tends to occur at about 150°C. For example, the Jurassic Kimmeridge Clay source rock of the North Sea is immature on the flanks of the basin, generates undersaturated black oil at intermediate depths and gas condensate deeper in the basin, and is thought to be generating dry gas where the source rock is most deeply buried. The distribution of oil and gas in Jurassic reservoirs (i.e., where great migration distances have not been involved) reflects the maximum temperatures recorded by the source rock, which are also the temperatures recorded today.

For gas-prone sources, little fluid is produced and the first major product of maturation is abundant gas, with a low molecular weight (Fig. 3.43).

Maturation in the reservoir

Oil that has migrated from the source rock and is subsequently trapped will also be influenced by increases in temperature, ultimately maturing to a mixture of gases, dominated by methane (CH_4). Gas cracking in the reservoir is characterized by the occurrence of associated bitumens (pure hydrocarbons with large aromatic ring structures resulting from the loss of hydrogen during methane production; i.e., dehydrogenation).

Bitumen and tar mats

Note that bitumens can also be produced by late influx of gas into an oil-filled reservoir (known as de-asphalting) or by bacterial degradation and/or water-washing of oil in a shallow reservoir (generally at below 80°C). Bitumens due to the influx of gas, also known as tar mats, are produced because increased gas saturation leads to a lower solubility of dissolved

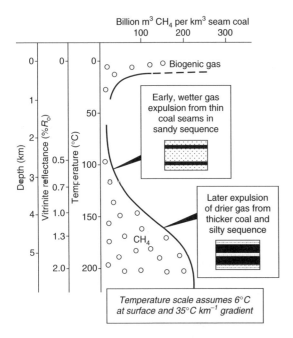

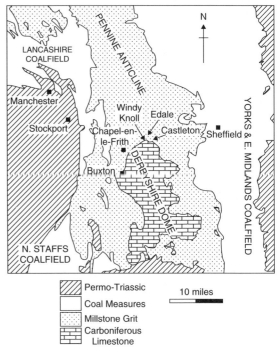

Fig.3.43 Maturation from a gas-prone source rock (1 km³). The curve is based upon changes in the H : C ratio with rank in Upper Carboniferous seam coal. (From Cornford 1990.)

Fig.3.44 A geologic map of the Peak District (North Derbyshire, English Midlands) and surrounding areas. (Modified after Stevenson & Gaunt 1971.)

heavy compounds. Bitumens formed due to thermal cracking are found in the Jurassic Norphlet Formation of the Gulf of Mexico (Ajdukiewicz 1995). Tar mats thought to be due to de-asphalting are found in several Middle Eastern oilfields (including Ghawar, the world's largest conventional oilfield in reserves discovered to date) and in the oilfields on the Alaskan North Slope, including Prudhoe Bay, the largest oilfield in North America.

3.8 CASE HISTORY: EDALE NO. I WELL, ONSHORE UK

3.8.1 Introduction

For this, our first case history, we travel back in time more than 60 years. The location is England. The date is 1938, a year in which exploration activity onshore UK was high (Lees & Cox 1937). The impetus for this activity was the UK's need to find its own oil. The British government wished to avoid the risk of losing overseas oil supplies to enemy submarine attacks, as had happened during World War I (Kent 1985).

Much of the exploration activity took place in the English Midlands (Fig. 3.44), where there was evidence of petroleum seepage, commonly in the form of bitumen. Moreover, both the anticlinal structures and the Lower Carboniferous limestone of the English Midlands were likened to the trap geometries and reservoirs of the Persian (Iranian) oil province (Lees & Cox 1937; Kent 1985). It was within this exploration setting that Steele Brothers (of London) drilled several unsuccessful oil exploration wells. In this case history, we examine one of those wells, Edale No. 1. We look at the failure of this well compared with the subsequent success of the East Midlands oil province in general.

3.8.2 The geologic background

The Late Paleozoic stratigraphy of the English Midlands comprises Dinantian limestones overlain by Namurian shales and sandstones. Westphalian sandstones, shales, and coals succeed the Namurian sedimentary rocks. A major unconformity separates these Carboniferous strata from the overlying Permo-Triassic red beds. The stratigraphy is now known in considerable detail (Fraser et al. 1990). Each of the major sedimentation episodes described above has been interpreted in terms of syn-rift (late Devonian and Dinantian), post-rift (Namurian and Westphalian A, B, and C), and inversion (Westphalian D and Stephanian) megasequences (Fig. 3.45). This depth of knowledge was not available to the geologist (Geoffrey Cotton) who mapped the Edale Anticline in 1937, for petroleum exploration purposes. Cotton could not call on seismic or the data derived from hundreds of wells drilled in the English Midlands since his time. However, he was aware that within the central Pennines there were significant thicknesses and lithological variations in the predicted reservoir (Carboniferous limestone) and seal (lowermost Namurian mudstones; Stamp 1923).

The structure of the English Pennines is dominated by a large, open, NNW–SSE anticline. Smaller-scale structures north of the Derbyshire Dome and in the adjoining East Midlands commonly have east–west or ESE–WNW trends. The Edale Anticline, the first fold structure to the north of the Derbyshire Dome, has an east–west trend (Fig. 3.46).

3.8.3 Petroleum geology

The exploration concept used in the late 1930s was simple but effective. The key elements required were a closed anticlinal structure (Fig. 3.46), a proven working source, and of course a reservoir. The following extract is taken from an internal (Steele Brothers) company report, written by the geologist Cotton (Gluyas & Bowman 1997). It is what we would today call a prospect evaluation memorandum:

There can be little doubt that the Edale shales are entirely suitable as a source rock the presence of oil in the bullions {carbonate concretions} I think proves the point. That oil has been expressed from the shales into the limestones is shown by the elaterite {bitumen} of Windy Knoll amongst other places.

The Edale Anticline is a first class structure with good closure and a large flat crestal area {the Edale Anticline lies between what we would call today the source kitchen and the seep at Windy Knoll}.

The only doubt is whether or not there is a suitable reservoir rock under the Edale Shales. The white limestone to the south is well bedded and jointed and obviously capable of containing and giving up fluids but there is some doubt whether this limestone will be present. The Edale Shales are believed to overlap the limestone and the limestone is known to be absent from the series in the north. Pot holes and caves have, however, been followed for some distance to the north of the outcrop near Castleton showing that the limestone continues for some distance below the shales but it is

not certain that the limestones will be present under Edale. If the limestones are present they should be struck at very shallow depth.

If the white limestones are absent there is always the possibility that other beds suitable as reservoir rock may be present.

3.8.4 Well evaluation: 1938

Edale No. 1 was plugged and abandoned as a dry hole. Shows of both oil and gas were recorded during the drilling of the well, but they were not present in commercial quantities. Additional Steele Brothers company archives give a little insight into why the well might have been considered a failure in 1938. The following extract is taken from a letter written by Professor W.G. Fearnsides in early January 1938, and addressed to the Steele Brothers exploration department. Fearnsides was an expert on the geology of the area around Edale. At the time of the letter, Edale No. 1 had penetrated about 420 ft, significantly deeper than the expected top pay:

As there is now a dry hole at 422 feet I think we may continue to hope for a more interesting fluid in joints traversing the limestone conglomerates and massive bedded limestones which lie ahead.

The massive limestones that were expected were never found. A further 300 ft of drilling only revealed thin limestones, interbedded with mudstones. Cotton had expressed concern that the reservoir might be missing. He was correct. The well was drilled outside the play fairway (Section 4.6). Instead of penetrating shelfal or reefal limestones, the well had penetrated basinal limestone turbidites (Gutteridge 1991). This is about as far as Fearnsides, Cotton, and Steele Brothers could have taken the story in 1938.

3.8.5 Well evaluation: 1990

Some 50 years after Edale No. 1 was drilled, new work on the source rocks and the local burial history has revealed additional information as to why the well failed.

Fraser and Gawthorpe (1990) and Fraser et al. (1990) have shown that for the central part of the English Pennines, petroleum generation almost certainly occurred before trap formation. They cite field evidence; minor gas shows and numerous small bitumen deposits, but no live oil. They take this to indicate that all the oil is residual and degraded. Such data might have been available to Cotton and his co-workers, but it is unlikely that their significance could have been appreciated.

Fraser et al. (1990) agreed with Cotton as to the source potential of the Namurian shales. These (Edale) shales are now known to have generated the petroleum for all of the oil pools

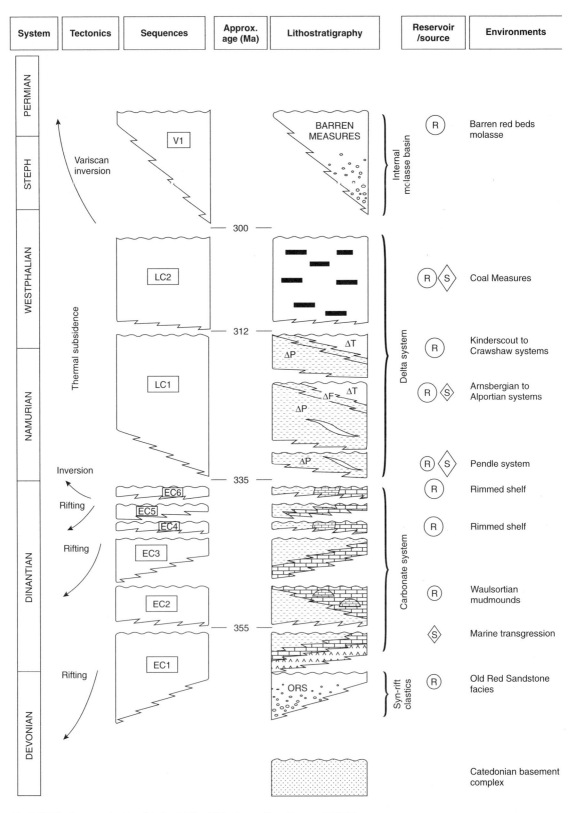

Fig.3.45 The chronostratigraphy of the English East Midlands area. (From Fraser et al. 1990.)

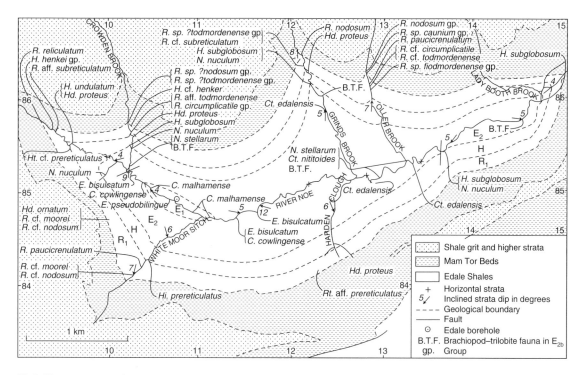

Fig 3.46 A geologic map of the Edale Anticline. This particular map is reproduced from Freeman et al. (1971), but approximately replicates the map produced by Cotton for Steele Brothers & Company.

in the English Midlands. However, unbeknown to Cotton, the Edale shales were buried to within the oil window during the Namurian and Westphalian, whereas the Edale anticlinal trap probably formed at the end of the Carboniferous, during the Variscan orogeny. Thus petroleum generated in the source kitchen to the north migrated through the Edale area without being trapped. Fraser argues (pers. comm., October 1996) that even in the absence of a conventional reservoir, oil might have been trapped in fractured basinal limestone had migration occurred after trap formation.

The data of Fraser et al. (1990) make it easy to understand why Edale No. 1 failed to find oil. However, history tells us that the Steele Brothers did not reach the same conclusion as Fraser et al. (op. cit.). Alport No. 1, which lies 7 km NNE of Edale No. 1, was drilled a few years later than the Edale well. It failed for the same reasons.

3.8.6 The English East Midlands petroleum province

Although the Edale and Alport exploration wells failed, a discovery was made at Eakring in the English East Midlands in 1939. The discovery was made not in Lower Carboniferous

limestone, but in Upper Carboniferous sandstone. Although small by global standards, the Eakring Field and the subsequent local discoveries were the major UK oil province for 30 years, until the discovery of North Sea petroleum. Alas, the hope of Lees and Cox (1937) that the English Midlands could become an oil province like Persia were not realized.

3.9 CASE HISTORY: THE PEARL RIVER MOUTH BASIN, OFFSHORE CHINA

3.9.1 Introduction

The Pearl River Mouth Basin is on the northern edge of the South China Sea, and between the islands of Taiwan and Hainan. It remained unexplored until the 1980s. Less than ten years from the opening of the basin for exploration, the two largest fields found to date had been discovered. The two fields are not located where most of the exploration companies expected to make discoveries. This case history allows us to examine a modern example of frontier exploration in an area hitherto little known to Western science.

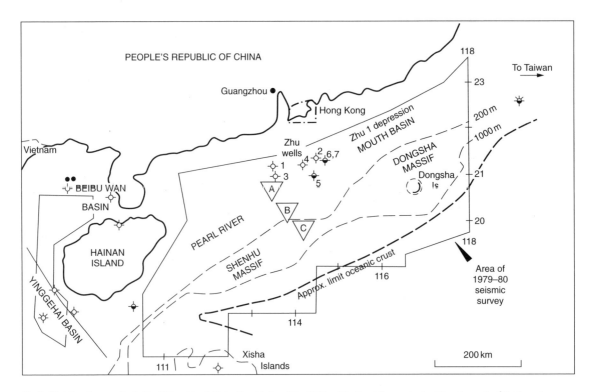

Fig.3.47 A location map for the Pearl River Mouth Basin, People's Republic of China. The Zhu 1 Depression (sag) is to the north of the Dongsha Massif. Degrees of latitude and longitude are shown along the margins of the geophysical survey area. (Modified from Tyrrell & Christian 1992.)

The People's Republic of China has a long history of successful oil exploration and production. However, until the early 1980s production came from onshore wells. The country did not have a developed offshore oil industry. Moreover, it did not have significant foreign investment in its oil industry. The "opening" of the People's Republic of China began in the 1970s, with meetings between Western heads of states and the Chinese rulers in Beijing. By 1979–80, about 40 companies were participating in joint geophysical studies across a large tract of the South China Sea, which overlies the Pearl River Mouth Basin. Some 60,000 km of 2D seismic, gravity, and magnetic data were acquired over an area of about 240,000 km². In preparation for the First License Round of 1982, the People's Republic of China had drilled seven stratigraphic test wells (Zhu 1–7). Several more wells had been drilled north of Hainan Island and a shallow test was drilled on one of the Xisha Islands (Fig. 3.47). Wells had also been drilled to the east of the license areas in Taiwanese waters. Thus the Zhu wells were at least 400 km (250 miles) from wells known to Western companies.

3.9.2 The regional geology and basin analysis

The WSW–ENE basin was initiated during Late Cretaceous to Early Paleocene rifting, a process which led to the opening of the South China Sea. The rift developed as a series of half-graben ("sags" in local terminology) grouped into three major depressions (Zhu 1, 2, and 3). The breakup unconformity with its coincident erosion occurred during the middle Oligocene. From late Oligocene to Holocene times, the basin underwent thermal subsidence.

The basement for the Pearl River Mouth Basin is formed by Mesozoic-age granitic and metamorphic rocks. The basin fill is dominated by clastic sequences. Syn-rift sedimentation was nonmarine: fluvial sands and lacustrine and fluvial muds were deposited. However, there appeared not to be a common pattern between sags. Zhu 1 contains a well-developed lacustrine section; Zhu 3 has a well-developed lacustrine section; and Zhu 2 appears not to have a significant lacustrine section but has, instead, a sand-dominated syn-rift interval. The post-rift sequences above the unconformity comprise paralic

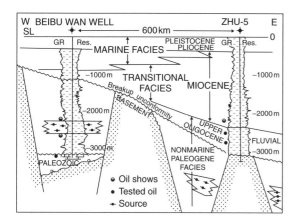

Fig.3.48 The general stratigraphic relationships in the Pearl River Mouth Basin as understood in 1982, prior to exploration drilling. (From Tyrrell & Christian 1992.)

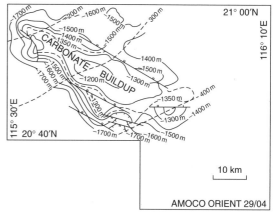

Fig.3.50 The top carbonate structure of the Liuhua 11-1 prospect on the crest of the lower Miocene carbonate platform, as mapped by Amoco geophysicists for the 1982 bid round. Water depths are shown by the southwest-trending heavy, dashed lines. (From Tyrrell & Christian 1992.)

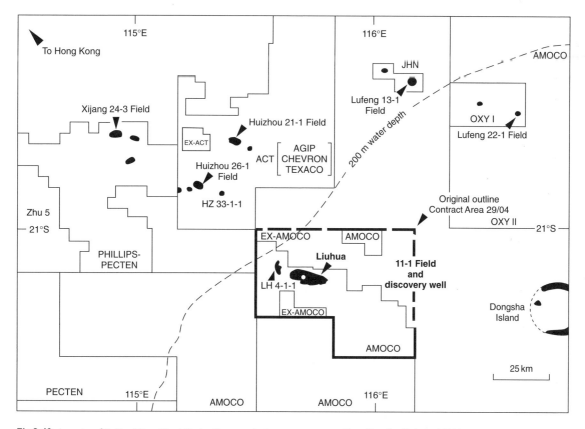

Fig.3.49 A portion of the Pearl River Mouth Basin adjacent to the Amoco contract area. (From Tyrrell & Christian 1992.)

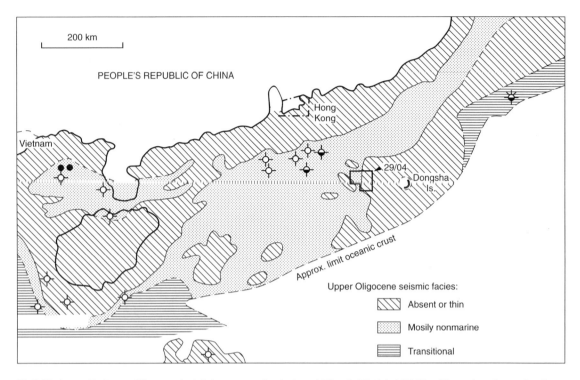

Fig.3.51 A generalized, upper Oligocene seismic facies map as produced prior to drilling the Whale in 1982. The widespread, sand-prone, largely nonmarine "seismic facies" overlies Paleogene graben with lacustrine source rocks. These sandstones were postulated to provide the long-distance conduit into the Amoco Contract Area (29/04). (From Tyrrell & Christian 1992.)

sediments, grading into marine sediments. The basal sand (mid–late Oligocene to earliest Miocene) within the post-rift section is extensive. Further south, the Dengsha and Shenhu massifs were drowned by the same overall transgression. In the wake of the transgression, a carbonate shelf system developed on the massifs, ultimately leading to reef development on Dongsha. In the middle Miocene, southward-directed progradational sequences developed over the central part of the license area (Tyrrell & Christian 1992). The progradational sequences fine from their source in the north, the proto Pearl River, southward. The Late Miocene to recent deposits are mud dominated.

Much of the information on the lithology and depositional environment for the basin-fill megasequences came from analysis of the Zhu wells. Two of the wells (Zhu 5 and Zhu 7) tested waxy crude, reservoired in paralic sandstones of upper Oligocene – lower Miocene age. This oil was typed as lacustrine in origin, and was linked to the Late Eocene Wenchang Formation (Cai 1987). The Wenchang Formation occurs deep within the sags. Later analyses of this source rock gave $P_2 (S_2)$ values of 8–19 kg t^{-1} and hydrogen index (HI) average of 660 (Todd et al. 1997). This source rock is known now to reach a maximum thickness of about 150 m in the Zhu 1 sag (Fig. 3.48).

3.9.3 Exploration history: 1983–5

Exploration proper began in 1983, with BP drilling the first well in the Enping License (south and west of the Zhu wells). This well, along with most of the wells drilled in the first few years after exploration began, was targeted at a large, simple anticline formed in the post-rift sequences by late Tertiary compression. Like so many of the wells to follow, Enping 18-1-1 had oil shows. Moreover, it contained abundant carbonate cement in parts of the post-rift interval. This carbonate contained isotopically light carbon, typical of organic matter or oil oxidation. It was postulated at the time that much oil could have escaped to the sea floor during Miocene times. Could it be that man had arrived some 20 million years too

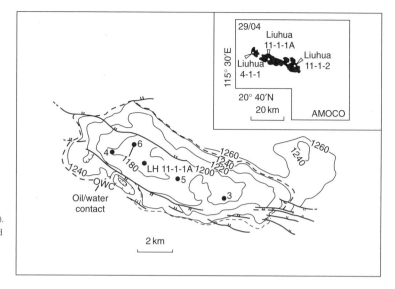

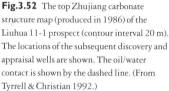

Fig.3.52 The top Zhujiang carbonate structure map (produced in 1986) of the Liuhua 11-1 prospect (contour interval 20 m). The locations of the subsequent discovery and appraisal wells are shown. The oil/water contact is shown by the dashed line. (From Tyrrell & Christian 1992.)

late? Had oil been generated, migrated, leaked, and been degraded long ago?

Many consortia had bid extensive exploration programs in order to secure the acreage, yet the results from the first couple of years' exploration were disappointing. Interest in the second licensing round, work for which was initiated in 1985, was significantly less. Conventional wisdom, from analysis of similar areally restricted, initially lacustrine basins elsewhere in Southeast Asia, dictated that most oil would be found within a 20 km radius of the kitchen area (Todd et al. 1997). Companies put much work into trying to identify sites of possible terminal lakes and hence source kitchen areas, in the early syn-rift sequences. However, for many companies, a satisfactory model of source distribution did not emerge.

In terms of the key elements — source, seal, trap, reservoir, and timing — reservoir alone was not an issue. The post-rift sequences of the Pearl River Mouth Basin are sand-rich and sandstones were also found within the syn-rift interval. A working lacustrine source (or sources), though proven by the occurrence of shows in the wells, had not been characterized in terms of volume or fecundity. As a direct result of the abundance of sand in the proto Pearl River Delta, seals are only poorly developed in the Oligo-Miocene section — at least in the northern part of the licensed area. Although they were clearly present and mappable, traps may have formed too late to capture migrating oil.

3.9.4 Exploration success: 1985–7

In January 1985, the ACT group (Agip, Chevron, and Texaco) announced the discovery of oil in their HZ 33-1-1 well. It flowed at 2839 bopd from two zones. Significantly, the well lay about 10 km (6 miles) south of the southern edge of the Huizhou (Zhu 1) sag and hence significantly south of any possible lacustrine source rock. This piece of news was greeted enthusiastically and, outside of ACT, probably nowhere more so than amongst the geoscientists at Amoco (Tyrrell & Christian 1992). Amoco had not been successful in obtaining exploration acreage in the first exploration round. Like so many of the exploration companies working the area at the time, they had considered the main targets to lie within anticlines overlying the source kitchens. The Amoco geologists were, however, acutely concerned about the apparent paucity of seals and consequent short oil columns in the Zhu wells. They considered poor seal quality and thus seal failure to be a major cause for concern within the mid-Miocene progradational sequences targeted. The specific reason for joy amongst the Amoco geologists was that the discovery had been made to the south and outside of the mid-Miocene play fairway, and hence in an area of poorer reservoir but better seal development at the mid-Miocene level. The reservoir was believed, then, and has proven to be since, the basal sandstone above the breakup unconformity (latest Oligocene or earliest Miocene). Moreover, the discovery was *en route* to the largest closure mapped in the area.

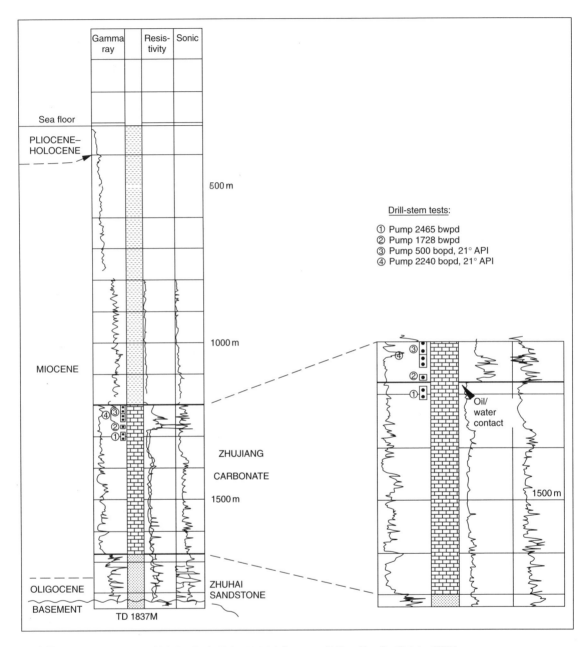

Fig.3.53 A well log and generalized lithology for the Liuhua 11-1-1A discovery well. (From Tyrrell & Christian 1992.)

Amoco negotiated for and obtained the license on the 29/04 Contract Area southeast of the HZ 33-1-1 discovery in November 1985 (Fig. 3.49). It contained the prospect known to Amoco geologists as the Whale, because of its size and shape (Fig. 3.50). Amoco were not the only company to map closure at this location. Nor were Amoco the only company to identify the lithology correctly as limestone in the form of a Miocene, carbonate reef complex on top of the Dengsha Massif. However, Amoco could well have been the only company to develop a model whereby the structure could be filled with oil generated in a basin at least 75 km to the northwest. Certainly, Amoco recognized the concept to be high-risk. The discovery at HZ 33-1-1 dramatically reduced that risk. The key aspect was that Amoco had tried to map the extent of the basal sandstone lying above the breakup unconformity, using seismic facies and continuity analysis (Fig. 3.51). It was by any standards an extensive sand complex, linking the source area in the Zhu 1 sag with the Whale prospect. The Zhu wells and some of the new exploration wells had proved vertical migration from the Wenchang source within the sag kitchens. HZ 33-1-1 proved that substantial lateral migration could occur once oil entered the basal sandstone on top of the breakup unconformity.

In January 1987, Amoco drilled LH 11-1-1A and struck oil within the Miocene reef complex (Fig. 3.52). About 75 m (246 ft) of net pay delivered 21° API oil at 2240 bopd (Fig. 3.53). The oil-bearing interval lay over a further 401 m (1316 ft) of carbonates and an underlying basal sandstone of 149 m (489 ft). The migration and seal models had been proven correct. Further drilling showed the Liuhua 11-1 to be a giant accumulation, and production began in 1996.

3.10 CASE HISTORY: OFFSHORE FALKLAND ISLANDS

3.10.1 Introduction

The aim of this case history is to examine frontier exploration as practiced at the end of the 20th century in the area around the Falkland Islands. The Falkland Islands lie in the South Atlantic (c.52°S, 59°W) and comprise two main islands and hundreds of smaller ones. The nearest landfall in South America is about 480 km (300 miles) from the archipelago. The total land area is 12,173 km^2 (4700 square miles). Port Stanley, the capital and only significant town, lies on the northeast coast of East Falkland. The designated exploration area completely surrounds the islands and occupies some 400,000 km^2 (154,400 square miles).

The Falkland Islands are a dependent territory of the UK. Development and execution of the first exploration license round was carried out by the Oil Management Team based in Port Stanley and the London-based Oil Licensing Administration. Technical support was provided by the British Geological Survey. Their data and analysis form the greater part of this case history.

The Falklands Islands exploration area was opened for investment in October 1995. The designated exploration area has been divided into quadrants that are subdivided into blocks (Fig. 3.54). Each quad occupies one degree of latitude and one degree of longitude. The designated exploration area comprises 82 quads and part-quads. There are 30 blocks in

each quad. For the first licensing round 19 tranches of acreage were offered. Each tranche holds between 6 and 30 blocks. The tranches cover parts of three of the four major sedimentary basins surrounding the Falklands (Richards 1995). License fees during the first five years of a license were set at US$30 per square kilometer per year. Thereafter the fees escalated substantially, although discounts were available as concessions against the number of wells drilled. This format was designed to encourage companies not to "sit" on acreage. Fees payable on any production will take the form of royalties and corporation tax based on profits.

On October 28, 1996 the Falklands Islands government announced that it had awarded licenses in the First Licence Round to five groups comprising 12 companies. Forty-eight blocks in seven tranches were awarded, covering an area of 12,800 km^2. The first well, operated by Amerada Hess, was drilled early in 1998.

3.10.2 The database

At the time of licensing, there were no wells drilled in the designated exploration area, although 17 wells had been drilled in the Argentinian part of the Malvinas Basin, and several more, relatively shallow, wells had been drilled by the Deep Sea Drilling Project (DSDP) on the Maurice Ewing Bank, more than 500 km due east of the Falklands (Ludwig & Krasheninnikov 1983).

The commercially available seismic database at the time of licensing is shown in Fig. 3.55. The total seismic line length, comprising a variety of vintages, was of the order of about 50,000 km in 1995. At this time there were few seismic lines across the Falklands Plateau Basin. Both proprietary and nonproprietary seismic have since been acquired over the licensed areas. Gravity and magnetic data were also available.

3.10.3 The regional geology and basin analysis

The Falkland Islands lie at the western end of the Falklands Plateau. The area is a displaced terrain, having been attached to southern Africa before the breakup of Gondwana. Both paleomagnetic evidence and similarities between the onshore geology of the islands and that of southern Africa have been used to aid the reconstruction (Taylor & Shaw 1989; Marshall 1994).

Four sedimentary basins have been identified, using gravity, magnetic, and seismic data, in the designated exploration area of the Falkland Islands. These are the Malvinas Basin, the North Falkland Basin, the Falkland Plateau Basin, and the South Falkland Basin (Fig. 3.56). The North Falkland Basin is separate from the other three interconnected basins. The

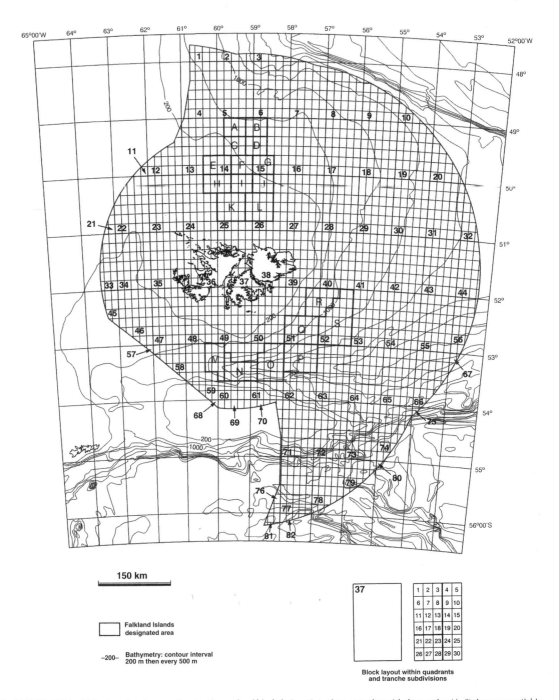

Fig.3.54 The Falkland Islands exploration area, showing the quad and block designation scheme together with the tranches (A–S) that were available in the first exploration licensing round. Tranches A–L lie in the North Falkland Basin, while M–S are within the South Falkland and Falkland Plateau Basins. (From Richards 1995.)

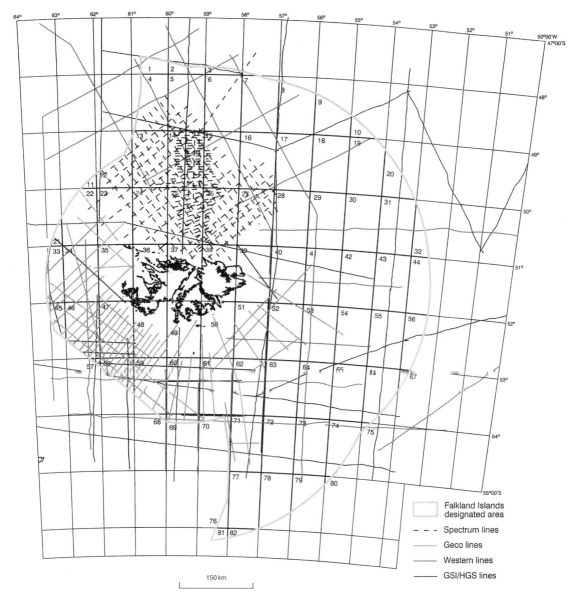

Fig.3.55 The location of the commercial seismic data available within the Falkland Islands Designated Area. (From Richards 1995.)

basins are largely Mesozoic in age, and probably overlie Devonian and Carboniferous sediments similar those seen at outcrop on the islands.

All four basins are likely to have been initiated by east–west extension as the Gondwanan supercontinent fragmented. As such, the basins are essentially north- to south-oriented rifts. However, each has been modified to a degree by either strike-slip faulting associated with the major wrench systems in the area or compressional tectonics and semiregional inversion. The effects of transpression are particularly clear in the southern part of the North Falkland Basin, where wrench tectonics have led to large-scale folding and the generation of flower structures.

The Falklands Plateau Basin is located to the east of the is-

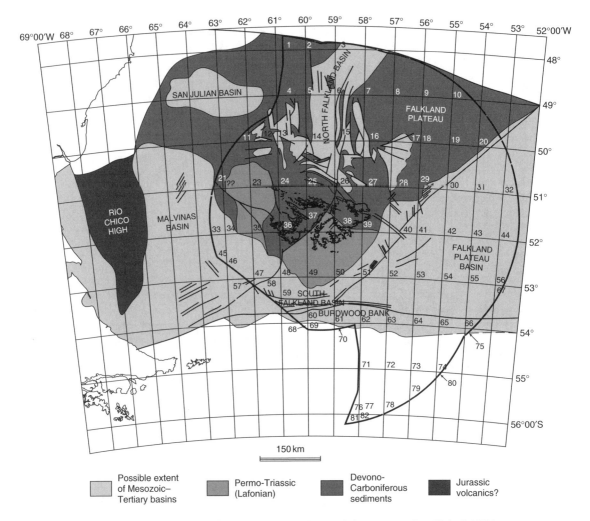

Fig.3.56 A simplified geologic sketch-map of the Falkland Islands Designated Area and adjacent regions. (From Richards 1995.)

lands. It is limited to the west by Devonian–Carboniferous rocks of the Falkland Plateau. The line of the faulted contact can be seen clearly on gravity and magnetic data (Fig. 3.57). This is probably due to deep-seated magnetic and basaltic rocks along the margin (Richards et al. 1996). A steeply sloping feature, the Falkland Escarpment, forms the northern boundary to the basin. The Burdwood Bank occurs to the south and Maurice Ewing Bank to the east.

DSDP boreholes on the Ewing Bank have sampled the easternmost margin of the basin. The basement here is Precambrian gneiss. The oldest rocks encountered overlying this basement are of Callovian age (Middle Jurassic), and these rest directly on the metamorphic Precambrian rocks.

The presence of Devonian, Carboniferous, and Permo-Triassic sediments in the basin itself has been inferred by Richards et al. (1996) on the basis of the seismic data linking the DSDP boreholes with the known geology onshore Falklands. A thick succession of Late Jurassic to Aptian (Cretaceous) mudstones rich in organic matter has been recovered from some of the DSDP boreholes on the Ewing Bank (Cromer & Littlejohn 1976; Deroo et al. 1983).

The South Falkland Basin lies to the southeast of the Falkland Plateau and Malvinas Basins. The southern edge of the basin is the boundary between the South American and Scotian Plates, a sinistral, transpressional system with northeasterly-directed thrusts (Craddock et al. 1983).

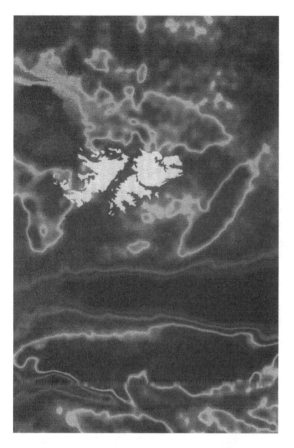

Fig.3.57 A gravity anomaly map from the Falkland Islands and surrounding areas. The strong linear feature east of the islands is the faulted western boundary of the Falklands Plateau Basin. (Reproduced courtesy of Nigel Press Associates.)

The stratigraphy of the basin has been described from seismic data and by extrapolation from wells either hundreds of kilometers to the west in the Malvinas basin or hundreds of kilometers to the east (DSDP wells on the Ewing Bank). Wedges of sediment believed to be of Jurassic to Early Cretaceous age have been interpreted as syn-rift sequences (Richards et al. 1996). The Cenozoic appears to be thin, while the basement is possibly Devonian–Carboniferous. Early Jurassic volcanic rocks may also be present.

The Malvinas Basin has yet to be licensed in the area under Falkland Islands jurisdiction. We will discuss it no further in this case history.

The North Falkland Basin consists of two major elements, a north–south rift system called the North Falkland Graben

and a series of smaller basins to the west of the graben (Fig. 3.58; Richards & Fannin 1997). Both features were controlled by east–west extension overlying Paleozoic thrust terrain (NW–SE). In the northern part of the North Falkland Graben, a mid-basinal high further separates the rift into eastern and western depocenters.

The basement is believed to be Devonian quartzites, sandstones, and shales, like those found on the Falkland Islands. Syn-rift sequences have been identified as Late Jurassic to Early Cretaceous. In the eastern depocenter, the syn-rift occurs between about 5 s and 2.3 s TWT (two-way time) on seismic sections. Early post-rift, topography-filling sequences occur between 2.3 s and 1.8 s. This interval is probably of Valanginian to Aptian age, and by analogy with the Ewing Bank evidence much of the interval may have been deposited under anoxic conditions. In the same early post-rift interval, both the eastern and western depocenters contain evidence in their northern part of infill by strongly progradational sequences (Fig. 3.59). These have been interpreted as deltaic (Richards & Fannin 1997). Late post-rift sequences infilled the remaining accommodation space during the thermal subsidence phase of the basin formation process (Fig. 3.60). By analogy with surrounding areas, this interval is likely to be dominated by marine mudstones of late Cretaceous to Paleogene age.

3.10.4 Petroleum geology

The key elements required for a basin to become a petroleum province are source, seal, trap, reservoir, and timing (Chapter 1). At this the earliest stage of frontier exploration, the most important questions to answer are connected with the petroleum source rock. If there is no petroleum source rock, the outcome of exploration is self-evident. If a source rock is present but is immature, then again there will be no petroleum to be found. If a source rock is present but is gas-prone, little or no oil will be found. Given the distance between the Falklands and potential markets, gas, even in large quantities, is probably without value. The same issue over gas arises if the source rocks, though oil-prone, are largely overmature. Given the importance of the source question, we will address it in detail.

Late Jurassic to Early Cretaceous mudstones rich in organic matter were recovered from the DSDP boreholes on the Ewing Bank. Here the mudstones were immature, with maximum TOC (total organic carbon contents) of about 6%. The organic material extracted from cores at DSDP site 511 is of marine origin and is largely oil-prone kerogen of Types I and II (Deroo et al. 1983). Royden and Keen (1980) con-

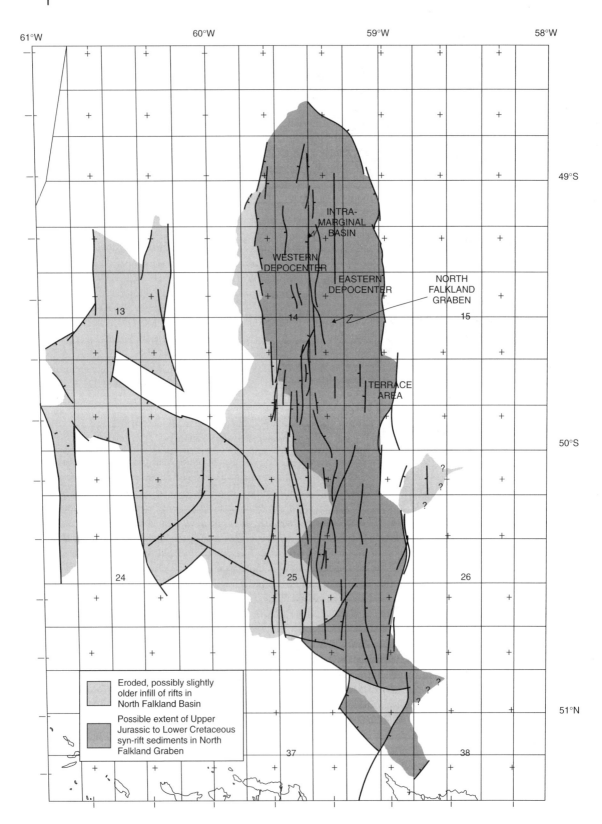

Fig.3.58 A simplified geologic map of the North Falkland Basin graben. (From Richards & Fannin 1997.)

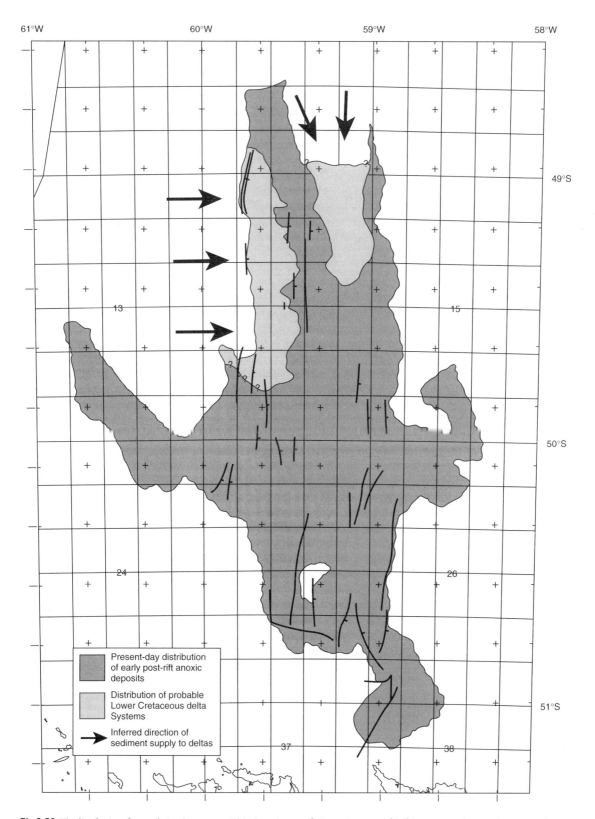

Fig.3.59 The distribution of progradational sequences within the early post-rift (Lower Cretaceous) fill of the western and eastern depocenters of the North Falkland Basin. (From Richards & Fannin 1997.)

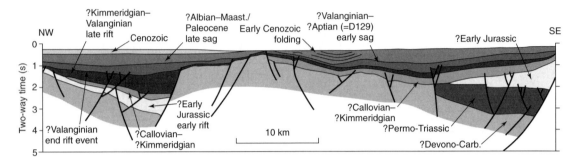

Fig.3.60 A geoseismic cross-section of the North Falkland Basin, showing pre-Jurassic tilted fault blocks, a syn-rift sequence (possibly early Jurassic), and thermal subsidence (sag) fill of Late Cretaceous and younger sediments. (From Richards 1995.)

structed a thermal model for the Falklands Plateau Basin to the west of the DSDP sites. They conclude that oil-prone kerogen, if present, was likely to be mature for oil generation at burial depths between 3 km and 4.3 km. Moreover, on the poor-quality seismic data available to Royden and Keen (1980), they identified a "pre-Albian" reflector, coincident with the potential source, at about 4 km over much of the area.

Indications of a working oil-prone source rock have also been identified within the Argentinian segment of the Malvinas Basin, where noncommercial quantities of both oil and gas have been recorded from exploration wells. In addition, a photograph of a presumed natural oil slick within the designated exploration area was published in the *PESGB Newsletter* (Petroleum Exploration Society of Great Britain) in February 1997. However, the specific location was not published and the publication postdated the awarding of the licenses.

From the above evidence, it might seem reasonable to assume that the Falklands exploration area contains mature, oil-prone source rock. However, let us reexamine the evidence carefully:
• Nowhere in the entire Falklands exploration area, nor in the adjacent Argentinian basins, are there commercial oil discoveries.
• Oil shows have been reported in the Malvinas Basin, but the Falklands segment of this basin has not been licensed.
• Immature, oil-prone source rock of Late Jurassic/Early Cretaceous age is known from the Ewing Bank some 500 km east of the Falklands Plateau Basin license area.
• These same-age sediments could be present at depths and temperatures sufficient for oil generation in each of the three basins licensed, but there is no evidence to indicate that such sediments would be of comparable source quality to the sam-

ples obtained from DSDP boreholes. Indeed, such sediments may be neither oil- nor gas-prone.
• The validity and utility of the photographed seep (slick) has yet to be proven.

The presence of possible trapping geometries is easy to infer from the basin histories and current seismic coverage. In the rift-dominated areas, the main trap types are likely to be rotated fault blocks in the pre-rift and four-way dip-closed highs in the post-rift sequences overlying the tilted fault blocks. Traps with a significant stratigraphic component may occur as wedges within the syn-rift, with updip shale-out.

The thrusted areas of the South Falkland Basin are likely to contain anticlinal trapping geometries, and the inverted portions of the North Falkland Basin may contain horst traps and four-way dip-closed structures in overlying draped sequences.

Satisfactory mapping of closure in such traps will be determined by their size and by the quantity of seismic data available from which to map them. Richards (1995) quotes a minimum line spacing of 10 km for the seismic data then available. Although this is clearly a coarse grid, large structures should still be mappable. Moreover, it will take large structures with large oil reserves to be economic in this area.

The Paleozoic and earliest Mesozoic sediments that outcrop on the Falkland Islands contain sandstones, which might make suitable, if possibly low-quality, reservoirs. Sandstones of probable Oxfordian age were recovered from DSDP site 330 in the Falklands Plateau Basin. A 300 m thick reservoir was also penetrated in an exploration well in the Argentinian San Julian Basin. Indeed, it had oil shows. Elsewhere, the presence of reservoir sandstones has been inferred from the sequence stratigraphic relationships observed on the seismic data. The two prograding wedges seen in the northern part of the North Falkland Basin are obvious candidates for reservoir development, both in the wedges themselves and in lowstand fans that may have been shed from them.

If it is indeed present, the presumed Late Jurassic/Early Cretaceous source interval is likely to form an effective regional seal. Similarly, transgressive systems tracts in the post-rift section are likely to form effective regional seals.

It seems possible from the work of Royden and Keen (1980) that much of the postulated source rock could be at peak generation today. If so, then provided that sufficient oil has already been generated, structures that are observable today could have been charged. It would, however, be sensible to model the source maturation and migration processes for the area under study to validate this speculative statement. In the areas that have suffered uplift, particularly the inversion area in the south of the North Malvinas Basin, it would be wise to be a little more cautious about the source maturation and migration story. Uplift and cooling of a source rock will stop oil generation. Moreover, uplift of a charged reservoir can lead to gas exsolution as the pressure declines. The gas will migrate to the top of the trap and can displace oil beyond the spill point. For example, this singular uplift process has caused all of the major traps in the Hammerfest Basin (northern Norway) to be flushed, by gas, of their oil charge.

3.10.5 The results of exploration

Amerada Hess spudded the well 14/9-1 in April 1998. The well was drilled in the North Falkland Basin, in the eastern depocenter and through the prograding wedge in the post-

rift section. At the time, leaked information that oil shows had been encountered led, in mid-May 1998, to a rapid rise in the value of shares for the participating companies. This was followed by an almost equally rapid decline in value once the well was "abandoned with shows" but without testing.

Despite the fact that the well failed to make a commercial discovery, the presence of shows was clearly important from the point of view of demonstrating that the North Falkland Basin does contain a working source rock. Despite the promise delivered by the results from the first well, the subsequent wells drilled in this first exploration campaign failed to find commercial quantities of oil and the licenses were relinquished. The second licensing round was announced in 2000. Might a second drilling campaign find commercial quantities of oil in the Falkland Basins?

FURTHER READING

Allen, P.A. & Allen, J.R. (1990) *Basin Analysis, Principles and Applications.* Blackwell Scientific, Oxford.

Emery, D. & Myers, K.J. (1996) *Sequence Stratigraphy.* Blackwell Science, Oxford.

Hunt, J.M. (1979) *Petroleum Geochemistry and Geology.* W.H. Freeman, San Francisco.

Van Wagoner, J.C., Mitchum, R.M., Campion, K.M., & Rahmanian, V.D. (1990) *Siliciclastic Sequence Stratigraphy in Well Logs, Cores, and Outcrops: Correlation of Time and Facies.* American Association of Petroleum Geologists Methods in Exploration Series, No. 7.

CHAPTER 4

EXPLORATION AND EXPLOITATION

4.1 INTRODUCTION

The phase of frontier exploration in a basin (Chapter 3) has proved successful. Imagine that several consortia have made significant petroleum discoveries, and the newly discovered fields are undergoing appraisal. What happens with regard to exploration elsewhere in the basin? The answer is simple. Once a commercial discovery has been made in a basin, the pace of exploration will increase dramatically. Those companies that have had the early success will try to increase their acreage holdings in areas that they believe will deliver similar petroleum discoveries. At the same time, other companies will try to emulate the successes of the fortunate few. Importantly though, those companies with early successes will be aware of one or more of the seal, trap, and reservoir combinations that work. Much effort will be put into stopping the unsuccessful companies from obtaining such knowledge, by ensuring that company data are kept secure. In the office, code names may be used for exploration wells, while at the well site important discoveries may go untested for fear of alerting other companies to the success, since testing commonly involves flaring. Mobilization of equipment for flaring, together with the flare itself, will of course be obvious to the informed observers from rival companies.

The petroleum exploration industry abounds with stories and tales pertaining to the early phases of exploration in a basin. Many such tales involve accounts of the lengths to which geologists will go to guard their newly discovered petroleum (Moreton 1995).

As the exploitation phase develops, different petroleum plays (reservoir/seal combinations) will be proven to work. Some of the new plays will arise by accident, the serendipitous discovery. Other plays will also be developed as the geoscientists working the basin begin to gain a better understanding of the seal development, reservoir distribution, trap configurations, and petroleum migration pathways. In this chapter,

we investigate the critical components of exploration success: seal, reservoir, petroleum migration, and trap. We also examine the concepts of plays and play fairways, and their extension into "yet to find" statistics.

Exploration wells are drilled on prospects and prospects are developed from leads. We examine both in this chapter, along with risk and uncertainty.

4.2 THE SEAL

4.2.1 Introduction

We can define a seal — the fundamental part of the trap, which prevents the petroleum from migrating onward through the rock — as a bucket without holes. The bucket is full of porous rock, with the pores in the rock full of fluid, both water (generally bound to the rock surface) and petroleum. Moreover, the bucket is upside down, because petroleum is buoyant and will continue to rise upward through the rock until it is arrested by the seal. The most common subdivision of seals distinguishes between seals in which petroleum is unable to force its way through the largest pores (membrane seals) and seals in which petroleum can escape only by creating fractures (hydraulic seals). Any lithology can act as a seal to a petroleum accumulation. However, attributes that favor a rock as a seal include small pore size (seals tend to be formed in fine-grained rocks), high ductility, large thickness, and wide lateral extent. Of course, the physical properties of the water and petroleum are important too. Water salinity, petroleum density, and interfacial tension between petroleum and water are the most important, and these properties will change according to changing pressure and temperature conditions (related to depth of burial, for example).

The most common lithology that forms a petroleum seal is mudrock, often referred to as shale. Mudrocks form approximately 60–70% of all sedimentary rocks. Mudrocks can be composed of either carbonate or siliciclastic minerals (or both), and mudrock sequences are often thick (>50 m) and laterally continuous (>1.0 km²). Examples of mudrock seals are found in all deltaic settings (including the Gulf of Mexico, Niger, and Nile Delta petroleum provinces) and many interior, rift and passive continental margin basins (including the northwest Australian shelf, northwestern Europe, the North Sea, and Southeast Asia). Halite can form a more effective seal but is a less common lithology, found only where conditions of high evaporation of sea water have taken place. The Upper Permian rocks of northwestern Europe contain Zechstein halite that is known to have trapped gas for long periods of geologic time. Halite forms part of the sealing lithology in many of the large Middle Eastern petroleum accumulations.

4.2.2 The membrane seal

When petroleum is trapped beneath a cap-rock seal, there is a buoyancy pressure (P_b). The magnitude of the buoyancy pressure is related to the contrast in density between the water (ρ_w) and the petroleum (ρ_p), and its height (h) above the free water level (FWL) at which both are in equilibrium (normally close to the petroleum/water contact). The relationship can be written as follows:

$$P_b = (\rho_w - \rho_p)h$$

where P_b is expressed in units of pressure (e.g., bars, psi, or MPa) and the fluid densities are expressed as pressure gradients (e.g., bars m⁻¹, psi ft⁻¹, or Pa m⁻¹), and is shown graphically in Fig. 4.1. The maximum petroleum column is controlled by the capillary entry pressure of the petroleum into the largest pores in the seal. The capillary entry pressure (P_d) of a water-wet rock is given by the equation

$$P_d = 2\gamma \cos\theta / R$$

where γ is the interfacial tension between the water and the petroleum, θ is the contact angle, and R the radius of the largest pore (Fig. 4.2). The interfacial tension and contact angle change with increasing temperature and pressure, and relate to fluid type and density. These properties are routinely established from laboratory experiments on rocks, and the procedure involves injecting the pores with mercury and converting to a petroleum–water system at *in situ* conditions using standard equations. The seal capacity now determines the height of a petroleum column that can be trapped beneath

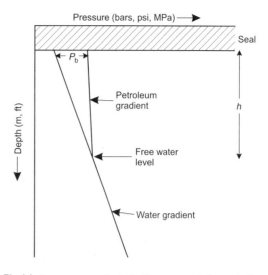

Fig.4.1 A pressure versus depth plot illustrates a typical water gradient (aquifer) supporting a petroleum column, whose steeper gradient leads to a pressure difference, or buoyancy pressure (P_b), at its maximum beneath the seal.

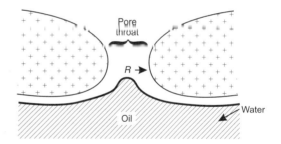

Fig.4.2 A schematic illustration of a pore throat between two grains. The radius of the pore throat and the buoyancy pressure, plus the interfacial angle and surface tension between the oil and water, determine whether oil can migrate through the pore throat or remains trapped beneath.

it, and the seal will be breached when the buoyancy pressure (P_b) exceeds the seal capillary entry pressure (P_d).

4.2.3 The hydraulic seal

When the capillary entry pressure of the rock is extremely high—for example, in evaporites or very tight mudrocks/shales—then the failure of the cap-rock seal is controlled by the strength of the rock that, if exceeded, creates natural tension fractures. The rock will fracture when the pore pressure is greater than both the minimum stress and the tensile strength of the rock. Structural geologists describe the stresses in rock in terms of three orthogonal components of stress (Fig. 4.3), one oriented vertically (S_v) and the other two oriented horizontally ($S_{h\,min}$ and $S_{h\,max}$). In relaxed sedimentary rocks found in an extensional basin or a young delta, S_v is usually greater than both $S_{h\,min}$ and $S_{h\,max}$, so the minimum stress (σ_3) is horizontal. Under these conditions, the rock will fracture by creating vertically oriented fractures that will propagate horizontally. In rocks under horizontal push, or compression, the vertical stress (S_v) is the minimum stress (σ_3) and the rock will fail by the opening of horizontal fractures. This condition is quite rare.

To calculate the hydraulic seal capacity it is necessary to know the magnitude of the minimum stress, which we will consider to be $S_{h\,min}$, since this is the condition for relaxed sediments such as those on passive margins (such as the Gulf of Mexico, for example) and in intracratonic basins, such as the North Sea. Quantitative estimation of $S_{h\,min}$ is not possible, but a measure of the magnitude of the minimum stress comes routinely from both leak-off test pressures (LOPs) in wells when casing is run, and from hydraulic fracturing when production demands enhanced permeability. A compilation of data in Breckels and van Eekelen (1982) shows that (1) the hydraulic fracture or "mini-frac" produces a better measure of minimum stress magnitude than LOPs, and that (2) there appears to be a relationship between $S_{h\,min}$, depth, and the magnitude of overpressure (pressure above hydrostatic). In other words, the estimation of hydraulic fracturing has to take into consideration the pressure state of the rocks; and, in general, the higher the pressure, the higher is the corresponding $S_{h\,min}$.

4.2.4 Faults

Faults can act as both conduits (migration pathways) and seals, depending on the hydraulic conditions, the rock properties of the faults, and the properties of the rocks juxtaposed across the fault. The consideration of faults as seals follows the same reasoning as for cap-rock seals above; that is, the sealing capacity of a fault relates to its membrane strength and its hydraulic strength. Membrane fault seals fail when the pressure

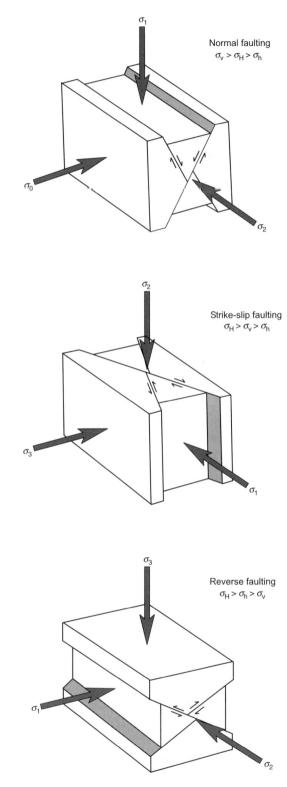

Fig.4.3 The relative magnitudes of the three principal stresses – one vertical (v) and two horizontal (H, h) – acting in a rock mass, and the associated direction of shear, for normal, strike-slip, and reverse-faulting regimes. The figure is based on the Andersonian classification of states of stress.

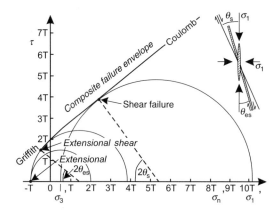

Fig.4.4 A Mohr diagram, used to determine when and how rocks will fail under shear and tension. Small differential normal stresses lead to failure due to the creation of tension (Griffith) cracks. (Redrawn after Sibson 1996.)

of the petroleum exceeds the entry pressure of the largest pores along the fault plane. Hydraulic fault seals fail when the fault is opened mechanically by high pore pressure that exceeds the minimum stress. Faults can be induced to move (shear) when the pore pressure exceeds the shear resistance along the fault, which can be lower than the minimum tensile stress, especially where there is a big difference between σ_1 and σ_3 (Fig. 4.4).

The main processes that act to reduce permeability, and hence increase seal efficiency, along and adjacent to fault planes include clay smear, cataclasis, and cementation of authigenic minerals such as quartz and dolomite. Clay smear is most effective when there is a high proportion of clay-rich rocks within the faulted section, such as in a deltaic environment of sand, silt, mud, and coal. Cataclasis is the process of grain-size reduction due to grinding of fault rock within the fault plane: the more movement along the fault, the greater the likely grain-size reduction, although the stress history is also important. When faults act as pathways for migrating fluids, particularly upward movement when temperatures are reducing, there is the potential for precipitation of authigenic minerals within the fault rock and in the adjacent rocks. Cements in faults often show indications of episodic flow conditions, suggesting cycles of fluid and pressure buildup and release (Byerlee 1993).

Faults can be effective barriers to flow, and hence create lateral seals within traps, as well as impeding migration of petroleum *en route* to a conventional trap. Evidence from some North Sea fields shows that migration of petroleum into traps tends to occur at only one or two points. For example, in the

giant Troll gasfield (Horstad & Larter 1997) there are two filling points, with mixing of oils over time. A fossil oil/water contact in the Troll Field is used as evidence of an earlier configuration of trapped oils, with barriers preventing oil from reaching the larger area of the present-day field. Hence fill-up of the field depends on continued supply and migration of the petroleum within the reservoir, limited by any lateral barriers to flow. Consequently, in a highly faulted trap, the distribution of petroleum throughout the trap depends on the pressures within the evolving (growing) petroleum column relative to the seal capacity of each fault. Differential fill-up is a feature of compartmentalized reservoirs. Some faulted compartments may be devoid of petroleum if the sealing capacity of the faults is not exceeded, even though high-quality reservoir may be present within the defined structural closure of the field (Section 5.4).

4.2.5 Trap fill

If there is an effective seal that prevents petroleum loss, petroleum traps can be filled to capacity. Alternatively, there may be less petroleum than they can accommodate; that is, the trap is under-filled. An under-filled trap can occur when the seal leaks before the structure is filled to capacity, or because there is an insufficient supply of petroleum from the source. Alternatively, a combination of both mechanisms will lead to under-filling, since seal leakage, petroleum migration, and trap filling are dynamic processes (that is, rate dependent). In some cases there appears to be more petroleum volume than can be contained within the limits of the seal; that is, the trap is over-filled. Over-filling may be a result of continuous recharge. This seems to have happened in the Eugene Island 330 Field in the Gulf of Mexico, where the total volume of petroleum extracted already exceeds the volume of petroleum that could be trapped. The explanation given for over-filling is an active petroleum charge up a deep-seated fault that can access petroleum from deeper reservoirs, and/or an active petroleum source rock. Apparent over-filling of some chalk fields in the Norwegian Central Graben may be attributed to diagenetic influences on chalk reservoir quality, preventing lateral migration at the structural spill points.

The control upon the volume of petroleum contained in the trap is principally governed by the capacity of the seal coupled with the supply of petroleum from its source. Sales (1993) suggested a classification of traps based on leakage versus spillage (Fig. 4.5), but this classification assumes that there is enough petroleum to fill the structure:

• Class 1 traps are those in which the seal strength is high enough that there is no leakage before petroleum fills to the spill point (structural closure of the trap);

• Class 2 traps are those in which the seal strength to oil allows complete filling to spill, but where the higher pressure gradient of the gas column causes leakage prior to fill;

• Class 3 traps are those where oil and gas columns will exceed the seal strength before the trap is filled.

When traps have enough seal strength to be filled to the spill point (Class 1), "Gussow's principle" may apply (Gussow

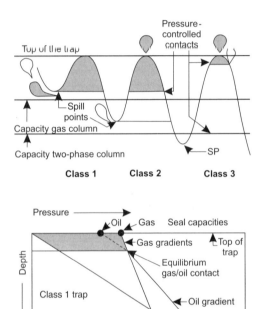

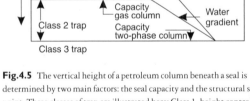

Fig.4.5 The vertical height of a petroleum column beneath a seal is determined by two main factors: the seal capacity and the structural spill point. Three classes of trap are illustrated here: Class 1, height controlled only by the spill point, with pressure below the seal capacity; Class 2, height jointly controlled by spill point and seal capacity; Class 3, height controlled by the seal capacity and above the spill point. (Redrawn after Sales 1993.)

1954). Ordinarily, oil is the first petroleum fluid to migrate from an oil-prone source rock and it will displace water downward from the crest of the trap where its buoyancy is arrested by the seal. Subsequently, gas will migrate into the trap, if conditions for gas generation are favorable, and gas will displace the oil downward. The first petroleum fluid to spill will be oil, later followed by gas when all the oil is displaced and the trap is filled with gas. The oil and later gas will migrate to other traps up dip from the spill point of the original trap in a "fill and spill" arrangement, with the fluid composition determined by the displacement timing of the fluids beyond the spill point. In this way, gas will fill traps at deeper levels than oil (Fig. 4.6).

4.2.6 The pressure seal

Seals relate to the trapping of petroleum, and may result in no-flow conditions for the petroleum over geologic time. The term "seal" has also been used to refer to conditions where very low flow conditions for water lead to the buildup of pore fluid pressures well above the hydrostatic, or normal, pressure for a column of water. We might refer to these as "pressure seals," although the term must be kept separate, and should not be used to indicate no-flow conditions over geologic time, as suggested by Hunt (1990). The term "dynamic seal" has been proposed in recognition of the interplay between pressure generation by some process, and the pressure leakage or dissipation by either Darcy porous media flow or by hydraulic leakage by microfractures and faults.

Pressure seals are recognized where there are variations in pore fluid pressures, excluding those related to the density of the pore fluids. Pressure seals can occur naturally over geologic time, or due to depletion in oil and gas production. The natural causes relate to mechanisms such as stress, fluid volume change, and hydraulic head (Osborne & Swarbrick 1997), coupled with rock properties that retard fluid pressure equilibration. Both single and multiple pressure transition zones are observed on profiles of pressure against depth (Swarbrick & Osborne 1996), and the magnitude of pressure

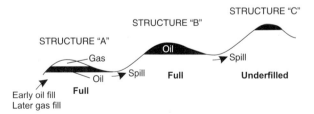

Fig.4.6 A conceptual diagram to show "fill and spill" during successive filling of shallower traps along a migration pathway.

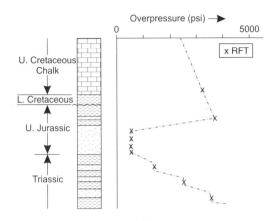

 (a)

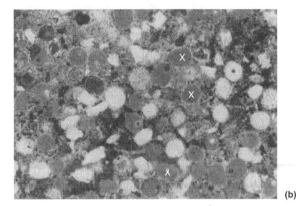 (b)

Fig.4.7 A pressure reversal (reduction of the amount of overpressure with increasing depth) in a North Sea well. The reversal was caused by sudden seal failure, which has drained the excess pressures from a thick Upper Jurassic sand and left the rocks above and below at various states of overpressure, depending on their permeability (and therefore their ability to drain into the lower-pressure sand).

Fig.4.8 (a) A scanning electron microscope photomicrograph of a porous (28%) and permeable (2200 mD) reservoir sandstone. The grains of the sandstone are loosely packed, there is little mineral cement, and the pores (x) are large and, most importantly, well connected one to another. The sample is from the aeolian Permian sandstones of the southern North Sea (Cleeton Gas Field). Field of view 2.7 mm × 2.0 mm. (Photograph by A.J. Leonard, reproduced courtesy of BP.) (b) A thin-section photomicrograph of a porous (30%) but impermeable (0.1 mD) sandstone. The grains of this sandstone were once opaline silica sponge spicules. During diagenesis of this sand the spicules dissolved, creating new, secondary pore space (X). However, the silica reprecipitated in the original pore space between the grains, as microcrystalline quartz cement. The pore throats are intercrystalline and minute. Oil and gas can get in, but will not come out in the time available to man. Such spicule-rich sandstones are common within the Late Jurassic of the central North Sea. Individual spicules are 100–150 μm long. (Photograph by J.G. Gluyas.)

change across them is highly variable. Large changes in pore pressure sometimes take place over narrow stratigraphic intervals (Fig. 4.7). A large pressure difference across the seal indicates low permeability, which is likely at values less than 10^{-21} m^2 (10^{-7} mD). By contrast, long transition zones can occur in thick, fine-grained rocks such as mudrocks at much higher permeability (e.g., 10^{-19} m^2 (10^{-5} mD)) if the sedimentation rate is high. Such sedimentation rates and rock properties are found in deep-water slope deposits supplied by active deltas (e.g., the Mississippi, Niger, and Nile Deltas).

4.3 THE RESERVOIR

4.3.1 Introduction

For a rock to be a petroleum reservoir, it need only be porous—capable of holding petroleum. For a rock to be an "economically viable petroleum reservoir"—that is, an oil- or gasfield—many other factors must be considered; some are geologic.

The rock must be permeable; that is, able to flow petroleum (Fig. 4.8). The volume of trapped petroleum must be sufficient to justify development and the reservoir not too compartmentalized. These elements are used to calculate oil- or gas-in-place estimates, which in turn are multiplied by a formation volume factor and a recovery factor to produce the reserves estimate—how much petroleum we expect to get out. A formation volume factor represents the change in volume of the oil that will take place when it is lifted from the high pressure and temperature of the reservoir and placed in the "stock tank" (Fig. 4.9). The equivalent term for gas is the gas expansion factor. The recovery factor is an estimate of the

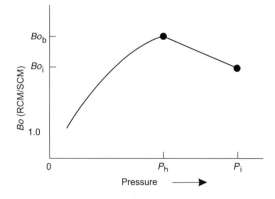

Fig.4.9 The formation volume factor (*Bo*) is a measure of the volume difference between reference conditions and reservoir conditions. In generating *Bo* the total system is considered (oil plus associated gas), not just the stock tank liquid phase. The volume change is a function of the partial molar volume of the gas in solution, the thermal expansion of the system with respect to temperature change, and compression of the liquid. No simple equations of state exist, and thus the liquid properties are either measured or estimated by using empirical data from analogous crude oils. Bo_i (initial) to Bo_b reflects the oil compressibility, while all the gas remains in solution at pressures above the bubble point (P_b). Beneath bubble point pressures, the curve reflects liquid shrinkage caused by gas exsolution. (From Archer & Wall 1986.)

percentage of petroleum that is likely to be recovered; this is rarely greater than 50% for oil and rarely less than 80% for gas (Table 4.1). In this section, we describe the essential characteristics of a reservoir (porosity, permeability, etc.), reservoir types, and their geometries.

4.3.2 Intrinsic properties

For the purposes of exploration, only a few intrinsic reservoir properties of a potential reservoir rock need be considered. All of the following properties pertain to the estimation of petroleum volume:

1 net to gross;
2 porosity;
3 permeability;
4 hydrocarbon saturation.

The question regarding whether any discovered petroleum would flow from its reservoir into the wellbore is only partially addressed in exploration, commonly because there are too few data on which to make a sensible decision. Additional reservoir properties are described in the appraisal and development chapters.

Net to gross

Net to gross is a measure of the potentially productive part of a reservoir. It is commonly expressed either as a percentage of producible (net) reservoir within the overall (gross) reservoir package or sometimes as a ratio. The percentage net reservoir can vary from just a few percent to 100% (Fig. 4.10). Care must be taken to recognize the difference between net to gross, the potentially productive part of a reservoir, and net pay, that which actually contains petroleum.

There are several ways in which net to gross can be defined. None are perfect. It is common to define net sand (or limestone) using a permeability cutoff. The exact value will depend upon the nature of the petroleum and the complexity of the reservoir but, as a rough guide, 1 mD is commonly used for gas and 10 mD is enough for light oil (for an explanation of the units, see the section below on Permeability). However, such detailed knowledge of permeability is only available when the reservoir has been cored or a petroleum flow test completed. For uncored intervals and uncored wells, we often rely on a combination of data on lithology and porosity from wireline logs (Chapter 2), calibrated to permeability data in a cored interval.

Fractured reservoirs present their own particular problems when it comes to prediction or even measurement of net to gross. Cores and specialist logs such as the formation microscanners or their equivalent sonic devices (Chapter 2) are critical for the identification of fractured intervals. Fractures or fracture sets can cross-cut different lithologies. Even hard, nonporous mudstones with porosity of a fraction of a percent may be productive intervals. For example, many of the large fields in the Zagros Province of Iran contain two highly productive, fractured carbonate reservoirs (Asmari, Tertiary and Sarvak, Cretaceous). Although the two reservoirs are commonly separated by several thousand meters of marl, the paired reservoirs are in full pressure communication via fractures through the marl. Extraction of oil from the upper reservoir, the Asmari, causes elevation of the oil/water contact in the lower, Sarvak, reservoir.

Porosity

Porosity is the void space in a rock (Fig. 4.11). It is commonly measured as either a volume percentage or a fraction (expressed as a decimal). In the subsurface this volume may be filled with petroleum (oil and gas), water, a range of non-hydrocarbon gasses (CO_2, H_2S, N_2), or some combination of these. Here we discuss porosity values and types. Measurement is covered in Chapter 2.

Table 4.1 Intrinsic properties and recovery factors for reservoirs within UK and Norwegian oil- and gasfields. (From Beaumont & Foster 1990; Abbots 1991; Gluyas & Hichens 2003.)

Field	Age/Formation	Reservoir	Net to gross (%)	Porosity (%)	Permeability (mD)	Fluid	Petroleum saturation (%)	Expected recovery factor (%)
Alwyn North	Jurassic/Brent	Sandstone	87	17	500–800	Oil	79	45–56
Alwyn North	Jurassic/Statfjord	Sandstone	65	14	330	Oil	80	35
Auk	Permian/Zechstein	Fractured dolomite	100	13	53	Oil	60	Field av. 19
Auk	Permian/Rotliegend	Sandstone	85	19	5	Oil	40	Field av. 19
Brae South	Jurassic/Brae	Sandstone	75	12	130	Oil	80	33
Britannia	Cretaceous/Britannia	Sandstone	30	15	60	Condensate	68	60
Buchan	Devonian/Old Red	Fractured sandstone	82	9	38	Oil	10–68	15–18
Cleeton	Permian/Rotliegend	Sandstone	95	18	95	Gas	83	75
Cyrus	Paleocene/Andrew	Sandstone	90	20	200	Oil	65	16
Ekofisk	Cretaceous/Chalk	Limestone (fractured Chalk)	64	32	<150	Oil	92	40
Eldfisk	Cretaceous/Chalk	Limestone (Chalk)	62	30	2	Oil and gas	70	20
Forties	Paleocene/Forties	Sandstone	65	27	30–4000	Oil	85	57
Fulmar	Jurassic/Fulmar	Sandstone	94	23	500	Oil	79	69
Frigg	Eocene/Frigg	Sandstone	95	29	1500	Gas	91	75
Heather	Jurassic/Brent	Sandstone	54	10	20	Oil	65	23
Leman	Permian/Rotliegend	Sandstone	100	13	0.5–15	Gas	51	83
Piper	Jurassic/Piper	Sandstone	80	24	4000	Oil	90	70
Ravenspurn South	Permian/Rotliegend	Sandstone	39–77	13	55	Gas	<90	58
South Morecambe	Triassic/Ormskirk	Sandstone	79	14	150	Gas	75	93
Scapa	Cretaceous/Valhall	Sandstone	—	18	111	Oil	77	26
Staffa	Jurassic/Brent	Sandstone	76	10	10–100	Oil	69	13
West Sole	Permian/Rotliegend	Sandstone	75	12	—	Gas	60	74

Typically, many reservoirs have a porosity in the range 20–30% but the full range varies from a few percent in some fractured reservoirs to around 70% in some types of limestone reservoir (Table 4.1).

Not all pores are alike: there are big pores and little pores, pores with simple shapes, and others with highly complex 3D morphologies. Knowledge of the size and shape of pores and the way in which they are interconnected is important, because it is these factors that will determine the permeability of the rock.

For sands and sandstones, many authors use a simple three-fold description of porosity: intergranular, intragranular, and "micro." The terms are largely self-explanatory. Intergranular porosity occurs between grains. Individual pores in a clean

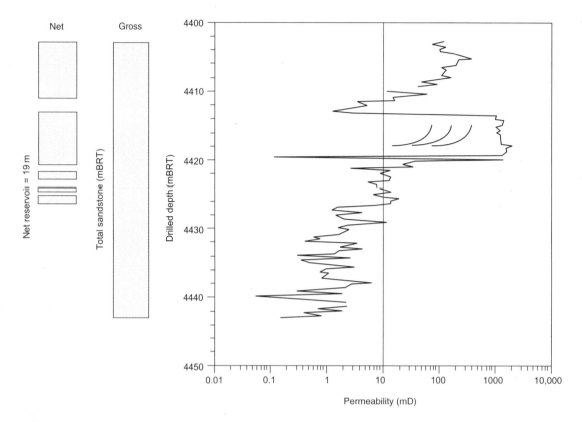

Fig.4.10 Net to gross is the ratio between reservoir rock capable of flowing petroleum and the gross reservoir interval. It is commonly defined using a permeability cutoff. The example shown here is from an oil-bearing, Jurassic, marine sandstone in the central North Sea. Thirty meters of sandstone was cut in one core, but only 19 m of the sandstone had a permeability (to air) greater than the chosen permeability cutoff of 10 mD. Thus the net to gross is here 63%.

sand will occupy approximately 40% of the total volume (grains plus void). For coarse sands the pores are larger than in fine sands. In most sands and sandstones the intergranular porosity is primary, a residuum of that imparted at deposition. Some intergranular porosity may be created in sandstones by the dissolution of mineral cements such as calcite (Schmidt & McDonald 1979). Conversely, intragranular porosity is largely secondary in origin, created on partial dissolution of grains in the sandstone. Minerals such as chert and feldspar commonly have pores within them. Any oil trapped in such secondary pores is unlikely to be released from the reservoir during production. Microporosity simply means small pores, those associated with depositional or diagenetic clay or other microcrystalline cements.

The development of porosity in limestones and dolomites is much more variable than that for sandstones. Both rock

Fig.4.11 Porosity. (a) Intergranular porosity (X) between quartz grains with quartz overgrowths, Ula Formation, Norwegian North Sea; thin-section photomicrograph, field of view 650 μm × 450 μm. (b) Intragranular porosity (X) within feldspar grain, Ula Formation, Norwegian North Sea; thin-section photomicrograph, field of view 1.3 mm × 0.9 mm. (c) Microporosity (arrowed) between illitized kaolinite crystals, Brent Group sandstone, UK North Sea; scanning electron microscope photomicrograph, field of view 25 μm × 17 μm. (d) Intergranular porosity (X) in limestone, beach rock, Bahamas; the rock has been partially cemented by aragonite. (From Bathurst 1975.) (e) Biomoldic porosity (x) within dasycladacean (D) and mollusk (M) molds, Pennsylvanian limestone, Texas; field of view 5 mm × 4 mm. (From Dickson & Saller 1995.) (f) Intercrystalline porosity (X) within dolomite, Zechstein dolomite, Dutch North Sea; the periphery of each pore is oil stained; thin-section photomicrograph, field of view 1.3 mm × 0.9 mm. (g) Cavernous porosity, Ste. Genevieve Limestone, Indiana; human for scale. (From Palmer 1995.)

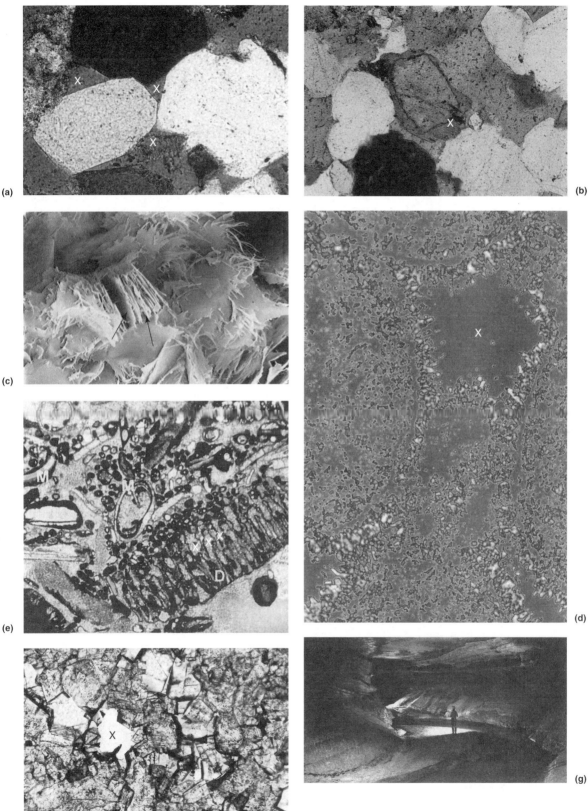

(a)

(b)

(c)

(d)

(e)

(f)

(g)

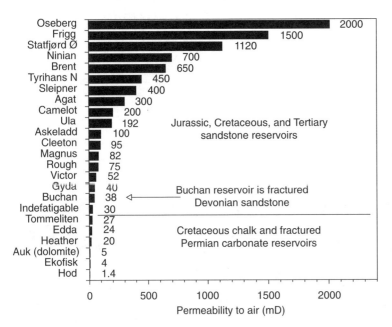

Fig.4.12 Average permeability for various producing fields on the UK and Norwegian continental shelves. Many of the low-permeability reservoirs are still able to deliver in excess of 10,000 bbl d⁻¹ per well because of low-viscosity oil, high-pressure drawdown (the pressure differential between the formation and the wellhead) and heterogeneous permeability distribution. For example, the Gyda reservoir contains high-permeability sandstone layers (>1000 mD) and most of the chalk fields produce through fracture systems. (Data sources are Spencer et al. 1987; Abbots 1991; Gluyas et al. 1992; Oxtoby et al. 1995.)

types are much more prone to mineral dissolution and precipitation than sandstones. This, coupled with the often varied suite of shell and other bioclastic materials in the carbonates, makes for a wealth of pore types: intergranular, intragranular, intercrystalline, intracrystalline, biomoldic, vuggy, fracture, and cavernous (Fig. 4.11). The size range for pores is also much greater for limestones than for sandstones: from micropores (a few μm) in individual oolite grains to giant cave systems (Bathurst 1975).

Permeability

Permeability is an intrinsic property of a material that determines how easily a fluid can pass through it. In the petroleum industry, the darcy (D) is the standard unit of permeability, but millidarcies (1 millidarcy (mD) = 10^{-3} D) are more commonly used. A Darcy is defined as a flow rate of 10^{-2} m s⁻¹ for a fluid of 1 cp (centipoise) under a pressure of 10^{-4} atm m⁻². Permeability in reservoir rocks may range from 0.1 mD to more than 10 D (Fig. 4.12). Permeability measurements made at the Earth's surface are commonly a factor or more greater than those in the subsurface. As a consequence, a pressure correction must be made to restore the value of permeability to reservoir conditions. This intrinsic rock property is called "absolute permeability" when the rock is 100% saturated with one fluid phase.

Water, oil, and gas saturation

It is rare in nature to find a reservoir entirely oil- (or gas-) saturated. More commonly, the pore system contains both oil and water. The proportions of each phase are usually expressed as percentages, linked to the abbreviations S_w for water, S_o for oil, and S_g for gas. Water and petroleum saturations are not constant across a reservoir. They vary in response to position in the oil column, and the permeability and mineralogy of the rock (Table 4.1). Oil and water saturations will also change as petroleum is produced.

4.3.3 Reservoir lithologies

Sandstone and limestone (including dolomite) are the most common reservoir lithologies. Sandstones dominate as important reservoirs in the USA (including Alaska), South America, Europe, Russian Asia, North Africa, and Australia. Limestones form the dominant reservoirs in the Middle East. They are also important in the Far East, western Canada, and some of the former Soviet states.

Sandstones, limestones, and dolomites of any age can make fine reservoirs. However, most of the best reservoirs in the world are relatively young. Commercial petroleum fields are more common in Cenozoic and Mesozoic sediments than in Paleozoic reservoirs (Fig. 4.13). Precambrian-age reservoirs

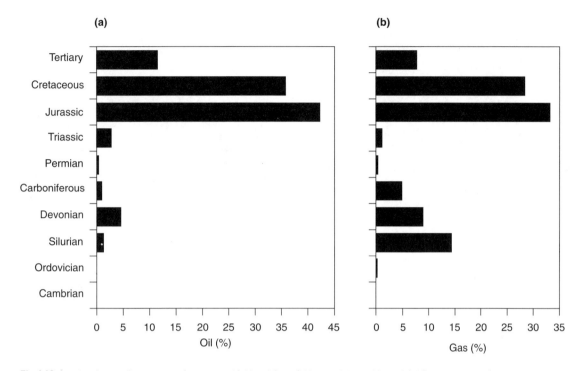

Fig.4.13 The distribution of reservoir ages for giant (a) oilfields and (b) gasfields around the world. (Modified from Tiratsoo 1984.)

Table 4.2 Fractured reservoirs.

Lithology	Field	Location	Reference
Sandstone	Buchan	North Sea	Edwards (1991)
Limestone	Asmari and Sarvak reservoirs	Zagros Trend, Iran	Alsharhan & Nairn (1997)
Limestone/dolomite	Lisburne	North Slope, Alaska	Belfield (1988)
Limestone/dolomite	La Paz	Lake Maracaibo, Venezuela	Mencher et al. (1953)
Granite	Sideki	Egypt	Helmy (1990)
Precambrian basement metamorphic rock	Clair (in part)	West of Shetland, UK	Coney et al. (1993)
Cherty mudstone	Point Arguello	California, USA	Mero (1991)
Cherty mudstone	Salym	Siberia, Russia	Zubkov & Mormyshev (1987)

are rare. There is no intrinsic reason why old rocks are more or less likely to be reservoirs than younger ones; it is simply that older reservoirs have had greater chance to be involved in tectonism or cementation. Both can lead to destruction of the reservoir.

In addition to sandstones and limestones, there are many other reservoir types. Fractured rock can form important reservoirs. The fractures alone may form the total pore volume of the reservoir. Alternatively, the fractures may help to drain petroleum from the intervening lower-permeability rock. Many different lithologies can form fractured reservoirs. Examples are given in Table 4.2.

4.3.4 The reservoir: sandstone depositional systems

Sediments, including those that may one day form a petro-

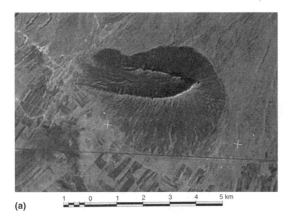

Fig.4.14 Alluvial fan development adjacent to a small remnant of a synclinal core, Sarvestan Valley, Fars Province, Iran. The syncline plunges westward. Alluvial fans are developed at the base of the scarp slopes on the north, east, and south sides of the exposure. An alluvial fan system has also developed along the axis of the fold, spilling out to the northwest and west of the fold. The radial nature of individual fans is clearly seen on the Landsat image, as is the lateral coalescence of fans along the scarp edges. The courses of ephemeral streams can be seen on the fan surfaces. The depositional angle of the fans lying at the foot of the southern and eastern scarp slopes can be seen on the conventional photograph. (Landsat image courtesy of Nigel Press and Associates.)

leum reservoir, can accumulate in many environments on the Earth's surface, from terrestrial settings to marine settings. The following discussion can be thought of as a journey, perhaps one that a sand grain would take, from mountain to sea bed.

Alluvial fans

Alluvial fans tend to develop along the front of mountains, along the sides of major valleys, or at the sides of glacier ice (Stanistreet & McCarthy 1993; Leeder 1999). High-energy upland streams and rivers lose energy quickly as they escape the confines of mountain valleys and, as a consequence, they drop their sediment load as soon as they reach flatter open land. The fans commonly develop into low-angle half-cones (Fig. 4.14). Their size is largely dependent upon the climate in which they develop, arid fans tending to be smaller than wet fans.

Deposition of sediment on the fan is intermittent, linked directly to periods of high rainfall in the mountains. At the head of the fan, flow tends to be channelized, whilst it is commonly more sheet-like on the lower parts of the fan. Sorting of sediment is commonly very poor in the channelized portions, although it increases down fan.

Individual alluvial fans are commonly small. Most are only a few kilometers across and the largest rarely reach more than a few tens of kilometers across. As a consequence, reservoirs

that are developed in alluvial fan sequences are commonly small, unless the fans are amalgamated along a fault front and/or a large thickness of potential reservoir rock accumulates as faulting creates accommodation space. Due to the nature of the depositional system, flash floods, ephemeral streams, and ponds, the sediment dumped on an alluvial fan tends to be very poorly sorted. The reservoir quality of these systems tends to be poor, in part due to rapid variation of facies and a high percentage of fines (clay- and silt-grade sediment).

A well-described example of an alluvial fan reservoir system is the Triassic of the Roer Valley Graben (The Netherlands). Here, the complex and rapid changes in depositional environment of the reservoir sandstones add considerably to an already high-risk exploration play (Winstanley 1993). Giant fields in alluvial fan complexes are even rarer; the Quiriquire Field in eastern Venezuela has produced 750 million barrels of oil from 678 wells in about 60 years. The oil in place has been estimated to be about 4 billion barrels (Salvador & Leon 1992). Here, the reservoir is composed of silts, sands, boulder beds, and interbedded mudstones of Pliocene age. The fan complex was shed southward from the actively growing coastal mountain range of eastern Venezuela. Today, it is buried at a depth of between 1000 ft and 3000 ft. Reservoir continuity in the field has been a problem, with production of heavy grade (av. 17° API) oil coming largely from lenticular sand pods.

Aeolian dunes

Aeolian dunes develop within deserts. Their classic shape, that of a crescent with the tips pointing down wind, is just one of a number of shapes that can form, others being linear or star shapes. Some dunes do occur alone, but more often they occur in massive sand seas. Modern examples of such seas commonly exceed 100 km² (Wilson 1973). A composite thickness of some 500 m has been recorded for the Permian aeolian sandstones of the southern North Sea, although this is made up from the preserved remnants of many dune fields stacked one on another. Aeolian dune sands often occur in association with fluvial (wadi) and sabkha deposits at the basin margins, and lake deposits, possibly ephemeral playa lakes, at the basin center (Fig. 4.15).

Aeolian sandstones are uncommon as reservoirs, because

(a)

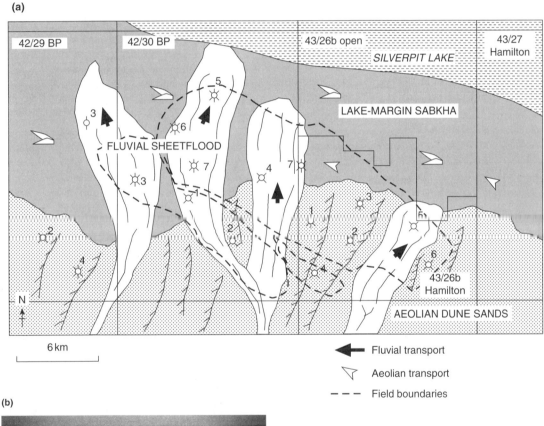

(b)

Fig. 4.15 (a) The margin of an aeolian dune field, showing the relationship between aeolian, fluvial, sabkha, and lacustrine sedimentation. The figure is a schematic representation of the paleogeography during the middle Early Permian around the Ravenspurn North Field, UK North Sea. (Modified from Ketter 1991b.) (b) The northern margin of the Erg Oriental (Tunisia). At this locality, the dune height reaches a few tens of meters; further south and west into the sand sea proper, the total thickness of the stacked dunes is measured in hundreds of meters. (Photograph by J.G. Gluyas.)

they have a low preservation potential—eventually the sea arrives and washes them away! In exceptional circumstances, commonly associated with catastrophic flooding of a subaerial basin, the sands are preserved.

The Permian Rotliegend Sandstone of Europe (Glennie 1986) and the Jurassic Norphlet Sandstone of the Gulf of Mexico (Mancini et al. 1985) are important examples. The Rotliegend sandstones form the main gas reservoirs of the southern North Sea, adjoining areas onshore The Netherlands and Germany, and further east into Poland. To date, some $4.1 \times 10^{12} \, m^3$ ($145 \times 10^{12} \, ft^3 = 1$ trillion cubic feet) of gas have been found (Glennie 1986). The Norphlet play is smaller, but still boasts some $0.2 \times 10^{12} \, m^3$ ($4.5 \times 10^{12} \, ft^3$) of gas (Emery & Robinson 1993).

The reservoir quality of many aeolian sandstones may be excellent. This is true of both the Rotliegend (Gluyas & Leonard 1995) and the Norphlet. They contain well-sorted rounded grains, and provided that they can avoid cementation such reservoirs will be both porous and permeable. Fluvial and sabkha deposits are commonly lower-quality reservoirs than the aeolian sandstones. This is largely due to differences in the way in which the rocks are sedimented. The fluvial and particularly sabkha deposits tend to be much less well sorted. Additionally, the quality of both fluvial and sabkha deposits is commonly reduced further via precipitation of minerals as ground waters evaporate.

Lakes

Lakes are common features of terrestrial sediment systems, but like aeolian systems the preservation potential of lacustrine deposits is low. Hence reservoirs developed in lacustrine sediments are uncommon, except in parts of Southeast Asia. In the Early Tertiary of parts of both China and Thailand (Min 1980; Ma et al. 1982), large rift-generated lakes formed the termini of sediment systems. Here oil and gas are reservoired in lacustrine fans. Such fans share many similarities with their marine counterparts (discussed later in this section). However, the volumes of lacustrine fans are small and, correspondingly, the field sizes are also small; a few tens of millions of barrels being the norm for reserves.

Fluvial systems

River systems connect the sites of sediment production (erosion) to areas of (coastal) deposition. For river sediments to be preserved and thus be potential petroleum reservoirs, deposition must occur in areas of net subsidence. When this happens, river complexes commonly form extensive and thick sand

(and mud) bodies (Leeder 1999). There are many examples of large petroleum accumulations within such sand bodies.

Rivers have a variety of forms. Their main channels may be braided or meandering, or anything between the two end members. For convenience, we will discuss the geometry of these two end members. However, the types of river do share some aspects in common. For example, rivers—be they meandering or braided—tend to migrate back and forth laterally through time and within a specific area. This leads to the development of a "belt" in which sands deposited by the channels will accumulate.

Braided rivers are commonly very sand-rich. As a consequence, the net to gross of petroleum reservoirs developed in such rocks is commonly very high (0.7–1.0). Thicknesses are commonly in the range of many tens of meters. The giant Prudhoe Bay Field (initial reserves estimate 9.5 billion barrels; Tiratsoo 1984; Case history 4.10) on the North Slope of Alaska occurs in Triassic sandstones that were deposited in a braided fluvial system. Fluvial sandstones are also important reservoirs in Europe. The Clair Field (6 billion barrels in place; Coney et al. 1993) west of Shetland (UK continental shelf) occurs largely in Devonian/Carboniferous braided fluvial sandstones. Similarly, the main part of Europe's largest onshore oilfield (Wytch Farm, UK, 240 million barrels reserves; Bowman et al. 1993; Fig. 4.16) also occurs within braided-fluvial sandstones.

The second category of fluvial systems, meandering rivers, can also form extensive sand bodies that may occur as petroleum reservoirs (Fig. 4.17). Net to gross in meander systems is typically less than that in braided systems (0.4–0.6). Locally, however, higher and lower ratios can occur due to the way in which (sand) channel stacking is controlled. For example, small-scale or incipient syn-depositional faulting may lead to a stacking of channels (Fielding 1984). As with the braided systems, massive oil- and gasfields can be developed in the sandstones of ancient meander systems. The Ness sandstones of the North Sea's Brent Province are lower delta plain deposits of meandering river complexes. They hold a substantial proportion of the petroleum in the Brent Province (Mitchener et al. 1992).

Deltas

On its journey from the mountains, our sand grain has is now nearing the sea. Where they are rich in sand, deltas and marginal marine complexes can form important petroleum reservoirs. Often, the best sandstones in such complexes are of excellent reservoir quality (e.g., well-sorted shoreface sandstones). However, because of the intimate mixture of many

Fig.4.16 An outcrop panel for the Otter Sandstone (Sherwood Sandstone equivalent), illustrating the internal architecture of a braided river deposit. (From Bowman et al. 1993.)

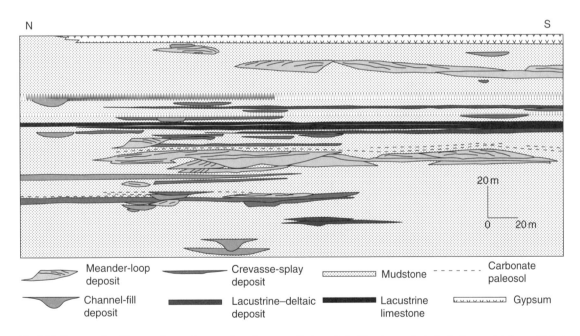

Fig.4.17 The reservoir architecture of a meander system in section, perpendicular to the bedding, Loranca Basin, Spain. Contrast this figure with that for the braided system in Fig. 4.17. (From Cuevas Gozalo & Martinius 1993.)

different lithotypes, abundant barriers and baffles to fluid flow often complicate reservoirs that have developed in such settings.

The geometry of delta systems and the associated marginal marine lithologies is controlled by the interplay of wave, tidal, and river energy. The well-known triangular diagram (Fig. 4.18) developed by Galloway (1975) and Elliott (1978) illustrates the dominant process in modern deltas. Figure 4.19 (Reynolds 1994) shows the relationship between delta type (river-, wave-, or tide-dominated), size, and the geo-

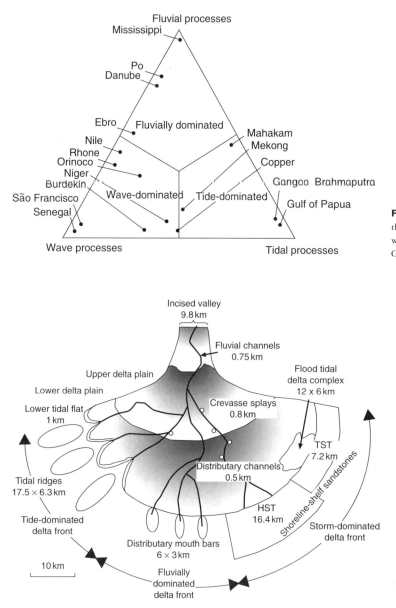

Fig.4.18 A delta ternary diagram, showing the distribution of modern deltas in terms of wave, tidal, and fluvial influences. (From Galloway 1975.)

Fig.4.19 The average size, shape, and location of sand bodies in wave-, tidal-, and fluvially influenced reservoirs. (From Reynolds 1994.)

metry of the sand bodies that develop. A delta is not the only coastal site for reservoir development. Sand, and thus potential reservoir deposition, can still be important, as beaches, intertidal flats, barriers or spits, and in lagoons.

Modern deltas that overlie older Cenozoic delta systems have been a common target for oil exploration. This is because the depositional systems can generate a total petroleum system of reservoir, source, and seal rocks (Chapters 3 & 4).

Moreover, the growth faulting (Section 4.5.3) often associated with delta development commonly creates trap geometries (Weber & Daukoru 1988).

Major petroleum provinces with deltaic reservoirs are the Tertiary of the US Gulf Coast (Galloway et al. 1982), the Niger Delta (Evamy et al. 1978), the southern Caspian Sea (O'Conner et al. 1993), the Mahakam Delta (Duval et al. 1992), and much of the Brent Province of the North Sea

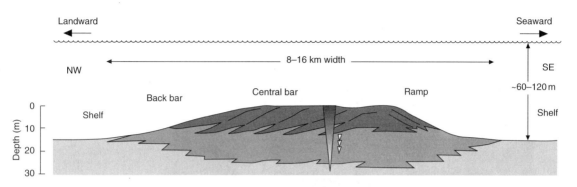

Fig.4.20 A schematic cross-section of Upper Cretaceous offshore sandstone bars, Duffy Mountain Sandstone, Colorado, USA. The cross-section is normal to the bar trend. The bar construction and migration were products of an interaction between fair-weather processes (?oceanic or tidal currents) and storm events (wind- or wave-generated currents). (From Boyles and Scott 1982.)

(Morton et al. 1992). In South America, many of Venezuela's large oilfields occur in delta–marginal-marine–shallow-marine systems (Hedberg et al. 1947).

Shallow-marine systems

Shallow-marine sandstones can form ideal petroleum systems. This is because they can commonly accumulate in association with a source rock, which may also act as seal. Compared with deltaic and marginal-marine deposits, shallow-marine sandstones are relatively simple and homogeneous. Net to gross can be very high and, unless faulting breaks sandstone bodies, communication between wells is good. There are no fixed rules as to "how thick." The thickness may vary from a few meters to a few hundred meters. It depends on the interplay between sediment supply and the creation of accommodation space through a combination of subsidence and sea-level rise and fall. This interplay will also control the degree to which the sand is diachronous, which will in turn determine how easy or difficult it is to correlate the sand well to well. As with deltaic systems, tidal and wave/storm energy are key controls on the development and internal architecture of sand bodies (Johnson & Baldwin 1986).

Shallow-marine sandstones of Cretaceous age form important reservoirs in the Midwest of the USA (Shannon; Williams & Stelck 1975; Fig. 4.20) and in the Western Canada Basin (Viking; Hein et al. 1986). Late Jurassic age shallow-marine sandstones are also important in the central North Sea (Bjørnseth & Gluyas 1995). Here the shallow-marine sandstone reservoirs are interbedded with the Mandal Mudstone (Norway) and Kimmeridge Clay (UK) source rocks. The shallow-marine, Jurassic/Cretaceous Toro

Sandstone is a singularly important reservoir interval in Papua New Guinea (Denison & Anthony 1990).

Submarine fans

Our sand grain is now nearing the end of its journey. It has not been captured in any of the terrestrial or shallow-marine systems, and it is on its way to deep water and a final resting place within a submarine-fan system (Fig. 4.21). Such systems commonly develop at the base of slope within sedimentary basins. Sand and mud are transported via mass flow, gravity-induced, processes from the shallow to deep water (Stow 1986). Some sediment may not make it all the way down to the basin, but be trapped on the slope in small intra-slope basins (Cook et al. 1982).

The composition of a submarine-fan deposit in terms of sandstone or mudstone will depend upon the sediment source area. You can only get out of a fan what you put into it. Fans can vary dramatically in size, from a few kilometers across (Navy Fan, offshore California) to thousands of kilometers across (Bengal Fan, Indian Ocean). Most of the largest fans tend to be rich in mud and poor in sand. Here, net to gross may be 0.1 or less. Even so, because of their large size, areas with higher than average sand concentrations can be viable reservoirs. For example, the Mississippi has shed a largely mud load into the Gulf of Mexico for the past few million years; nonetheless, there is sufficient sand in the system (Weimer 1990) to make the Plio-Pleistocene an attractive exploration target (Braithwaite et al. 1988).

Sand-rich fan systems (net to gross commonly >0.6), despite being considerable smaller than their mud-prone counterparts, can themselves be huge. Fan size is highly unlikely

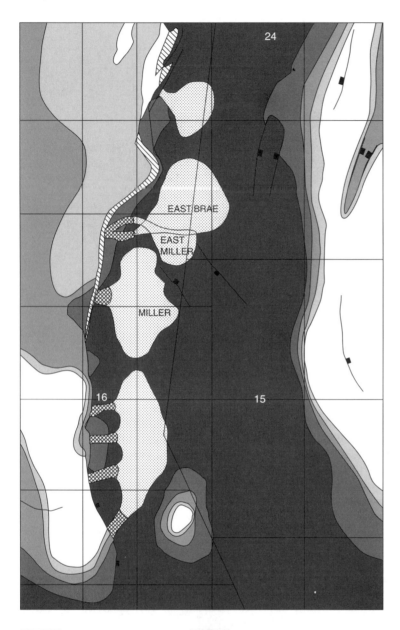

	Eroded/not deposited		Drowned slope
	Fault		Pelagic basin
	Shoreface		Feeder system
	Drowned shelf		Basin-floor fan

Fig.4.21 The summarized paleogeography for the late Kimmeridgian of the South Viking Graben, North Sea, illustrating basin-floor submarine fans and their feeder systems. (From Rattey & Hayward 1993.)

to be the limiting factor on petroleum pool size. The Paleocene Forties Fan of the North Sea is hundreds of kilometers in diameter (Whyatt et al. 1991). Collectively, the Paleocene sandstones of the North Sea are the reservoir within many fields (Forties, Nelson, Maureen, Arbroath, Montrose, Balmoral, McCulloch, etc.; Gluyas & Hichens 2003). The Viking Graben of the North Sea was also the site of substantial sand-rich, submarine-fan sedimentation during the Late Jurassic (Rattey & Hayward 1993). The sediments shed into the basin during this time now host a large number of oilfields, condensate fields, and gasfields, including the giant Magnus and Miller/Brae fields (Shepherd 1991; Cherry 1993; Garland 1993). Outside of the North Sea, submarine-fan reservoirs are important in the Gulf of Mexico, California, and offshore West Africa (Anglo, Nigeria, and Gabon) and Brazil. Deep-water drilling offshore Angola has led to the discovery of several giant oilfields in sand-filled slope canyons generated from the ancient Congo Delta. Exploration continues in ever-deeper water, with the expectation that petroleum accumulations will be discovered in the basinal fans.

4.3.5 The reservoir: limestone and dolomite

While on its journey from source to final resting place deep in the sea, our sand grain failed to visit all of the possible environments in which potential petroleum reservoirs could form. We have yet to examine carbonate systems. Limestones and dolomites form some of the largest petroleum reservoirs in the world. Many of the biggest occur in the Middle East. Other areas in which carbonate reservoirs deliver large quantities of oil and gas are western Canada, Mexico, Texas (USA), Norway (central North Sea), Poland, Kazakhstan, western and southeastern China, Iran, and Libya.

The range of carbonate depositional environments likely to produce significant petroleum reservoirs is more restricted than that for clastics. Almost all of the major petroleum reservoirs in carbonates accumulated as shallow-marine sediment. The important exceptions are relatively deep-water pelagic chalks (the Ekofisk complex of the North Sea; D'Heur 1984) and the similarly deep-water resedimented reservoir carbonates of the Poza Rica Trend in Mexico (Enos 1988). Exclusively nonmarine carbonate-producing environments are of little consequence for petroleum reservoirs. As such, we limit further discussion to marine settings. For further information on carbonate sedimentology, the reader is referred to Tucker and Wright (1990). One additional general point that differentiates carbonate reservoirs from clastics is the role played by "life" in producing carbonate sediment and carbonate frame-

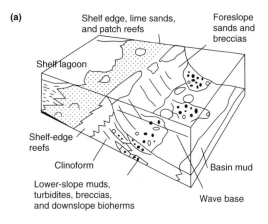

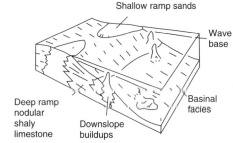

Fig.4.22 Block diagrams of (a) a rimmed carbonate shelf with a landward shelf and (b) a carbonate ramp homocline. (From Tucker & Wright 1990.)

works (reefs). Since carbonate sediments are products of living and dead organisms, part of their form has changed in parallel with evolution.

The gross subdivision commonly used for carbonate marine environments is ramp versus platform (Fig. 4.22). A ramp slopes gently (<1°) to progressively deeper water (Burchette & Wright 1992). A platform is a flat area with shallow water and steep or precipitous sides. The term "carbonate shelf" can be used to describe the shallow-water parts of either platform or ramp. Ramps tend to form open shelves that are strongly influenced by waves together with tidal and oceanic currents. Coarse-grained detritus can be abundant within such environments. Protected shelves comprise a shallow sea floor (~10 m), trapped within fringing barriers such as reefs, islands, or shoals, from deep (abyssal) water. The internal parts of protected shelves are commonly low-energy environments, with only the fringes affected by significant wave energy. Currents generated by both winds and tides can produce local, high-energy depositional environments and thus coarse-grained detritus. Modern examples of open shelves include western Florida and the Yucatan of the Caribbean, North Australia, and the Persian Gulf. The Bahamas and the Great Barrier Reef (eastern Australia) are examples of protected shelves.

Like their clastic counterparts, there is a clear link between the reservoir potential of a carbonate body and the environment in which the host sediment accumulated. High-energy ooid and shell shoals can and often do make excellent reservoirs. Framework reef complexes are also prime reservoir targets. However, unlike siliciclastics, carbonates can undergo almost complete transformation on weathering and/or diagenesis, to produce reservoirs from former seals and seals from former potential reservoirs. Some of the diagenetic and weathering processes will be discussed later in this chapter. We use a similar order for description of carbonate environments and reservoirs as used earlier for clastics; from shoreline to deep sea.

Shelfal and ramp carbonates

The petroleum found in the largest and most prolific oilfields in the world occurs within the primary, intergranular porosity of shelfal, marine limestones (Ghawar, the largest oilfield in the world, with estimated reserves of 83 billion barrels, and the other Middle and Upper Jurassic fields of Saudi Arabia; Stoneley 1990). Packstones and grainstones of Cretaceous age also form important parts of reservoirs in the Persian Gulf and the United Arab Emirates. For example, the Zakum oilfield of Abu Dhabi is developed in Cretaceous grainstones and packstones of the Thamama Group (Alsharhan 1990). Within the Zakum Field, oil occurs in low- to high-energy lithofacies, deposited in lagoons, shoals, and platform edge deposits. Burchette and Britton (1985) described the Mishrif Formation (late Cretaceous of Abu Dhabi), deposition of which occurred in a ramp setting. They demonstrated that the reservoir lithofacies were deposited on the uppermost slope, as shoals and biostromes. The areas available for reservoir development within such carbonate-dominated shelves are large in comparison to even the largest fields. Only in exceptional circumstances, therefore, is the depositional system likely to be a limit on field size.

Shelfal carbonates are also important reservoirs in other parts of the world. Carbonate sand bars composed of skeletal grainstone form the reservoir for more than 150 oil- and gasfields in the Permian Basin of West Texas and southeast New Mexico (Craig 1990). The reservoir quality of the grainstones was originally high. This was then further enhanced by a combination of dolomitization, mineral leaching, and fracturing. The trend covers an area of about 20,000 km^2. Shelfal carbonates were also the target for some of the earliest oil exploration onshore UK. Lower Carboniferous limestones in the English Midlands were first penetrated by a successful oil exploration well in 1919 (Case history 3.8). Intermittent drilling up until the late 1980s failed to make any more commercial discoveries, although it did substantiate the viability of the Carboniferous limestones (and dolomitized limestones) as potential reservoirs. The main problem seems to have been one of delivering sufficient petroleum to charge the reservoirs.

Reefs

The foregoing discussion has concentrated on shelf carbonates in either ramp or platform settings, although platform margin reefs also form important reservoirs. Individual reefs and reef complexes are the key component of the Devonian play fairway in the Western Canada Basin (Alberta). Such complexes hold about 90% of the total oil in place (>15 billion barrels) within the play fairway (Prodruski et al. 1988; Fig. 4.23). Platform fringing algal bindstones also hold a substantial part of the oil in place within Liuhua 11-1, the biggest oilfield (15 km × 4 km) offshore China (Tyrrell & Christian 1992; Case history 3.9). Cretaceous, frame-building, rudist bivalves are the prime reservoir around Mexico's famous Golden Lane Trend. The so-called Golden Lane "Atoll" is an old Cretaceous platform (c.100 km N–S by c.50 km E–W) fringed by many petroleum fields (Guzmán 1967).

Reef size is clearly important when considering the volume of reservoired petroleum. At the lower end of the size range, pinnacle reefs (defined as having a greater height than diameter) may be only a few hundred meters across and patch reefs a few kilometers across. Fringing reef complexes are commonly much larger, being many tens of kilometers long and several kilometers wide (Mexico's Golden Lane; Guzmán 1967). *En echelon* vertical stacking of reef complexes, as in the Western Canada Basin, may produce even wider belts (Barss et al. 1970). If we take the example of a pinnacle reef some 500 m in diameter and 500 m in height, inclosed in sealing basinal muds, an estimate of the recoverable volume of oil (reserves) can be made. The following factors will be used for the calculation:

- Reservoir flat lying and full to spill (area = $(500/2)^2 \times \pi$; height = 500);
- net to gross (n:g) = 0.8;

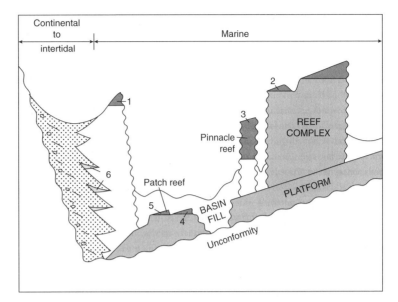

Fig.4.23 About 90% of the Devonian oil reserves of western Canada occur in transgressive-phase reef complexes. The reservoir distribution commonly controls the trap geometry. This schematic figure shows the reservoir and trap types. (From Prodruski et al. 1988; reproduced with the permission of the Minister of Public Works and Government Services Canada, 2001, and courtesy of the Geological Survey of Canada.)

Trap types:
1 Updip termination of large reef complex
2 Channel within reef complex
3 Pinnacle reef
4 Updip termination of platform
5 Patch reef on platform
6 Subtle facies change in reef complex

• porosity average 25% ($\Phi=0.25$);
• matrix permeability high (about 100 mD), with additional conductive fractures;
• water saturation average 25% ($S_w=0.25$, $S_o=0.75$);
• light oil (about 35° API);
• formation volume (shrinkage) factor (fvf) 1.3;
• recovery factor (rf) 40% based on initial gas expansion followed by water injection;
• there are approximately 6.29 barrels in 1 m³.
Therefore

standard barrels of oil originally in place (STOOIP) = area × height × n:g × F × S_o/fvf × 6.29 = 71 mmstb

and

reserves = STOOIP × rf = 28 mmstb

With potential reserves at 28 million barrels, the pinnacle reef would have to be close to market and inexpensive to drill for it to be an economically viable target. For example, such a reef would probably be drilled in Texas but not in Alaska.

Deep-water carbonates

Reservoirs in deep-water carbonates are not so common as those in shallow-marine systems. Nevertheless, there are a few important instances in which deep-water carbonates form major reservoirs. The carbonate equivalent of clastic submarine-fan systems is to be found in detritus shed from platform carbonates into deep water. The Poza Rica Trend of onshore eastern Mexico (Enos 1988) consists of oilfields that have developed in carbonate turbidites and other resedimented deposits that are believed to have been shed from the adjacent carbonate platform. That adjacent platform is the world famous Golden Lane Trend of Mexico. Within the Poza Rica Trend, the reservoir has been much improved following dissolution of abundant and formally aragonitic rudist bivalve shells.

The Upper Cretaceous pelagic carbonates, the Chalks, of northern Europe act as reservoirs for an estimated 350×

10^6 Sm^3 (in "standard cubic meters"; equivalent to 2.2 Bbbl, "billion barrels") of recoverable oil and 275×10^9 Sm^3 (9.7 Tcf, "trillion cubic feet") of recoverable gas in the Norwegian Sector of the North Sea (Spencer et al. 1987). Much of this oil and gas occurs in the fields associated with the Ekofisk complex, which was the first commercial discovery in the North Sea, and the first giant oilfield to be discovered offshore Norway (D'Heur 1984). The Chalks comprise skeletal debris of the nanoplanktonic coccolithospheres, which can be highly porous, but commonly permeability is low due to the small pore-throat size. Locally, both resedimented intervals and/or fractures can lead to large improvements in reservoir quality.

The journey of our sand grain, be it quartz or skeletal carbonate, is complete. The discussion of carbonate reservoirs is incomplete. Both dolomitization and karstification can lead to the creation of reservoir (porosity) where once there was tight rock. Dolomite is calcium magnesium carbonate ($CaMg(CO_3)_2$) and calcite calcium carbonate ($CaCO_3$). The chemical processes will be discussed in the following chapter, and the resultant reservoirs here.

Dolomite

Producing dolomite reservoirs occur in most continents and reservoirs range from Precambrian (Tong & Huang 1991) through Paleozoic to Mesozoic and Tertiary in age (Sun 1995). Zenger et al. (1980) estimate that about 80% of the recoverable petroleum in carbonate-hosted reservoirs of the USA occurs in dolomite and only about 20% in limestone. The same ratio probably applies to the producible reservoirs in the Permian Zechstein of Europe (England to Poland; Taylor 1981), while older carbonate plays in Europe and Russian Asia are almost wholly dolomite. Dolomitization and karstification control the productivity of carbonate reservoirs (Late Devonian to Permian) of the Timan–Pechora Basin on Russia's northern coast (Gérard et al. 1993). For the dolomites, porosity ranges from 8% to 23% and permeability up to 2 D, while adjacent limestones are tight.

Many dolomite or dolomitic limestone reservoirs are intimately associated with evaporitic rocks. Both shallow-water tidal flat/lagoonal and deep-water, basin-evaporite associations are common. The Upper Jurassic dolomite reservoirs of the Arabian platform and the Cretaceous dolomites of West Africa are well-described examples of the shallow-water association (Brognon & Verrier 1966; Wilson 1985; Baudouy & LeGorjus 1991). Dolomite reservoirs in peritidal-dominated carbonate are typical of geologic periods without major continental glaciations, such as the late Precambrian–Middle Ordovician, Late Silurian, and Late Triassic–Early Jurassic

Fig.4.24 Tower karst within Paleozoic limestones, Zhoaqing, Guangdong Province, China. The precipitous hills contain vuggy and cavernous porosity within an otherwise low- (near-zero) porosity limestone. (Photograph by J.G. Gluyas.)

periods (Sun 1995). During such periods, small sea-level fluctuations lead to major inundation of shallow platforms. Examples of dolomitic reservoirs that are not associated with evaporites occur in the Silurian of the Anadarko Basin, USA (Morgan 1985), the Devonian of the Alberta Basin, Canada (Phipps 1989), and the Carboniferous of the Illinois Basin, USA (Choquette & Steinen 1980).

Karst

Karstified limestones and dolomites represent the second major group of carbonate reservoirs not directly linked to depositional environment. Karst is a product of mineral dissolution (Fig. 4.24). It develops where carbonates are exposed to meteoric water, often linked with episodes of sea-level fall. Karst features are—caves, collapse breccias, dissolution enhanced joints and fractures, and vugs—well known to geologists and geographers alike. The degree to which karst will create porosity is a function of the volume of water flushed through the system and the susceptibility of the rock to penetration of the water through preexisting matrix porosity and fracture systems.

The presence of karst does not guarantee a petroleum reservoir. The karstic surface may not survive reburial. Infilling of the karstified surface by muds deposited by the transgression, which finally floods the exposed carbonate, may destroy reservoir potential totally.

Large dissolution features can also be created by circulation of hot basinal (mineralizing) brines. Porosity created by this process is commonly referred to as thermal karst, and oil production is known from such thermal karst.

A number of examples have already been cited where karstification has contributed to the viability of a carbonate reservoir. They include the Liuhua Field in the South China Sea (Tyrrell & Christian 1992; Case history 3.9), the Permian reservoirs of Texas and New Mexico (Craig 1990), and parts of the Upper Permian in the Zechstein Basin in Europe (Clark 1980). The first two oil-producing fields to be developed in the North Sea (Argyll and Auk, each with reserves of about 100 mmbbl) have karstified and brecciated, Zechstein dolomite reservoirs (Robson 1991; Trewin & Bramwell 1991).

Perhaps the most spectacular of the karst reservoirs are the so-called buried hills. The buried hills are in fact remnant topography (irregular unconformity surfaces) that has been buried beneath younger sediments. Many of the onshore Chinese basins contain buried hill reservoirs. Tong and Huang (1991) describe examples from the Damintun depression in northern China, where oil is reservoired in a combination of dissolution-enhanced fractures in Proterozoic dolomites and fractured metamorphic rocks. The Casablanca Field offshore Spain contains about 80–90 mmbbl of reserves in a karstified Upper Jurassic limestone "buried hill." The porosity in the reservoir is only about 3%. However, weathering during the Tertiary increased the porosity and permeability in a zone 90–150 m (300–500 ft) beneath the surface of the limestone, through the creation of vugs and the enhancement of fracture porosity (Watson 1982).

4.3.6 Fractured reservoirs

Open fractures contribute in a variety of ways to both the viability and performance of reservoirs. A fracture system may contain all of the pore volume for the reservoir as well as controlling the permeability, or provide permeability for a porous but otherwise low-permeability reservoir. Finally, open fractures can enhance the permeability of an already permeable reservoir. Conversely, closed fractures and faults with clay smear or nonreservoir-to-reservoir juxtaposition will increase the compartmentalization in a reservoir. In the following discussion, we concentrate our efforts on reservoirs where open fractures either constitute the reservoir porosity or provide permeability to an otherwise unproducible reservoir. Field segmentation due to fracturing and faulting is reviewed in Section 5.4.

Open fractures, and dissolution-enlarged fractures and vugs, contribute almost all of the pore space in the Albion–Scipio and Stoney Point fields of the Michigan Basin, USA (Hurley & Budros 1990). The field geometry is completely defined by the reservoir geometry. The reservoir is about 50 km long, 1.6 km wide, and up to 150 m thick (orig-

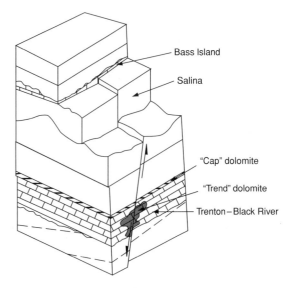

Fig.4.25 A schematic diagram showing the structural-stratigraphic interpretation of the Albion–Scipio Trend. The unconformity on top Bass Island removed any relief generated movement after the Salina Fault. The whole of the trap is defined by the fractured and dolomitized portion of the Trenton–Black River limestone. Vertical scale about 900 m. (From Hurley & Budros 1990.)

inal oil in place = 290 mmbbl, original gas in place = 276 bcf). The fractured, porous, and permeable Ordovician dolomite reservoir is set in tight limestone (Fig. 4.25). Fracturing and associated dolomitization are associated with left-lateral wrench faulting in the underlying Precambrian crystalline basement. The dissolution and dolomitization processes were probably associated with the introduction of hot saline fluids; that is, thermal karst.

The storage volume in open fractures in the giant Amal Field (Libya, 1.7×10^9 bbl reserves) is much greater than that contributed by the tight Cambrian quartzite matrix. Similarly, the Furbero Field in the southwestern part of the Poza Rica Trend (Mexico) is reservoired in baked and fractured shales, created by the intrusion of gabbro dykes (Tiratsoo 1984).

There are a large number of fields in which fracture volume does not contribute significantly to total hydrocarbon volume, but where fractures have a significant impact on reservoir producibility. Enhancement of producibility may occur because open fractures connect thin and otherwise poorly connected reservoir layers. Alternatively, the large surfaces of the fractures may allow oil to be drained from a low-permeability reservoir.

The Oligocene–Miocene Asmari Limestone reservoir of

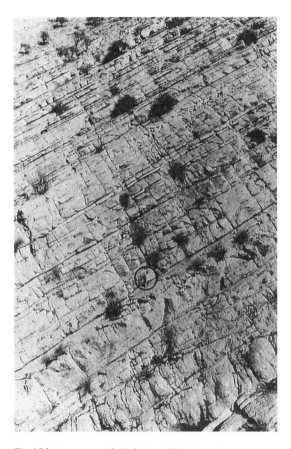

Fig.4.26 The southwest flank of Kuh-e Pabdeh-Gurpi, Zagros Mountains, Iran: upper beds of Asmari Limestone, showing the bedding plane distribution and related variations in fracture density (man circled for scale). (From McQuillan 1973.)

the Zagros fold belt in Iran produces from fractures that tap into oil with matrix porosity (Fig. 4.26). The productivity of the fields that have Asmari Limestone reservoir is phenomenal. In the mid-1970s, the output from the eight largest fields averaged more than 5 million barrels a day from only 378 producing wells (Tiratsoo 1984). Narrow 70–100 μm hairline fractures within the uppermost 1–5 m of the Lisburne Field reservoir (Alaska, USA) have been estimated by Belfield (1988) to effect an increase in production up to ten times that to be expected from the matrix permeability alone. Rapid inter-well pressure responses during an interference test on the field indicated that although the fractures are extensive, their volume is small in comparison with that of the pore space in the matrix. The host reservoir at Lisburne is layered limestone and minor dolomite of Carboniferous age.

4.4 MIGRATION

4.4.1 Introduction

Migration is the process (or processes) whereby petroleum moves from its place of origin, the source rock, to its destruction at the Earth's surface. Along the route, the petroleum's progress may be temporarily arrested and the petroleum may "rest" on its journey within a trap. The process of migration may be divided into three stages (Fig. 4.27):
• Primary migration—expulsion of petroleum from the source rock;
• secondary migration—the journey from source rock to trap;
• tertiary migration—leakage and dissipation of the petroleum at the Earth's surface.

4.4.2 Primary migration

Many different mechanisms have been proposed to account for the way in which petroleum moves out of a source rock and into a reservoir or carrier bed. The various proposed mechanisms may be divided into two broad categories: those in which petroleum or precursor petroleum products are assisted in their escape from the source rock by water; and those mechanisms that invoke migration of petroleum as a separate phase, independent of any associated water movement. Indeed, immobilizing or removing the water becomes an important prerequisite of the mechanisms that involve migration of petroleum as a separate phase.

Primary migration, if water-controlled, could include the following mechanisms: solution of oil in water, solution of gas in water, solution of nonoil organic molecules in water, micellar formation, and emulsion of oil in water. Movement of the solution/emulsion would then occur through one or more of the following: diffusion, convection, meteoric water movement, compaction-induced water movement, and release of water during clay dehydration.

The other possibility, that primary migration occurs as a discrete petroleum phase, can also be explained by a variety of mechanisms. The petroleum could occur in several forms: as oil as a single phase, as a solution of oil in gas, or as a solution of gas in oil. The transport mechanisms could then include migration within the kerogen network of the source rock or migration within the mineral network of the source rock. The driving force for discrete petroleum-phase migration would be the pressure differential between the source and carrier bed, acting with or without capillary imbibition at the source rock/carrier bed boundary.

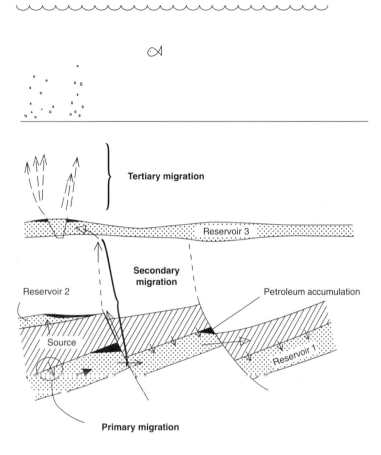

Fig.4.27 A sketch of the three stages of migration: primary migration out of the source rock and into a trapped reservoir (reservoir 2) or into a carrier bed (reservoir 1); secondary migration in carrier reservoir 2 and up faults into reservoir 3; and tertiary migration (dissipation) from reservoir 3 to the surface.

It was not until the final decade of the 20th century that a consensus began to emerge as to which of the above groups of mechanisms was the most likely explanation for primary migration. Most researchers now favor hypotheses in which petroleum is expelled from the source rock as a separate phase within a water-wet rock matrix (England & Fleet 1991). The important elements in the arguments against petroleum migration in aqueous solution are that the solubility of most petroleum constituents are too low to transport the required masses, and that a solution process would enrich reservoired petroleum in the soluble aromatic compounds, whereas the reverse is observed. The other possible water-controlled processes are also difficult to envisage. Natural soaps are needed to promote micelles, which are rare and, like emulsions, would be essentially immobile in the source rock if they formed, due to their large size relative to the small pore-throat size in the source rock.

Methane is one exception to the general statement that the solubility of the major constituents of petroleum in water is low. Methane is soluble, particularly under the conditions of high temperature and pressure found in the deep subsurface. For example, many shallow gas deposits migrated as dissolved gas, with exsolution of the gas occurring in the near subsurface at low pressure and temperature.

BP Research did much work on primary migration during the 1980s. Some of the important breakthroughs in understanding the primary migration process were hastened at the end of 1982, when the workers at the research center discovered that core had accidentally been cut in an actively expelling source rock in one of the BP exploration wells. UK continental shelf well 16/7b-20 was the discovery well for the giant Miller Field (Gluyas et al. 2000). The expected top reservoir was at about 3900 m (below rotary table). Shows were encountered at about that depth and a core-barrel run in hole. Four cores were cut in a heterolithic sequence of thin (cm–dm) sandstone turbidites and dark gray to black source-

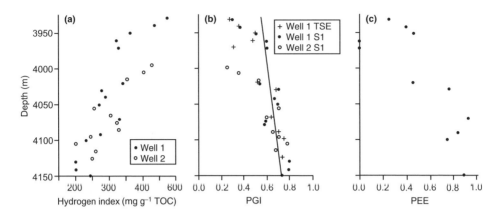

Fig.4.28 (a) The average hydrogen index (mg g^{-1} TOC) as a function of depth, Kimmeridge Clay Formation, Miller Field, UK continental shelf. The hydrogen index was measured every 2 m: the points shown are averages for each 10 m interval (five samples), plotted at the mid-point of the range. (b) The average petroleum generation index as a function of depth (calculated using both S_1 pyrolysis and TSE methods for well 1). (c) Average petroleum expulsion efficiency (PEE) as a function of depth for well 1, calculated using TSE data. (From Mackenzie et al. 1987.)

rock shales of the Kimmeridge Clay Formation. Most of the turbidite beds were oil stained. Thick sandstones comprising the producible reservoir *sensu stricto* were encountered and cored at 4047 m (below rotary table). Further cores were cut at an equivalent stratigraphic position in the first appraisal well (16/7b-23). They too were mudstone dominated.

Mackenzie et al. (1987) report the results of an elegant set of experiments performed on these Kimmeridge Clay Formation cores, and designed to elucidate the primary migration process. The key pieces of information were as follows:

• The hydrogen index (HI) decreased systematically in both wells over the sampled range (3850–4150 m, Fig. 4.28);
• The petroleum generation index (PGI) increased systematically over the same depth range (Fig. 4.28);
• Petroleum expulsion efficiency (PEE) increased systematically like PGI (Fig. 4.28).

Taken together, the three observations are powerful arguments to support the hypothesis that in the sampled area the maturity of the source rock, and hence petroleum generation, increase with depth and temperature.

A more detailed set of experiments was then carried out on mudstone and sandstone couplets. Analyses were made at the bases of thick mudstones, at their interface with sandstones, and on relatively thin mudstones (2–3 m) sandwiched between sandstones. The results are presented graphically in Fig. 4.29. In general, these detailed experiments demonstrated that the margins of mudstones are especially depleted in C9 and heavier components, while the reverse was true for C5–C8 components. The edges of mudstones

had either similar or enhanced concentrations of the light components.

In order to link these observations on the distribution of petroleum components with a mechanism for primary migration, further observations were made upon the porosity and pore-size radii of the mudstone. Typically, the porosity was in the range of 2–3% and the pore radii were 3–4 nm. Such pore radii are approximately five times greater than most of the component molecules of the petroleum. From these data, Mackenzie et al. (1987) said there was no reason to invoke fractures for expulsion to occur. This particular issue remains contentious. Other workers would argue that primary migration is effected through microfractures produced as a direct result of the volume, and hence pressure, increase associated with petroleum formation.

Mackenzie et al. (1987) went on to point out that their results could best be explained by pressure-driven flow of a petroleum-rich phase as the main expulsion mechanism for source rocks. Specifically, they demonstrated that petroleum was first expelled when the volume of generated petroleum approximately matched the volume of the pore space within the mudstone. In other words, the mudstone was almost fully saturated with petroleum before expulsion occurred. This tended to support earlier observations on lean source rocks. Cooles et al. (1986) noted that source rocks with less than 5 kg t^{-1} of initial petroleum yield will not achieve sufficient saturation for expulsion to occur.

Pepper (1991) has taken the primary migration story one step further. He was concerned about the differing behavior of

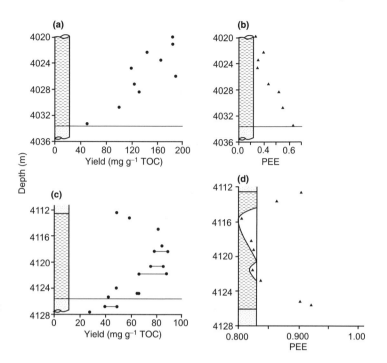

Fig.4.29 (a) The total soluble extract yield relative to TOC (mg g⁻¹ TOC) for the base of a thick mudstone in well 1. (b) The total soluble extract yield relative to TOC (mg g⁻¹ TOC) for a thick mudstone between two sandstones in well 2. (c) The petroleum expulsion efficiency for the thick mudstone in well 1. (d) The petroleum expulsion efficiency for the thick mudstone in well 2. (From Mackenzie et al. 1987.)

different source rocks with similar labile kerogen contents. The simple kinetic model dictates that such source rocks should deliver petroleum of similar compositions over similar temperature ranges. Yet this is commonly not the case. Instead, Pepper (1991) proposed that the expulsion behavior of a source rock was controlled by the relative proportions of petroleum-generative kerogen and petroleum-retentive kerogen (Figs 4.30 & 4.31). For example, humic coals would tend to expel petroleum products of higher maturity and at higher temperatures than a clastic source rock containing the same labile kerogen components. The observations expressed by the data in Fig. 4.31 are of course completely opposite to the proposal made ten years previously; that is, that kerogen could act as a conduit for petroleum escaping from the source.

Gas expulsion may occur in a similar fashion to that of oil, albeit at higher temperatures. Clearly, the volume increase associated with gas generation is massive, be it directly from kerogen or from thermal decomposition of previously formed oil. Pressure-driven expulsion will occur either through the existing pore network or through induced fractures. During gas generation, previously generated, short-chain liquid hydrocarbons may become dissolved in the gas and expelled with it. This mechanism has been used to explain the production of condensate from Type III kerogen in overpressured mudstone (Leythauser & Poelchau 1991).

4.4.3 Secondary migration

Secondary migration takes petroleum from the source location to trap or traps via carrier beds. The defining aspect of secondary migration is that it concentrates or focuses the petroleum. On escape from the source rock, petroleum is dispersed over a large area. By the time it reaches the relatively restricted area of a trap, it can occupy in excess of 90% of the pore volume in the reservoir. Secondary migration is temporarily arrested once the migrating petroleum enters a trap. Disruption of the trap or over-filling of the trap can lead to remigration of the petroleum to a higher structural level under the same secondary migration process. Such secondary migration ends when the petroleum approaches the Earth's surface.

The medium through which the petroleum travels during secondary migration is also quite different from that of the source rock. The pore size and thus the permeability in a carrier bed — be it a sandstone, a carbonate, or a fractured lithology — is much larger than that in a source rock. The movement of petroleum in carrier systems is very similar to its movement from reservoir to wellbore during petroleum production. In consequence, the mathematics used by reservoir engineers to describe the production process can be applied to secondary migration.

The driving mechanism for secondary migration is the

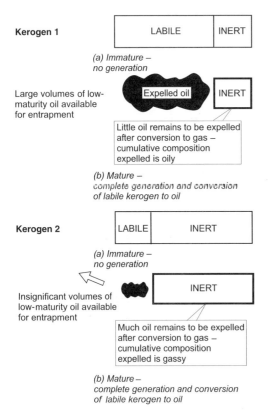

Fig.4.30 An outline of Pepper's (1991) hypothesis linking petroleum expulsion efficiency to the abundance and nature of the retentive kerogen within a source rock. (From Pepper 1991.)

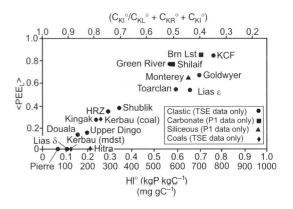

Fig.4.31 Average petroleum expulsion efficiency plotted against hydrogen index (HI°) for 20 source rocks from around the world. There is a high degree of positive correlation between the expulsion efficiency and the initial organic matter quality, irrespective of the absolute initial petroleum potential. The kerogen-linked petroleum expulsion hypothesis predicts that rocks with very different organic matter contents will have similar expulsion efficiencies provided that they share a common organic matter type. (From Pepper 1991.)

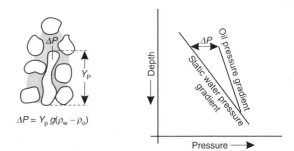

Fig.4.32 Buoyancy as a driving force for secondary migration. Buoyancy is the pressure difference between a point in the petroleum column and the surrounding pore water. It is a function of the density difference between the petroleum and the water and the height of the petroleum column. A large buoyancy pressure may develop at the tops of large, low-density (gas) petroleum columns. Pressure measurements throughout the petroleum column define the petroleum pressure gradient. This intersects the hydrostatic gradient at the petroleum/water contact. Y_p = height of petroleum column, g = acceleration due to gravity, ρ_w = subsurface density of water, ρ_p = subsurface density of petroleum. (From Allen & Allen 1990.)

density difference between the petroleum (less dense) and water (more dense). The density difference is expressed through the buoyancy force generated by the pressure difference between a point in a continuous petroleum column and the adjacent pore water (Fig. 4.32):

$$\text{buoyancy force } \Delta P = Y_p g(\rho_w - \rho_p)$$

where Y_p is the height of the petroleum column, g is the acceleration due to gravity, ρ_w is the subsurface density of water, and ρ_p is the subsurface density of petroleum.

In hydrostatic situations—that is, where no water movement is involved—we need only consider this buoyancy force. Under conditions of water flow (hydrodynamic) through a secondary migration route, the movement of petroleum will be modified. The interaction between the buoyancy driving the petroleum and the flowing water may retard or enhance migration. Indeed, the direction as well as the rate of petroleum flow can be modified. Such directional modification in

a hydrodynamic system can lead to a different distribution of oil pools than would be generated in a hydrostatic system. Moving water will also affect the distribution of petroleum in an individual trap, and may lead to tilted fluid contacts.

We have spoken of the buoyancy driving force for petroleum. The restricting force to petroleum migration is capillary injection pressure. A slug of petroleum migrates from pore to pore in a carrier bed, squeezing through the intervening pore throats. The force required to move petroleum through a pore throat is a function of the radius of the pore throat, the interfacial tension between the petroleum and the water, and the wettability of the rock–petroleum–water system:

$$\text{displacement pressure} = \frac{2\gamma\cos\theta}{R}$$

where γ is the interfacial pressure between petroleum and water (dyne cm^{-1}), θ is the wettability, expressed as the contact angle of the petroleum/water interface against the rock surface (degrees), and R is the radius in the pore (cm).

Interfacial tension is a function of the compositions of water and petroleum, not the rock. The interfacial tension is lower for high-gravity, volatile oils than it is for lower-gravity oil. Interfacial tension also decreases as temperature increases.

The interfacial tension between gas and water is commonly higher than for oil–water systems. This means that displacement pressures are higher for gas relative to oil in the same rock. The opposite is commonly true with respect to buoyancy. It is commonly much greater for gas than for oil. This combination of factors, all of which change as a function of temperature or pressure and hence depth, but at different rates, does lead to unexpected consequences at shallow depth. For example, gas may be trapped beneath a mudstone at shallow depth, but the same mudstone will fail to trap oil (Fig. 4.33).

So far, we have only discussed the physics of secondary migration, and in particular how petroleum leaves one pore and enters another. The macroscale comprising carrier beds, reservoirs, and basins is of course much more complex, although the lessons learned at the microscale allow us to understand the way in which petroleum migrates from source to trap.

The buoyancy effect means that petroleum will tend to rise within the sediment column. The capillary effect dictates that, in the absence of other forces, petroleum will migrate from a small pore to a large pore. Furthermore, petroleum (and water) will attempt to equilibrate with respect to pressure. That is, flow can be induced by pressure differential (either overpressure or hydrodynamics).

It is possible to estimate the likely migration directions from source bed to reservoirs by mapping the orthocontours of the likely carrier systems (Fig. 4.34). Orthocontours are simply lines constructed on a map at right angles to the contours. Instead of displaying areas of equal height (or depth), they depict lines of maximum dip. The buoyancy effect dic-

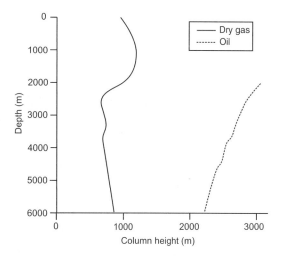

Fig.4.33 The decrease of oil column height with depth (contrary to popular opinion), despite the fact that the mudstone seal compacts with burial to give smaller pore throats. This is due to the reduction in both interfacial tension and oil density with depth (temperature). Temperature and pressure effects on gas/water interfacial tension also produce an unusual curve for the gas column height as a function of depth. Near the surface, the pore-throat radius of the compacting mudstone decreases faster than the interfacial tension, and the sealing capacity increases. Below this, to about 2500 m, the decrease in interfacial tension is more important and the seal capacity decreases. The curve explains why shallow gas pockets are so common, gas is trapped most efficiently between 500 and 1500 m. The seal capacity is a function of many factors, all of which vary with depth at different rates. As a consequence, simple trends in petroleum column height should not be expected. (Pers. comm., C.J. Clayton 1990.)

tates that the rising petroleum will follow such orthocontours. Clearly, such an exercise must be attempted on the geometry of the carrier bed(s) as it was during the phase of petroleum migration. This clearly leads to attempts to reconstruct the basin history in terms of deposition, structuring, and source-rock maturation. The product of these complexities, coupled with the difficulty of mapping the subsurface, means that while a general understanding of migration directions is possible for most basins, it is commonly very difficult to make an accurate assessment as to whether individual petroleum prospects lie on or off migration routes. This can be further exacerbated if the basin is believed to have been subjected to a phase of hydrodynamic flow that modified the distribution of the petroleum migration pathways.

The capillary effect controls how much of a carrier bed becomes petroleum saturated. Rarely are carrier beds of uniform grain-size distribution. Thus petroleum will tend to migrate along the coarsest, high-permeability pathways (Fig. 4.35).

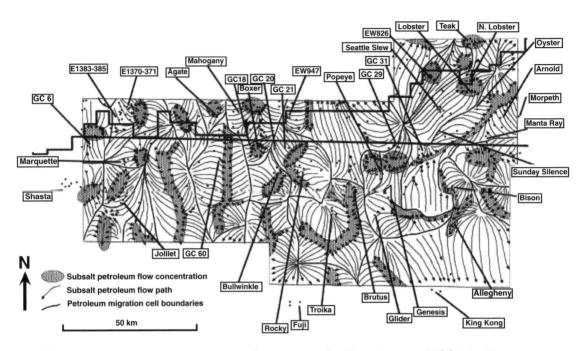

Fig.4.34 Reconstructed subsalt petroleum migration pathways from the Ewing Bank and Green Canyon areas, Gulf of Mexico. The arrows (orthocontours) show flow paths up the dip of the interpreted base salt. Zones of flow concentration are highlighted, with flow cells highlighted in heavy lines. The locations of fields and discoveries are shown to illustrate their correspondence to original subsalt petroleum migration concentrations. (From McBride et al. 1998.)

This may occupy 10% or less of any particular formation. Open fractures have the same effect as coarse beds. Petroleum will exploit them. Temporarily open fracture systems are commonly invoked as the mechanism whereby migrating petroleum "jumps" upward in the stratigraphy of a particular basin.

Although the calculations performed by England et al. (1991) indicate that only a small portion of a carrier bed is exploited by migrating petroleum, the total volume of the rock through which petroleum passes from source to trap can become large if the migration distances are large. This has consequences for the quantity of petroleum "lost" during migration. Migration losses are defined as the volume of petroleum generated minus the volume of petroleum trapped (including petroleum that reaches the surface). Such petroleum ends up in minitraps within the carrier bed. Such minitraps can vary in size between individual dead-end pores and small-scale structural closures on the top of the carrier bed. It is probable that there may be many millions of these pore-sized traps along any migration route.

England et al. (1991) have estimated that total migration losses may be as little as 5%, based on typical migration distances. They point out that there is a finite distance over which petroleum will migrate before there is insufficient input of petroleum from the source to generate output of petroleum into the trap. How far petroleum can migrate is a question that has taxed explorers for decades. Very long migration distances, well in excess of 100 km, have been calculated for oil accumulations in some foreland basins. In these instances, development of the basin continuously "chases" the petroleum to the foreland. In contrast, 20 km is often used as a rule of thumb for petroleum generated in small lacustrine basins although, indeed, sometimes such rules of thumb can be spectacularly wrong (Case history 3.9).

The rate at which petroleum migrates can be calculated using Darcy's Law:

$$q = -\frac{k}{\mu}\frac{\mathrm{d}\theta}{\mathrm{d}z}$$

where q is the volume of flow rate ($\mathrm{m}^3\,\mathrm{m}^{-2}\,\mathrm{s}^{-1}$), k is the permeability, μ is the viscosity ($\mathrm{Pa\,s}^{-1}$), and $\mathrm{d}\theta/\mathrm{d}z$ is the fluid potential gradient.

Fig.4.35 Petroleum migration along high-permeability sandstone beds within a stacked sequence of turbidite sandstones and siltstones. The migration route was exposed during the excavation of a road cutting in Ecuador. (Photograph by M. Heffernan.)

Typical permeability values are as follows:

sandstones $10^{-12}-10^{-15}\,\mathrm{m^2}$ (1 D–1 mD)

limestones $10^{-14}-10^{-17}\,\mathrm{m^2}$ (10 mD–10 µD)

From these data, it is possible to calculate that the migration rate for petroleum in sandstone is 1–1000 km per million years, and in limestone 0.01–10 km per million years (England et al. 1991). One further aspect needs to be considered when secondary migration processes are being evaluated. So far, we have concentrated our description on the physical process of migration, how the petroleum moves through the pore space. In addition to the movement of petroleum through the rock, there are commonly chemical changes to the composition of the petroleum. Such changes are commonly divided into three types.

Phase changes will occur in petroleum as a result of its mi-gration upward to regions of lower pressure and temperature. This is most important for high-temperature and -pressure condensates, but any oil will exsolve some gas if the pressure in the formation drops below the bubble point. The residual petroleum and generated gas are then likely to behave differently with respect to subsequent migration.

At low temperatures (<70°C) and in regions where there is significant water flow, petroleum may be degraded by bacterial action or by water-washing. The bacterial process follows a systematic loss of the n-alkanes, branched alkanes, isoprenoids, alkylcyclohexanes, and polycyclic alkanes. This progressive destruction of the petroleum leads to increases in the pour point and viscosity of the oil and a lowering of the API gravity.

The physical process of migration can lead to chromatographic separation of the components of the petroleum, the rock itself being the stationary phase. This process has been referred to as geochromotography.

4.4.4 Tertiary migration

Tertiary migration includes leakage, seepage, dissipation, and alteration of petroleum as it reaches the Earth's surface. In Chapter 3 we examined the products of such processes, particularly insofar as they might be useful as direct petroleum indicators in a poorly understood basin. To recap, the products of seepage may be gas chimneys in the shallow sediment, gas hydrate layers and mounds, cemented pock marks and mud volcanoes, effects on vegetation and live oil, and gas seepage at the surface. For more information on these products, the reader should refer to Chapter 3. Here, we examine briefly the tertiary migration processes.

The physical processes that drive tertiary migration are the same as those that operate during secondary migration. Buoyancy drives the petroleum to the surface. This may be helped or hindered by overpressure gradients or hydrodynamics. Perhaps the only major difference that can be used to separate tertiary migration from secondary migration is the rate of petroleum supply. Trap failure, through capillary leakage, hydraulic fracture, or tectonism, supplies petroleum into a new carrier system much more rapidly than does a maturing source rock.

The direction of tertiary migration can be vertical, horizontal, or some combination of both. However, there are some simple guidelines that will help the explorer to understand what is happening in his or her basin of study. Vertical migration is promoted by the presence of thick mudstones, rapid deposition (causing high overpressures), and a paucity of extensive dipping carrier beds (sandstones or carbonates).

Tilted sandstone and carbonate beds will promote lateral migration and focusing of petroleum. This in turn will lead to seepage over much more restricted areas, as the petroleum breaks out of a carrier bed rather than seeping pervasively through a seal. As a gross generality, foreland basins are characterized by well-developed lateral migration and highly focused areas of leakage (e.g., the Faja of Venezuela, Wolf Lake, and the Athabasca tar sands of Canada). Using a similarly gross generality, rift basins and passive margins tend to have better developed vertical migration (e.g., the Gulf of Mexico and the North Sea).

4.5 THE TRAP

4.5.1 Introduction

There are many ways in which petroleum may be trapped. The criteria whereby the migration of petroleum to the Earth's surface can be arrested (albeit temporarily) have already been discussed. Much has been written about traps, both how to find them and how to classify them. Various approaches have been used for trap classification. North (1985) summarizes much of what has been written and discusses the merits of geometrical and genetic classifications. His preference for use of a very simple, high-level, geometric classification, of convex and nonconvex traps, works well if you, as the explorer, are prepared to use the simplest of exploration techniques; searching for the "big bumps." Here we use the three-fold classification as adopted by Allen and Allen (1990) and by volumes in the Treatise of Petroleum Geology Atlas of Oil and Gas Fields series, published by the American Association of Petroleum Geologists. The three trap categories are structural, stratigraphic, and hydrodynamic (Table 4.3). This classification is a pragmatic one. It stimulates the explorer to try to understand his or her basin before populating it with traps. However, it does not pretend to present a perfect subdivision of the world's traps.

Structural traps may be generated through tectonic, diapiric, compactional, and gravitational processes (Fig. 4.36). In broad terms, the range of structural traps in a basin can be deduced from knowledge of the basin evolution. Structural traps occur within the Zagros fold belt of Iran, the El Furrial Trend (transpressional anticlines) of eastern Venezuela, the wrench systems of California, the tilted fault blocks of the Brent Province (North Sea), the diapiric (salt-associated) traps of the Texas Gulf Coast, the gravity-generated rollover anticlines of the Niger Delta, and the compaction-generated anticlines over basement highs of the Arabian Shield. Almost the entire world's discovered petroleum is in structural traps.

Table 4.3 Structural and stratigraphic traps. (Modified from Allen & Allen 1990.)

Structural	Tectonic	Extensional Compressional
	Diapiric	Salt movement Mud movement
	Compactional	Drape structures
	Gravitational	Listric fault generation
Stratigraphic	Depositional	Pinchouts (dunes, bars, reefs, channels, etc.) Unconformities (erosional subcrop, karst, etc.)
	Diagenetic	Mineral precipitation Mineral dissolution (thermal karst, dolomitization) Tar mats Permafrost Gas hydrate crystallization

Stratigraphic traps, sometimes referred to as subtle traps (Halbouty 1982), are formed by lithological variations imparted to a sediment at deposition or generated subsequently by alteration of the sediment or fluid through diagenesis (Fig. 4.37). Petroleum provinces with dominant stratigraphic traps include the Oficina Trend of Venezuela, in which much oil is trapped within fluvial channel sands, the Tertiary submarine-fan sands of the North Sea, the carbonate reef complexes of the Western Canada Basin, and the sub-unconformity traps of giant fields in the Lower 48 (East Texas) and Alaska (Prudhoe) fields of the USA. Diagenetic traps can be caused by mineralization, tar-mat development, permafrost, or gas hydrate formation. Much of the world's remaining undiscovered petroleum resource at conventional water depths and onshore will be found in stratigraphic traps.

Hydrodynamic traps are uncommon (Fig. 4.38). They are caused by differences in water pressures associated with water flow, which creates tilt on the contacts between hydrocarbons and water. Consequently, the fluid contacts in hydrodynamic traps do not correspond directly to structural maps of the reservoir. They are favored in, but not exclusive to, foreland basins.

The trapping mechanism for many fields is commonly a combination of structural and stratigraphic elements or, more rarely, structural elements and hydrodynamic

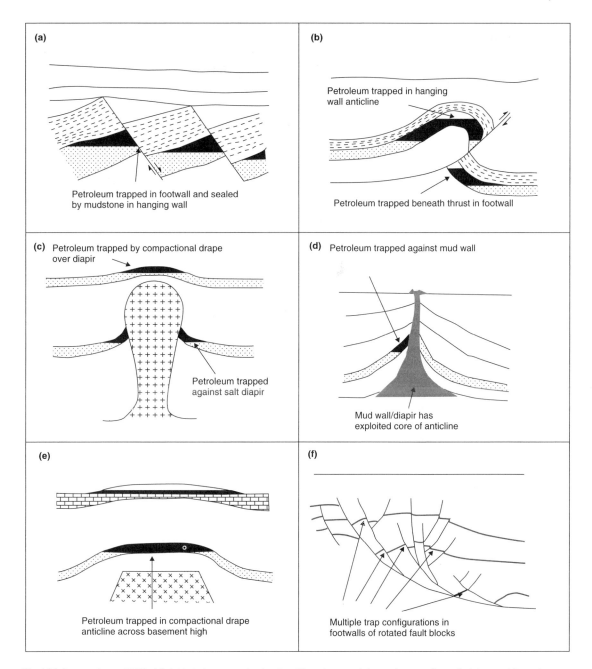

Fig.4.36 Structural traps. (a) Tilted fault blocks in an extensional regime. The seals are overlying mudstones and cross-fault juxtaposition against mudstones. (b) A rollover anticline on a thrust. Petroleum accumulations may occur on both the hanging wall and the footwall. The hanging-wall accumulation is dependent upon sub-thrust fault seal, whereas at least part of the hanging-wall trap is likely to be a simple four-way dip-closed structure. (c) The lateral seal of a trap against a salt diapir and a compactional drape trap over the diapir crest. (d) A trap associated with diapiric mudstone, with a lateral seal against the mud-wall. Traps associated with diapiric mud share many features in common with those associated with salt. In this diagram, the diapiric mud-wall developed at the core of a compressional fold. (e) A compactional drape over a basement block commonly creates enormous low-relief traps. (f) Gravity-generated trapping commonly occurs in deltaic sequences. Sediment loading causes gravity-driven failure and produces convex-down (listric) faults. The hanging wall of the fault rotates, creating space for sediment accumulation adjacent to the fault planes. The marker beds (gray) illustrate the form of the structure, which has many favorable sites for petroleum accumulation.

(a)

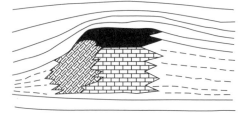

(b)

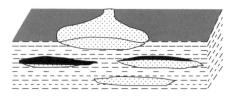

(c)

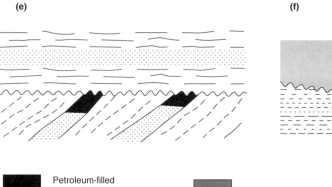

(d)

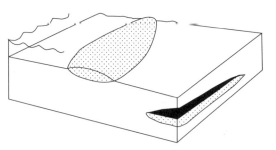

(e)

(f)

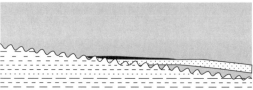

 ■ Petroleum-filled
reservoir

▦ Limestone–water-
bearing

▒ Sandstone–water-
bearing

▨
      } Nonreservoir lithologies
▬

Fig. 4.37 Stratigraphic traps. (a) A "reef" trap. Oil is trapped in the core of the reef, with fore-reef talus and back-reef lagoonal muds acting as lateral seals and basinal mudstones as top seals. (b) A pinchout (sandstone) trap within stacked submarine-fan sandstones. The upper surface of the cartoon shows the plan geometry of a simple fan lobe. The lateral, bottom, and top seals are the surrounding basinal mudstones. (c) A channel-fill sandstone trap. The oil occurs in ribbon-shaped sandstone bodies. The top surface of the cartoon shows the depositional geometry of the sandstone. The total seal may be provided by interdistributary mudstones, or by a combination of these and marine-flooding surfaces. (d) A shallow-marine sandstone bar completely encased in shallow-marine mudstone. The upper surface of the cartoon shows a prolate bar. (e) A sub-unconformity trap. The reservoir horizon is truncated at its updip end by an unconformity and the sediments overlying the unconformity provide the top seal. The lateral and bottom seals, like the reservoir interval, predate the unconformity. (f) An onlap trap. A basal or near-basal sandstone onlaps a tilted unconformity. The sandstone pinches out on the unconformity and is overstepped by a top-seal mudstone.

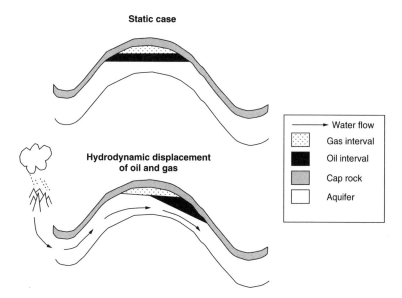

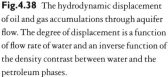

Fig.4.38 The hydrodynamic displacement of oil and gas accumulations through aquifer flow. The degree of displacement is a function of flow rate of water and an inverse function of the density contrast between water and the petroleum phases.

conditions. We will not specifically examine combination traps in detail, for that would be repetitious. However, a brief look at the Brent Province in the North Sea is worthwhile, since it serves to illustrate how subtle changes in the trap geometry between fields can lead to a range of trap definitions in what is clearly a single petroleum province.

The traps in this "Brent Province" illustrate the difficulty in creating a rigid nomenclature. The Brent Province comprises westerly-dipping fault blocks created during Late Jurassic rifting of the North Viking Graben. The prime reservoirs in the area are the Middle Jurassic, Brent Group deltaic sandstones. The rifting process led to the development of a number of trap configurations. The simplest features are the rollover anticlines, such as in the Hutton Field. Elsewhere in the Brent Group, sediments were lifted above sea level and erosion ensued. Such erosion combined simple planation of the rift shoulders and mass wastage of the easterly-directed scarp faces. Figure 4.39 combines three cross-sections across the simple structural closure of the Hutton Field (Haig 1991), the relatively simple stratigraphic (unconformity) closure of the Brent Field (Struijk & Green 1991), and the combined structural and stratigraphic elements of the Ninian Field (van Vessem & Gan 1991). Thus the Brent fields are divided between structural and stratigraphic traps despite sharing a common origin. Nonetheless, the division of the Brent traps into different types does force the geoscientist to try to understand what makes individual traps work (or fail), and it makes whoever searches for the next one that much the wiser.

4.5.2 Migration and trap formation

The timing of trap formation relative to that of petroleum generation and migration is critical. Clearly, if it is to be viable, the trap has to form at the same time or earlier than petroleum migration. In simple rift basins and passive margin basins, calculation of both the timing of trap formation and that of petroleum migration may be relatively straightforward. There are now a reasonable number of commercially available computer software packages which allow calculation of the structural and thermal histories of a basin, given input of some well data, seismic configurations, and source-rock characteristics.

Similar calculations for foreland and transpressional basins are likely to be much less precise, because such compressional systems commonly involve the destruction or severe modification of early-formed structures. This may lead to multiple phases of petroleum remigration. The destruction process may include that of the source rock.

4.5.3 Structural traps

Traps formed by compressive tectonic processes

Compressive tectonic regimes commonly lead to the development of large-scale contractional folds and thrusts. Such contraction is most common at convergent plate boundaries and transpressional strike-slip plate boundaries (wrench systems). Contractional folds form the trap geometries for a large

(a)

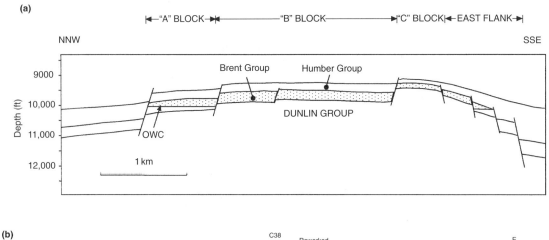

(b)

(c)

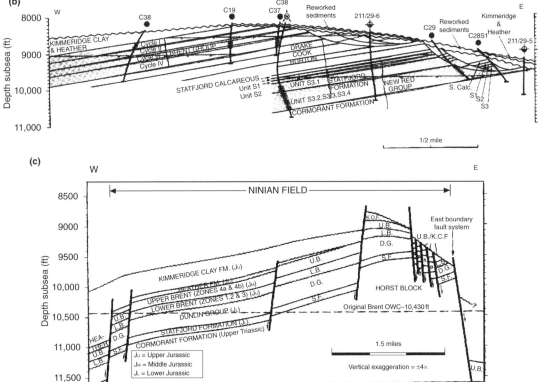

Fig.4.39 Cross-sections of the Brent Province field. (a) A geologic interpretation of a dip-oriented seismic line across the "B" block of the Hutton Field. The trap is structural and reliant on fault seal. (From Haig 1991.) (b) A cross-section of the Brent Field. The trap is stratigraphic and wholly dependent upon a sub-unconformity seal beneath Heather Formation or Kimmeridge Clay Formation mudstones. (From Struijk & Green 1991.) (c) A cross-section of the Ninian Field. The crestal block of the Ninian Field is fault-sealed, while the flanks have unconformity seals. (From van Vessem & Gan 1991.)

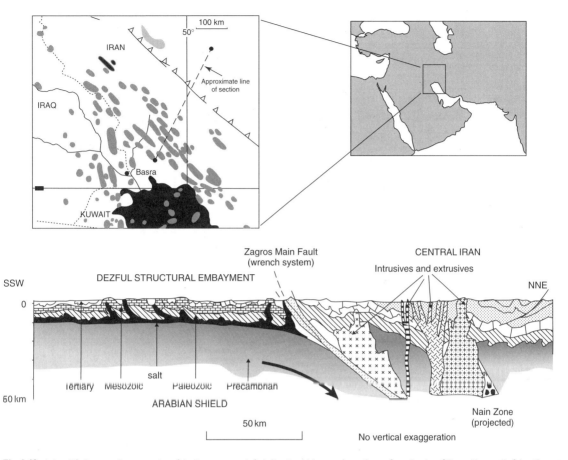

Fig.4.40 A simplified map and cross-section of the Zagros orogenic belt (Iran), which extends southwest from the city of Qum. Almost all of the oil-and gasfields lie within the area of the Dezful Structural Embayment, which is some 600 km long (NW–SE) and 200 km wide. Traps are compressional anticlines. (Modified after Beydoun 1991.)

part of the world's most important petroleum province, the Middle East. Here compression over the past few millions of years has led to the development of major anticlines in the Zagros Mountains of Iran. The Zagros folds have amplitudes of 3 km and axial lengths of 10–200 km (Tiratsoo 1984). These huge structures often have several pools that have developed in crestal culminations along their length (Fig. 4.40).

A second example of large-scale compressional folds and associated thrusts forming a major oil province is the El Furrial Trend of eastern Venezuela. The anticlinal traps of the trend were developed during convergence of the Caribbean and South American Plates during the Miocene and Pliocene. The trend was discovered in the early 1980s, soon after Venezuela nationalized its oil industry. The discovery was made by drilling deeper than the "traditional" plays (Neo-

gene deltaic sandstones; Aymard et al. 1990) and into folded and thrusted Early Tertiary and late Mesozoic strata, deposition of which predated formation of the foreland basin (Carnevali 1992). Most of the traps are large ramp anticlines (Fig. 4.41). They have oil columns with an average depth of 400 m (1300 ft) and reservoirs formed from high net to gross shallow-marine sandstones. For the four major fields in the trend (Fig. 4.41), reserves have been estimated at 3.9 Bbbl (oil) and 34 Tcf (gas).

The same tectonic trend continues from Venezuela into Trinidad, where compressional features occur in very young Miocene and Pliocene sequences. However, the trap geometries differ from those within the Early Tertiary of eastern Venezuela; they are more complex. The complexity is due to a combination of sedimentation simultaneous with

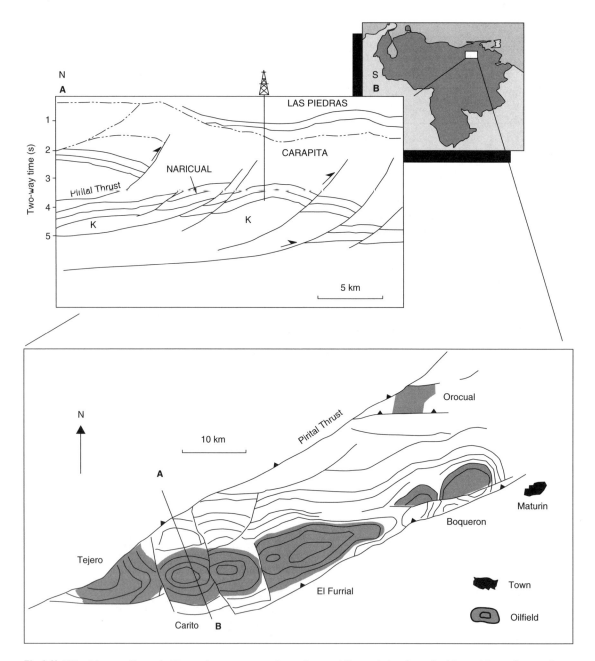

Fig.4.41 El Furrial, eastern Venezuela. The petroleum traps occur as large rollover anticlines on the hanging walls of thrusts. Most are four-way dip-closed structures, while some have a dependency on a fault component along their southeastern thrust margins. (Compiled from Aymard et al. 1990; Carnevali 1992.)

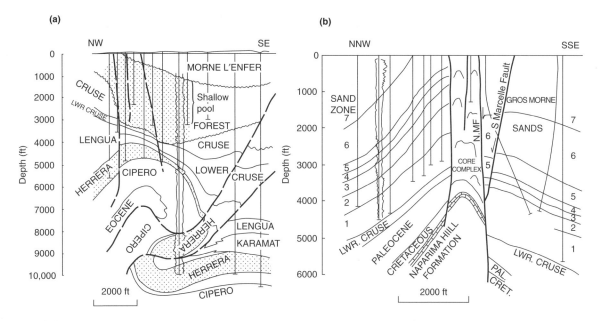

Fig. 4.42 Trinidadian trap geometries. The thrust-related anticlinal structures within the Plio-Miocene of southern Trinidad are much tighter than the stratigraphically older, rollover anticlines of the El Furrial Trend in eastern Venezuela: (a) Penal Field; (b) Beach Field. Within the cores of the anticlines, much of the compression has been accommodated via mobilization of mud and associated mud diapirism. (From Ablewhite & Higgins 1965.)

compressional folding and thrusting, and to post-depositional diapiric movement of thick middle to late Tertiary mudstones (Fig. 4.42; Ablewhite & Higgins 1965). In North America, thrust-linked rollover anticlines form the major trap type in the Wyoming–Utah thrust-belt fields (Lowell 1985; Fig. 4.43) and the southern foothills of the Alberta Basin in Canada.

Compressional anticlines are also a common feature within both transpressional and transtensional strike-slip basins. Anticlinal culminations are commonly aligned *en echelon* with the long axes of the anticlines, oblique to the sense of movement (Fig. 4.44; Harding 1985). Compressional anticlines may also form as a result of inversion on old extensional faults. Typical examples occur in the Central Sumatra Basin of Indonesia (Gluyas & Oxtoby 1995), the Rotliegend Province of the southern North Sea (the Leman Field; Hillier & Williams 1991) and the Maui Field offshore New Zealand (Abbott 1990; Fig. 4.45).

Traps formed by extensional tectonic processes

Traps formed by extensional tectonics are very common in rift basins. Here we adopt the approach of Allen and Allen (1990) in discussing only structural traps produced by extension of the basement. Structural features generated by gravity effects will be examined in the following section.

The East Shetland Basin of both the UK and the Norwegian Sectors of the North Sea contained around about 15 billion barrels of recoverable oil (Bowen 1991; Fig. 4.46). Much of this oil was trapped in tilted fault blocks formed during Late Jurassic rifting. In the pre-rift section, oil is trapped in the sandstones of the Middle Jurassic Brent Group, together with other sandstones of both Jurassic and Triassic age.

Outside of the North Sea, traps formed through tectonic extension are important in the Gulf of Suez, the Haltenbanken area, offshore mid-Norway (Spencer & Eldholm 1993), and in the pre-rift sections of the Gippsland Basin, Australia (Rahmanian et al. 1990).

Traps formed by diapiric processes

The specific gravity of salt (halite) is about 2.2 g cm^{-3}; that of fully consolidated rock is about 2.5–2.7 g cm^{-3}. In consequence, salt is buoyant relative to most other sediments and sedimentary rocks. Over geologic timescales, salt is also able to deform plastically. The onset of salt movement may be caused by a variety of initial conditions or subsequent processes. Salt may have been deposited heterogeneously, in-

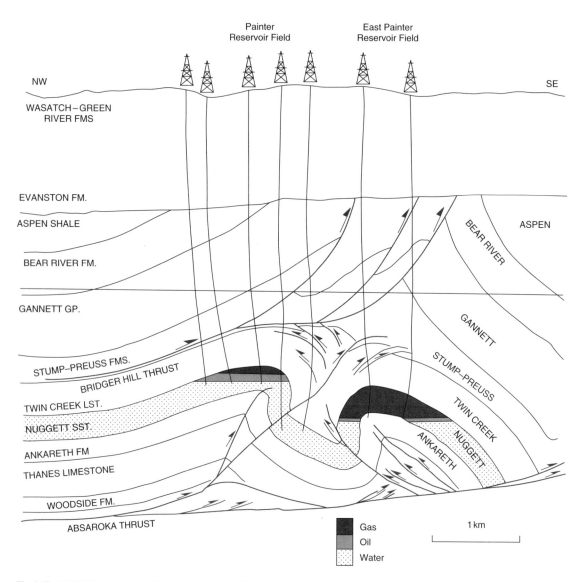

Fig.4.43 A NW–SE cross-section of the Painter Reservoir and East Painter Reservoir Fields (Wyoming–Utah thrust belt, USA). The traps occur as rollover anticlines associated with parasitic thrusts on the main Absaroka Thrust. An analysis of both petroleum-bearing and dry structures in the thrust belt has led to the conclusion that, for the play to work, the reservoirs in the hanging wall of the thrust must be in contact with the Cretaceous-age source rocks in the footwall. (From Frank et al. 1982.)

terspersed with clastics; the overburden may be heterogeneous with respect to density, or tectonic processes may disrupt the homogeneity of the salt layer or overburden. Layers of salt may then aggregate into swells and eventually pillows. Subsequently, the salt diapir may rise through the overburden. The terminal stage of diapirism occurs once the diapir becomes detached from its host salt layer.

Processes that are very similar to those associated with salt diapirism can occur in association with muds. Rapidly deposited muds are commonly water-rich, overpressured, and in consequence highly mobile. Mud lumps (Niger Delta), shale walls, diapirs, and mud volcanoes (Trinidad, Azerbaijan) are all products of mass mud movement.

Clearly, diapiric movement of both salt and mud can create

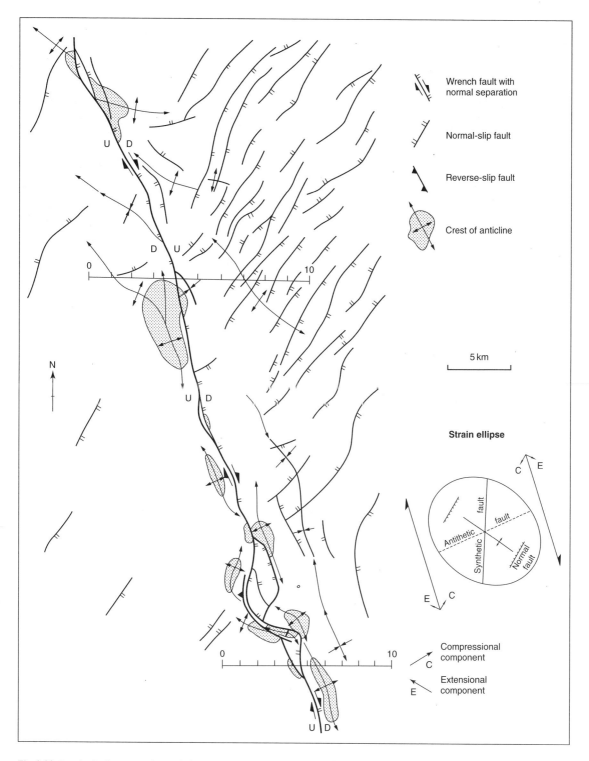

Wrench fault with
normal separation

Normal-slip fault

Reverse-slip fault

Crest of anticline

5 km

Strain ellipse

Compressional
component

Extensional
component

Fig.4.44 Anticlinal culminations along and adjacent to a major right-lateral wrench fault, Andaman Sea. (From Harding 1985.)

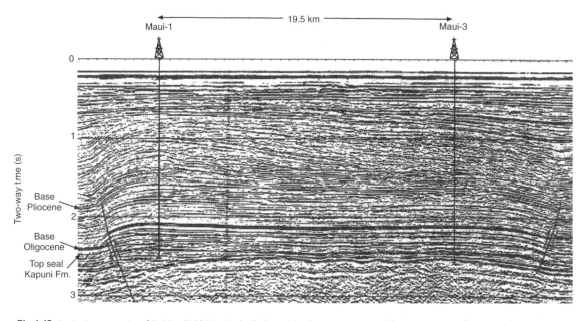

Fig.4.45 A seismic cross-section of the Maui Field, New Zealand. The anticlinal structure was created by inversion on the field margin faults. (From Abbott 1990.)

anticlinal structures that could form petroleum traps. However, the opportunities for trap formation in association with salt (or mud) movement are much more diverse than simple domal anticline formation. Halbouty (1979) lists nine distinct trap geometries associated with penetrative diapirs (Fig. 4.47). In addition, trap configurations can also develop in the areas of salt withdrawal. The "turtle" structure anticline develops via increased sedimentation in areas of salt withdrawal. Later, as salt continues to feed the diapir, the structure flounders and flips into an anticline (Fig. 4.48).

The factors that control diapirism are poorly understood. The induration state of the overburden is one factor that seems to exert some influence. North (1985) contrasts the large-scale, low-amplitude salt swells of Saudi Arabia and Kuwait with the penetrative features of diapirs in Texas and Louisiana (USA). In the Middle East there is an abundance of well-indurated competent carbonates in the overburden, while in the US Gulf Coast the overburden comprises poorly consolidated muds and sands. North (op. cit.) concludes that the salt has not been able to punch its way through the limestone overburden. Such a hypothesis might explain some of the differences in salt pillow/diapir geometries, but it fails to explain all, since Iraq contains salt piercement structures at the surface (Wyllie 1926).

Greater Burgan (Kuwait), the second largest oilfield in

the world (>75 Bbbl reserves; Brennan 1990) developed over a large low-amplitude salt swell. The area of the field is about 300 square miles, so large that it took 13 years (from discovery in 1938) for the development teams to recognize that a dome rather than a homocline was being developed! The structure is a north–south elongate dome. Flank dips are only a few degrees and the dominant structural elements are simple radial faults. The prime reservoir is the Cretaceous Wara Formation, also known as the Burgan Sandstone. However, there are a further 11 productive and nonproductive reservoirs (tar, heavy and light oil, and gas), ranging in depth from 60 m to 3000 m and in age from Pleistocene to Jurassic. The salt is believed to be the Infracambrian Hormuz Salt Series that began, in the area of Burgan, to move toward the end of the Jurassic. All formations younger than the Jurassic thin over Burgan, which indicates that although halokinesis was continuous, it failed to keep pace with subsidence.

Similar, simple anticlinal dome traps typify the Cretaceous Chalk fields of the Central Graben of the Norwegian North Sea (Hancock 1984). As with the Middle Eastern examples, the key controlling structures are Zechstein (Late Permian) salt pillows. Collectively, the Norwegian Chalk fields contain in excess of $340 \times 10^6 \, m^3$ (>2.1 Bbbl) of oil and condensate, and $250 \times 10^9 \, Sm^3$ of gas (Spencer et al. 1987). The largest of

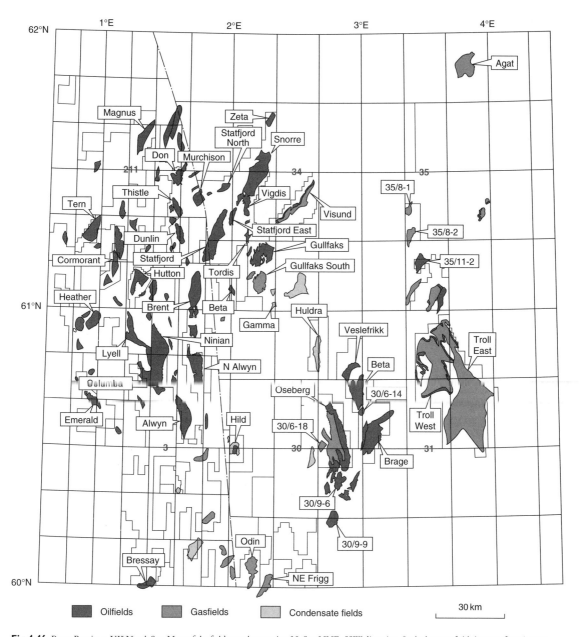

Fig.4.46 Brent Province, UK North Sea. Most of the fields are elongate in a N–S or NNE–SSW direction. Such elongate field shapes reflect the trap configuration as being tilted fault blocks. The primary reservoir intervals vary between fields, although most contain some Middle Jurassic (pre-rift) Brent Group reservoirs. In Snorre, the reservoir is Triassic fluvial sandstone, and in Magnus Upper Jurassic (syn-rift) submarine-fan sandstones are the main productive horizon. Troll, although reservoired in Upper Jurassic sandstones, is on the continental shelf, away from the rift province. Its peculiar shape is a function of the low-relief structure and near-horizontal reservoir. The near-equant Agat Field is reservoired in post-rift Cretaceous sandstones.

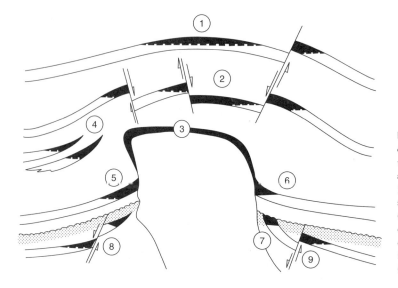

Fig.4.47 A sketch section showing the common types of petroleum traps associated with salt domes. 1, An anticlinal drape over a dome; 2, a graben fault trap over a dome; 3, porous cap rock; 4, a flank pinchout and sandstone lens; 5, a trap beneath an overhang; 6, a trap uplifted and buttressed against a salt diapir; 7, an unconformity; 8, a fault trap downthrown away from a dome/diapir; 9, a fault trap downthrown toward a dome/diapir. (From Halbouty 1979.)

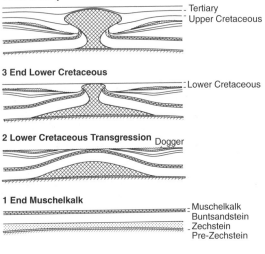

Fig.4.48 Turtle-structure anticlines form from the inversion of sediments that accumulate in sinks generated by salt withdrawal (2). As the salt diapir continues to grow, the areas of former sedimentation flounder on the subsalt and in so doing become positive structures (4). (From Jackson 1995.)

the accumulations, Ekofisk, accounts for about 70% of the oil and 50% of the gas.

Traps associated with diapirs rather than swells tend to be much smaller in aerial extent than the giants described above. They also tend to be much more structurally complex, commonly containing both radial and concentric fault patterns. For example, the Machar Field (STOOIP about 228 mmstb) of the central North Sea (Foster & Rattey 1993) is roughly circular in outline, with a diameter of about 4 km (Fig. 4.49). Reservoir rock occurs in both Cretaceous Chalk, which overlies much of the Zechstein salt diapir, and Paleocene submarine-fan sandstones, which onlap the flanks of the diapir. Steeply dipping reservoir units, structural complexity, and a seismically poorly defined reservoir interval all combine to make imaging of this field exceedingly difficult.

Traps formed by compactional processes

Differential compaction across basement highs, tilted fault blocks, carbonate shelf rims, reefs, or isolated sand bodies in mud can lead to the development of relatively simple anticlinal traps. Some of the most simple are also the largest. The world's largest field, Ghawar (Stoneley 1990) in Saudia Arabia, contains almost 100 billion barrels of reserves. The oil occurs in Jurassic carbonates draped over and compacted around a north–south-trending basement high. At the other end of the size range, compactional drape traps over Devonian pinnacle reefs occur in the Peace River Arch area of Alberta, in western Canada (Prodruski et al. 1988).

The Balmoral Field in the central North Sea is an example of drape over a tilted fault block. The field is an elongate anticline some 6.5 miles long (NW–SE) by 2 miles wide (Fig. 4.50; Tonkin & Fraser 1991). Reserves have been estimated at

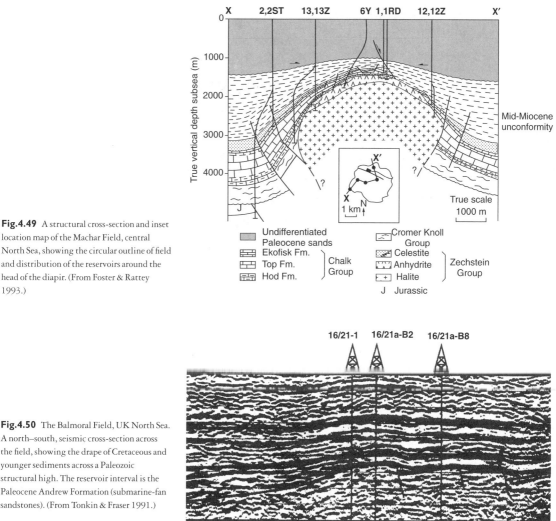

Fig.4.49 A structural cross-section and inset location map of the Machar Field, central North Sea, showing the circular outline of field and distribution of the reservoirs around the head of the diapir. (From Foster & Rattey 1993.)

Fig.4.50 The Balmoral Field, UK North Sea. A north–south, seismic cross-section across the field, showing the drape of Cretaceous and younger sediments across a Paleozoic structural high. The reservoir interval is the Paleocene Andrew Formation (submarine-fan sandstones). (From Tonkin & Fraser 1991.)

about 68 million bbl, based on a 45% recovery factor. The reservoir is the Paleocene Andrew Formation, which is compacted over rotated fault blocks containing well-indurated Devonian sediments.

Elsewhere within the North Sea, compaction induced, four-way dip-closed structures are common around Paleocene, Eocene, and Jurassic sand-rich, submarine-fan lobes. The trap that supports the giant Frigg Field ("Gas Initially In Place," GIIP, $235 \times 10^9 \, Sm^3$; reserves $175 \times 10^9 \, Sm^3$; Brewster 1991) was formed by compactional drape around sand-rich fan lobes. The bird's foot geometry of the field outline reflects the shape of the depositional system. The similarly giant Forties Field ("STandard barrels of Oil Initially In Place," STOOIP, 4.3 Bbbl, reserves 2.5 Bbbl; Wills 1991) owes part of its closure to differential compaction between the reservoir sand and its laterally equivalent mudstone. For Forties, the closure has been accentuated by further compactional drape over the Forties–Montrose ridge. Similar compactional processes have formed the trap and accentuated the depositional geometry of the reservoir in the East Brae Field (South Viking Graben, North Sea; Fig. 4.51).

Traps formed by gravity processes

Traps formed by gravity-driven processes are particularly important in large recent deltas. The best-described examples

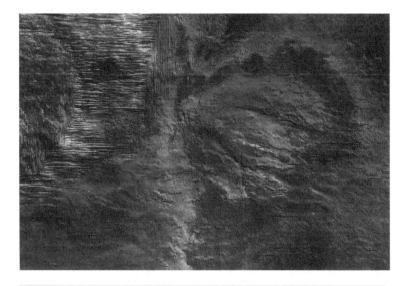

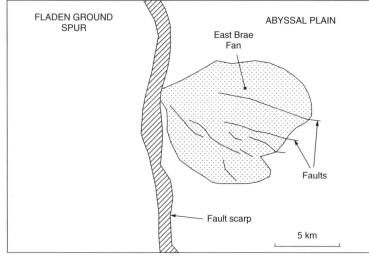

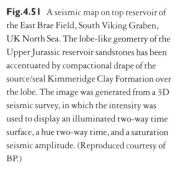

Fig.4.51 A seismic map on top reservoir of the East Brae Field, South Viking Graben, UK North Sea. The lobe-like geometry of the Upper Jurassic reservoir sandstones has been accentuated by compactional drape of the source/seal Kimmeridge Clay Formation over the lobe. The image was generated from a 3D seismic survey, in which the intensity was used to display an illuminated two-way time surface, a hue two-way time, and a saturation seismic amplitude. (Reproduced courtesy of BP.)

are from the US Gulf Coast (Lopez 1990; Fig. 4.52) and the Niger Delta (Weber & Daukoru 1975). The gravity structures form independently of basement tectonics (either extension or compression). They owe their existence to shallow detachment (commonly at a few kilometers below datum) along a low-angle, basinward-dipping plane. The drive mechanism is provided by the weight of sediment deposited by the delta at the shelf/slope break or on the slope itself.

In the Niger Delta the detachment planes are highly mobile muds, while in the Gulf Coast (Mississippi Delta) detachment occurs on both muds and the Louanne Salt

(Jurassic). The key detachment surfaces are commonly listric, concave-up, and concave basinward in plan view. The main faults are commonly large, being tens of kilometers from tip to tip.

Displacements on the listric faults are driven by the weight of deposited sediment. In consequence, they tend to grow in magnitude during sedimentation (Bischke 1994; Fig. 4.52). Hence the term "growth fault" is commonly used to describe such structures. Because fault planes are commonly concave-up, accommodation space at the head of the fault is created by rotation of the footwall. Thus the oldest sediments within the

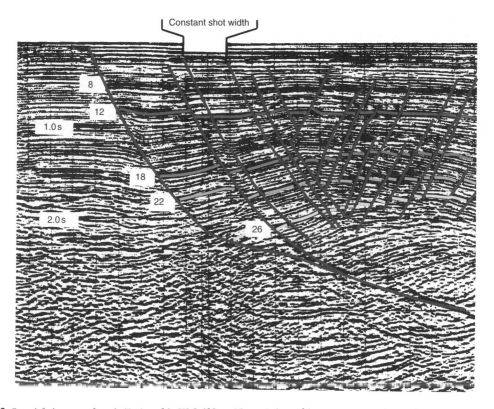

Constant shot width

Fig.4.52 Growth fault patterns from the Tertiary of the US Gulf Coast. The vertical axis of the seismic section is displayed as two-way time (TWT) in seconds. The complex synthetic and antithetic faulting masks the stratigraphic relationships and petroleum fields developed in such sediments would be highly segmented. Correlation surfaces have been highlighted. Surface 18 is a sequence boundary onto which overlying sediments onlap. The sequence between boundaries 22 and 26 thickens dramatically toward the master fault. (Simplified from Bischke 1994.)

hanging-wall sequences are commonly steeply dipping. Displacement on growth faults is often large (>300 m in the Niger Delta).

Although gravitational collapse is extensional and thus creates space in the proximal parts of the shelf–slope system, compression occurs at the toes of the listric growth faults. This too may generate anticlinal traps.

Gravitational process can also lead to the generation of other trapping geometries. Mud lumps, ridges, and diapirs may be induced (see the preceding section), and surface topography on the delta slope can generate conditions that are appropriate for the formation of stratigraphic traps.

4.5.4 Stratigraphic traps

Traps formed by depositional pinchout

From the top to the bottom of a systems tract, each deposi-

tional environment is capable of producing a juxtaposition of permeable and impermeable sediments which might one day form a stratigraphic trap for petroleum. Many examples of this type of trap have already been discussed in the reservoir section of this chapter. In practice, the reservoir geometry becomes the trap geometry. Examples include aeolian dunes encased in lacustrine mudstone, sand-filled fluvial channels cut into mud-rich overbank deposits, shallow-marine bar sandstones surrounded by marine shales, carbonate reefs isolated by inclosing marls, and submarine-fan sands trapped within the domain of pelagic mud. Rittenhouse (1972) managed to amass a collection of over 30 possible stratigraphic traps based on pinchout. We will illustrate just a few.

The Paradox Basin (Colorado and Utah, USA) contains a large array of small oil (a few mmbbl) and gas (a few bcf) fields that occur as stratigraphic pinchout traps (Baars & Stevenson 1982). Here, Devonian reservoirs are present within shallow-marine bar sandstones and Carboniferous reservoirs within carbonate mounds. Baars and Stevenson (1982) report that the distribution of the stratigraphically trapped fields within the basin was considered to be random. Given that it was not possible to adequately image the reservoirs using seismic, their distribution was considered to be unpredictable, until — that is — it was recognized that the preexisting Paleozoic structural style controlled the distribution of the carbonate mounds. The Paleozoic structures are mappable on seismic, and this had led to the discovery of at least one oilfield without the aid of serendipity!

Although the Paradox Basin traps have proved difficult to find, when found they are of relatively simple shape. Their geometries are either prolate bar forms or more equidimensional carbonate mounds. Pinchout traps formed in deltaic (paralic) settings are often much more complex in outline, and because potential reservoir sandstones are commonly discontinuous, multiple pools (clustered fields) are common. On land or in areas where lease areas are small, fields reservoired in paralic sandstones may have multiple names. This is because the field will be independently discovered at a number of locations, and only after years of outward-stepping development drilling will the competing companies realize that they are all developing the same reservoir. This point is further illustrated in the case history on Trinidad at the end of Chapter 6. The Fyzabad – Forest Reserve pool covers six independently named fields. Fyzabad and Forest Reserve, being the largest portions, have now subsumed the other names.

The highly complex shape of many pinchout trap fields in paralic reservoirs reflects the complex interplay of sand geometries in channels, mouth bars, crevasse splays, and so on (Fig. 4.53). Petroleum distribution in individual sands may have a much simpler and thus understandable distribution (Fig. 4.54).

Stratigraphic traps and in particular pinchout traps have another interesting property — interesting at least to those

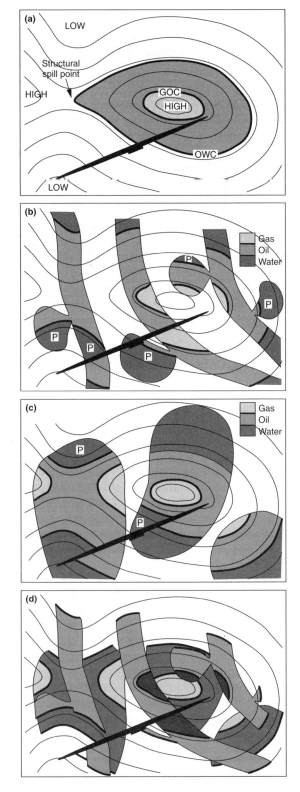

Fig. 4.53 Paralic field shapes are commonly of a highly complex shape, because of the interaction between the structure and sedimentary bodies of differing architectures. The complexity is multiplied because individual paralic sandstones tend to be stacked. There are commonly multiple fluid contacts, and gas-, oil-, and water-bearing sands can be stacked vertically. The four examples shown are: (a) a field shape on a simple faulted anticline, where the reservoir interval is much larger than the structure; (b) the same structure as in (a), but with reservoirs developed in channel and crevasse splay sands that are smaller in area than the structure; (c) the same structure as in (a), but with mouthbar sands, which again have a smaller areal extent than the structure, as the reservoir; (d) diagrams (b) and (c) combined, with the channel and mouthbar sands at different levels. (Pers. comm., A. Reynolds 1994, reproduced courtesy of BP.)

Fig.4.54 The Oficina Field, eastern Venezuela. The isopach, structural contour, and fluid distribution are for a single sand (the "L" sand). By 1947, the field was thought to have 160 individual reservoirs in 37 productive sands across 22 field segments. At the time, "L" was the most important. By 1990, the whole field had produced over 600 mmbbl. The "L" sand comprises two connected channels, with petroleum trapped against cross-cutting faults. The structure dips to the NNE and there are no spill points. The field geometry, like so many in the Oficina Trend, is controlled by sand-body geometry and cross-cutting faults. Wells (black dots) outside the productive area of "L" target other reservoirs. (Modified after Hedberg et al. 1947.)

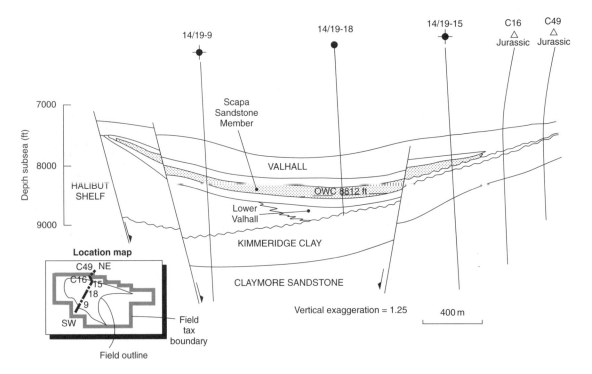

Fig.4.55 A NE–SW cross-section through the Scapa Field of the Witch Ground Graben, UK continental shelf. The trap geometry is a plunging syncline and the stratigraphic trap is caused by the pinchout of the Cretaceous reservoir sandstone. (Modified from McGann et al. 1991.)

petroleum geoscientists who might wish to exploit an area "already known to be played out." By definition, pinchout stratigraphic traps do not require any particular structure to complete closure. In consequence, trapped petroleum can and does occur in synclinal settings. The Scapa Field in the Witch Ground Graben area of the North Sea contains about 200 mmbbl STOIIP (c.63 mmbbl reserves) in a plunging, synclinal trap. The trap is little faulted and almost all of the trapping elements are updip pinchouts (Fig. 4.55; McGann et al. 1991).

The same situation exists in the Mutiara/Sanga Sanga Trend on the Mahakam Delta in Kalimantan. The trend has a successful exploration history that dates back to 1897. However, our interest focuses on the Samboja Field discovered in the late 1930s and, on the same trend, the Pamaguan Field discovered in 1974. Both are dip-closed anticlines. Step-out drilling into the syncline separating the two fields failed to find any water. Indeed, the well sunk at the axis of the syncline found only oil and gas. The crests of the anticline and the syncline are vertically separated by about 6000 ft, and the Mutiara Field contains over 80 different petroleum-bearing paralic sandstones (Safarudin & Manulang 1989). Overall,

the petroleum volume in the synclinal part of the trap is about 1.5 times as much as that in the anticlinal parts of the trap.

Traps formed by unconformities

Attenuation of the updip portions of potential reservoir interval by an unconformity can create truly massive traps with enormous petroleum catchment (drainage) areas. The largest oilfield in North America is Alaska's Prudhoe Bay (about 25 Bbbl of liquid and more than 20 Tcf of gas in place; Szabo & Meyers 1993; Fig. 4.56). The field lies over 100 km to the northeast and updip of the axis of the Colville Trough, from which the oil migrated (Hubbard et al. 1990). A relatively simple homoclinal dip links the field to the source area. The updip end of the Triassic fluvial reservoir sand is eroded by a Lower Cretaceous unconformity. Upper Cretaceous Colville Mudstone overlying the unconformity combines with faulting to seal the field.

East Texas, the largest oilfield in the US Lower 48, is also a stratigraphic trap. The productive Woodbine Sandstone reservoir, with its initial reserves of about 6.8 billion barrels, is sandwiched between two unconformities. The sand rests

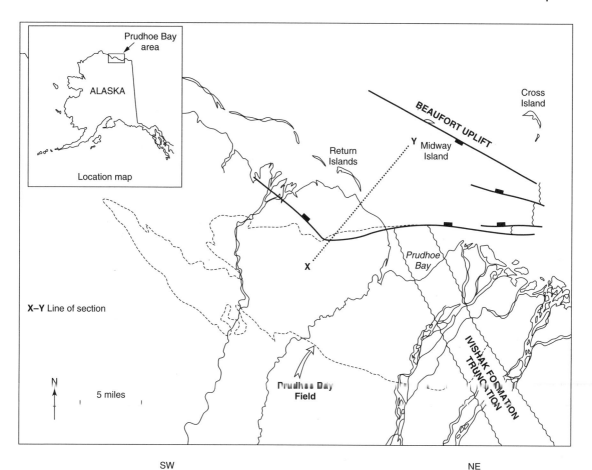

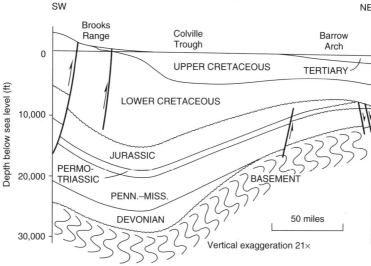

Fig.4.56 A map and cross-section of the Prudhoe Bay Field, Alaska. Petroleum occurs within the Permo-Triassic Ivishak sandstones, trapped beneath sub-Jurassic and sub-Early Cretaceous unconformities. (Adapted from Melvin & Knight 1984.)

upon the Washita Group mudstones and is itself truncated beneath the Austin Chalk. The field, some 40 miles long and 5 miles wide, is a simple homoclinal dip to the west (Willis 1988).

Troll is the largest offshore gasfield in Europe (Fig. 4.57), with about $1700 \times 10^9 \, Sm^3$ of gas and approximately $600 \times 10^6 \, Sm^3$ of medium-gravity oil (Hellem et al. 1986). Like Prudhoe Bay, it lies at the periphery of a basin (the Viking Graben). The western limb of the field dips gently toward the basin and the petroleum drainage area is large. The Upper Jurassic Viking Group sandstones that form the reservoir are eroded at the field crest. As such, field closure relies on sub-unconformity sealing beneath Tertiary mudstones, and faulting elsewhere.

Each of the unconformity traps described above depend on a combination of trapping mechanisms, which rely in large part on a planar or gently folded unconformity. Unconformities do of course come in a variety of shapes. The most spectacular of the unconformity-bounded traps have already been described earlier in this chapter. They are those commonly referred to as "buried hills." Such hills are residual topography of a one-time land surface. A sea or lake subsequently inundated the areas and the hills were buried beneath mudstone. Thus it is the unconformity surface that has the trapping geometry. Buried hill plays in northern China are well-described (Zhai & Zha 1982; Tong & Huang 1991). However, as Zhai and Zha (1982) point out, the term "buried hill" has slipped into common usage amongst Chinese geologists, such that it may be used to describe any elevated feature in the subsurface, such as a horst block. The term has much more restricted usage outside of China, where it refers to true paleogeomorphological traps.

Traps formed by diagenetic processes

Mineral precipitation tends to reduce reservoir quality (Section 5.6). In exceptional circumstances, porosity impairment may be so severe that it is destroyed completely. Such tight rocks may act as seals and trap petroleum. Here, diagenetic traps formed either by mineral precipitation or phase change (water to permafrost, oil to tar, and water and methane to gas hydrate) are described.

Mineral cements are known to form top, lateral, and even bottom seals to reservoirs. Examples in carbonate systems are more numerous than those in clastic systems. Already described, earlier in this chapter, is the Albion–Scipio Field of Michigan (USA). All surrounding rock to the trap is thoroughly cemented limestone and dolomite. Indeed, the vuggy and cavernous dolomite reservoir was created from this

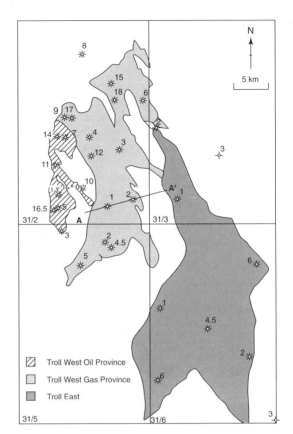

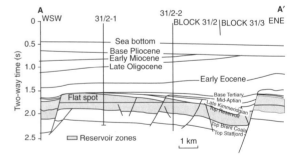

Fig.4.57 The crest of the giant, gas and oil Troll Field (Norwegian North Sea) subcrops a base Tertiary unconformity. (From Horstad & Larter 1997.)

impermeable rock by mineralizing fluids associated with an underlying, deep-seated wrench system. A comparable situation exists for many of the carbonate-hosted oilfields of Abu Dhabi; porosity only exists where there is oil. Areas that at one time must have been the aquifers to the oilfields have been thoroughly cemented. For a few fields, such cementation has allowed trap integrity to be maintained despite tilting of the field after petroleum accumulated.

An unusual situation exists in the Pleasant Valley Field of California (USA). Kaolinite-cemented sandstones act as seal in an otherwise unclosed trap (Schneeflock 1978). Quartz cement is also known to form cap rock. The Norphlet play of the Gulf of Mexico has as its reservoir Jurassic, pure quartz, aeolian sandstone. In many wells the upper tens (rarely hundreds) of meters of the sandstone are completely cemented by quartz, while the porous and permeable reservoir below is chlorite-cemented (Emery & Robinson 1993). The Norphlet play is commonly very deep (>20,000 ft). As such, it must have been a very brave geologist who held the exploration manager and drilling superintendent at bay as the first well in this play plunged deeper and deeper into a quartz tombstone above top pay.

Tar mat seals are common in the shallow subsurface. They also act as cap rock for the largest single accumulation of heavy (viscous) oil in the world; the Faja or Orinoco Tar Belt of southeastern Venezuela (James 1990). Uncertainty over the size of this accumulation is large, but current estimates suggest that there may be 1.2 trillion barrels of oil in place, with a mere 250 billion barrels recoverable using today's technology. Tar also forms a major seal component for Venezuela's largest — and the world's fourth largest — conventional field, Bolivar Coastal, on the eastern side of Lake Maracaibo (estimated reserves >30 Bbbl; Tiratsoo 1984). In the east of Venezuela, tar seals much of the 4 billion barrels of oil originally in place in the Quiriquire Field. Tar seals and tar sands are also common within the Western Canada Basin and the Californian basins.

Gas trapped beneath permafrost forms large fields in the northern part of the West Siberia Basin, adjacent to the Kara Sea (North 1985). In cold regions, gas (methane) is also trapped as clathrate (gas hydrate, see Chapter 3). Clathrates comprise pentagonal dodecahedra of 20 hydrogen-bonded water molecules. Methane is held within the dodecahedron cages (Huheey 1975). Such clathrates could have formed a Glacial Epoch precursor to the permafrost traps of West Siberia. They are also common within the shallow subsurface sediments beneath cold seas. The Deep Sea Drilling Project (DSDP) has mapped many occurrences around the world.

4.5.5 Hydrodynamic traps

Large-scale movement of water in the subsurface has already been examined in part in Chapter 3. Here we examine the influence of water movement on trapped petroleum. The idea that moving water could and would control the distribution of both oil and gas traps was first advocated in 1909. The hypothesis had a number of advocates until the 1930s, when the number of publications on the topic dwindled and the anticlinal theory of petroleum accumulation reassumed its position as the only favored theory. Twenty years later, Hubbert (1953) resurrected and advanced the ideas on hydrodynamic trapping. Indeed, he went as far as to suggest that normal anticlinal trapping was a special case; static water rather than dynamic water.

It is possible to demonstrate the concept of hydrodynamic trapping in the laboratory simply by flowing oil and water through an inverted U-tube. However, in nature, hydrodynamic flow paths and trapping geometries are of course much more complicated than those in the laboratory. In consequence, it is difficult to prove that many of the reported occurrences of hydrodynamic traps are indeed hydrodynamic traps. Differential heights of oil columns on different limbs of an anticline might be due to inefficient transmission of oil through a fault system; or facies variations, with oil being fed from one side only.

Those traps with undoubted hydrodynamic credentials tend to be in foreland basins, where subsurface reservoir units commonly crop out in adjacent mountain belts. The outcropping reservoir units are recharged with meteoric water and the hydraulic head drives the flow through the basin. Two of the best documented examples are in the Frannie Field of the Big Horn Basin, Wyoming (Adams 1936; Fig. 4.58) and the East Colinga Extension Field of San Joaquin Valley, California (Chalmers 1943). In both instances, there is sufficient information to map the tilted oil/water contacts, to rule out the possibility of significant permeability barriers in the systems, and to explain the water flow in terms of the adjacent topography and subsurface structure.

4.6 PLAY AND PLAY FAIRWAY

4.6.1 Play

In this short section, we examine "play" and "play fairway." The concepts of play and play fairway are useful insofar as they combine the sub-regional source, reservoir, and seal elements in exploration but omit the trap. The idea is to be able to define a part of a basin in which petroleum could be trapped, rather than to define a specific prospect. Thus a petroleum play comprises a seal and reservoir combination (Fig. 4.59),

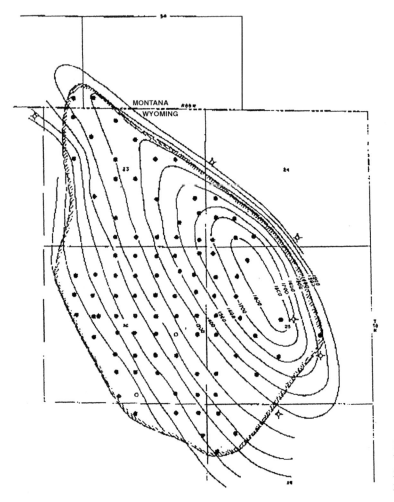

Fig.4.58 The Frannie Field, Big Horn Basin, Wyoming. The oil/water contact has been proven to be tilted on the basis of well penetrations. (From Hubbert 1953.)

coupled to a mature source rock. A play may be proven, in which case at least one oil- or gasfield exists with the defined seal and reservoir (Table 4.4). A play may also be hypothetical, a conceived possibility of a seal and reservoir pair, but one in which a petroleum accumulation has yet to be found. Importantly, a play can be proven in a basin without implying that every trap within that play will contain a petroleum pool. For example, the Cretaceous Austin Chalk of Texas is a proven play. However, only a portion of the exploration wells drilled for this play have been successful. Some have failed because the wells were not drilled on valid traps. Other exploration wells have failed because petroleum never filled the trap.

4.6.2 Play fairway

A useful development from the concept of the play is the play fairway. It is the area within which the play can be expected to work. Since it is an area, the play fairway can be represented on a map (Fig. 4.59). The area outside a particular play fairway is the area in which the reservoir and seal combination is known or believed not to exist. This could be because either one or both were not deposited, or because they were deposited and then eroded. The area of a play fairway may also be modified to exclude those parts that could not have received a petroleum charge. Within the area bounded by the play fairway, there will be areas of greater and lesser risk of the play working. It is commonly difficult to quantify such a risk. In consequence, many companies use a qualitative description of the risk within the play fairway: high, medium, and low. Again, maps can be constructed to depict this risk, and such maps are commonly referred to as common risk segment maps (Fig. 4.60). The maps may be annotated with descriptions of the particu-

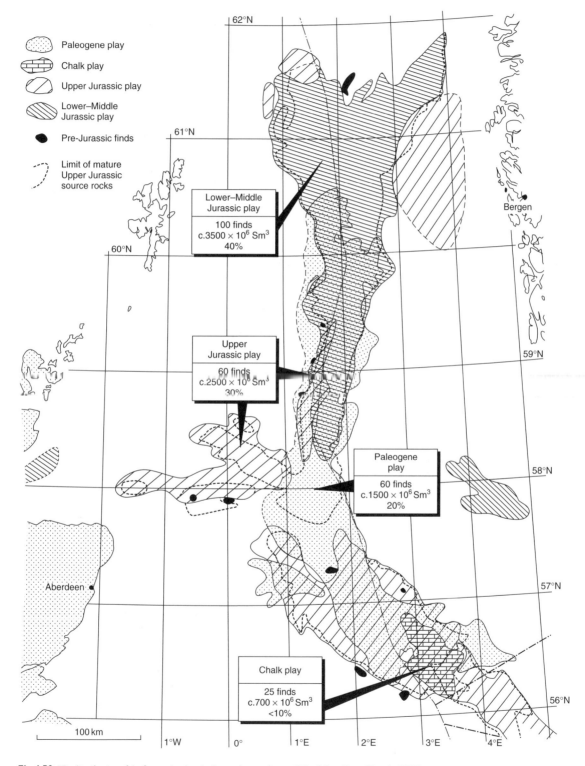

Legend:
- Paleogene play
- Chalk play
- Upper Jurassic play
- Lower–Middle Jurassic play
- Pre-Jurassic finds
- Limit of mature Upper Jurassic source rocks

Lower–Middle Jurassic play
100 finds
c.3500 × 10⁶ Sm³
40%

Upper Jurassic play
60 finds
c.2500 × 10⁶ Sm³
30%

Paleogene play
60 finds
c.1500 × 10⁶ Sm³
20%

Chalk play
25 finds
c.700 × 10⁶ Sm³
<10%

Bergen

Aberdeen

100 km

Fig.4.59 The distribution of the four main plays in the northern and central North Sea. (From Glennie 1998.)

Table 4.4 Central and northern North Sea plays. (Modified from Johnson 1998.)

Stratigraphic unit	Play		Fields
	Reservoir	**Regional seal**	
Paleogene	Eocene fan	Eocene mudstone	Frigg, Harding, Gryphon, Tay
	Eocene channel/canyon fill	Eocene mudstone	Alba, Chestnut, Fyne
	Paleocene fan	Paleocene or Eocene (Balder) mudstone	Forties, Balmoral, Maureen, Heimdal, Andrew, Nelson
	Paleocene submarine channel fill	Paleocene mudstone	MacCulloch, Kappa (discovery)
	Paleocene shelf-edge sandstones	Eocene (Balder) mudstone	Bressay, 9/3 (discovery)
Late Cretaceous	Chalk	Paleocene mudstone	Ekofisk, Eldfisk, Tor
Early Cretaceous	Apto–Albian submarine-fan (Kopervik and Captain) sandstone	Lower Cretaceous mudstone	Britannia, Captain, Goldeneye
Early Cretaceous	Hauterivian–Barremian sandstone and conglomerate	Lower Cretaceous (Cromer Knoll) mudstone	Scapa, Spey, North Area Claymore
Late Jurassic	Kimmeridgian–Volgian submarine-fan sandstone	Kimmeridgian–Lower Cretaceous (Cromer Knoll) mudstone	Magnus, Brae, Miller, Ettrick, Borg, Main Area Claymore
	Callovian–Volgian shelfal sandstone	Kimmeridgian–Lower Cretaceous (Cromer Knoll) mudstone	Piper, Beatrice, Hugin, Scott, Fulmar, Clyde, Ula, Gyda
Middle Jurassic	Aalenian–Callovian paralic sandstones	Upper Jurassic (Humber Group) to Upper Cretaceous mudstone and marl	Sleipner, Brent, Bruce, Beryl, Cormorant, Alwyn North, Gullfaks (Brent reservoir), Statfjord (Brent reservoir)
Early Jurassic and Triassic	Rhaetian–Liassic paralic to shelfal sandstones	Lower Jurassic mudstone	Cook and Statfjord reservoirs in Brent Province fields
	Scythian–?Norian fluvial sandstones	Jurassic–Cretaceous mudstone	Marnoch, Snorre
Paleozoic	Zechstein carbonates, Rotliegend, Upper Carboniferous, and Devonian sandstones (subcrop play, fractures commonly important)	Upper Jurassic–Upper Cretaceous mudstone and chalk	Auk, Argyll, Stirling, Buchan, Main Area Claymore, Ettrick (Zechstein reservoir)

lar risk. As exploration continues within a basin, the play fairway maps for a particular play will become better defined, as will the description of the risk. Common risk segment displays on play fairway maps can be used to denote high-grade areas for exploration. The combination of maps from two or more plays can be used to reduce the exploration risk still further.

4.7 LEAD AND PROSPECT

4.7.1 Introduction

Exploration wells are drilled into prospects. Prospects are volumes of rock in the Earth's crust that are believed, but not proven, to contain the four key components outlined in this chapter and in Chapter 3: a valid trap, an effective seal, a reservoir, and a petroleum source rock that has generated and expelled petroleum into the trap. Once an exploration well penetrates a prospect, it ceases to be a prospect. It becomes either a proven petroleum (oil or gas) field, or more likely a dry hole (which means that it lacks petroleum).

In this section, we examine how prospects are defined, evaluated, and ranked. The prospect commonly is the precursor to a plan for an exploration well. This too is investigated, as is the process that occurs when a well is drilled but fails to find a petroleum pool.

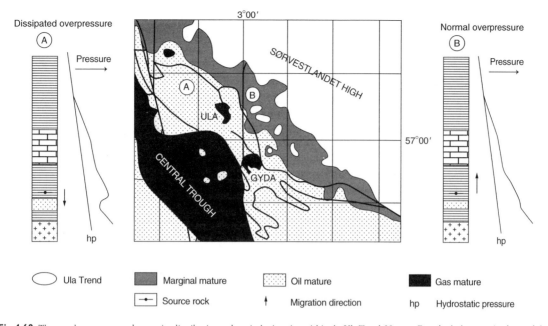

Fig.4.60 The petroleum source-rock maturity distribution and vertical migration within the Ula Trend, Norway. From both the maturity data and the migration direction data, it would be safe to conclude that the Sørvestlandet High is an area in which the chance of finding a petroleum pool within a Jurassic sandstone reservoir is low (high risk). (From Bjørnseth & Gluyas 1995.)

4.7.2 Lead, prospect, and prospect evaluation

A lead is nothing more than an ill-defined prospect. The definition may be inadequate in any of the key components of an effective reservoir, a petroleum charge, or trap integrity. Notice here that the choice of words — "effective reservoir," "petroleum charge," and "trap integrity" — differentiates the prospect or lead from the play and play fairway. The difference is important. For a prospect or lead, the explorer is trying to assess risks at a specific location. Work beforehand will have already defined the play and play fairway. The play may be proven or unproven, but the prospect is defined independently in terms of the three prospect risk components. For a proven fairway, the play risk is unity. For an unproven fairway, the play risk is less than unity. However, so as not to double-risk a prospect, the approach used when considering a prospect in an unproven fairway is to consider the individual prospect risk on the basis of a working play. The total risk is then derived by multiplying the play risk by the prospect risk.

Since prospects are nowadays most commonly defined on the basis of a trap mapped using seismic data, it follows that most leads are so-called because the trap has not been adequately mapped, or cannot be adequately mapped given the data available. The boundary between what is a lead and what is a prospect is open to individual interpretation; a seller's "prospect" may only be a buyer's "lead."

Having mapped a trap, it is commonly possible to conceive of ways in which the trap may not be valid. There may be uncertainties surrounding the conversion of the seismic data from two-way time to depth. An apparent trap in time may not be a trap in reality if the varelocity field across the area of the prospect is variable (Fig. 3.8). Alternatively, a trap may be divisible into different areas, each of which carries a different risk with regard to integrity. This situation most commonly occurs where both folding and faulting define a trap, or where structure and stratigraphy are needed to define a trap.

The prospect map in Fig. 4.61 shows a trap where the upper part has four-way dip closure. However, the deeper part of the prospect relies upon a cross-fault seal to maintain its integrity. In consequence, it is possible to assign different risks to different parts of the trap. The same is true for the prospect sketched in Fig. 4.62. The integrity of this trap relies upon shale-out of the reservoir sandstone. It is highly unlikely that such a trap would be drilled during the early stages of exploration in a basin. However, once the distribution of reservoirs in the play fairway is understood, such a prospect might be drilled, particularly if the area were covered with 3D seismic data.

The risk on effective reservoir relates to an assessment of the porosity and permeability of the prospective reservoir; that is, its capability to hold and produce petroleum. In order to predict porosity before drilling, it is common to plot

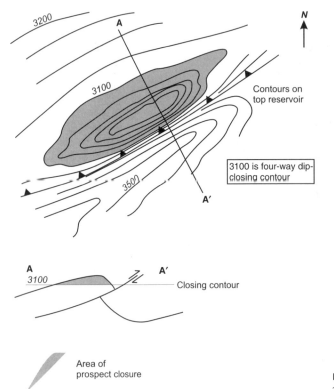

3200

A

3100

N

Contours on
top reservoir

3100 is four-way dip-
closing contour

3500

A'

A   A'
3100

Closing contour

Area of
prospect closure

Fig.4.61 A sketch of a prospect with four-way dip closure. The deepest closing contour is 3100 m.

porosity against depth (or some other parameter) for existing discoveries (Gluyas 1985) and calculate a best-fit line. The measured property (depth) is then used to predict the unknown (porosity). This can work, sometimes. Alternatively, an attempt may be made to map the property in known wells and hence either interpolate or extrapolate to the new prospect. This too can work sometimes. Permeability is rarely properly or independently assessed at the exploration stage. It is often assumed that if the reservoir is porous, it is also likely to be permeable. This may not be true. A fuller treatment of the porosity and permeability of reservoirs is given in Chapter 5.

The petroleum charge is commonly the most difficult component to assess accurately during the prospect evaluation process. While, in a gross sense, it may be straightforward to plot the petroleum migration direction from mature source to basin margin, it is rarely possible to be sure that the prospect under examination is on the migration pathway for the liberated petroleum. Earlier in this chapter we showed that buoyancy and capillary forces control secondary migration of petroleum. Subtle changes in grain size, open fractures

and faults, and nuances in the structural fabric between source and trap will each influence the migration direction of petroleum. So-called "migration shadows" are common in many petroleum provinces. In such situations, a perfectly valid trap and reservoir pair is dry, the petroleum having migrated past the trap without entering it.

In the preceding paragraphs, we have discussed the factors that control the presence or absence of a prospect. However, an assessment of presence or absence alone (risk) is insufficient information when it comes to determining whether a prospect will be drilled. It is also necessary to evaluate the expected petroleum in place, and to further estimate how much might be recovered. The process of determining oil or gas in place and reserves before a well is drilled is called "prospect evaluation" (Case history 4.10). The component parts to prospect evaluation are values for gross rock volume above the petroleum/water contact, net to gross, porosity, petroleum saturation, formation volume factor, and recovery factor. These are of course exactly the same data that are required during appraisal of a discovery in order to assess the size of the discovery (Chapter 5).

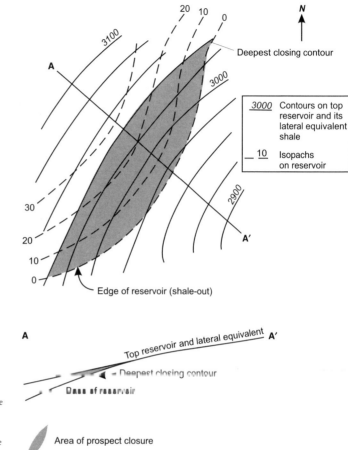

Fig.4.62 A sketch of reservoir shale-out. There is no closure on the top reservoir horizon or its lateral equivalent surface within the continuous shale sequence to the southeast of the reservoir wedge. Closure is defined by the interplay between the zero-thickness sandstone edge and contours on the top of the sandstone reservoir.

4.7.3 The prospect inventory

As the name suggests, a prospect inventory is a list of prospects. The list may pertain to a particular license or to a particular company's licenses in a specific basin. The list may cover all of the known prospects in a basin. Alternatively, it may be a company's entire portfolio. The purpose of the prospect inventory, whatever the grouping, is to allow either summation of the prospects and/or ranking of the prospects. Summations may be used to compare licenses or basins. For example, a company may calculate that it has seven prospects with an estimated 400 million barrels recoverable in its North Sea license portfolio, but only three prospects with an estimated 300 million barrels recoverable in its deep-water Gulf of Mexico licenses. It could use such data, along with the risk of failure on the prospects (Section 4.9), to determine where to place resources. Similarly, the seven prospects in the North Sea portfolio could be ranked in terms of size, risk, value, or risked value, and these data could be used to determine the order in which prospects are drilled.

4.7.4 Well prognosis

A well prognosis is an important part of any drilling campaign, whether for exploration, appraisal, development, or production. A well prognosis is the geologic section expected to be penetrated by a well. The key aspect from a petroleum geoscience perspective is the identification of the depth to top prospect and base prospect. However, prognoses on the elevation and characteristics of key horizons in the overburden will help to improve the estimation of the depth to top reservoir. Once these key horizons are penetrated during drilling, adjustments can be made to the depth to prospect prediction.

Well prognoses have a greater utility than to the petroleum

geoscientist alone. They are also of critical importance to drilling engineers. A drilling engineer will look to the petroleum geoscientist to provide information on hazards to drilling, such as shallow pockets of gas (usually methane), hard formations, sticky formations, dramatic pressure transitions, and possibly poisonous gas (H_2S). Although the drilling and monitoring safety equipment allows detection of gases and pressure changes, the well prognosis will provide valuable information on the likely presence and position of such hazards. The geologic description of the overburden will also influence the planning of a well in terms of what type of drill bit is used and how many might be needed (hard versus soft formations). The required geometry of the well — vertical, deviated, high-angle or horizontal — coupled with the lithology will also control the drilling method.

A geophysicist and geologist working together commonly produce a well prognosis. The geophysicist will provide the key horizons and their depths, and the geologist will supplement the depth description with information on lithology and possible hazards. In some instances, it may be possible to derive relative pressure data from the seismic analysis and so improve the description. While every attempt will be made to ensure that the well prognosis is accurate, there are likely to be some uncertainties with every new well. This can be expressed on the prognosis; as, for example, a range of depths for a particular horizon or descriptions illustrating the uncertainty about the presence of a particular lithology.

Clearly, a well prognosis is developed before and during the planning of a well. However, it is not a static document. Information gathered during the drilling of a well can be used to update the document. The information can come from analysis of the cuttings generated during drilling. It may be useful to have a biostratigrapher analyze the microfossils or palynomorphs at rig site, so as to determine the stratigraphy of the well. Wireline logs run either between drilling sections or as "logging while drilling" are used by geologists to determine the accuracy of the prognosis and adjust the drilling plan.

4.7.5 Failure analysis

Most exploration wells fail to find petroleum. That emphasizes the fact that petroleum exploration is a risky business. Even the most prolific petroleum provinces deliver success rates of less than 1 in 2. The petroleum industry as a whole has been very poor at analyzing its failures. All too often, failures have been dismissed as "to be expected given the risk." Yet, quite clearly, there is a big difference between a well that fails for the expected reason and one that fails for some other reason. In the first instance, the geoscientists could justifiably

say that they understood the petroleum system. This statement cannot be made if the well fails for an unexpected reason. In two of the case histories reported at the end of this chapter (Alaska and eastern Venezuela), exploration wells failed for the reasons predicted.

An example of a well that failed to find petroleum for unexpected reasons is the Bright prospect well, drilled in UK continental shelf Block 20/3a in the mid-1990s. The well was designed to appraise a discovery made a few years earlier in turbidite sandstones of Jurassic age. Before the well was drilled, a considerable amount of work was done on the likely seismic response at the prospect location. This involved modeling the effects of different lithologies (sandstone and mudstone) and different fluids (water, oil, and gas) within porous sandstone. A seismic amplitude anomaly was used to help define the prospect and tie it to the older discovery well. The prognosis was for a sandstone reservoir filled with low-gravity oil. The main uncertainty was seen to be the thickness of the pay. The well was drilled and the expected sandstone was missing, as of course was the oil. The well was dry. Clearly, an analysis of this failure could have provided insight into what was controlling the seismic response and hence improved the prognosis for the next well. Indeed, the operator outlined such a study. However, ultimately the partnership decided that the failure had defined the prospect as uneconomic and no further work was done. In this instance, such a decision can be justified in commercial terms. However, if at some time in the future the area around the Bright prospect becomes the focus of renewed attention, then the well's failure to find petroleum must be investigated properly.

4.8 YET TO FIND

4.8.1 Introduction

Most petroleum exploration companies wish to have a view on how much petroleum remains to be found in a given area (basin, sub-basin, license, etc.) before committing significant expenditure. On the basis of a "yet to find" study, a company can decide to enter a new basin or license or exit from the same. The two key products of such a study are the number of petroleum pools that are expected to be found and their size (volume) distribution. Such statistical data will be further qualified with descriptions of the likely petroleum phase (heavy oil, light oil, or gas), the range of plays, depths to targets, and reservoir type and quality. The following three subsections examine different approaches to this same problem, of determining how much petroleum remains to be found.

4.8.2 Areal richness and prospect summation

Superficially, the simplest approach to calculating the yet-to-find volume of petroleum in a basin is first to calculate how much petroleum has been generated. From such a figure, the quantity of petroleum already found and that assumed to be lost to the surface or on migration routes is then subtracted. It will be appreciated that such an approach is subject to large errors. The only figure that can be established with reasonable confidence is the quantity of oil already found. We examined earlier in this chapter the steps involved in calculating the quantity of generated petroleum. It involves estimation of the organic richness of the source rock and then determination of its maturation and expulsion history. Some estimate of the petroleum lost on the migration routes may be possible, but it is extremely difficult to quantify how much has escaped *en route* to the surface of the Earth. In consequence, yet-to-find estimates based upon petroleum generation and migration criteria are likely to be order-of-magnitude estimates. Nonetheless, this approach may be the only one available in new basins, which lack the historical data that can be used to calculate a more accurate yet-to-find figure. In its simplest form, this geochemistry-driven approach does not attempt to place petroleum in prospects, and aside from the definition of the basin or sub-basin under study there is no attempt to specify the location of the yet-to-find petroleum.

In areas where some exploration has taken place, it may be possible to use a modification of the geochemical method. Clearly, while it may be possible to estimate the quantity of generated petroleum, it is not easy to then calculate how much of it might have been trapped. However, if part of a basin has been explored, the quantity of petroleum trapped in the explored part is reasonably well known. It is then a trivial matter to calculate the areal richness of the basin from the sampled portion. This figure, commonly quoted as millions of barrels per square kilometer, may then be applied to the unexplored part of the basin.

This approach was used in the early 1990s to estimate the yet-to-find volume of petroleum in the unexplored part of the Eastern Venezuela Basin (Case history 4.11). The eastern part of the Eastern Venezuela Basin lies beneath swamp and the distributary channels of the Orinoco River. It remained unexplored long after the more accessible parts of the basin had been explored. In Trinidad to the east, the basin statistics were well known, as they were in the El Furrial and associated trends to the west and south. The areal richness of the explored parts of the basin is about 14 million barrels for each square kilometer. This translates to about 7 billion barrels for the area beneath the swamp. With this areal richness approach, like the petroleum generation approach, there is no attempt to define specifically where the petroleum is to be found.

In well-explored and well-understood basins, it is common to use a different approach, which is usually referred to as "prospect summation." In this method, the likely petroleum in place and the reserves are calculated for all the mapped prospects in one's own licenses, competitors' licenses, and unlicensed acreage. The reserves in all are then summed to give a grand total of reserves and numbers of prospects. Historical data on the finding rate can be used to calculate the risked reserves for the summed prospects.

At first sight, such a prospect summation approach may appear to be the most rigorous and the one most likely to deliver an accurate estimation of the "yet-to-find" petroleum volume. However, prospect mapping, even in well-explored basins, can be problematic. Bain (1993) published a stunning example of prospect summation. Bain (op. cit.) worked for the UK licensing authority during the 11th Licence Round. For one particular block on offer, 16 companies made license applications. These companies had mapped prospects in the block based on stratigraphic pinchout of Tertiary sandstones. Each company would have had its prospect summation yet-to-find for the block. Collectively, the 16 companies covered the entire block with prospects (Fig. 4.63)! Clearly, the prospectivity had been overestimated by most of the applicants. Indeed, it is common for yet-to-find figures derived from prospect summation to be high. Geologists are very good at recognizing possible trapping geometries, but less proficient at assessing risk. Ideally, yet-to-find figures for a particular area should be calculated using several different methods, the outcomes compared, and the possible errors identified.

4.8.3 Pool size distribution

The various methods used to calculate the volume of petroleum yet to be found have been outlined above. The approach using geochemical data requires at very least a rudimentary understanding of source-rock deposition, subsequent maturation, and expulsion of petroleum. The prospect summation approach requires detailed knowledge of at least part of a basin. It is often the case that when a company first enters a new basin, such data may not be readily accessible. This has certainly been the case over the past decade, at a time when Western companies have gained access to formerly closed oil industries in South America, the former Soviet Union, and most recently the Middle East. In these areas and elsewhere, the most commonly available data are estimates of field re-

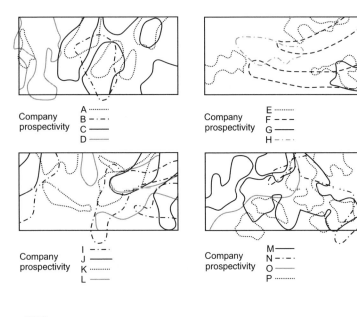

Company prospectivity
A ....
B —·—·—
C ———
D ———

Company prospectivity
E ....
F — — —
G ———
H —··—··

Company prospectivity
I —·—·—
J ———
K ....
L ———

Company prospectivity
M ———
N —·—·—
O ———
P ....

Fig.4.63 In the UK continental shelf 11th Licence Round, 16 companies applied for the exploration block shown in this figure. The primary targets were Tertiary sandstones in stratigraphic traps. On each of the four block outlines are four prospectivity maps, one for each company. Together, they cover the whole block, a clear case of where some companies' interpretations must have been wrong. (From Bain 1993.)

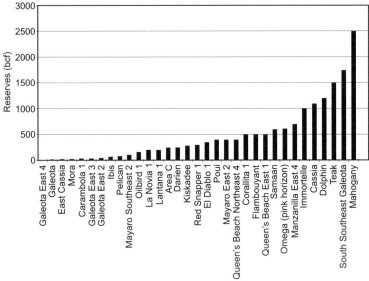

Fig.4.64 The field (reserves) size distribution for the Colombus (gas) Basin, offshore eastern Trinidad. Where possible, it is preferable to plot gas or oil in place distributions rather than reserves distributions, because different recovery factors will be used by different companies to calculate reserves figures. However, in practice for gasfields there is little difference between production strategies, and hence recovery factors, since most are produced via pressure depletion.

serves size (Campbell 1997). At the technical level, we are interested in oil or gas in place rather than reserves, but rarely are such data available. Moreover, although the reserves data are commonly flawed, or even downright wrong for individual fields, much information can be gained from studying the distribution of the data.

Regardless of whether pool size data be considered on a global, basinal, or individual play scale, the distribution pattern is much the same. There are a few giant pools, quite a lot of middle-sized pools, and many small pools (Fig. 4.64). Where reserves figures alone are published, the abundance of small fields is often underestimated, since small fields are not actively explored for, and when found are ignored because they are uneconomic to produce (Fig. 4.65). This hyperbolic, logarithmic, or geometric distribution of pool sizes is what one might intuitively expect from nature—a few big traps and numerous small ones.

The expectation that nature will fit its pool size distribution to one of these mathematical curves has been used in many instances to generate the yet-to-find in many partially explored basins. Such a rigorous mathematical treatment of pool size distribution data seems persuasive. From such an analysis it is not only possible to derive a single yet-to-find volume for a whole basin, but also to predict the pool sizes of the fields to be found. However, the precise choice of curve and of intercepts has a dramatic influence on the total yet-to-find figure, as well as on the prediction for the individual pool sizes. In practice, the range of uncertainty remains large. Despite such caveats, plotting pool size distribution data for partially explored basins remains a useful analytic method, particularly when combined with subsequent work designed to better understand the distribution of pool sizes.

The Iranian Sector of the Persian Gulf is a fine example of where pools seem to be missing from the pool size distribution. The data are shown in Fig. 4.66, from which it is clear that there is a large gap between the size of the largest fields (>2500 mmbbl boe, where "boe" means "barrels of oil equivalent") and the upper size limit of the smaller fields (250 mmbbl boe). It is possible to speculate that fields in the size range 250–2500 mmbbl boe remain to be found.

4.8.4 Creaming curves and destruction of value

The largest, simplest structurally trapped pools of petroleum tend to be found earliest in the exploration history of a basin. As the number of exploration wells on an individual play or within a whole basin increases, so the average volume of oil proven by each discovery diminishes. At some point in the exploration history of a basin, the cost of exploration will exceed the value derived from the discoveries. The point at which this happens will be determined by many factors. The complexity of the geology, the difficulty of the operation (e.g., offshore versus onshore), proximity to market, the local fiscal regime, and of course the oil (or gas) price will all contribute to defining this point. When this point is passed, a company indulges in exploration with the high probability that it will lose money or destroy the value of its earlier discoveries.

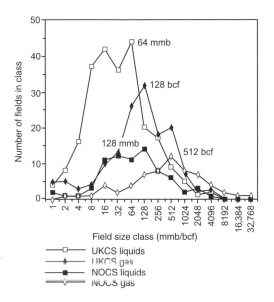

Fig.4.65 The pool size distribution for both oil and gas accumulations in the UK and Norwegian North Sea. The data are plotted using a geometric distribution of size categories. It is widely accepted that, due to sampling bias, large numbers of smaller pools are omitted from such plots. Small pools are not explored for, and even when found may not be recorded in statistical tables designed to capture economically successful exploration. (Reproduced courtesy of BP.)

Fig.4.66 The reserves size distribution for Iranian petroleum fields in the Persian Gulf. Unlike the mature area of the North Sea, there is no progression from many middle-sized fields to few very large ones. This leads the exploration geoscientist to suspect that fields in the reserves range 250–2500 mmbbl may yet be found. Moreover, when the data used to compile this graph were examined further, only one field was recorded as a stratigraphic trap. Thus even before necessary examination of geologic data, the geoscientist is likely to suspect that there are opportunities to discover more stratigraphically trapped petroleum within the Iranian Persian Gulf.

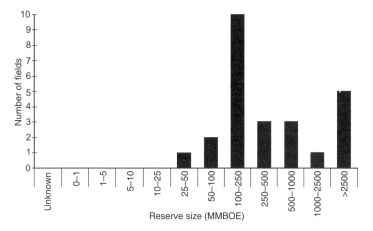

A very simple way of illustrating this process of diminishing returns is to plot the cumulative volume of petroleum discovered against time or the number of wildcat wells drilled. For a mature exploration play, the resultant curve will be hyperbolic, with the more recent discoveries adding little to the cumulative volume of discovered petroleum. These hyperbolic curves are commonly referred to as "creaming curves." The name comes from drawing an analogy with taking the cream (the best bit) from the top of the milk. For an immature exploration area, the end gradient of the curve is steep. Naturally, basins will reach exploration senility at different times after different numbers of wells have been drilled.

A detailed examination of a creaming curve for a basin may deliver additional information. It may, for example, be formed from a series of stacked hyperbolas, each of which corresponds to the discovery and subsequent exploitation of a different play (Fig. 4.67). Alternatively, discontinuities on the curve may indicate where changes in technology have opened a new part of a basin. This is the case for the Berkine Basin in Algeria, where improvements in seismic technology have allowed better resolution of the tilted fault block plays that lie beneath thick, modern sand dune fields (Fig. 4.68).

Creaming curve analysis is therefore a powerful tool that

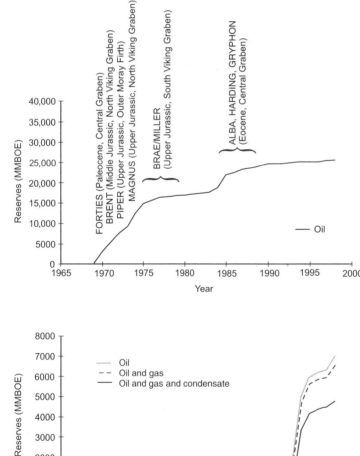

Fig.4.67 The creaming curve for oil accumulations in the UK North Sea Graben system, with reserves plotted against date of discovery. The steepest part of the creaming curve occurred during the first six years after the initial discoveries. During the period 1970–5, the major Paleocene play of the Central Graben, the Brent play of the North Viking Graben, and several of the Upper Jurassic plays were discovered. Although new discoveries and a few lesser plays were added between 1975 and 1983, the volume of discovered oil was much less than in the initial phase of exploration. The step in the creaming curve centered on 1984–5 corresponds to oil discoveries in hard-to-find (other than by accident!) stratigraphic traps within Eocene sandstones.

Fig.4.68 The creaming curve for the Berkine Basin, Algeria (equivalent to the Ghadames Basin, Libya). The massive increase in the quantity of petroleum found in the mid-1990s compared with the small quantities found in the preceding 25 years has been linked directly with advances in seismic technology. Before 1993, the seismic images produced from beneath the thick, modern sand dunes in the Sahara were poor. This curtailed exploration activity in the area, despite the fact that it was known to be an oil province.

enables the explorer to determine whether an area is "played out" — whether it is likely to be devoid of additional economically viable petroleum pools in the known plays. The explorer may then choose to encourage his or her company to exit the basin or target a new unexplored play.

4.9 RISK AND UNCERTAINTY

4.9.1 Introduction

Exploration for petroleum is commonly described as a risky business. This is an accurate statement, because the most likely outcome for most exploration attempts is failure. However, the rewards for success are usually, although not always, large. Moreover, despite the odds being somewhat better than the average lottery, the stake money, or "cost to find out," is substantially larger. When it comes to success, the analogy with gambling on the outcome of a horse race or other sporting event, card game, or casino activity breaks down further, for there is much uncertainty about the size of the win. The exact value of the success may not be known for many tens of years after the discovery. That is, the value may not be known until the field becomes (economically) depleted, the last barrel of oil sold, and the field abandoned.

In this short section, we examine both risk and uncertainty as applied to petroleum exploration, appraisal, development, and production.

4.9.2 Risk

Risk may be defined as the probability of a specific or discrete outcome. For example, a single toss of a coin may deliver a head or a tail. The probability of each outcome is definable before the event. There is a 50% probability of obtaining a head and a 50% probability of obtaining a tail. Another way of expressing this is to say that the probability of obtaining a head (or tail) is 1 in 2. In other games of chance, the number of possible outcomes may be larger, and in consequence the probability of obtaining a specific outcome may decrease. For example, the probability of rolling a "three" on a normal cubic die is 1 in 6, while the probability of hitting a "three" in roulette is 1 in 37.

Clearly, it is possible to consider more than a single event. To return to our coin-tossing example, if two sequential tosses are made the possible outcomes are as follows: head & head, head & tail, tail & head, and tail & tail. The probability of each outcome is 50% × 50% = 25%. Naturally, if the order is unimportant, head & tail = tail & head, and thus the proba-

bility of obtaining one head and one tail is 50%. A sequential series of risked events can be represented diagrammatically on what is commonly known as a decision tree (Fig. 4.69).

Before examining the probability (risk) of success in petroleum geoscience, we will examine our simple coin-tossing experiment one stage further. Some of the readers of the above paragraphs will already have realized that the 50% probability assigned to heads and tails is only an approximation. There is an exceedingly small but finite possibility that the coin will land and remain on its edge, displaying neither head nor tail. Although the probability of such an outcome is exceedingly small, we highlight it to emphasize the difference between a 2D mathematical puzzle — "heads or tails," in which there are only two possible outcomes — and the real world, in which there are three outcomes. Petroleum geoscience is a rather more complicated system than "real-world coin-tossing." In consequence, there are commonly many more possible outcomes for a particular event than heads or tails. The different outcomes may not have equal probabilities and, importantly, there may be outcomes that are not recognized.

Earlier in this chapter, we examined the representation of play risk in terms of maps, and the use of prospect risk in terms of prospect ranking and management. However, nothing was mentioned regarding the value of risk to assign to a particular play or prospect. In a well-explored area, with abundant information on the drilling success rate, the application of prospect risk may be no more than adoption of the historical figures. For example, the success rate for discovering oil in the Brent Province of the North Sea is 1 in 3 (also reported as a 33% chance of success) and therefore the risk on a new prospect could be given a value of 1 in 3. Such a risk may be deemed unacceptably high in the Brent Province, where today virtually all the exploration is directed toward reserves growth for existing fields. In this instance, an attempt could be made to reduce the risk (increase the chance of success) by studying both the failed wells and the successful wells in the surrounding area, to determine what components failed and what worked. If, for example, such a study revealed that 80% of the failed exploration wells were the result of drilling on a nonexistent trap due to inappropriate depth conversion from the seismic, then quite clearly such data for the new prospect would require careful examination.

The approach to risk assignment in areas with few exploration wells is much less analytic than used above. The team working the area needs to describe the range of possible outcomes for the component parts of both play and prospect, remembering that the prospect risk needs to be independent of the play risk. In other words, the prospect needs to be considered on the basis of the play working. It is then up to the mem-

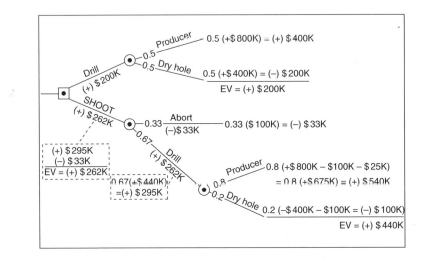

Fig.4.69 Decision tree analysis for a simple development well problem. The decision nodes are squares and the chance nodes circles. The problem here was one of deciding whether to drill a development well on the periphery of a field with or without additional seismic data. There was concern that the planned location could have been structurally low, and that with new seismic data it may have been possible to drill the well in a more favorable location. There was also uncertainty about the reservoir quality, although it was deemed that this could not be addressed without drilling. The costs in this example were as follows: cost of seismic data, $100,000; cost of deferring production to allow acquisition and interpretation of seismic data, $25,000; cost of a dry hole, $400,000; and mean present value of a producing well $800,000, including the cost to drill. There were two options, to drill without new seismic data (A) or with new seismic data (B). In each instance, there were two possible outcomes, success or a dry hole. However, the probability of success was improved through the use of the seismic data. A 50:50 chance of success without seismic data was improved to a 67% chance of success of ensuring that the well was in a favorable structural position. This resulted in an increase in the chance of success to 80% for a profitable producer. Clearly the preferred option is B, because the estimated present value was $262,000 compared with $200,000 for the option without seismic data. However, if the cost of seismic acquisition was doubled to $200,000, then the preferred outcome would be to drill without new data. (From Rose 1992.)

bers of the team, plus guests, to estimate the probability of occurrence of the outcomes. The mean, or robust mean, of the outcomes may then be used as the risk values for play and prospect. Personal bias is always present in such a decision, but the use of several people—or even lots of people—to deliver the risk values tends to obviate the optimism (or pessimism) of the explorer who is working directly with the data.

The committee voting approach works well in many situations. However, by its very nature of averaging it tends to over-risk events that in reality have a very high probability of occurrence and under-risk events with an extremely low chance of occurrence. Such an effect needs to be taken into consideration when data falls at the extremes.

4.9.3 Uncertainty

Uncertainty may be defined as the range within which the true value of a parameter lies. For example, once an oil discovery has been made by an exploration well, we can be certain we have an accumulation, but we may be uncertain about the size of the accumulation. Uncertainty about the size of an accumulation turns out to be the end product of uncertainty sur-

rounding each of the parameters that go into the calculation. For example, there will be uncertainties regarding the trap volume, the position of fluid contacts, the net to gross, the porosity, the oil saturation, and more. Moreover, the degree of uncertainty will vary between parameters.

In order to calculate the uncertainty surrounding an end product such as oil in place, reserves, or net present value (NPV), it is necessary to estimate the uncertainty associated with the input parameters and propagate these component uncertainties through the calculation. Uncertainty is commonly displayed as a probability distribution, in which the *x*-axis is the value of a parameter and the *y*-axis the probability of occurrence of that parameter (Fig. 4.70). There are many different shapes to probability distribution graphs (Fig. 4.71). The full choice of distributions to represent a parameter is unlimited. However, because in the oil industry we often know very little about our systems, there is a tendency to use simple distributions: ones in which we can define minima, maxima, and most likely occurrences. For example, the Viboral case history at the end of this chapter (Case history 4.11) examines the uncertainties associated with oil in place.

The uncertainty associated with a product can be calculated from the inputs in a variety of ways. Formulas exist for adding, subtracting, multiplying, and dividing probability levels, and for propagating uncertainties through calculations (Davis 1986). Alternatively, Monte Carlo simulations can be performed. In a Monte Carlo simulation, many thousands of calculations are performed (user-defined), in which the input parameters are chosen randomly from the defined distribution for that parameter. This allows calculation of the output distribution, from which statistical measures can be derived (mean, median, deciles, skew, etc.). In these Monte Carlo simulations it is also possible to introduce correlation between input variables. As an example, consider a prospect

in which the structure is well defined but the petroleum-phase risk is great. In this instance, the gross rock volume in the oil leg and the gross rock volume in the gas leg would be negatively correlated. Such a negative correlation could be used in the Monte Carlo simulation. In an alternative situation in which the gas/oil contact is defined from a well but the structure is ill-defined from seismic, a positive correlation between gross rock volume for oil and gas could be used. This captures the situation in which the oil and gas volumes both vary sympathetically with the trap volume. Examples of uncertainties and how to deal with them occur throughout the following chapters.

4.10 CASE HISTORY: NORTH SLOPE, ALASKA

4.10.1 Introduction

Alaska contains the largest oilfield in North America, Prudhoe Bay. It was discovered in 1967 and it is expected to deliver well over 50% of the estimated 20 billion barrels of reserves attributed to this prolific petroleum province. The license area that contained the Prudhoe acreage also covered the area of what is now the Kuparuk Field. By 1990, daily production from Kuparuk made it the second largest field in North America.

The lease sale of the area in which both Prudhoe and Kuparuk were discovered delivered $6.1 million in winning bids. A further 44 lease sales took place in the two decades fol-

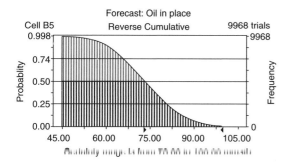

Fig.4.70 A probability distribution (exceedance curve) for a hypothetical oil accumulation based upon a Monte Carlo simulation using distribution data for net rock volume, porosity, oil saturation, and formation volume factor. Tabulated percentiles and mean are also given. (Constructed using Crystal Ball™ software.)

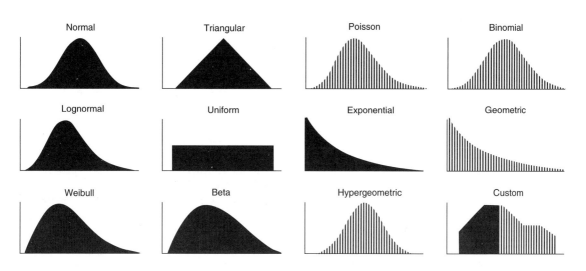

Fig.4.71 A gallery of possible possibility distributions that can be used in Monte Carlo simulations. (From Crystal Ball™ software.)

lowing discovery of the two fields and these sales attracted $5 billion in winning bids. Despite this high outlay, few of these 44 leases have delivered significant discoveries. Clearly, companies indulging in the expensive bid process and equally expensive exploration on the Alaskan North Slope were in severe danger of making substantial financial losses or destroying the value of the early discoveries. With this in mind, BP undertook a regional review of the prime, Ellesmerian 1 play fairway in the mid-1980s. The purpose was to determine what oil remained to be found and hence develop an exploration strategy for subsequent lease sales.

The area for this Ellesmerian 1 study was large, approximately 300 miles east–west by 150 miles north–south. The quantity of data available for analysis was similarly substantial, with 210,000 km (130,000 miles) of 2D seismic data and 225 exploration wells, of which 107 had extensive geochemical data. Even some outcrop data were available. The main study team comprised five people working on the project for a year. Specialist stratigraphers and geochemists provided support services. Hubbard et al. (1990) published much of the geologic story reproduced here. S.P. Edrich, R.P. Rattey, and M.E. Bushey were important contributors to the internal company documentation of the study.

The component parts of the study were to map the distribution and thickness of reservoir intervals and petroleum source formations. This was followed by calculation of the quantity of petroleum generated as a function of time. Migration pathways were assessed, as were the probable migration losses. The study area was then divided into segments that carried similar exploration risks. The yet-to-find petroleum volumes were then calculated, the outcome of which was used as a basis for the exploration strategy.

4.10.2 Reservoir distribution and quality

The main reservoir in the Ellesmerian 1 play fairway is the Triassic Ivishak Formation (Fig. 4.72). It comprises a complex suite of alluvial fan and fan deltas overlain by fluvial sediments. These were shed southward from a northern landmass. The coarse clastics form a belt about 100 km wide, stretching from the north coast of Alaska southward (Fig. 4.73). Net to gross is high throughout (>80%) and there is evidence to suggest that the landmass from which the clastics were shed was high relief in the east and lower relief in the west. The northern edge of the formation is determined by a combination of depositional pinchout and erosion. The southern edge of the formation interdigitates with contemporaneous marine-shelf deposits.

The cleanest sandstones have a porosity in the range

15–25% and lie toward the northern, proximal edge of the alluvial system. Cementation is not extensive, although some porosity has been lost to siderite precipitation. Dissolution of chert grains beneath the Beaufortian unconformity has improved reservoir quality (Melvin & Knight 1984).

4.10.3 The petroleum charge system

The volume of petroleum available to charge the system was estimated by considering the volume, quality, maturity, and expulsion efficiency for all the known and probable source rocks within the area (Fig. 4.74). Calculations were performed as time series to enable estimation of the petroleum flux available as the basin evolved. To model the sequential maturity of each of the source rocks, it was necessary to develop a heat-flow history for the basin and a geologic (burial and uplift) history. Heat-flow data were borrowed from modern-day basins analogous to the basin development stages of the North Slope, which comprise stable continent, failed rift, and foreland events. Both seismic data and vitrinite reflectance data were used to calculate the burial and uplift history for the area.

The oldest significant oil source rock is the Shublik Formation. It generated and expelled almost all of its oil and much of its gas during the first two time slices modeled, the Lower Brookian (115–95 Ma) and the Middle Brookian (95–50 Ma). Today, only a tiny area remains mature for oil generation. The remainder of the formation is in the gas window or is overmature. The volume of the oil estimated to have been generated from the Shublik across the basin was 250–600 billion barrels, 58% of which was generated in the Lower Brookian, 38% in the Middle Brookian, and only 4% in the Late Brookian.

A similar calculation was made for the other source rocks in the area. The next stage was then to determine the most likely secondary migration routes for each source by time slice. The Ellesmerian 1 play fairway has the most efficient and lowest-risk petroleum charge system in the area (Fig. 4.75). A large part of the Shublik source kitchen overlies directly the Ivishak reservoir. Downward migration of petroleum from source to reservoir occurred over large areas. Direct evidence of this comes from the Ivishak itself. The uppermost 100 m of the formation commonly contains shows of oil even outside mappable closures (traps). The second charge system is from the Lower Kingak LVU (Low Velocity Unit) into the top Sag River Sandstone. The distribution of the Kingak/Sag River couplet is much the same as the Shublik/Ivishak couplet and they may be considered together. The third charge system for the Ellesmerian 1 play fairway is the most localized and least

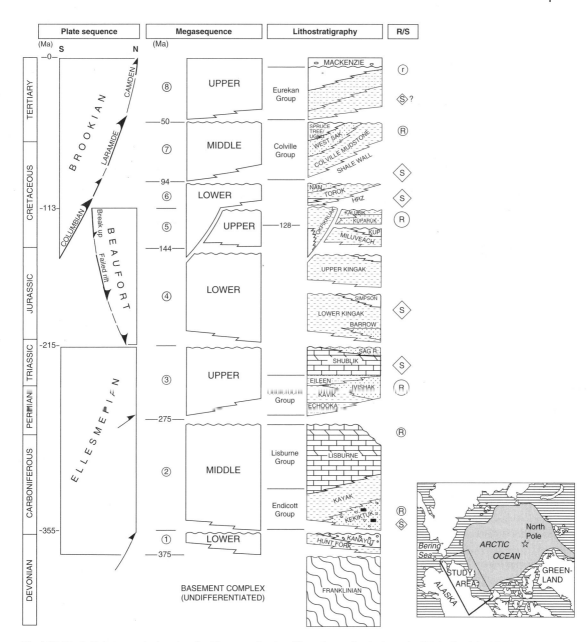

Fig.4.72 North Alaska summarized stratigraphy. All ages are given in millions of years, after the timescale of Harland et al. (1990). Megasequences are named and numbered from 1 to 8, as indicated by the numbers in circles to the left of the central column. The HRZ and Shale Wall lithostratigraphic units are hemipelagic deposits at the base of the Torok and Colville Mudstone terrigenous deposits respectively. The "S" symbol within a diamond denotes a petroleum source rock and the "R" symbol within a circle denotes a reservoir rock. (From Hubbard et al. 1990.)

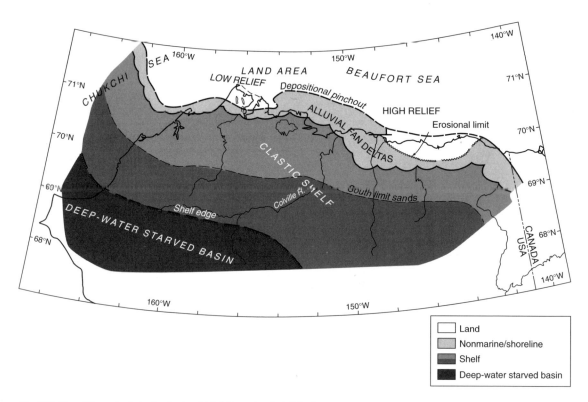

Fig.4.73 Upper Ellesmerian facies distribution, Ivishak Formation, Early Triassic.

important. It comprises direct downward charging of the Ivishak from the HRZ (High Radioactive Zone) where the base HRZ unconformity overlies the Ivishak. This situation occurs over a small area on the Prudhoe High.

The next stage of the project was to compare the petroleum volume and chemistry with the calculated yields of the five main source rocks. The maturity and expulsion calculations indicate the most important source rocks to be the Brookian HRZ and Shale Wall Formations. Together, these have generated about 50% of the total petroleum in the basin. However, geochemical analysis of the oils in the Alaskan fields showed that Shublik sourced oil accounted for 70% of the trapped petroleum. This was an important conclusion, for it showed that trapped petroleum was not simply a function of generated petroleum but, rather, in this petroleum-rich system, of the migration route. The simple Shublik–Ivishak system was much more efficient at trapping petroleum than the other systems. In consequence, any traps relying on the Kingak, HRZ, or Shale Wall sources must carry a substantial migration risk.

The basin analysis information was used to examine the likely migration routes and entrapment possibilities for petroleum generated in each of the source rocks and for each of the time slices. In the Lower Brookian (115–95 Ma) time slice, most of the oil generated from all source rocks was focused toward the northwest corner of the fairway. Although traps were present at this time, most were at temperatures of 65°C or less. In consequence, oil reaching such traps is likely to have been extensively biodegraded. During the Middle Brookian (95–50 Ma) interval, a combination of continued burial coupled with local isostatic uplift in the extreme west and thrusting in the east changed the migration pathways. Petroleum was focused toward the central and eastern part of the fairway for entrapment on the Prudhoe High. By this time, burial had taken traps to depths (temperatures) at which biodegradation of oil was not likely to be a factor. Subsequent to the Middle Brookian interval, trap formation continued in the far west (the Beaufort Sea area), but most of the new oil was generated far to the east. Thus none of the source rocks were able to charge the new traps. Finally, in the past few million years it is probable that some gas has been generated from Kekikituk coals. This gas may have charged the gas

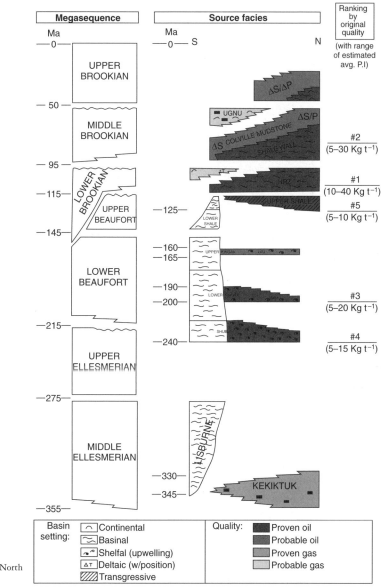

Fig.4.74 Source rock systems on the Alaskan North Slope.

caps in the Prudhoe Bay and Endicott fields, and promoted de-asphalting of the oil in the fields, with the consequent production of tar mats.

4.10.4 The distribution of the top seal

The next step in the evaluation of the Ellesmerian 1 play fairway was to map the distribution of the top seal. Over much of the area, the Ivishak Formation is overlain by the Shublik mudstones, which act as both as source and seal. However, on the Barrow Arch the Shublik and some of the Ivishak were eroded during the Early Cretaceous phase of rifting. The eroded Ivishak lies below what is known as the "128 Ma unconformity." Thin but extensive sheet sandstones (the Beaufortian Kuparuk sandstones) overlie this unconformity, so ensuring that much of the oil drained from this part of the

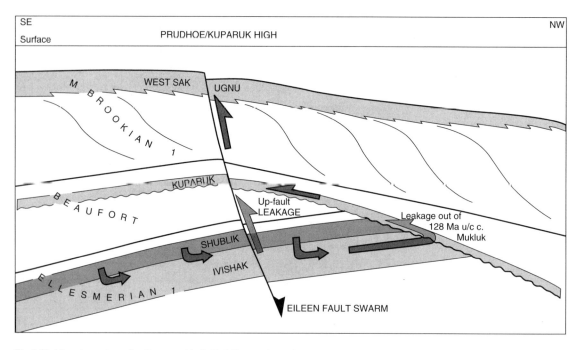

Fig.4.75 Migration pathways for oil generated in the Shublik source interval.

Ellesmerian play fairway (Fig. 4.75). Leakage of oil through these sandstones has been used to explain the failure of the Mukluk well in the early 1980s. Mukluk was famous before it was drilled because the prospect was of comparable size to Prudhoe Bay. When drilled, it was found to contain only residual oil and thoroughly oil-stained Kuparuk thief sandstones on top of the Ivishak. It now seems probable that oil that passed through the Mukluk area eventually pooled in what is now the Kuparuk Field in the younger Beaufortian fairway. Lesser and partial leakage also occurred through faults on top of the Prudhoe High.

4.10.5 Drilling statistics and common risk segments

The Ellesmerian 1 play fairway has been thoroughly explored. Eight fields have been found, of which Prudhoe Bay is by far the largest. It will ultimately produce substantially more than 10 billion barrels of oil. The other fields will typically deliver a few hundred million barrels each. All eight fields have been discovered within the same area, on or close to the regional Prudhoe/Kuparuk High.

A combination of the geologic analysis and the drilling statistics allows the Ellesmerian 1 play fairway to be divided into four common risk segments (Fig. 4.76). Common risk segment 1 is the area of the Prudhoe/Kuparuk High. Here the reservoir risk is moderate, governed by the effectiveness of the Ivishak Sandstone reservoir. High-quality reservoir occurs in the proximal parts of the alluvial–fluvial systems tract, and where dissolution of labile grains has increased porosity and permeability. For the same area, the charge risk is low because of the juxtaposition of mature Shublik source rock with the reservoir. The top seal is the key risk, because of the breaching that occurs beneath the "128 Ma unconformity," with the high probability of thin thief sands at this horizon. In consequence, only prospects that have well-defined four-way dip closure at the basal Beaufortian level have a low top-seal risk.

Common risk segment 2 lies immediately west of CRS 1. It contains no proven reserves. The reservoir and top-seal risks are the same as for CRS 1. However, the charge risk is high. The basin analysis work on this area showed that many of the traps had developed in the past 35 Ma. We know from the same analysis that this interval postdated the main phases of Shublik/Kingak source-rock maturation and expulsion. In other words, the traps formed after migration.

Common risk segment 3 lies west of CRS 2. It too contains no proven reserves. It shares a similar reservoir risk to segments 1 and 2 and has a similar charge risk to that of segment 2; traps probably developed too late to capture migrating oil.

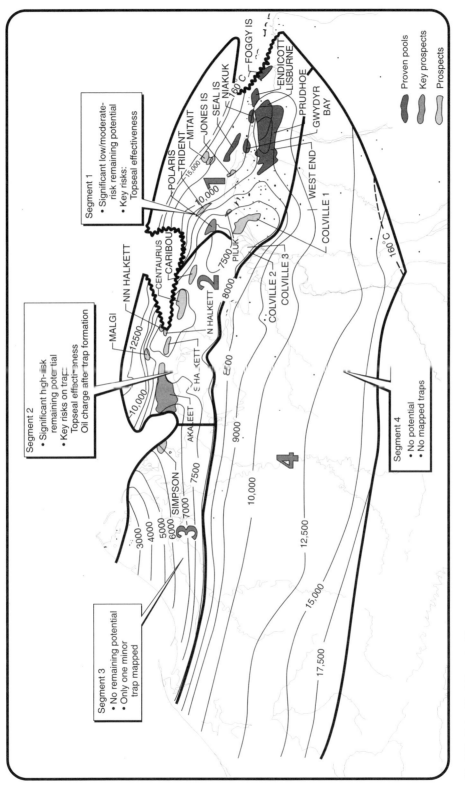

Fig.4.76 Alaskan North Slope play fairways.

In addition to the major charge risk, it also has a major seal risk. The Shublik Mudstone is only poorly developed in the west, where it grades into a sandy shelf facies.

Common risk segment 4 is the largest area, lying due south of segments 1–3. It contains no proven reserves. It has also only been lightly explored. The area is characterized by low-angle homoclinal dip to the south. Here the reservoir risk is high. The area covers what was the distal part of the Ivishak alluvial–fluvial system. This area is also distant from the zone of sub-unconformity, porosity enhancement. The burial depth is greater than for CRS 1, and this in itself is likely to be associated with a reduction in reservoir quality due to compaction and cementation of the remaining Ivishak sandstones. The charge and top-seal risks are low. In consequence, the area is one of high risk, due to the reservoir issue. However, if a different reservoir were to be identified, a new petroleum play could be opened.

4.10.6 Conclusions

The product of this regional review of North Alaska was a recommendation to perform no further exploration in the Ellesmerian 1 play fairway. This led ultimately to the relinquishment by BP its acreage in the western and central Beaufort Sea. This was achieved by early 1987.

4.11 CASE HISTORY: VIBORAL, EASTERN VENEZUELA

4.11.1 Introduction

In this case history, we examine an exploration opportunity in eastern Venezuela. The likely oil in place is estimated, as are the prospect risks. The background to this case history is that Venezuela nationalized its oil industry in 1975 (Betancourt 1978). By the early 1990s, the Venezuelan government was considering allowing foreign and private companies back into its oil industry. The Quiriquire Field (Section 4.5) was offered for reactivation in the First Reactivation License Round. No bids were received. It was again offered in the Second Reactivation License Round. This time, exploration acreage to the south of Quiriquire was included in the license area. The additional area covered what could be an eastern extension of the prolific El Furrial Trend, discovered by Lagoven (a state oil company) in the 1980s (Carnevali 1992). There was much interest in the new offer, and subsequent to bidding the unit was awarded to Maxus (a US exploration company) and Otepi (a local Venezuelan company). Early in 1994, Maxus invited potential partners to farm into the license.

For the purposes of this case history, we consider only the Viboral prospect. The other leads identified in the block and the potential value of the existing Quiriquire shallow and deep fields are ignored. The following data were available:

• A location map (Fig. 4.77);
• a description of the basin setting and local tectonics (Fig. 4.78);
• a map of TWT to top reservoir (Fig. 4.79) — we take it as a given that the mapping was accurate and the horizon identified as top reservoir was indeed top reservoir, and we should add that this was not certain at the time of the farm-in;
• data for a time-to-depth conversion (Fig. 4.80);
• a description of the reservoir in the El Furrial Trend and the time-equivalent interval in Quiriquire Deep to the north;
• porosity and permeability data from El Furrial, Quiriquire Deep, and potential analog reservoirs in Colombia (Fig. 4.81);
• a description of the regional seal;
• a description of the source rock for El Furrial (Fig. 4.82) and data sufficient to make estimates of maturity and phase (to an approximation — gas has no value in eastern Venezuela). From these data, oil (or gas) in place (STOOIP) may be calculated and estimates of the risk associated with trap, source, and reservoir made. It is even possible to consider component risks associated with all three parameters. For the trap, this could include examination of the risk associated with a four-way dip-closed structure or one partially dependent upon a fault seal. For the reservoir, both presence and effectiveness (porosity and permeability) need to be considered. Component source risks include those attached to presence, organic matter type and maturity in the source kitchen, and phase in the reservoir.

4.11.2 The problem

The location

The Quiriquire License is in eastern Venezuela, a few miles east and north of the oil town of Maturin. The northern part of the license is hilly, while to the south are plains (Fig. 4.77).

The tectonic setting

Eastern Venezuela is dominated by a neoformed coastal mountain range (thrust belt). The east–west range developed on the northern margin of a foreland basin that occupied much of what is now eastern Venezuela (Eva et al. 1989). The foreland basin developed in the Oligocene (Carnevali 1992; Fig. 4.78). Compression has given way to dextral transpression, with the Caribbean Plate moving eastward relative to South America at a few centimeters each year.

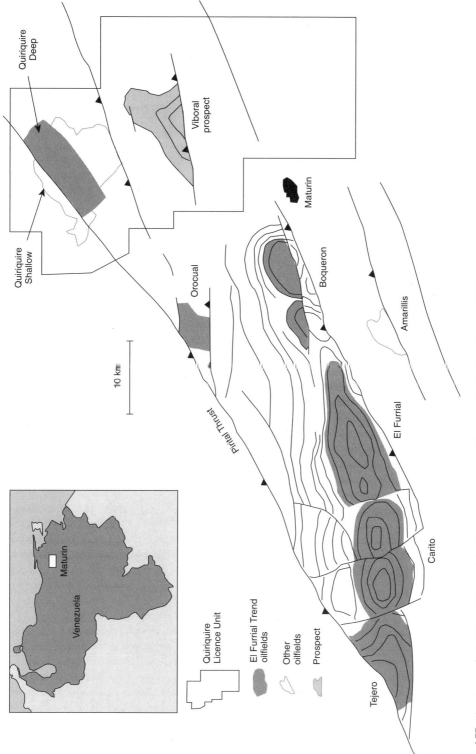

Fig.4.77 A location map for the El Furrial Trend and the Viboral prospect in eastern Venezuela. (Modified from Carnevali 1992.)

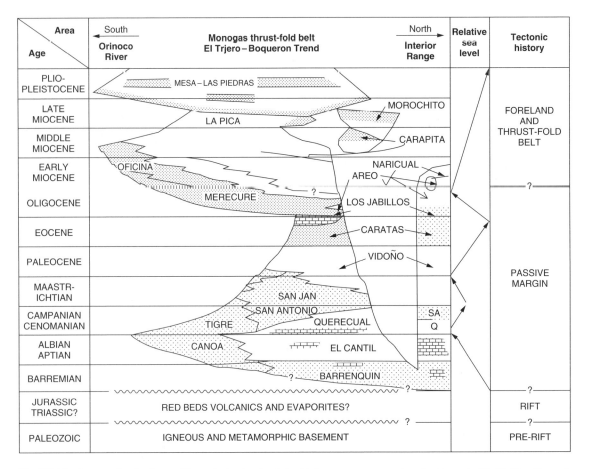

Fig.4.78 The chronostratigraphy of eastern Venezuela, showing the relationship between source, reservoir, and seal intervals. (Modified from Carnevali 1992.)

The prospect map

The trap was mapped on a coarse seismic grid with a typical line spacing of 1.5–2.0 km (Fig. 4.79).

Depth conversion

Data to enable calculations of depth from the seismic map are given in Fig. 4.80. The data are taken from the seismic stacking velocities across the Viboral prospect and from check shot data (from deep wells in the El Furrial Trend).

The reservoir

All the potential reservoir horizons are sandstone. At the prospect level (depth), those which could form the reservoir are the Oligocene Narical, the Areo or Los Jabillos formations, the Cretaceous San Juan Formation, the Eocene Caratas or Paleocene Vidoño formations.

The quality of the El Furrial reservoirs is high (Fig. 4.81). The sandstones are clean, quartz arenites. Most are medium-grained and cemented only by quartz. Average porosity is about 16% and average permeability 500 mD. Locally, the permeability may be as high as 3500 mD. Specific poroperm data for the El Furrial reservoirs were not available. As such, porosity and permeability data from analogous, quartz-cemented quartz arenites may be used. The available analog data are from the Cusiana Field in Colombia (Cazier et al. 1995), the Miller Field of the North Sea (Gluyas 1985), and Tertiary sandstones from the Paris Basin (Cade et al. 1994).

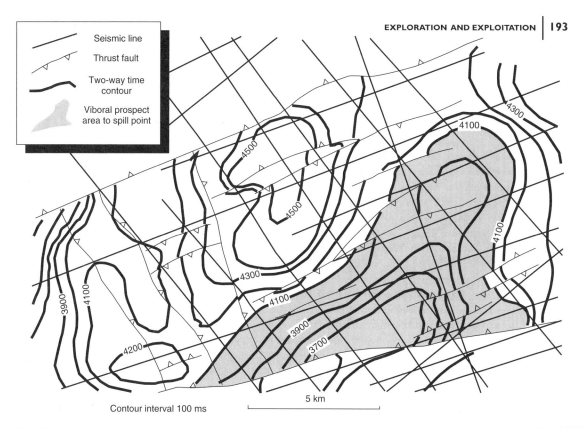

Fig.4.79 A simplified map of the Viboral prospect. The contours are in milliseconds two-way time. (Mapping by Spencer Howe, November 1993.)

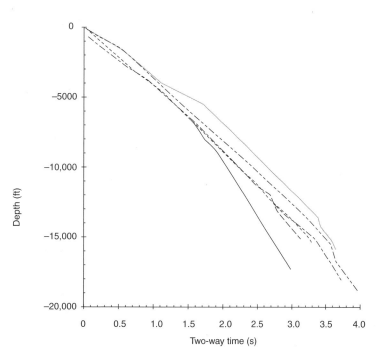

Fig.4.80 Time-to-depth conversion for wells in the El Furrial Trend. (Constructed by Mark Shann, November 1993.)

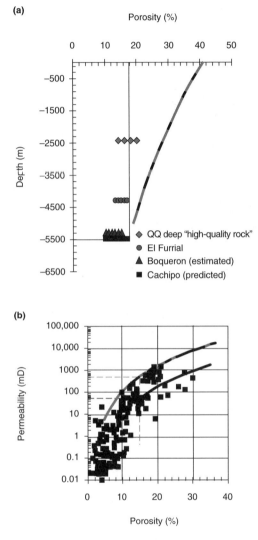

(a)

Fig.4.81 (a) The porosity–depth relationship for the El Furrial Trend. (b) The porosity–permeability relationship for the El Furrial Trend and similar quartz-cemented sandstones from Colombia, the North Sea, and the Paris Basin.

The seal

Mudstones can be mapped regionally and are well developed throughout an area much greater than that of the Quiriquire License.

The source

The primary source rocks for the Eastern Venezuela Basin are the Upper Cretaceous Querecual and San Antonio Formations, which belong to the Guayuta Group (Fig. 4.78). The Guayuta Group comprises calcareous and siliceous shales, deposited in a deep, anoxic marine basin. The average thickness is about 610 m (2000 ft) and the maximum thickness at outcrop is 1040 m (3400 ft). Aymard et al. (1990) suggest that up to 55% of the thickness could be considered source rock, and such source potential intervals have between 0.25% and 6.6% of total organic carbon with a hydrocarbon yield of 454 mg of hydrocarbons per gram. About 200 km² of the thrust slab that contains the Viboral prospect lies at depths where the temperature is in excess of about 110°C. At this temperature, the source rocks (estimated at 200 m thick) will generate and expel petroleum (Cooles et al. 1986).

4.11.3 The solution

The trap

It is clear from the mapping (Fig. 4.79) that the trap is a hanging-wall anticline on the back of a thrust. The trap is three-way dip-closed, with the southern margin fault closed. There is no indication of an independent four-way dip-closed portion of the trap. The top seal is provided by the thick mudstones of the Miocene Carapita Formation. The area of the trap is about 49 km² (12,000 acres). Depth to crest, calculated by combining the depth map with the velocity depth data, is 5500 m ± 500 m (18,000 ± 1650 ft). Vertical closure for the prospect, based on the same velocity transform, is about 1000 m (3280 ft), with a range of 500–1500 m (1650–4920 ft).

The critical factor for this trap is the fault seal on the southern margin. Such a trapping mechanism is known to be effective for the Quiriquire Deep Field (Fig. 4.77), but has failed elsewhere in the trend. Given this uncertainty, the probability of success for the trap was estimated at 0.7.

The reservoir

An important consideration when assessing the reservoir distribution in the area is that the north–south compression has telescoped the facies belts. Today, the Viboral prospect and Quiriquire Deep are only 15 km apart. During the Lower Tertiary, these locations were probably more than 25 km apart.

In order of significance, the potential reservoirs are as follows:

• The Oligocene Naricual and Los Jabillos Formations.

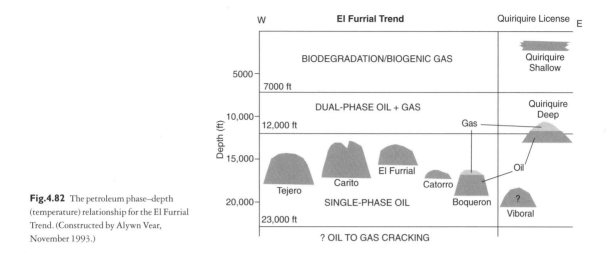

Fig.4.82 The petroleum phase–depth (temperature) relationship for the El Furrial Trend. (Constructed by Alywn Vear, November 1993.)

Table 4.5 Possible reservoir intervals within the Viboral prospect.

Formation	Thickness (m)	Evidence
Naricual/Los Jabillos	50	Distal location of Quiriquire Deep = 25 m
		Boqueron contains about 100 m
		Orocual has a reported gross of 140 m
San Juan	70	El Furrial and Orocual net 200 m
		Distal Quiriquire Deep = 20–100 m
Caratas and Vidoño	0	Gross interval estimated at 100 m, but with a net of 0 m because of poor quality. The formations are absent in El Furrial and Boqueron, present in Orocual, and about 500 m thick in Quiriquire Deep

These regionally extensive transgressive sandstones are productive in both the El Furrial Trend and in Quiriquire Deep.
• The Cretaceous San Juan Formation. These are widespread shallow-marine sandstones, proven as productive in El Furrial, Orocual, and Quiriquire Deep.
• The Eocene Caratas and Paleocene Vidoño Formations. These shallow-marine sandstones are absent in the El Furrial Trend but present in Orocual and Quiriquire Deep.
The depositional environment for all of these sandstones is shallow-marine inner shelf on the then-passive margin of the South American Plate. The estimated thickness for the reservoir package in Viboral was 120 m, with a range of 90–250 m (Table 4.5). Net to gross is in the range of 0.3 (minimum), through 0.5 (most likely) to 0.7 (maximum).

Porosity for the reservoir in Viboral was estimated to be in the range of 10–17%, with an average of 14%. Permeability was estimated to be between 20 and 200 mD. The estima-tions were based on the few local data conditioned to the analog data trends. The permeability estimate is dependent upon encountering medium-grained quartz arenites (Fig. 4.81).

The critical factors linked with reservoir risk were determined as the presence of thick sandstones, given the paucity of such in Quiriquire Deep, and the reservoir quality of the sandstones at 5500 m, deeper than any of the other El Furrial reservoirs. The probability of success was estimated at 0.6.

The source

The Upper Cretaceous Querecual Formation is the main source rock for the whole of the Eastern Venezuela Basin. There may also be some contribution from the Miocene Carapita Formation. The migration path was most likely southward and upward to Viboral, from within the deeper parts of the thrust sheet carrying the Viboral Anticline. The

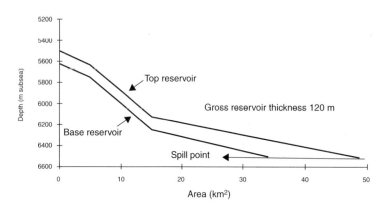

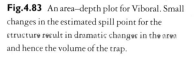

Fig.4.83 An area–depth plot for Viboral. Small changes in the estimated spill point for the structure result in dramatic changes in the area and hence the volume of the trap.

Querecual Formation commonly has a minimum petroleum yield of 15 kg t^{-1}. In the region of Viboral it is estimated to be about 200 m thick. The area is today at maximum burial and within the oil window.

The combined thickness and richness of the Querecual Formation could generate 60 mmbbl km^2 from the oil kitchen within the Viboral drainage area (about 200 km^2). Thus the total potential petroleum flux in the Viboral thrust sheet was about 12 billion barrels. We had no specific formula for calculating how much of the generated petroleum might reach Viboral. However, El Furrial has a STOOIP of 4300 mmbbl and a drainage area of 250 km^2. Clearly, if El Furrial is a valid analog, there is easily sufficient petroleum to fill Viboral.

The petroleum phase is a separate issue. The Querequal source contains Type II kerogen that is highly oil prone (Chapter 3). Moreover, many of the El Furrial Trend pools contain undersaturated oil. The depth of the Viboral prospect is also in the range in which a single-phase fluid might be expected; too deep for gas exsolution and too shallow for in-reservoir cracking of oil to oil and gas (Fig. 4.82). Despite such positive indications, there remains some residual risk that part, or all, of the prospect will contain gas. One concern is theoretical; that late- (hot-) generated gas from the source rock will have displaced some of the oil. The second concern is practical; the nearby Boqueron Field contains a gas cap that occupies 10% of the trap volume (Russo 1993). The origin of the Boqueron gas cap was not known at the time of prospect evaluation. Thus the probability of success was estimated to be 0.8 for a working, oil-prone source, capable of expelling sufficient petroleum to fill Viboral. Formation volume factors for the El Furrial Trend are typically 1.4, with a range of 1.25–2.0.

The combined risk factor for the trap (0.7), reservoir (0.6),

and source (0.8) delivers an overall probability of success of 0.34. Thus there was a probability of about 1 in 3 of finding an oil pool of the specified volume. Although 1 in 3 indicates that failure is more likely than success, such a figure is low compared with many exploration prospects.

The volume of petroleum

So far, we have only considered the risk of finding an oil pool. We now need to calculate its size. Data have already been provided for gross reservoir thickness, net to gross, porosity, and formation volume factor. Chapters 1 and 5 contain information on petroleum saturation and typical recovery factors.

Bulk rock volume can be calculated in a number of ways from the data provided. The prospect map would nowadays be produced on a workstation using a mapping package. This is unlikely to be readily available to the readers of this book. Thus we suggest the following. The trap area can be estimated by overlaying the Viboral prospect map on graph paper on a light table. For each depth interval, squares are then counted to estimate the area. Indeed, this is how the area of the Viboral prospect was calculated when first the farm-in was offered. Another method is to copy the map onto a sheet of paper. Measure the area of the paper and weigh the paper on an accurate set of scales. Now cut the paper into depth (time) intervals and weigh these. The weights of the pieces of paper may now be converted to areas by reference to the area and weight of the whole sheet.

The prospect area and volume can be represented on an area–depth plot (Fig. 4.83). Such plots graphically illustrate the sensitivity of prospect volume to slight changes in spill points derived from different interpretations of the seismic data or time–depth relationship. From the data here, together with estimates of petroleum saturation and the recov-

Table 4.6 Input parameters for the Viboral prospect oil-in-place calculation.

Parameter	Value	Mean	Variance	Kappa
Bulk rock volume ($\times 10^6$ m^3)		5836.58	1504.38	0.2578
Minimum	2152			
Mode	5837			
Maximum	9522			
Net to gross (fraction)		0.500	0.0816	0.1633
Minimum	0.300			
Mode	0.500			
Maximum	0.700			
Porosity (fraction)		0.137	0.0143	0.1049
Minimum	0.100			
Mode	0.140			
Maximum	0.170			
Average oil saturation (fraction)		0.870	0.0122	0.0141
Minimum	0.840			
Mode	0.870			
Maximum	0.900			
1/(Oil formation volume factor) (stb rb^{-1})		0.667	0.0624	0.0935
Minimum	0.500			
Mode	0.700			
Maximum	0.800			
Recovery factor (fraction)		0.300	0.0816	0.2722
Minimum	0.100			
Mode	0.300			
Maximum	0.300			

ery factor, it is possible to calculate a reserves figure of about 450 mmstb. The exact figure will depend on the choice of oil saturation (87%) and recovery factor (30%). One cubic meter contains 6.2898 barrels of oil at standard conditions.

We have now calculated a single, deterministic figure for oil reserves in Viboral, and given its risk of only 1 in 3 there may be sufficient data to make a farm-in offer. However, it is commonly very useful to have some estimate of uncertainty in reserves, since there is only one certainty about the reserves calculated above; that is, the figure is incorrect. Various approaches can be used to estimate uncertainty. We illustrate the use of the parametric method. Table 4.6 contains minimum, mode, and maximum data for the parameters needed to calculate the reserves. These data are given above. The information derived from these inputs includes the following:

$$\text{mean, } \mu = \frac{(\text{minimum} + \text{mode} + \text{maximum})}{3}$$

variance (for a triangular distribution),

$$\sigma^2 = \frac{(a(a-b) + b(c-a) + c(c-a))}{18}$$

where a is the minimum, b is the most likely value, and c is the maximum;

$$\text{kappa} = \frac{\text{variance}}{\text{mean}}$$

$$\text{impact on uncertainty} = 100 \times \frac{\text{kappa}^2}{\Sigma(\text{kappa}^2)}$$

Tables 4.7 and 4.8 contain two important pieces of information besides the mean reserves (436 mmstb). These are the impact on uncertainty of the key input parameters and the probability exceedance curve.

From Table 4.7 and Fig. 4.84, clearly 70% of the uncertainty in the estimated reserves comes from only two parameters, the recovery factor and the bulk rock volume. It may not be possible to do much about the uncertainty that surrounds the recovery factor before drilling the prospect and sampling any petroleum discovered. However, it may be possible to remove some of the 30% uncertainty in the bulk rock volume by shooting a 3D seismic survey over the prospect and better defining the trap. The cost versus

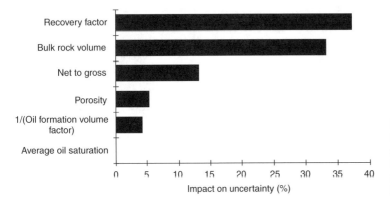

Fig.4.84 The impact on uncertainty for the estimated unrisked reserves for Viboral. Seventy percent of the uncertainty in reserves is carried by only two parameters, recovery factor and bulk rock volume. (Reserves calculation by Andrew Evans, November 1993.)

Table 4.7 An impact analysis for the Viboral prospect.

Parameter	Mean	Variance	1 + kappa^2	Impact on uncertainty
Bulk rock volume	5836.577	1504.38	1.066	33.29
Net to gross	0.500	0.08	1.027	13.36
Porosity	0.137	0.01	1.011	5.51
Oil saturation	0.870	0.01	1.000	0.10
1/(Formation volume factor)	0.667	0.06	1.009	4.38
Recovery factor	0.300	0.08	1.074	37.12
Reserves (mmstb)	436.439	194.99	1.200	—

the benefit of such a survey must first be assessed before a decision is taken. The power of the uncertainty analysis is that it allows identification of the major risks to delivery of a successful project. Knowledge of what the major uncertainties are allows either collection of data to reduce the uncertainty or planning of interventions to mitigate against the worst effects of the downside case, should it prove to be a reality.

The second important piece of data is the reverse cumulative frequency curve (Fig. 4.85 & Table 4.8). Should Viboral prove to be an oil discovery, there is a 90% probability that the reserves will exceed 231 mmbbl and a 50% probability that they will be greater than 399 mmbbl. Having such data helps the decision process when it comes to ranking investment options.

4.11.4 Conclusions

This case history illustrates two points. The first is the sort of approach that can be taken to prospect evaluation. Some of

the data for the prospect evaluation can be taken direct from the prospect. Other data need to be derived from analogs. The second point is that much uncertainty surrounds the estimates for the various parameters that are needed for reserves estimation. By including those uncertainties in the reserves calculation, options are created. Thus it will be possible either to buy more information to reduce the uncertainty (e.g., shoot a 3D seismic survey) or plan intervention strategies to avoid the worst of the downside case.

4.11.5 Postscript

An exploration well was drilled on the prospect during the first half of 1996. The trap was present, and within the depth range expected. Migration had delivered liquid petroleum from source to trap. However, in agreement with the greatest anticipated risk, the reservoirs were either absent (Tertiary) or of very poor quality (Cretaceous). Commercial petroleum flow rates were not obtained.

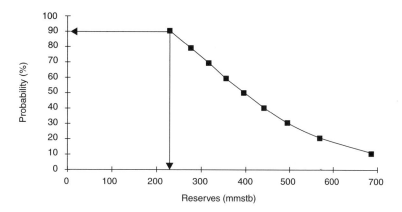

Fig.4.85 The reserves probability distribution for Viboral.

Table 4.8 Reserves estimates for the Viboral prospect (average and probabilistic results).

	Reserves (mmstb)
Minimum	17
Mean	436
Maximum	2566
Variance	195
Mode (log normal)	332
Median (log normal)	399
P_{90}	231
P_{80}	278
P_{70}	319
P_{60}	358
P_{50}	399
P_{40}	444
P_{30}	498
P_{20}	571
P_{10}	688

4.12 CASE HISTORY: SAN JOAQUIN, CALIFORNIA, USA

4.12.1 Introduction

Between 1900 and 1911, there were 40 or more professional geologists and geologic engineers working for the major oil companies in California. For its time, this was a greater concentration of petroleum geologic skills than anywhere else in the world (Blakey 1985). At much the same time, California was the most productive area in the world. With such a long and successful history, it is hardly surprising that by the 1980s the exploration opportunities for prospecting companies were a little more restricted than they had been three-quarters of a century earlier.

In this case history, we examine exploration possibilities in a thoroughly explored and largely depleted basin, the San Joaquin Basin. It is one of many basins within the tectonically active, wrench-controlled basins of mid-Tertiary age in the borderlands of California. The San Joaquin is also the basin that has produced the most oil, and still produces the most today. Moreover, the remaining reserves are also estimated to be greater than those of the other explored Californian basins. In 1990 the estimated total reserve base for the basin (oil produced and to produce) was about 12.7 billion barrels (Fig. 4.86).

Most, possibly all, of the obvious structural traps were drilled before or during the 1960s. Giant fields abound, but today exploration targets are more modest. Today, a new discovery with 30 million barrels of recoverable oil is about the level of expectation. Moreover, today such reserves volumes are likely to be found within subtle stratigraphic traps. The challenge facing the explorer of recent decades has been to predict the distribution of suitable reservoirs in appropriate stratigraphic trapping configurations. BP worked on this problem during the mid-1980s and we take the information for this case history from internal company documents (Rogers 1990).

4.12.2 The basin geometry and megasequence development

Sediments belonging to two megasequences, separated by a major regional unconformity, fill the San Joaquin Basin (Fig. 4.87). The Cretaceous megasequence, also called the Great Valley Sequence, comprises the deposits of deltas and subma-

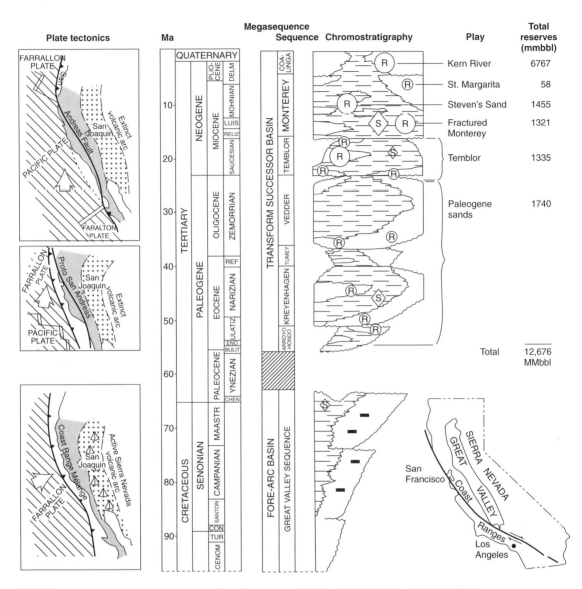

Fig.4.86 The stratigraphy, tectonic nomenclature, and petroleum resource definition within the San Joaquin Basin, California.

rine fans. It is up to 3000 m (10,000 ft) thick. These Cretaceous strata were deposited in a fore-arc basin that separated the active volcanic island arc of the Sierra Nevada to the east from an accretionary wedge in the west. The wedge lay above a subduction zone that carried the oceanic Farrallon Plate beneath the continental North American Plate.

Subduction ceased in the early Tertiary and was replaced by a strongly oblique, dextral strike-slip system. It was this major plate reorganization that produced the regional un-

conformity. The Tertiary sequence that developed above the unconformity did so in massively fragmented terrain. Along the length of the wrench system, rhombochasms opened, subsided rapidly, and were destroyed and uplifted with equal rapidity. In the San Joaquin Basin, the Tertiary megasequence may be divided into seven sequences, each separated by a local angular, erosional unconformity. BP targeted the Steven's Sandstone within the Middle Miocene Monterey sequence.

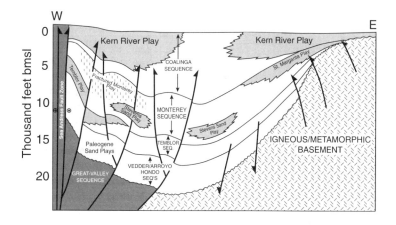

Fig.4.87 A sketch section across the San Joaquin Basin, California, showing the relationships between the petroleum plays.

4.12.3 The database

The data from 4500 exploration and development wells lies in the public domain. Data from about 450 of these were used to analyze the distribution of Steven's Sandstone. Given the dramatic lateral changes that occur within the lithology of the basin-fill sequences, high-quality, biostratigraphic control is essential for correlating between wells. Of the 450 wells used in the study, biostratigraphic data was available for 76. Moreover, about 25 of these wells had fully cored sections. The quantity and quality of the seismic data were also excellent. About 4000 km (2500 miles) of 2D seismic data on a 1.5 km × 1.5 km (0.9 miles × 0.9 miles) or closer grid were available. The exploration success rate and the resultant pool size distribution were well understood.

4.12.4 The Steven's Sandstone stratigraphic play

The Steven's Sandstone accumulated as a series of submarine-fan lobes. As stated earlier, almost all the structural traps in the Steven's Sandstone have already been drilled. What remains are the downflank stratigraphic pinchouts of the lobes into the surrounding basinal siliceous mudstones, a prolific source rock called the Monterey Chert (Fig. 4.88). Thus the most important technical issue that faced the team investigating this play was to identify the pinchout edges of individual sand bodies, be they channels or lobes. The approach to this problem was to construct maps showing the lithofacies distribution within individual sequences. This required use of the high-quality biostratigraphic data, their integration with the wireline log and seismic data to produce well-to-well correlation panels, and then ultimately the maps.

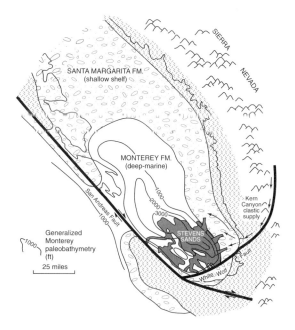

Fig.4.88 The distribution of the Steven's Sandstone reservoir and the Monterey source-rock intervals.

Given the large well and field database available for the area, it was already known that the productivity of individual wells was strongly dependent upon the lithofacies in which the well was completed and the depth from which production was obtained. The coarser-channel and channelized lobe deposits are more porous and permeable than the unchannelized lobe and thin, low-density turbidite sandstones. Their pro-

Table 4.9 Porosity and permeability of the mid-fan lithofacies of the Steven's Sandstone.

Depth (ft)	Depth (m)	Porosity (%)	Permeability (mD)
8000–10,000	2400–3000	16–22	30–250
10,000–12,000	3000–3600	12–18	1–30
12,000–14,000	3600–4200	12–16	0.5–10.0
14,000–16,000	4200–4900	10–14	0.5–5.0

Table 4.10 Initial production rates from the Steven's Sandstone.

Lithofacies	Initial production (bbl d^{-1})	
	Mean	Standard deviation
Channel	649	557
Channelized lobe	250	171
Lobe	112	67
Turbidite	6	5

ductivity is less (Table 4.9). Similarly, diagenesis has reduced the quality of the more deeply buried sandstones. Thus the lithofacies maps were then overlain with data derived from core analysis and well test information. The product was a series of maps, one on each sequence, showing the structural configuration, lithofacies distribution, and reservoir quality (Table 4.10).

4.12.5 The outcome of exploration

Even given the quantity and quality of the seismic and well data, the exploration program was seen as a risky adventure. Two prospects were identified in pinchout edges. These were risked at 1 in 8 and 1 in 12. A further prospect identified by an amplitude versus offset (AVO, Chapter 5) anomaly, and thought to be a gas condensate accumulation, was assigned a risk of 1 in 12. Only one prospect was seen as low risk (1 in 2).

This prospect was considered to be a redrill of bypassed oil from an older drilling and production campaign.

The four wells were drilled in the mid-1980s, a period when two discoveries were made in the Steven's Sandstone play (in 1985 and 1986). Unfortunately, neither of these discoveries fell to BP. Failure analysis for the four wells revealed that the stratigraphic predictions were correct. The sandstones, and thus the reservoir intervals, were at the well-site locations as predicted. Three of the wells failed because of lateral, and thus updip, seal failure. That is to say, transition from the Steven's Sandstone into its lateral equivalent, the Monterey Chert, and mudstone was less abrupt, as had been anticipated. The fourth and lowest-risk well failed because the reservoir found was of insufficient quality for production.

FURTHER READING

Allen, P.A. & Allen, J.R. (1990) *Basin Analysis, Principles and Applications*. Blackwell Scientific, Oxford. See Chapters 10 & 11.

England, W.A. & Fleet, A.J. (1991) *Petroleum Migration*. Special Publication No. 59, Geological Society, London.

Glennie, K.W. (Ed.) (1998) *Petroleum Geology of the North Sea, Basic Concepts and Recent Advances*, 4th edn. Blackwell Science, Oxford.

Halbouty, M.T. (ed.) (1982) *The Deliberate Search for the Subtle Trap*. American Association of Petroleum Geologists Memoir 32, 57–75.

Leeder, M.R. (1999) *Sedimentology and Sedimentary Basins From Turbulence to Tectonics*. Blackwell Science, Oxford.

Reading, H.G. (1986) *Sedimentary Environments and Facies*, 2nd edn. Blackwell Scientific, Oxford.

CHAPTER 5

APPRAISAL

5.1 INTRODUCTION

Petroleum has been discovered. The appraisal process is designed to determine the size of the pool and whether the petroleum accumulation should be developed. Appraisal commonly involves drilling wells and the acquisition of more seismic data. The data collected during appraisal are used to delineate the petroleum pool, to establish the degree of complexity of the reservoir, to characterize the fluids (petroleum and water), and to judge the likely performance of the field when in production. These technical assessments will be merged with economic criteria to establish whether the discovery has value and whether it can be developed commercially.

The outcome of an appraisal program will either be a development program, postponement of a development decision, or abandonment of the project. If the decision is to develop the petroleum pool, then the value of the project will depend directly upon the way in which the results of the appraisal program are used to design the development. Project value is controlled by reserves, production rate, operating expenditure (OPEX), capital expenditure (CAPEX), oil (or gas) price, and transportation costs. The reservoir and trap description derived during appraisal controls initial estimations of reserves, production rate, OPEX, and CAPEX (Fig. 5.1). However, the cost of appraisal is high and it must to be subtracted from the eventual value of the accumulation. Thus the petroleum geoscientist must be able to determine how much uncertainty to carry into the development program.

In an ideal situation, the uncertainty in reserves defined during exploration diminishes through the appraisal, development, and reservoir management (production) phases of a field's life (Fig. 5.2). Moreover, the range of reserves at sanction (development decision) should be large enough that subsequent reserves estimates do not fall outside the range. Typically, reserves estimates fall during field appraisal as the complexity of the petroleum accumulation becomes apparent. The reverse trend is often true of the reservoir management phase. Here, a combination of dynamic data, new technology, and facilities upgrades can lead to reserves growth (Chapter 6).

The importance of the appraisal program can be stressed by reference to two examples taken from the Brent Province within the North Sea. In the first example, the Magnus Field, reserves were significantly underestimated. Although the exploration department in BP had adequately captured the uncertainty in both the oil in place (550–2200 mmstb) and the reserves (220–880 mmstb), a single value for the reserves of 450 mmstb was calculated at the end of the appraisal program. The corresponding oil-in-place figure was estimated to be 1000 mmstb ± 15%. The facilities were designed and built to deliver the 450 mmstb. However, before a single barrel had been produced, the oil-in-place and reserves estimates were increased by 20%. Fifteen years after the development decision was made, the calculated oil in place was 36% higher than had been estimated at sanction and reserves were up by 61% (still within the original exploration range). Superficially, the spectacular reserves growth shown by Magnus is good news, and although such a situation is better than having to downgrade the reserves continuously, it is detrimental to project value. The facilities installed on Magnus were insufficient and the wells too few. The plateau production achieved was 155,000 bopd. This can be compared with a plan of 120,000 bopd. With the appropriate facilities, the rate could have been over 220,000 bopd.

The Thistle Field is an example of a field for which reserves were overestimated and, as a consequence, the facility build was too large and thus too costly (Brown et al. 2003). Sanction for the field development was obtained in 1974. At that time, the pool had been penetrated by only three wells and seismic coverage was limited to a 2D data set. Oil in place was estimated at 1350 mmstb. By 1992, the oil-in-place figure had been recalculated to be 820 mmstb. The corresponding reserves figures were 490 mmstb and 410 mmstb. The spectacular reduction in oil in place was not completely reflected by the decline in reserves, because it had proved possible to raise recovery from 36% to 50%. This increase in the recovery factor was achieved although the reservoir proved to be far more complex (segmented) than had been appreciated at sanction. However, many infill wells were required to deliver the increased recovery.

In this chapter, we examine the geoscience components within the process of appraisal. It will be clear to the reader that the scale and range of observation change from exploration to appraisal. No longer is the geoscientist concerned with the presence and maturity of the source rock or the risk

of migration, for the petroleum accumulation has been found. However, the ellipsoidal bump that proved an adequate description of the trap at the exploration stage is no longer sufficient. A detailed description of the trap envelope and its contents, the reservoir and petroleum therein, is needed.

Sections 5.2, 5.3, and 5.4 examine the definition of the trap envelope in terms of the bounding surfaces (fluid contacts and trap geometry) and internal segmentation. In Sections 5.5 and 5.6, we further analyze the reservoir architecture and reservoir characteristics on a pore scale. The theme of reservoir description is developed in Section 5.7, where the utility of seismic data is explored. The primary products of the appraisal program are estimates of oil and gas in place based upon the appraisal data. These data, together with reservoir architectural and a few dynamic data, are then used to populate a reservoir model (Section 5.8).

The three case histories are taken from the North Sea and the Middle East. All three examples concern discoveries that could be described as having a marginal possibility of being developed as commercially viable fields. This was a conscious choice, because it is necessary to make a real decision at the end of appraisal. For large, obviously commercial, fields, the decision to develop is usually straightforward (although the design of appropriate facilities may not be easy).

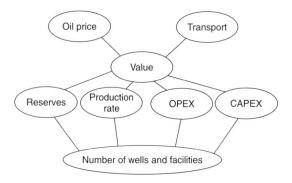

Fig.5.1 The interrelationship between reserves, production rate, operating expenditure, capital expenditure, and oil price. (From Dromgoole & Spears 1997.)

5.2 THE TRAP ENVELOPE

5.2.1 Depth conversion

Much of the data that we use to describe the subsurface —

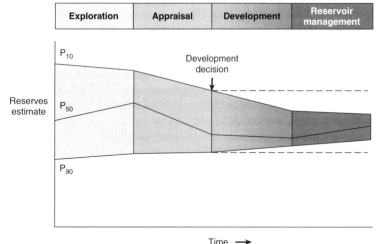

Fig.5.2 Temporal changes in reserves estimation during field development and production. In the ideal situation, the uncertainty in reserves should both decrease with time and stay within the initial reserves range estimate.

whether it be for appraisal, exploration, or production – are derived from seismic. Moreover, such seismic data are often displayed using two-way time, not depth, for the vertical axis. In making geologic interpretations using conventional seismic displays, we make a mental assumption that two-way time and depth are positively and simply correlated. In gross terms, this assumption is true on most occasions. The interpretation of geologic structures and geometries on normal seismic sections is perfectly valid. However, it is important to be aware that on some occasions the Earth's velocity field is not constant at a given depth. In consequence, obvious structures in the time domain may not exist in the depth domain, and real structures in the depth domain may not exist in the time domain and thus be invisible on seismic data (Fig. 3.12). The gross assumption that two-way time and depth are positively and simply correlated is also commonly inadequate for the purposes of well prognosis, well engineering, and evaluation of petroleum volume in a trap. In order to make each of these calculations, it is necessary to convert time to depth. The aim of this section is to examine what data are used in this process, the various methods that can be employed, the resultant uncertainties, and their effects upon the way in which the subsurface structure is interpreted.

Quite clearly, the relationship between depth and seismic two-way travel time is the velocity of seismic energy in the subsurface. The velocity is controlled by the mineralogy of the rock, the state of cementation, the pore fluid, and the porosity fabric. The seismic velocity is greater in well-indurated rocks such as limestone and dolomite compared with most sandstones. Soft sediments such as muds and coals are particularly "slow." Rock with fluid-filled pores transmits seismic energy more slowly than nonporous rock, and petroleum-filled pores reduce velocity more than water-filled pores. Small quantities of gas can cause dramatic velocity reductions. The pore fabric is also important; for example, a low-porosity but fractured rock commonly has a lower-velocity transmission than a rock with the same primary porosity. This is because the fracturing weakens the fabric of the rock and lowers its elasticity.

The degree of sediment compaction and cementation commonly increases with depth, and as a consequence seismic velocities commonly increase with depth. Exceptions to this generality occur within zones of fluid overpressure (Chapter 3). In such situations, the fluid supports part of the lithostatic load, porosity is relatively high, and thus fluid content is relatively high. The seismic velocity is therefore lower in overpressured intervals than in similar hydrostatically pressured intervals.

It is clear from the above paragraphs that heterogeneity in the lithology distribution, fluid distribution, and pressure (overpressure) distribution will cause heterogeneity in the velocity field. Such velocity heterogeneity needs to be understood to make a detailed conversion between two-way time and depth.

The data used for depth conversion come from one or more of four sources: calibrated sonic logs, pseudo-velocities, stacking velocities, and regional knowledge. Pseudo-velocity is the term used to describe velocity data calculated from combining well depths with seismic arrival times. Velocity data may be calculated for a point (from a sonic log), as an average for an interval or as the average from the Earth's surface (Tearpock & Bischke 1991). Such velocity data may then be used to convert to depth using one or more of a time–depth curve, an average velocity calculation, a layer cake conversion, or ray tracing.

A time–depth curve derived from check shot and VSP data (Chapter 2) was used in the Viboral case history (Case history 4.11). A similar suite of well data from the Moray Firth area of the North Sea (Fig. 5.3) was used to convert time to depth for several oil pools reservoired in Oxfordian sandstones. The conversion algorithm was simply derived by taking an average of the time–depth curves calculated from the wells.

An average velocity calculation uses a simple algorithm for a whole well (Fig. 5.4). No attempt is made to incorporate systematic variations in velocity that can occur with depth. The technique is often applied when wells are without sonic logs, check shot data, or VSPs, or when such data are of suspect quality. For this method to be accurate, an extremely good correlation is required between the seismic section and the time data from the wells.

In the interval velocity method for time-to-depth conversion, the geology represented on the seismic section is divided into its constituent layers (sequences). The average velocity for each layer is then calculated and the resultant layer cake of average velocity data used to make the conversion (Fig. 5.4).

Ray tracing is commonly used to check the validity of time-to-depth conversions. The method uses the generated depth maps and velocity data to generate a synthetic seismic section. The generated section is compared with the real seismic data. If the match is poor, the input data are readjusted until an acceptable match is obtained. The refined input data may then be used for time-to-depth conversion elsewhere in the area of interest (McQuillan et al. 1979).

5.2.2 Mapping surfaces and faults

Subsurface maps are a fundamental part of petroleum

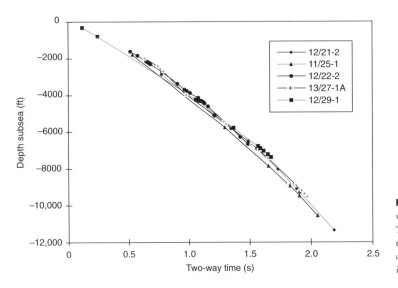

Fig.5.3 Time versus depth data for wells within the Inner Moray Firth, UK North Sea. The data are from check shot surveys in each of the wells. This curve would be used to convert a two-way time map derived from seismic data into a depth map.

geoscience. They are used at all stages of exploration, appraisal and development. Regional geologic maps were examined in Chapter 3, and the use of structural maps to interpret petroleum migration and prospectivity was covered in Chapter 4 (including the case histories). We choose to study mapping more closely in this chapter because at the appraisal stage the need for accurate and reliable maps increases disproportionately relative to the data available. Many of the criteria that are used to determine whether a petroleum pool will be developed or abandoned without development will be based upon subsurface maps, and yet the quantity of data available is likely to be limited to a few wells and a 2D or 3D seismic survey.

Maps can be produced for both surfaces and intervals (or their properties). Here, we look at surfaces. Intervals and interval properties are examined later in this chapter (Section 5.8), although the methodology for construction of the surface and interval maps is similar. Maps are most commonly constructed on stratigraphic (iso-geologic time) surfaces. These may be either conformable surfaces or unconformities. Time slice extractions from seismic data are also maps, and these can also form the basis for structural and stratigraphic interpretations. It is also possible to produce maps on fault surfaces. In such instances, it is common to plot the intersection of the stratigraphy on the fault surface for both the footwall and hanging-wall surfaces.

Cross-sections can also be constructed from maps. The construction of cross-sections can further aid the interpretation of both structure and stratigraphy.

Both structural and interval maps are 2D representations

of 3D data. The elevation, thickness, or property data are commonly portrayed as contours. Contours are lines that connect points of equal value. Relative contour spacing conveys much information about the properties of a surface (Fig. 5.5). Equally spaced contours indicate that a surface has constant slope (dip). Closely spaced contours indicate a more steeply dipping surface than more widely spaced contours. Therefore, in producing a map from point source well data or from seismic data, care must be taken during the contouring process to ensure that the map accurately conveys the interpretation intended.

There are a few important rules to consider when producing a contour map. A contour line cannot cross itself. It must also form a closed surface (although this may be beyond the bounds of the map). Contours appear to merge on vertical surfaces and appear to cross on overhanging surfaces, but in 3D space the contour lines lie one above another. To avoid confusion, contours on overhanging surfaces are commonly shown dashed (Fig. 5.6). Repeated contour lines of the same value indicate a change of slope (synclinal or anticlinal culmination; Fig. 5.7).

Tearpock and Bischke (1991) provide a simple guide to contouring, to enable easy construction and subsequently easy understanding by the reader. They suggest the following:

• Contour maps on surfaces should have a chosen reference. For depth or height maps, this is commonly mean sea level. For elevations below sea level, the contour values should be negative. For seismic time maps, the reference surface is also

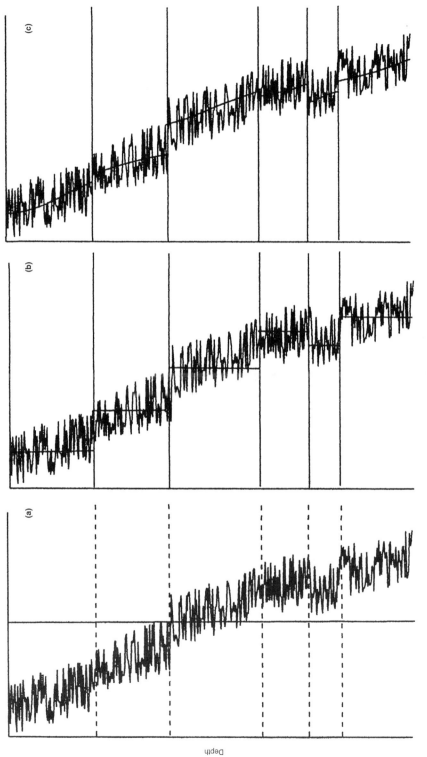

Sonic velocity

Depth

Fig.5.4 A synthetic sonic log for a well, showing three possible methods for calculation of velocity profiles that can then be used for depth conversion of seismic data. (a) The average velocity model; (b) the internal velocity model; (c) The instantaneous velocity model.

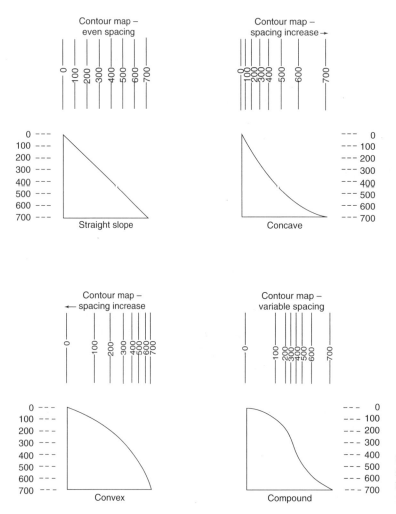

Fig.5.5 The spacing of contour lines is a function of the shape and slope of the surface being contoured. (From Tearpock & Bischke 1991.)

commonly sea level for marine surveys or a defined surface datum for land surveys, and the contours are lines of equal two-way time.

• The contour interval should be constant, although every fourth or fifth contour, known as an index contour, should be slightly thicker or bolder.

• Maps should have a graphic scale. This avoids problems caused by reproducing the map through enlargement or reduction.

• Hachured lines are commonly used to indicate close depressions.

• Contouring should be begun in the areas with the highest density of data and several contours constructed in the same area before venturing into areas of sparse data coverage.

• Contours should be smooth and the simplest contour solution that honors all points should be constructed.

There are four basic methods for contour construction: mechanical contouring, parallel contouring, equal-spaced contouring, and interpretive contouring. All are methods, not a depiction of reality, and in each case assumptions are made that may not be valid when comparisons are made with reality. In mechanical contouring, the contours are equally spaced between control points. This means that slopes remain uniform between control points and only change at control points. In parallel contouring, the contours remain parallel — or nearly so! Unlike the mechanical method, this method does not assume uniformity of slope or dip angle. Carried to its limit, this method can yield unrealistic maps, with undu-

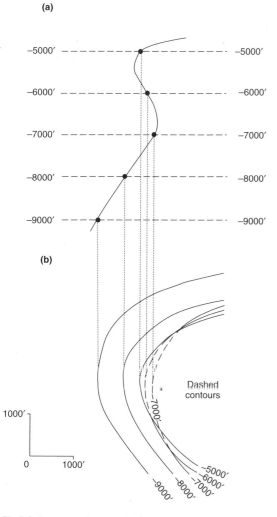

Fig.5.6 Contours on the underside of an overhanging structure (a) are shown as dashed on a map (b) so that they are clearly different from those on the upper surface. The contours shown here are in feet. (From Tearpock & Bischke 1991.)

imagination to develop the contouring. This method is the one used most frequently and most preferably. It also yields the most realistic geologic interpretation, and yet it is also the most subjective.

For maps produced from 2D seismic data, the data density is commonly greater than that produced solely from well data. Elevation data (two-way time) are transferred from lines interpreted on the seismic section onto shot point maps. In situations in which 3D seismic data are being interpreted, the data density is commonly so large that the map essentially makes itself as the geologic surfaces are interpreted. Indeed, on most workstation software it is possible to pick every 10 or 20 lines in a 3D seismic volume and then "zap" the surface. In this process, the software automatically picks the same reflectors on adjacent lines. In structurally complex areas such automatic picking may break down, since the continuity between lines is insufficient. With such a high density of data, it is commonly possible to produce a map with a similar level of detail to many topographic maps. Moreover, the high density of data may also allow mis-picked surfaces to be easily identified. Computer-aided contouring can suffer from the same problems as hand contouring, although because the choice of algorithm (method) is not always readily apparent to the operator, the detection of errors is rarely easy.

Faults are commonly an important part of overall trap geometry and internal segmentation of a petroleum pool. Care must be taken in construction of the fault surface, because the interpreted position of the fault will control the calculated volume of the trap or field segment. Care is also needed to define accurately the position of a fault surface from the perspective of drilling wells. Since faults are commonly boundaries within fields, wells are often drilled close to them. Production wells are commonly drilled close to fault block crests (Yaliz 1997) and injection wells for gas or water can be drilled at crestal or flank limits. A well drilled on the wrong side of a fault can be a massive waste of money. A detailed analysis of fault construction, fault balancing, and section restoration is beyond the scope of this book. For this, the reader is referred to Suppe (1985). Here, we will examine fault nomenclature, fault surface maps, and the representation of faults on (stratigraphic) horizon maps.

The components of a fault system are shown in Fig. 5.8. The footwall of a fault is the volume beneath the fault surface and the hanging wall is the volume above the fault surface. This nomenclature holds for both normal and reverse faults. The vertical separation (B–C) is the vertical component of the bed displacement, measured as the vertical difference between the projections of a marker horizon on either side of the fault onto that vertical plane. The vertical separation is what

lose highs and lows that lie outside the control points simply as a result of the contouring method. Equal-spaced contouring assumes a uniform slope or dip angle across whole flanks of structures. This method may also yield numerous highs and lows that are unconstrained by real data. As the name suggests, interpretive contouring is a method whereby the geologist has complete freedom within the constraints of the fixed data points to interpret the shape and spacing of the contours. No assumptions are made about bed dip or constancy of slope. Instead, the geologist will use his or her experience and

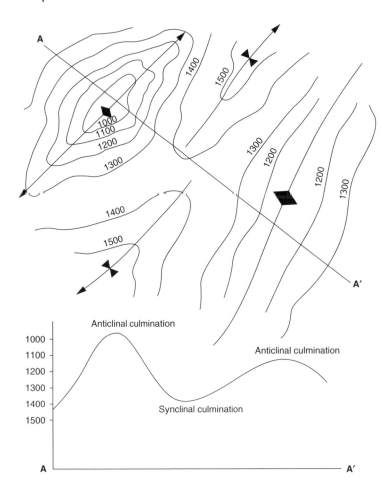

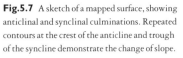

Fig. 5.7 A sketch of a mapped surface, showing anticlinal and synclinal culminations. Repeated contours at the crest of the anticline and trough of the syncline demonstrate the change of slope.

is measured in vertical wells and on vertical cross-sections. The missing section (or repeated section for a reverse fault) is called the "fault cut" when measured vertically. The "fault throw" (B–C) is the vertical component of the dip slip (B–D). The "fault heave" (C–D) is the horizontal component of the dip slip, measured orthogonal to the strike of the fault.

For maps constructed on stratigraphic surfaces, fault heave appears as holes in normally faulted terrain (Fig. 5.9) and as overlapping surfaces in reverse-faulted terrain. It is particularly common to find maps in which the gaps and/or overlaps have been ignored. Quite clearly, in these instances the geologist constructing the map has misrepresented the extent of the formations on either side of the fault surface and, by extension, has also misrepresented the volume of rock to either side of the fault. Representation of reverse faults is also problematic in much of the current generation of geocellular modeling software (Section 5.8). When using such software,

it is often necessary to construct a separate geocellular model to either side of a reverse fault. This is to capture accurately the volume and true extent of both the footwall and the hanging wall. To mitigate the shortcomings of such software, some geologists will choose to make their reverse faults vertical and produce a single geocellular model. Naturally, this compromises the volume calculations. Moreover, such a model cannot be used for well placements.

Fault surface maps are most commonly constructed to enable examination of the relative positions of sealing horizons and reservoir horizons on either side of the fault plane. Such Allan diagrams (Fig. 5.10) allow analysis of potential fault seal mechanisms and determination of spill points (Fig. 5.10).

5.2.3 Spill points

The spill point of a structure is the deepest closed contour on that structure. Spill points define the vertical limit of the trap. The volume of rock above the spill point is within the trap and that below the spill point is outside the trap (Fig. 5.11). Knowledge of the spill point for a particular trap will enable the vertical closure of the trap to be calculated. Thus spill point identification is an important part of petroleum field definition and ultimately calculation of petroleum in place.

Although both the definition and significance of a spill point are straightforward, the identification of spill points during both exploration and appraisal can be extremely difficult. The density of seismic and well data and its distribution may be insufficient to define adequately any spill point (Fig. 5.12). Some traps have a variety of possible spill points, the integrity of each depending upon different factors, such as four-way dip closure or fault closure (Fig. 5.13). Petroleum that is trapped stratigraphically in isolated sandstone or carbonate bodies may be without true spill points.

The relationship between the petroleum/water contact and the spill point is an important one. Much effort is put into understanding the relationship during appraisal. Fields may not be full to spill; that is, the petroleum/water contact may occur above the spill point. Equally, petroleum can be found below the mapped spill point. Of course, this indicates that

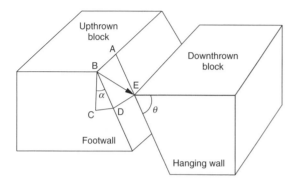

Fig.5.8 Fault nomenclature. AB, strike-slip; BD, dip-slip; BE, net slip; CD, heave; BC, throw; θ, dip of fault; α, hade of fault.

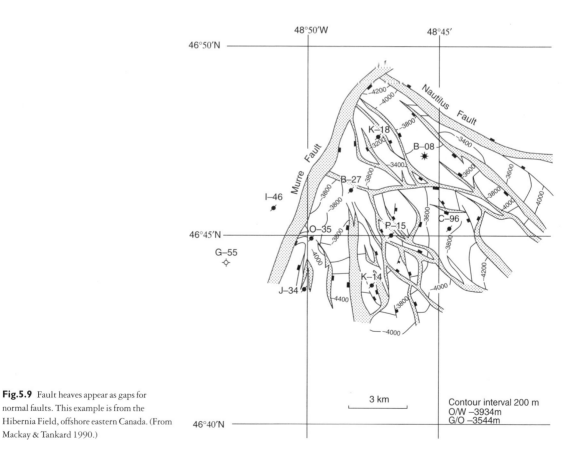

Fig.5.9 Fault heaves appear as gaps for normal faults. This example is from the Hibernia Field, offshore eastern Canada. (From Mackay & Tankard 1990.)

3 km

Contour interval 200 m
O/W −3934m
G/O −3544m

(a)

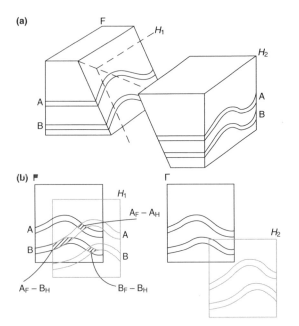

(b)

Fig.5.10 (a) A block diagram showing folded and faulted reservoir intervals A and B. The hanging wall is shown with two displacements, H_1 and H_2. (b) An Allan diagram showing the intersections of the reservoir intervals A and B from both the footwall and the hanging wall on the fault plane for displacements H_1 and H_2. The reservoir intervals are juxtaposed across the fault plane for displacement H_1 (shaded). Fluids could flow across the fault plane at the points of contact between the reservoirs. The displacement H_2 is too great for the reservoirs to be juxtaposed.

the mapped spill point is not the true spill point, which must occur deeper.

5.3 FLUID DISTRIBUTION AND CONTACTS

5.3.1 Fluid contacts and transition zones

A critical part of any appraisal program is determination of the fluid contacts (gas/oil, gas/water, and oil/water) in a petroleum pool. Until a well is drilled into the ground, such contact depths can only be estimated on the basis of a calculated spill point or a direct petroleum indicator from seismic. A common plan is to drill the exploration well on a prospect in such a position that it proves a volume of petroleum (updip) that is deemed to be economically viable. Following the success of that well, the first appraisal well will be drilled in such a position as to penetrate the expected petroleum/water contact and, in consequence, define the size of the pool (Fig. 5.14). Although this is a sensible strategy, there is a range of possible outcomes for a well, only one of which is penetration of an oil/water contact.

An appraisal well may penetrate petroleum and water intervals in different reservoir units separated by nonreservoir units. From such a well or wells it is possible to define "oil (or gas) down to" and "water up to" elevations. These elevations can also be referred to as "deepest known oil" and "highest known water." In such instances, the simplest explanation is that the oil/water contact lies between the water up to and oil down to. However, this need only be true if the oil- and water-bearing intervals are in pressure communication (Fig. 5.15).

If the appraisal process reveals different contacts in different parts of the field, then clearly the field must be compartmentalized. It can be either layered or faulted, or both. Segmented reservoirs are further investigated later in this section.

Fluid contacts can also be gradational. For an oil plus gas pool at its critical point, there will be a continuous transition from gas down into oil. This was the case for the giant Brent oilfield (UK North Sea) when it was first found. Pressure reduction during production has caused the fluid to no longer be at the critical point and a definable gas cap now exists. The contact between petroleum and water may also be gradational, although the causes are very different. In poor-quality reservoirs, an abundance of small pores and hydrophilic min-

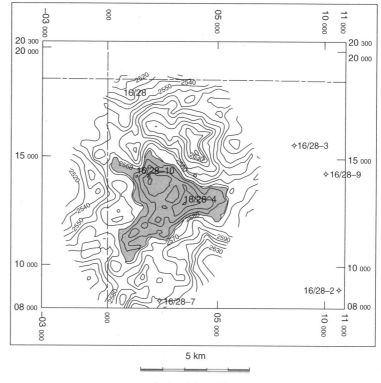

Fig.5.11 The Cyrus Field (UK North Sea) has a four-way dip-closed trap. The reservoir comprises a thick sequence of Paleocene submarine-fan sandstones. The structural spill point, equivalent to the deepest closing contour, is at 2550 m subsea. The oil/water contact is coincident with the mappable spill point. (Reproduced courtesy of BP.)

5 km

Contour interval 10 m
Depth in meters subsea

erals tends to promote high water saturations and gradational petroleum/water contacts. In gas reservoirs, considerable quantities of gas may be dissolved in the formation water underlying the gas leg. Although in this instance the contact may appear sharp, gas can be produced out of solution below the gas/water contact (when pressure is lowered).

The boundary between the petroleum-bearing interval and the water-bearing interval in a field, be it relatively sharp or gradational, is known as the transition zone (Fig. 5.16). Within the transition zone there is a downward decrease in oil saturation and a downward increase in water saturation. Although the transition is smooth, the interval can be divided into two distinct zones. Oil can be produced from the upper zone because the petroleum saturation is above the irreducible oil saturation. In the lower zone, the oil is immobile and only water can be produced from here. At a short distance below the observed oil/water contact is the free water level. The relationship between the free water level and the oil/water contact is best illustrated by a laboratory experiment using capillary tubes of different diameters, dipped into a dish of water (Fig. 5.17). The wetting phase (water) rises

highest in the narrowest tube. Thus, while the free water level is a property of a reservoir system, an oil/water contact is a local phenomenon that is dependent upon the capillary pressure threshold near the wellbore. Fortunately, in high-quality reservoirs, the difference in elevation between the oil/water contact and the free water level is small. However, in poor-quality reservoirs, care must be taken to ensure that differences in elevation in oil/water contacts, measured in different wells, do reflect differences in elevation of free water level and not simply variations in rock properties.

In the absence of a direct identification of a petroleum/water contact in a well, the depth to a contact can be calculated from measurements of pressure data in both the oil or gas legs and in the aquifer. The basic premise is that at the oil/water contact the pressure in the oil zone must equal that in the water zone. In reality, of course, this statement of equilibrium defines the free water level. The pressure in the water (P_W) at the oil/water contact is given by

$$P_W = X_{OWC}G_W + C_1$$

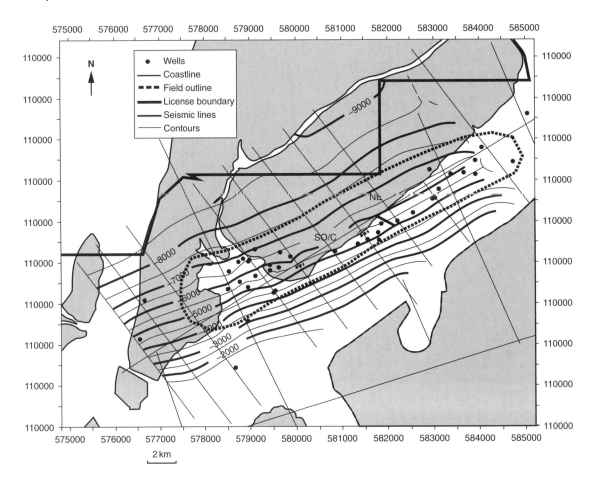

Fig.5.12 A structural map of the Pedernales Field, eastern Venezuela. The distribution of oil-bearing wells is shown by the field outline in heavy dashed line. From the 2D seismic data available when this map was constructed (1994), it was not possible to define spill points, or indeed the updip trap limit for Pedernales. (Reproduced courtesy of BP.)

where C_1 is a constant that represents any degree of overpressure or underpressure.

At depth X_{OWC}, $P_{W(OWC)} = P_{O(OWC)}$. Above the oil/water contact the pressure in the oil leg is simply a function of the pressure of the oil at the contact minus the density head of the oil. Thus estimation of an oil/water contact can be made by extrapolation of the oil and water gradients to the point of crossing (Fig. 5.18). The pressure data can be gathered from RFT analysis and well tests. The fluid density data can be measured from formation samples or calculated from wireline logs. The saturation information can be measured from both core and logs, while capillary pressure data are taken from core measurements.

5.3.2 Intra-field variations in petroleum composition

Almost all pools of oil and gas are heterogeneous with respect to the composition of their petroleum and associated water. An important part of the appraisal process is to determine how the compositions of petroleum and water vary both across a field and with depth in a field. Knowledge of compositional variation in a field will enable the geoscientist to understand how, and possibly when, the field filled with petroleum and the direction of filling. If the field is to be developed, the compositions of both petroleum and water together with their compositional ranges and spatial distribution will be used as a basis for the facility's design.

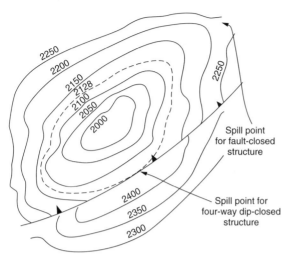

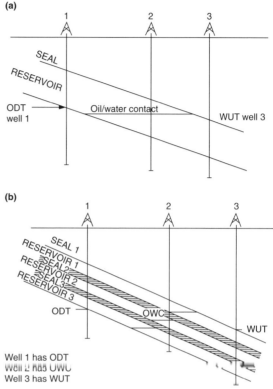

Fig.5.13 Spill points on four-way dip-closed and fault-closed structures. Four-way dip-closed contour, 2128 m subsea; the trap requires fault closure from 2128 m to 2200 m subsea.

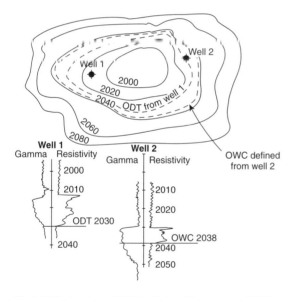

Fig.5.14 Exploration well 1 failed to find an oil/water contact (OWC). The base of the oil-bearing interval in the well was at the same depth as the base of the reservoir sandstone. This point is called the "oil down to" (ODT). Well 2 appraised the discovery made by well 1 and it penetrated an oil/water contact.

Well 1 has ODT
Well 2 has OWC
Well 3 has WUT

Fig.5.15 (a) The oil/water contact penetrated by well 2 lies between the ODT in well 1 and the "water up to" (WUT) in well 3. (b) Superficially, the well results displayed here are the same as those in (a): the oil/water contact penetrated by well 2 lies between the ODT in well 1 and the WUT in well 3. However, there are three reservoirs, each with different oil/water contacts. Unless additional data (such as pressure measurements) are taken in each of the reservoir intervals in each of the wells, the possibility that the field contains several different oil/water contacts would be missed using ODT and WUT data alone.

Variations in fluid composition also have implications concerning field segmentation (compartmentalization), reservoir simulation, satellite field development, and even unitization.

The properties that commonly vary across fields (Archer & Wall 1986) are the gas to oil ratio (GOR), the condensate to gas ratio (CGR), the oil viscosity, the gas composition, and the content of nonhydrocarbon compounds, such as nitrogen, carbon dioxide, and hydrogen sulfide.

An example of the way in which compositional differences can influence the utility of facilities comes from the Gyda Field in the Norwegian North Sea. The Gyda platform was

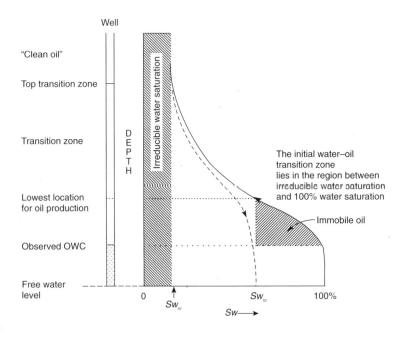

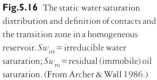

Fig.5.16 The static water saturation distribution and definition of contacts and the transition zone in a homogeneous reservoir. Sw_{irr} = irreducible water saturation; Sw_{ro} = residual (immobile) oil saturation. (From Archer & Wall 1986.)

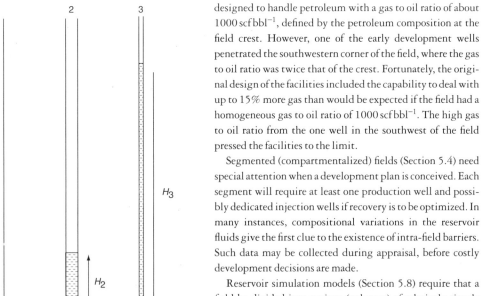

Fig.5.17 Capillary rise above free water level. (From Archer & Wall 1986.)

designed to handle petroleum with a gas to oil ratio of about 1000 scf bbl⁻¹, defined by the petroleum composition at the field crest. However, one of the early development wells penetrated the southwestern corner of the field, where the gas to oil ratio was twice that of the crest. Fortunately, the original design of the facilities included the capability to deal with up to 15% more gas than would be expected if the field had a homogeneous gas to oil ratio of 1000 scf bbl⁻¹. The high gas to oil ratio from the one well in the southwest of the field pressed the facilities to the limit.

Segmented (compartmentalized) fields (Section 5.4) need special attention when a development plan is conceived. Each segment will require at least one production well and possibly dedicated injection wells if recovery is to be optimized. In many instances, compositional variations in the reservoir fluids give the first clue to the existence of intra-field barriers. Such data may be collected during appraisal, before costly development decisions are made.

Reservoir simulation models (Section 5.8) require that a field be divided into regions (volumes) of relatively simple composition, because a reservoir simulator needs to start with an equilibrium pressure–fluid system. Black-oil simulators allow gas and oil mixing under equilibrium P–V–T assumptions. So although it is true that they cannot handle complex fluid gradients, they do routinely handle simple ones. Com-

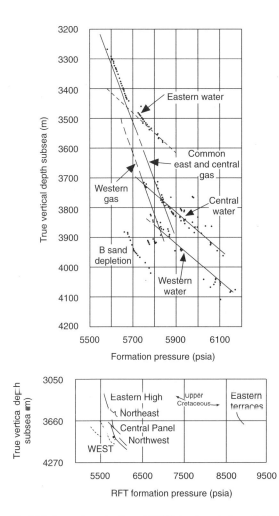

Fig.5.18 Repeat Formation Tester (RFT™) data from the Bruce Field, UK North Sea. The field is extensively segmented and displays parallel oil (against pressure) gradients and parallel water (against pressure) gradients in the different segments. (From Beckly et al. 1993.)

positional simulators are capable of handling more complex fluids; however, they are difficult to use.

Gas, condensate, and oil rarely have the same value. In fields where the gas to oil ratio, or the gas to condensate ratio, vary spatially, one part of a field may be more or less valuable than another part of a field. Where such different parts have different owners, fluid compositional differences can become a major part of unitization discussions (equity battles).

The major factors that can cause petroleum fluid variations in a field are filling of reservoirs from one direction, biodegra-

dation of oil in reservoirs, seal failure and vertical leakage, gravitational segregation, and thermal segregation.

For a field filling from a single direction, the end of the field closest to the maturing source rock will receive ever more mature petroleum as time progresses (Fig. 5.19a). The maturity of the petroleum will be marked by an increased gas to oil ratio and a decreased viscosity. This may be accompanied by changes in the content of nonhydrocarbon gases. Fields filled with petroleum from more than one direction and from more than one source rock will also display lateral variations in petroleum composition (Fig. 5.19b).

Cool (<70°C), shallow oil pools are prone to biodegradation. This can occur when bacteria-laden near-surface water comes into contact with the oil. The bacteria attack the short-chain hydrocarbons, and as a consequence the most heavily affected area of the field has the lowest oil gravity, the lowest gas to oil ratios, and the highest viscosity (Fig. 5.19c). The Quiriquire Field in eastern Venezuela (Chapter 4) is a classic example of a field in which petroleum composition variation is a product of biodegradation.

Fields that have been filled from other fields, either by spillage or seal failure, commonly show systematic variations in fluid composition (Fig. 5.19d). Such fractionation of petroleum is common in oilfields offshore southeastern Trinidad and the Gulf of Mexico (Thompson 1988). In both areas, the seals are poor, and in consequence gas and light hydrocarbons can leak upward, leaving heavier petroleum behind.

A static column of petroleum, such as that in an oilfield, is affected by the Earth's gravity. The denser components (molecules with more than six carbon atoms) sink and the less dense molecules, such as methane, rise. Such segregation can often be seen by changing gas to oil and condensate to gas ratios up and down a petroleum column. Creek and Schraeder (1985) illustrated such a phenomenon in the East Painter Field of Wyoming (Fig. 5.20).

The magnitude of gravitational segregation has proved difficult to model from thermodynamic considerations. In general, however, the process seems to be most pronounced in deep, hot, condensate- or gas-rich oil reservoirs. Under such near-critical conditions, gravitational segregation can lead to situations such as that in the Brent Field, UK continental shelf (Tollas & McKinney 1988), in which no distinct gas/oil contact can be observed. The petroleum changes gradually. Gas occurs at the top of the pool. This overlies condensate and this in turn overlies oil.

Thermal gradients may also induce compositional segregation within petroleum columns. However, the effects are difficult to model and in the real world difficult to differentiate from those of gravity. Modeling performed on the gas cap

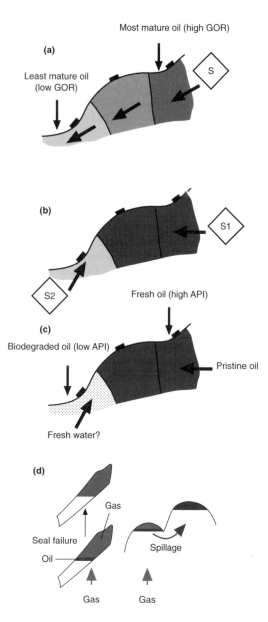

(a)

Most mature oil (high GOR)

Least mature oil
(low GOR)

S

(b)

S1

S2

Fresh oil (high API)

(c)

Biodegraded oil (low API)

Pristine oil

Fresh water?

(d)

Gas

Seal failure

Oil

Spillage

Gas   Gas

Fig.5.19 Intra-field petroleum variations. (a) Inherited differences due to filling from one source kitchen. The most mature, high-GOR, petroleum occurs nearest to the active source rock. (b) Inherited differences caused by the presence of two source-rock kitchens. (c) The effect of biodegradation on reservoired oil. Biodegradation removes the lighter components of the oil. (d) A cross-section illustrating seal failure and leakage causing compositional changes in reservoired oil. (Figure originally drawn by W.A. England.)

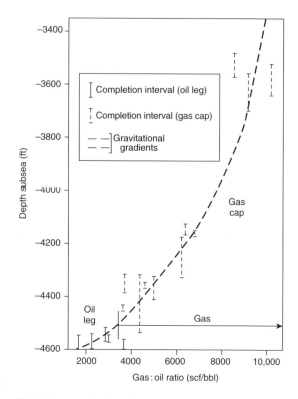

Depth subsea (ft)

Completion interval (oil leg)

Completion interval (gas cap)

Gravitational gradients

Gas cap

Oil leg

Gas

Gas : oil ratio (scf/bbl)

Fig.5.20 An example of gravitational segregation causing vertical changes in the gas to oil ratio (GOR), from the East Painter Field, Wyoming. (Pers. comm., W.A. England 1992.)

of the giant Prudhoe Field in Alaska suggested that thermal convection within the laterally extensive gas cap could have led to evaporation of the formation water from parts of the cap. This has produced anomalously low water saturations in some areas of the gas column (pers. comm., W.A. England 1992).

In a field without barriers, the rate of lateral mixing (diffusion and convective overturn) is insufficient to homogenize petroleum composition in all but the oldest fields. For a medium- to light-gravity oil, diffusion can homogenize a 100 m oil column in about 1 million years for molecules with up to 200 carbon atoms. In contrast, 40 million years would be required to homogenize methane content over a typical well spacing of 2000 m (Table 5.1), and larger molecules would require hundreds of millions of years. It is not that there is any physical difference between the horizontal and vertical diffusion processes but, rather, that the distances involved are larger and the time taken increases as the square of the distance.

Table 5.1 Order-of-magnitude diffusion rates for methane and molecules with 12 and 200 carbon atoms.

Distance (m)	Methane (Ma)	C_{12} (Ma)	C_{200} (Ma)
100	0.1	0.2	1
2000	40	80	400

Density-driven mixing operates much more quickly than diffusion mixing. If, for example, a field is filled from two directions by oil of different compositions and hence densities, the denser oil will sink to the bottom of the reservoir while the less dense oil will rise to the top. The rate of this process depends mainly upon the viscosity of the oils, although horizontal and vertical permeability are also important. For a light to medium oil, density-driven mixing can go to completion in 1 million years or less over typical inter-well distances of a few kilometers. More viscous oils will mix more slowly.

5.3.3 Intra-field variations in water composition

The composition of water, like that of oil, may not be constant in and around an oilfield. This is true for the water trapped within the petroleum leg of the field and that in the underlying aquifer. Variations in the chemistry of oilfield waters result from water and rock interaction (Warren & Smalley 1994). At the time of deposition the interstitial water in sediments will be fresh water, marine water, evaporitic water, or some combination thereof. Such water may begin to be buried with the sediment. However, invariably the original formation waters are lost and replaced by meteoric (originally rain) water. Such meteoric water can be characterized by its hydrogen and oxygen isotope ratios (Emery & Robinson 1993).

Although most formation waters are derived from meteoric water, few are fresh. The salinity of water and the specific composition of the dissolved salts are controlled by the extent to which the water reacts with the enclosing rock. The rate at which the formation water reacts with the host rock can be slowed by the presence of petroleum, simply because the water becomes trapped and immobile as the petroleum saturation increases. Naturally, such a process does not occur within the aquifer to a petroleum accumulation and, as such, the "water leg" can continue to evolve. In consequence, water trapped within the oil or gas legs of a petroleum accumulation may have a different composition and hence salinity than the surrounding aquifer. In a continuously evolving system, it is likely that the aquifer would be more saline than the water trapped in a petroleum leg. However, it is also possible that following accumulation of petroleum pools, wholesale exchange of aquifers can yield water legs that are less saline than associated water within the petroleum legs.

It is important to measure the salinity and composition of trapped water and aquifers for a number of reasons. Precipitation of mineral salts from formation waters can lead to damage of the reservoir and production pipework (Chapter 6). The salinity of oilfield formation waters is also utilized to calibrate the resistivity logs used for the determination of petroleum saturation (Chapter 2).

In an example from the North Sea, it was discovered late in the life of the field that the aquifer was more saline than the water trapped within the oil leg. Since the aquifer salinity had been used to calibrate the resistivity logs, the water saturations in the oil leg had been underestimated. That is to say, there was less oil in the field than had been calculated.

In the above paragraphs, we have examined the bulk chemical differences between the water chemistry of the oil and water legs in an oilfield. During the 1990s, the development of the technique known as residual salt analysis (RSA, Chapter 2) allowed characterization of formation waters in great detail using core.

The $^{87}Sr/^{86}Sr$ isotope ratio of the formation water can be measured from dried core chips simply by washing the evaporated salts from the chips with distilled water (Smalley et al. 1995). The strontium isotope ratio in the solution can then be measured very accurately using a mass spectrometer. The advantage of using an isotope ratio rather than a bulk chemical characterization is that isotope ratios are significantly less affected by disturbance of the water–rock system by drilling, coring, and extracting the samples from the Earth than is the bulk chemistry. The drilling and sampling process will affect the Eh, pH, pressure, and temperature of the formation water, all of which can lead to reaction, precipitation, and dissolution. These processes will affect the bulk strontium composition, but not the isotope ratio.

Once it became possible, using the RSA technique, to characterize the formation waters in great detail, variations in water chemistry within oilfields, and indeed within individual reservoir units, became apparent. The high degree of variability of the isotope ratio within oilfields has in some instances been used to explain the way in which the field filled with petroleum (Fig. 5.21). These same data can be used to describe field segmentation (Section 5.4).

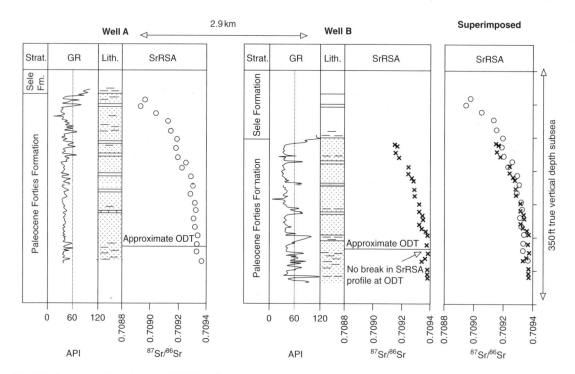

Fig.5.21 Strontium residual salt analysis (SrRSA) trends for two wells in the Arbroath Field, UK continental shelf. Core was cut across the whole of the oil-bearing reservoir interval and into the underlying water leg. The SrRSA profiles are almost identical when compared using the original flat-lying oil/water contact as a datum. The profiles do not match when compared using a stratigraphic datum. The similarity of the profiles when compared using the oil/water contact datum is best explained in terms of a progressive change in the $^{87}Sr/^{86}Sr$ isotope ratio of the formation water coincident with oil filling the field. Water progressively depleted in ^{87}Sr was rendered immobile as the field filled with oil. (From Mearns & McBride 1999.)

5.4 FIELD SEGMENTATION

5.4.1 Introduction

Rarely is a petroleum field a simple tank—one that could, given sufficient time, be drained with a single well. Most fields are segmented. Fields may contain barriers to flow of fluids either laterally along the stratigraphy or vertically across the stratigraphy. Some fields may be segmented in both ways. Moreover, barriers to flow may not be totally sealing. The flow restrictions may be no more than baffles. During appraisal of a field it will only be possible to examine barriers under virgin conditions. Some barriers may change their properties during production as fluid is extracted from the field and pressure changes occur.

In this section we will examine the geometry of baffles and barriers, and how to identify their presence and location.

5.4.2 Barriers to lateral flow

Faults are the most common barriers to lateral flow of petroleum from a field (Fig. 5.22), although not all faults seal, and even sealing faults may not always have sealed. Faults may seal under the following conditions:
• Where throw has juxtaposed reservoir and nonreservoir horizons;
• where fault zones are cemented;
• where clays have been smeared along the fault plane;
• where cataclasis of the rock along the fault plane has produced granulated zones of impermeable rock.

Faults are most commonly identified during interpretation of seismic data. More rarely, faults are observed from wireline log and core data. In order to determine whether a fault is likely to be a juxtaposition seal between segments of an oil- or gasfield, it is necessary to map the intersection between the footwall and hanging wall along the plane of the fault (Fig.

(a)

10 miles

(c)

(b)

Fig.5.22 Barriers to lateral flow. (a) A faulted sandstone bed. (Reproduced courtesy of BP.) (b) Cemented granulation seams associated with faults within the Triassic sandstones of the East Irish Sea and onshore northern England basins, Eden Valley, England (the coin in the center of the photograph is 2.5 cm in diameter). (Reproduced courtesy of BP.) (c) An intensity, hue, saturation (IHS) image of the top reservoir structure of the Kuparuk Field, Alaska. The main, detailed part of the image is covered by 3D seismic data, while the remainder of the area has only 2D seismic coverage. The intensity of the faulting has clear implications for segmentation within the Kuparuk Field. (From Dromgoole & Spears 1997.)

5.10). Alternatively, the stratigraphy of the faulted block can be plotted against displacement (Fig. 5.23). Such diagrams, developed by Knipe (1997), allow calculation of the sealing capacity of juxtaposition and clay smear faults.

Faults that have associated cemented zones are common but notoriously difficult to identify in the subsurface. Incipient, inactive, and active faults modify the permeability fabric of the host rock and induce stress anisotropy. Such modification causes, at the very least, flow diffraction for moving fluids

in the area of the fault. Indeed, many faults act as fluid flow conduits — a process that may itself cause cementation.

Barriers to lateral flow are also present as lateral shale-outs of reservoir lithologies. Such changes in lithology may not always be obvious from the limited quantity of data commonly available during the early stages of appraisal. This is particularly the case in reservoir systems with a complex reservoir architecture. Well-to-well reservoir correlation exercises may fail to recognize the true relationships between the reservoir bodies (Case history 5.10).

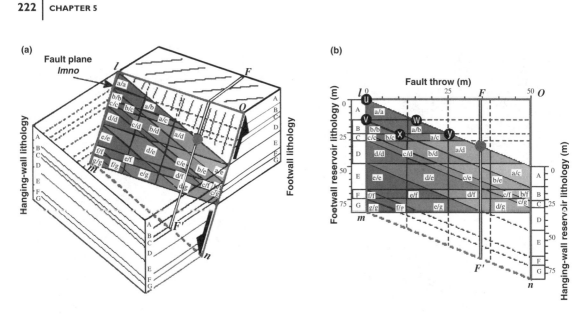

Fig.5.23 Juxtaposition diagrams. (a) A 3D fault with varying displacement along its length. The juxtaposition diagram is based on extracting the fault plane (lmno) and its intersections with the stratigraphy shown. (b) This juxtaposition diagram can be considered a "see-through" fault. Footwall stratigraphic units (e.g., unit C) intersect with the fault (the footwall cutoffs) from behind the diagram plane, and the hanging-wall cutoffs intersect from in front of the diagram plane. Faults can be shown as fault traces (e.g., FF′). Juxtapositions can be assessed by reading from a selected point on the fault trace (e.g., dot on fault FF′), looking first horizontally and left, back to the footwall stratigraphy (i.e., the upper unit, D, forms the footwall at this point), and then looking down and right to the hanging-wall stratigraphy (top unit A is juxtaposed in the hanging wall). Each of the triangle areas (e.g., uvw) and trapezoidal areas (e.g., vwxy) on the diagram represent areas on the fault plane with different juxtapositions. For example, the triangle uvw represents the area where unit A in the hanging wall is juxtaposed against unit A in the footwall. Trapezoid vwxy represents the area where juxtaposition of unit A in the hanging wall is against unit B in the footwall. (From Knipe 1997.)

5.4.3 Barriers to vertical flow

Here, we are concerned with barriers and baffles that impede flow upward and downward across the stratigraphy. This is commonly referred to as vertical flow, although clearly if the reservoir was vertical it would be necessary to take care with nomenclature to avoid confusion.

Barriers to vertical flow are essentially similar to those already described for horizontal flow. They include low-angle faults (thrusts), cemented layers, bitumen layers, and of course the natural stratigraphic composition of the gross reservoir interval (Fig. 5.24).

A significant difference between barriers to lateral flow and those to vertical flow is that the latter are commonly penetrated by wells. For that reason, it is often easier to identify potential barriers to vertical flow during appraisal.

5.4.4 Identification of flow barriers

We said in opening this section that seismic data are used to map faults, but that faults may or may not be responsible for segmentation of a petroleum accumulation. However, there are a number of tools that can be used to identify the presence of barriers before the field is committed to development and hence major expenditure. Conventionally, data on intra-field barriers have been gained from the analysis of well test information. A detailed discussion of well test analysis is beyond the scope of this book. However, the basic methodology is to examine the production behavior of a well as parameters such as pressure drawdown and choke size are changed. Measurements are made of production rate for all fluids (gas, oil, and water). The length of a production test can vary from a few hours to many months; if circumstances and cost permit, the longer the test the better as far as gathering useful information is concerned. A long test at an offshore location may not be feasible because of the cost of the drilling and test facility,

Fig.5.24 Barriers to vertical flow. (a) Cemented layers within the Jurassic Bridport sandstones, Dorset, UK. (b) A geocellular model of the Douglas Field, East Irish Sea Basin, showing a low-porosity (cemented) bitumen layer within otherwise porous and permeable Triassic reservoir sandstones. (Reproduced courtesy of BHP Petroleum Ltd.) (c) A low net to gross sandstone shale sequence, Middle Jurassic fluvial overbank deposits, Yorkshire, England. The shales between the sandstones would be barriers to flow for similar rocks in the subsurface.

and because the petroleum fluids cannot be exported and sold to pay for the test. Flaring may also be a problem because of environmental issues and associated taxes.

Probably more important than the test is the recovery of the reservoir afterwards. The speed and manner of the recovery tell the reservoir engineer a great deal about the local continuity of the reservoir, as well as the approximate distance to baffles and barriers. In order to gather these data, pressure meters are placed in the well and the pattern of pressure buildup recorded. In order to maximize the useful information from such tests, it can be important to "listen" to natural pressure changes in the reservoir before the system is disturbed by the test itself. This can help to filter out unwanted information, such as the effects of lunar and Earth tidal signals.

Most of the geologic tests designed to examine compartmentalization rely on analysis of some component of the

fluid system. Measurements can be made on either the petroleum or aqueous components. The basic premise of many such tests is that differences between samples taken from different wells in the same field or different reservoir intervals in the same wells may indicate the presence of barriers. The types of fluid differences that can be observed have been covered in Section 5.3 and the analytic methods introduced in Chapter 2.

5.5 RESERVOIR PROPERTY DISTRIBUTION

5.5.1 Introduction

Reservoirs and reservoir properties are heterogeneous when viewed on a range of scales from the pore to the play fairway

Fig.5.25 Heterogeneity at different scales, from basin to pore. (Reproduced courtesy of BP.)

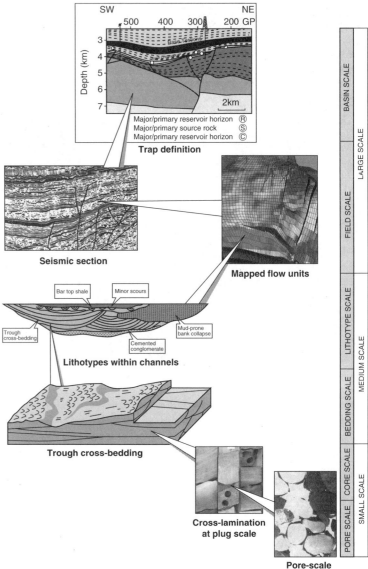

Trap definition

Seismic section

Mapped flow units

Bar top shale | Minor scours

Trough cross-bedding | Cemented conglomerate | Mud-prone bank collapse

Lithotypes within channels

Trough cross-bedding

Cross-lamination at plug scale

Pore-scale heterogeneity

Major/primary reservoir horizon ⓡ
Major/primary source rock ⓢ
Major/primary reservoir horizon ©

BASIN SCALE · LARGE SCALE · FIELD SCALE · LITHOTYPE SCALE · MEDIUM SCALE · BEDDING SCALE · CORE SCALE · SMALL SCALE · PORE SCALE

(Fig. 5.25). Field appraisal provides the opportunity to begin to understand the range of reservoir properties that exist within the discovered petroleum pool. In this section, we will examine the basic description of reservoirs, lithofacies, and lithotypes, and the way in which the lithofacies are joined (reservoir geometry and reservoir correlation).

5.5.2 Lithofacies and lithotypes

The concept of the "facies" was first used by geologists and miners in the 19th century (Reading 1986), who recognized that rock units with particular features were useful in correlating coal seams and ore bodies. Gressly (1838) was the first to use the term. Reading (1986) reports that the term and its usage have been the source of much debate. Here, we will limit ourselves to discussion of the derivative term "lithofacies," which is used to describe the physical and chemical characteristics of a rock. The criteria that can be used to describe a lithofacies include mineralogy, grain size, sedimentary structures, color, and fossil assemblages. Elsewhere in this book the terms (wireline) log facies and seismic facies are also used. These too are descriptive terms, used to indicate comparable wireline log responses and seismic responses respectively. We avoid terms such as turbidite facies, shallow-marine facies, and post-oregenic facies, all of which are to a greater or lesser extent interpretive.

A lithofacies is the basic unit of outcrop or core description. The physical characteristics of a lithofacies commonly allow some interpretation of sedimentary process, from which may be derived some indication of depositional environment. For example, a lithofacies comprising sandstone with climbing ripples indicates deposition from an unidirectional current. Thus the process is known, but such current activity could occur in a variety of depositional environments, from fluvial to deep-marine. Groups of lithofacies that occur together are commonly termed "lithofacies associations." Each lithofacies can be interpreted in terms of a sedimentary process and together the different sedimentary processes enable better definition of the depositional environment than does a single lithofacies. For example, a commonly associated group of lithofacies is laminated mudstones, graded and laminated sandstones, and chaotic mud-supported conglomerates. The individual processes for each of these lithofacies could be interpreted as deposition from suspension (mudstones), turbidity current (sandstones), and debris flow (conglomerates). Taken together, these processes are only likely to have occurred in a small number of possible depositional environments, such as deep lakes and deep seas.

In progressing from the description of lithofacies to the interpretation of process and eventually depositional environment, care must be take to understand the boundaries and thus the relationships between lithofacies. In core or outcrop, the boundaries between lithofacies can be gradational, sharp, or erosive. Only where boundaries are gradational can the vertical succession of lithofacies be unequivocally regarded as indicative of laterally coexisting depositional systems (Walther 1894). In instances in which the boundary between lithofacies is clearly erosional, it may be possible to interpret the degree of induration of the underlying lithofacies, but it may not be possible to determine what has been eroded or not deposited at the lithofacies boundary. Sharp boundaries between lithofacies could represent either conformable or erosional surfaces.

The interpreted lithofacies and lithofacies associations are the foundation for the reservoir description. The sedimentologist uses the lithofacies data together with analog information to construct a picture of the reservoir in terms of interlocking "geobodies." During the appraisal of a field, the description of the individual lithofacies may not change significantly as more wells and cores are added to those obtained from the exploration wells. However, the under-standing of the spatial relationships between lithofacies will develop significantly and the description will change from a statistical representation of lithofacies distribution to a more deterministic description.

Before considering reservoir geometries, we must first make a small detour. The lithofacies and lithofacies associations are the basic units for the geologic description of the reservoir; however, they may not be adequate descriptive elements for how the reservoir will behave during petroleum production. The derivative term "lithotype" has been invented to enable characterization of the reservoir in terms of how it is likely to perform under production. The main difference between the lithotype and the lithofacies is that the definition of the lithotype is based upon the permeability characteristics of the rock rather than the full suite of physical and chemical properties. Moreover, from a reservoir performance perspective, lithofacies with moderate to high permeability require more detailed characterization than those with low permeability.

In a typical lithotype subdivision of a reservoir, a single lithotype will have a range in permeability of one order of magnitude or less, but lithologies with less than about 1 mD permeability are commonly lumped together into a single lithotype. In practice, for lithofacies with moderate (1–100 mD) permeability there will be a one-to-one mapping between a lithofacies and a lithotype. For the best lithofacies in a reservoir, the permeability may be hundreds of millidarcies to tens of darcies. If this is the case, it is necessary to subdivide these lithofacies into several lithotypes of, say, 100 mD to 1 D and 1 D to 10 D. The lithotypes, rather than the lithofacies, will be used in the reservoir simulation.

5.5.3 Reservoir body geometry

The relationship between field size and reservoir body size and geometry was introduced in Chapter 4. The small size of individual alluvial fans and pinnacle reefs was recognized as the likely defining size for trap envelopes in such systems. In contrast, large-scale, sand-rich submarine fans were considered to be larger than the footprint of the largest of structurally trapped petroleum fields. The complex stratigraphic trapping geometries of paralic systems were seen as indicative of the interplay between sand-body geometry and trap configuration occurring on similar scales. The difference between exploration and appraisal is that we now need to examine the heterogeneity at the next scale down. At their most simple, the geometry of reservoir bodies can be considered to approximate to one of two basic forms: highly flattened oblate and prolate spheroids. The vernacular

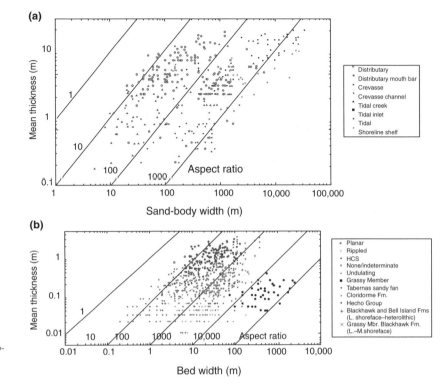

Fig.5.26 Thickness-to-width relationships (aspect ratios) for sandstones from different depositional environments. (a) Paralic sand bodies; (b) mid-lower shoreface and deep-marine beds. (Reproduced courtesy of BP.)

descriptions—pancakes and ribbons—probably convey their shapes better to the reader than do the mathematical terms.

Many geometrical parameters could be used to characterize such reservoir geometries. However, we must be careful to use only those that we can be expected to be able to measure and to simulate using analog data. In that respect, the two most commonly used are thickness and width. Thickness can be measured from core and perhaps wireline logs. Width cannot be measured from either of these, although in exceptional circumstances it may be possible to estimate width from seismic data. Whether or not width can be measured, it is useful to be able to understand the relationship between the two parameters for different geobodies deposited in different environments. The geobodies commonly referred to in such geometric classifications are sand bodies, beds, lithotypes, and shales. The sub-lithotype heterogeneity permeability correlation is discussed in the following section.

For different geobodies, there is commonly a correlation between thickness and width (Fig. 5.26). The numerical rela-

tionship between thickness and width for a given geobody is referred to as the aspect ratio.

Although they are important for reservoir description and reservoir modeling, length-to-width relationships are much more difficult to obtain and estimate. For example, it is commonly possible to make many measurements of channel thickness to channel width from outcrop, for ancient fluvial systems. Only in exceptional circumstances, where exposure is complete and structure conducive, is it possible to measure lengths. Modern analogs can be used in such circumstances, and there has been some success at flume modeling of braided fluvial systems.

Much work has been done over the past decade on the aspect ratios of sand bodies, beds, lithotypes, and shales, and from the statistical data generated it is possible to make a few generalizations. The aspect ratios for sand bodies and beds have a much greater range (typically between 10:1 and 10,000:1) than do those for lithotypes and permeability correlation (typically 1:1 to 20:1). This is not too surprising

Table 5.2 Aspect ratios for clastic sand bodies in fluvial, paralic, and shallow-marine environments.

Depositional system	Element (sand body, bed, or shale)	Aspect ratio (width to thickness)
Fluvial	Low-sinuosity or meandering channel fill at <10 m thick	3:1 to 70:1
Fluvial	Low-sinuosity or meandering channel fill at 5–100 m thick	70:1 to 1000:1
Fluvial	Braided and anastamosing	10:1 to 1000:1
Paralic	Channel fill (distributary, crevasse)	10:1 to 100:1
Paralic	Depositional lobes	100:1 to 1000:1
Paralic	Shoreline sand bodies	200:1 to 5000:1 (length to width, 10:1 to 50:1)
Shallow-marine	Middle shoreface, beds 0.05–2 m thick	1:10 to 1:100
Shallow-marine	Middle shoreface, beds >2 m thick	1:50 to 1:200
Shallow-marine	Lower shoreface	Poor correlation of thin beds <30 cm thick

when one considers that the lithotype, which is derived from the lithofacies, is commonly a product of a distinct physical process, rather than an agglomeration of processes. It also gives hope insofar that the bewildering array of sand-body geometries can be modeled by considering their much more predictable component parts. Aspect ratios of sand bodies, beds, and shales are given in Table 5.2.

Fewer data have been collected on deep-marine and carbonate systems than on fluvial to shallow-marine clastic systems. For deep-marine systems, at least the few data demonstrate aspect ratios that are specific to the system from which they were collected. The lack of global patterns could simply reflect the paucity of the numerical data currently available. Shale width to thickness ratios are shown in Fig. 5.27. Aspect ratios vary from 2:1 to about 10,000:1 for thicknesses between 0.01 m and 2 m. In Fig. 5.27, there appear to be distinct differences in aspect ratio and width for different depositional environments: fluvial, shallow-marine, and deep-marine.

5.5.4 Reservoir correlation

The concept and utility of stratigraphic correlation was introduced in Chapter 2 and some of the applications were examined in Chapter 3. The discussion so far has been one of understanding basins and play fairways. However, at the appraisal stage the problem of correlation can become a greater rather than lesser problem compared with what one might suppose. The need is to correlate penetrations of the reservoir in wells that are commonly only a few kilometers apart. The aim is to obtain a correlation between wells that will honor the flow units in the reservoir. The correlation will control the geometric arrangement of the reservoir units used in the reservoir simulator. The simulator will be run and the development scheme and well locations chosen on the basis of the results. Naturally, if the well-to-well correlation used in the simulator is inappropriate, then the reservoir will not behave as the simulation predicts. There may only be minor differences between the model and reality, or there may be large differences. This happened with the Clyde Field (Fig. 5.28), where there was sufficient mismatch that development of the field was suspended.

At the appraisal stage, biostratigraphy can remain an important tool, but in many instances identification of reservoir flow units will be beyond the resolution of biostratigraphic methods. In reservoirs that are barren of biostratigraphic material, chemostratigraphy or magnetostratigraphy may be used.

Sequence stratigraphy (Section 3.6.6), worked within a biostratigraphic, magnetostratigraphic, or chemostratigraphic framework commonly provides the most powerful reservoir correlation tool. The ease or difficulty encountered in using sequence stratigraphy to solve an appraisal-scale correlation problem depends upon both the well spacing and the depositional environment of the reservoir interval. If the well spacing is larger than the size of the individual elements that make up the reservoir, then accurate correlation will prove difficult. The same will be true if the reservoir interval is highly cyclic and the cycles are all much the same.

The fundamental approach to correlation based on sequence stratigraphy is to use the wireline log responses to define depositional trends. Care must be exercised when doing this to ensure that the trends are products of depositional changes and not due to fluid changes (petroleum versus

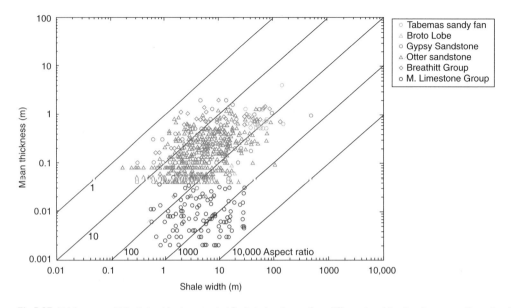

Fig.5.27 Thickness-to-width relationships (aspect ratios) for shales/mudstones from different depositional environments. (Reproduced courtesy of BP.)

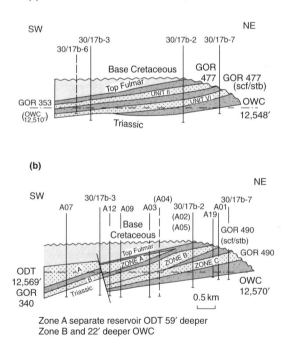

Fig.5.28 The Clyde Field, UK North Sea. Schematic cross-sections before (a) and after (b) the initial phase of development drilling, showing the reservoir zonation. (From Turner 1993.)

water). The suite of commonly identified patterns on gamma, SP, resistivity, and sonic logs is illustrated in Fig. 5.29. It is common, then, to identify maximum flooding surfaces and maximum regression surfaces. The wireline log responses are then used to define the systems tracts between the flooding and regression surfaces (Fig. 5.30). Each well is subject to the same treatment, and the combined well logs plus sequence stratigraphic interpretations are used as the basis for inter-well correlation (Fig. 5.31).

Correlation based upon rock properties using the above methods may be refined by incorporating data derived from the analysis of both petroleum and water (Sections 5.3 & 5.4).

5.6 RESERVOIR QUALITY

5.6.1 Introduction

At the pore scale, reservoirs can differ dramatically in their quality. Quality may be measured as either the capacity of the reservoir rock to hold fluids (porosity) or its capacity to transmit them (permeability). We have already covered the basics of porosity, permeability, and fluid saturations in Chapter 4. In this section, we examine intrinsic reservoir properties further. We also look at the post-depositional processes that can modify both porosity and permeability within potential reservoirs: compaction, cementation, and mineral dissolution.

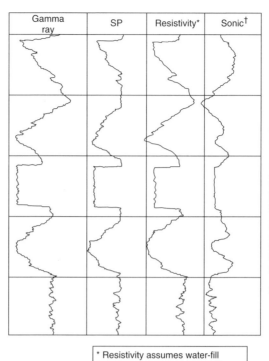

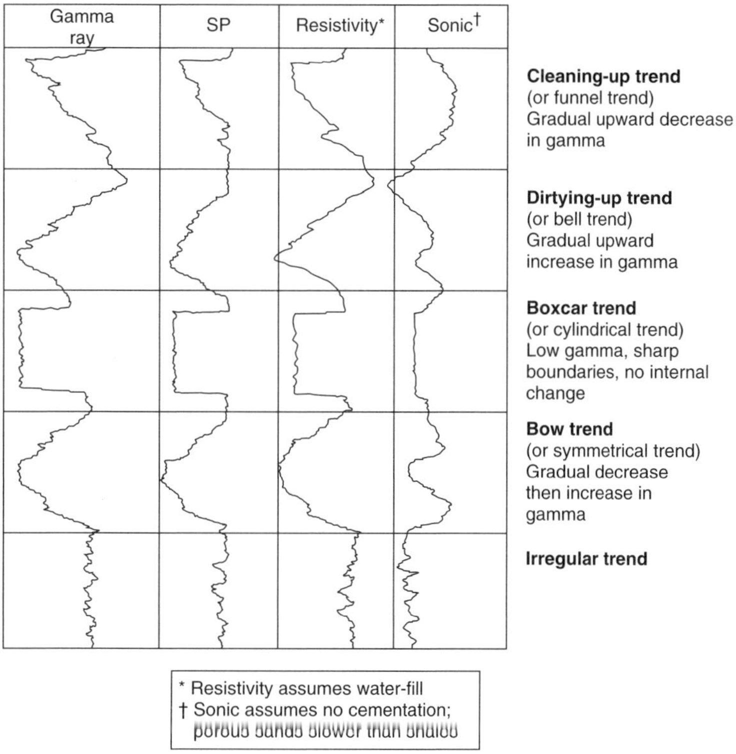

Gamma ray	SP	Resistivity*	Sonic†	
				Cleaning-up trend (or funnel trend) Gradual upward decrease in gamma
				Dirtying-up trend (or bell trend) Gradual upward increase in gamma
				Boxcar trend (or cylindrical trend) Low gamma, sharp boundaries, no internal change
				Bow trend (or symmetrical trend) Gradual decrease then increase in gamma
				Irregular trend

Fig.5.29 Idealized log trends, assuming saltwater-filled porosity. (From Emery & Myers 1996.)

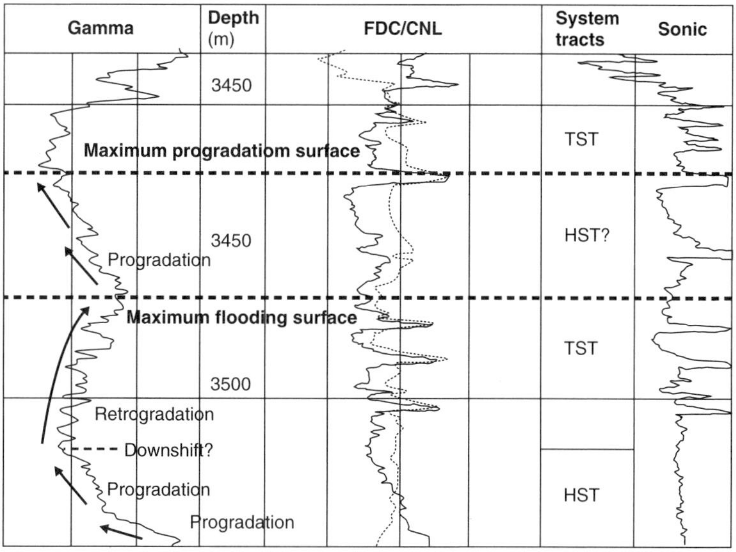

* Resistivity assumes water-fill
† Sonic assumes no cementation; porous sands slower than shales

Gamma	Depth (m)	FDC/CNL	System tracts	Sonic
	3450			
Maximum progradatiom surface			TST	
Progradation	3450		HST?	
Maximum flooding surface			TST	
Retrogradation	3500			
Downshift?				
Progradation			HST	
Progradation				

Fig.5.30 A well log through the Late Jurassic reservoir of the Ula Field, offshore Norway. The reservoir consists of a series of shallow-marine parasequences arranged in progradational and retrogradational stacks. The maximum flooding surfaces and maximum progradation surfaces are used to guide the reservoir stratigraphy. HST, highstand systems tract; TST, transgressive systems tract. (From Emery & Myers 1996.)

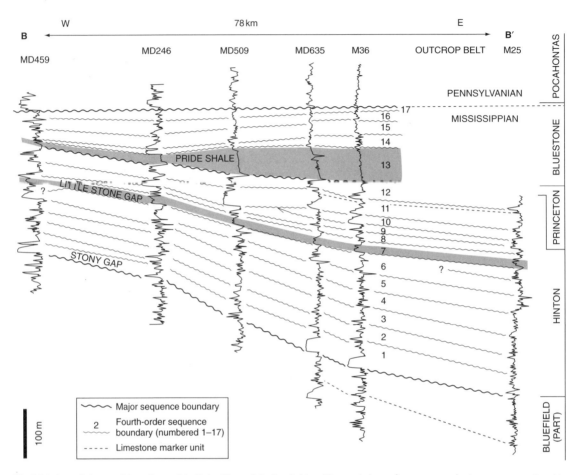

Fig.5.31 A correlation panel through part of the Carboniferous of the Appalachians. The correlation surfaces were erected using sequence stratigraphic principles. (From Miller & Eriksson 2000.)

5.6.2 More intrinsic reservoir properties

Porosity, permeability, grain density, and water saturation are measured on core plugs during routine core analysis. Such data, though of prime importance in determining the storage volume and flow potential of a reservoir, do not yield all of the information how a well or field will behave on production. Further data are generated from a combination of additional, special core analysis and from petrophysical analysis of the wireline logs. We are now at the interface between petroleum geoscience, reservoir engineering, and petrophysics. Nonetheless, it is useful to examine briefly some of these aspects of petroleum science, since they form part of an appraisal program.

Relative permeability

Absolute permeability to air is measured during routine core analysis. Such figures need to be corrected to reservoir conditions in terms of a compaction correction (Fig. 2.20), yet such corrected data are still only valid for single-phase flow. A consequence of there being more than one fluid in the pore system is that neither water nor oil will flow as readily as if there were only one phase (Fig. 5.32). This is known as relative permeability. As a conceptual device, think of relative permeability in terms of a group of joggers running down a street. If the street is empty, the joggers will be able to move along at a pace determined only by their fitness and capabilities. In street filled with busy, nonjogging phase (shoppers), the transit

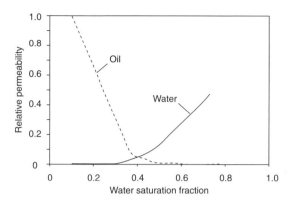

Fig.5.32 Relative permeability. The figure shows the dramatic decline in permeability to oil from 100% to about 5% as water saturation increases from about 15% to about 35%. The curves are for 22° API, 15 cp oil in fine-grained uncemented sand (about 25% porosity) from the Pedernales Field of eastern Venezuela.

time (the relative permeability to joggers) will increase massively as shoppers get in the way. In a similar way, the microscopic viscous drag forces between the oil, water, and mineral surfaces act to reduce the permeability to one of the phases. The importance of accurate determination of relative permeability is high. The relative permeability data obtained during appraisal will be used in the development design process. In detail, such data will be used to calculate the standoff (vertical separation) between production wellbores and petroleum/water or gas/oil contacts, and in calculations associated with the pressure drawdown that will be placed on production wells.

Relative permeability is measured in the laboratory through experimentation. The experiments are run at reservoir temperatures and pressures, using a combination of real or simulated formation water and oil. Sequences of oil and water are flushed through the samples of preserved core, and the relative permeabilities of the two phases at a variety of partial saturations are measured. Relative permeability is highly sensitive to both the detailed rock type (the lithofacies) and the fluid types. Moreover, the transformation of relative permeability data obtained on core plugs or whole core to reservoir-scale behavior carries large uncertainties.

The quality of the relative permeability data will be revealed during field production. Field-scale relative permeability data can be calculated once the undesired phase (usually water or gas) starts to break through into the (oil) production wells. Naturally, one's hope is that water break-

through will occur either when or later than predicted using the relative permeability measurements made on core during appraisal.

Wettability

Linked to the oil and water saturations (Chapter 4.4.3) is the concept of "wettability." In some reservoirs, the surface of the rock is water wet, with petroleum lying at the centers of the pores. A few reservoirs are oil wet. There are no firm rules as to what sort of reservoirs are oil or water wet. Indeed, most are now considered to have mixed wettability. The factors that influence wettability are the mineralogy of the reservoir and the presence or absence of natural surfactants within the petroleum (i.e., polar compounds, Chapter 2). As a guide, clastic reservoirs are usually water wet, while carbonate reservoirs may be oil or water wet. The distinction is important, because it has a dramatic effect on both the recovery factor and the process chosen for secondary recovery (Archer & Wall 1986). Sweep efficiency to water-flood tends to be better in water-wet systems than it does in oil-wet systems.

Visualization of wettability properties has become possible in recent years. Samples of preserved core are bathed sequentially in low-viscosity, hydrophobic, and hydrophilic resins. The resins are imbibed into the rock, where they replace the petroleum and water phases respectively. Once the resins have set, the rock can be sliced and polished to produce a conventional thin section or polished slab. These can then be viewed under a scanning electron microscope using back-scattered electron imaging. Since the resins are designed to have different mean atomic numbers, their distributions within the pore structure can be seen in terms of different gray levels on the back-scattered display. While the technique appears to be a powerful one, there is uncertainty as to whether the process of removing the rock sample from the Earth and then subjecting it to the preparation process appreciably affects its wettability and the distribution of the fluid phases.

Resistivity, cementation, and tortuosity factors

In Chapters 2 and 4, we introduced wireline logs and described the data that can be derived from them. Petrophysical analysis of wireline log data involves quantification of the properties outlined previously. The key numerical derivatives are the formation resistivity factor (F), the cementation factor (m), and the tortuosity factor (a).

The formation (resistivity) factor is the ratio between the resistance of a fixed volume of formation water and the same volume of formation water plus formation (i.e., nonconducting rock). The unit of measurement is the ohm-meter^2 per meter. This is commonly abbreviated to "ohm-m" and in practice the measurements are made in milli-ohms.

Naturally, the more any particular rock becomes cemented and loses porosity, so the more the formation resistivity factor increases. The link between F and porosity has been found from experimentation to be a power function corresponding to the form:

$$F_{\kappa} = \frac{a}{\Phi^{m}}$$

where m is the cementation factor and a is the tortuosity factor.

The value of m varies between about 1 for loosely consolidated rock and about 3 for well-indurated limestone. The factor a is simply a measure of the intercept of the relationship between m and F. It commonly lies in the range 0.6–2.0.

5.6.3 Controls on reservoir quality

When sands and carbonates are deposited, they are both highly porous and highly permeable. Sands commonly have between 40% and 50% pore space, and some carbonates may contain as much as 70% pore space. It will come as no surprise that the permeability of such systems is enormous (tens to hundreds of darcies). Few oil and gas reservoirs are quite so porous and permeable. Sandstone and carbonate reservoirs with more than 30% porosity are rare, and some reservoirs are still viable with matrix (as opposed to fracture) porosities as low as 10%.

Permeability may exceed 10 D in the best reservoirs, but in many reservoirs we have to make do with tens to hundreds of millidarcies. Gas reservoirs can still perform with a mean permeability of as little as 1 mD, while 10 mD is often accepted as a lower limit for light oil.

The processes whereby beautifully porous and permeable sands and carbonates are turned into tight sandstones, limestones, and dolomites are collectively known as diagenesis. Two processes are active: compaction and cementation. Compaction involves the shortening of the sediment column during burial: water is expelled and the overall pore space diminishes. Cementation involves little or no shortening of the sediment column but, rather, the mineral cements are brought in solution to the sandstone or carbonate and precipitated in the pore space, so reducing porosity (Gluyas & Coleman 1992). The two processes are combined in pressure dissolution. Most of the post-depositional diagenetic effects

downgrade reservoir quality. However, mineral dissolution can lead to the creation of secondary porosity that can yield an improvement in reservoir quality (Choquette & Steinen 1980; Emery & Myers 1990; Estaban 1991).

It is useful to discuss sandstones and carbonates separately as far as reservoir quality is concerned, for although we know that some geochemical processes are common to both rock types, we can more easily quantify and thus predict the effect on sandstones compared with what can be done for carbonates.

5.6.4 Compaction and cementation in sandstones

Sand compaction

Loose sand compacts easily. During the earliest stages of burial, much of the compaction is taken up by the rearrangement of grains. Simple burial combined with seismic shock will turn loose sand into consolidated sand. The amount of porosity lost will depend largely on how well sorted the sand is. In a poorly sorted sand, more porosity will be lost than in a well-sorted sand—the little grains filling in between the bigger ones (Vesic & Clough 1968). As burial continues, rough edges tend to be knocked off grains, so aiding greater compaction. At deeper levels (about 1–4 km), the sand begins to behave like a linearly deformable solid (Atkinson & Bransby 1978). Deeper still, plastic deformation is probably more common. The boundaries between where this process act will vary from sand to sand and basin to basin, and in some instances may be gradational.

The net outcome of all the above processes is that sands compact when stressed, but they de-compact very little when the stress is released. This means that if you find compacted but uncemented sandstone at the Earth's surface, you should be able to calculate the maximum stress that it suffered in any previous burial phase. Such a stress calculation can be used to provide an estimate of the maximum burial depth.

In the geologic literature, there are a large number of so-called compaction curves for sandstones (Baldwin & Butler 1985, and references therein). Alas, most of these curves are porosity–depth plots rather than porosity–stress plots. As such, the great swaths of data on these plots include, but do not differentiate, the effects of fluid overpressure and sandstone cementation. Experimental data are available on the way in which sands compact (Atkinson & Bransby 1978; Kurkjy 1988), and for clean quartzose or arkosic sandstones these data have been turned into a compaction equation by Robinson and Gluyas (1992a):

$$\Phi = 0.5 \exp\left(\frac{-10^{-3}z}{2.4 + 10^{-4}z}\right)$$

where z is in meters. In this equation, porosity is expressed as a fraction (i.e., <1) and the equation is calibrated to a normal hydrostatic pressure gradient. If the system is overpressured, the pressure borne by the grains is less than in a hydrostatic system and an effective depth must be calculated. As a simple rule of thumb for typical burial depths of 2–4 km, 1 MPa of overpressure is equivalent to about 80 m less burial. The equation is well tested, predicting porosity to within ±3% at 95% confidence limits (Fig. 5.33; Gluyas & Cade 1997).

Sands that contain easily squashed grains, such as glauconite or mica, or those rich in matrix clay lose porosity much more readily for an applied stress. Kurkjy (1988) quantified the effects of compaction on such sandstones using laboratory experiments.

Sand cementation

There are many minerals that can act as cements in sandstones. The most common and abundant are quartz, carbonates, clays, and zeolites. Evaporite minerals, other sulfates, sulfides, oxides, feldspar minerals, and other forms of silica occur widely, but rarely are they of volumetric importance in reservoir sandstone (Primmer et al. 1997; Fig. 5.34). The common habits of these minerals are shown in photomicrographs in Fig. 5.35.

In a review of about one hundred reports on diagenetic history, spread across the world and across basin types, Primmer et al. (op. cit.) observed the following patterns:
• Quartz-dominated diagenesis is the most common diagenetic style (40% of studies);
• the association between an early diagenetic grain coating of clay and minor quartz is more common than clay diagenesis alone;
• early or late diagenetic carbonates are the dominant cement phase in 10–15% of studies;
• significant quantities of zeolites are reported in 10–15% of studies;
• evaporites are rarely of significance as cements in sandstones.
The observations above give some clues as to the diagenetic processes that occur when a sand is buried. The assemblage of diagenetic minerals in a sandstone is dependent on:
• The composition of the deposited sand (e.g., a volcaniclastic sand is far more chemically reactive than a quartzose sand).
• The depositional environment. This controls the composi-

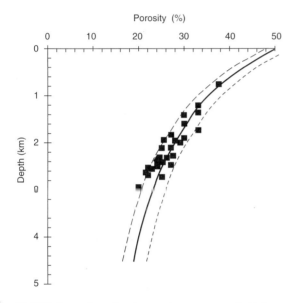

Fig.5.33 An experimental sand compaction curve compared with data on the porosity of uncemented, hydrostatically pressured sandstone reservoirs. (From Gluyas & Cade 1997.)

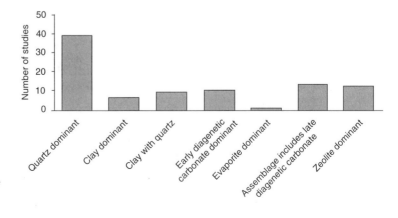

Fig.5.34 The main cements in sandstones. The data are taken from a survey of the diagenetic mineralogy of about 100 sandstone reservoirs. (From Primmer et al. 1997.)

(a)

(b)

(c)

(d)

(e)

(f)

(g)

tion of the initial pore water and thus the course of early diagenesis (e.g., a marine sandstone is much more likely to suffer calcite cementation soon after deposition than an aeolian sandstone, because the marine pore waters are rich in both dissolved calcium and bicarbonate).

• The composition of "nearby" lithologies. Sandstones are probably not closed systems during diagenesis (Gluyas & Coleman 1992), and as a consequence their diagenetic evolution commonly interacts with diagenetic processes in adjacent lithologies (e.g., evaporites, limestones, etc.).

• The burial history. The maximum burial depth, or the depth at which cementation occurs, will control the degree of compaction suffered by a sandstone. For sandstones that have been uplifted and exposed to erosion and freshwater flushing, there may be a local improvement in reservoir quality due to leaching (dissolution) of grains from the sandstone (Emery & Myers 1990).

• The thermal history can control both the presence and the absence of some diagenetic minerals and the volumes precipitated. Two prime examples are illite and quartz. Illite only precipitates in significant quantities above about 100°C (Small et al. 1992). In many basins, the quantity of quartz cement increases with depth. The rate of change of quartz cement with depth is greater in basins with a higher thermal gradient compared with those from basins with a lower geothermal gradient (e.g., North Sea = 35°C km^{-1} and Hammerfest = 60°C km^{-1}: Selley 1978; Rønnevik et al. 1983; Gluyas 1985).

In order to make some value of this knowledge in terms of understanding our petroleum reservoir, it is necessary to understand the quantitative effects of these cement minerals on permeability. The most exciting approach to this problem

that we have seen is that adopted by Bryant et al. (1993) and based on the work of Finney (1968). A bead pack containing 8000 equally sized spheres was used to simulate a sand. For the bead pack, the coordinate geometry of all the randomly packed beads was measured, and as a consequence the porosity and permeability could be calculated. Computer simulation was then used to mimic compaction, and then simple (syntaxial) cementation. The results of the initial studies using the monodisperse (perfectly sorted) bead system were almost identical to the porosity-to-permeability relationship for a real well-sorted, compacted, and partially cemented sand (from the Tertiary of the Paris Basin). The confidence generated by this result led the team to vary the style of the modeled cementation to mimic different morphologies adopted by quartz or clays or carbonate cements. The effects of less than perfect sorting were solved by incorporating the empirical sorting–permeability relationships published by Beard and Weyl (1973). Two examples are illustrated in Fig. 5.36.

The timing of cementation relative to petroleum migration is also important when it comes to assessing the reservoir quality of an undrilled prospect. It is now fairly widely recognized that the processes of so-called "late burial diagenesis" often take less than about 10 Ma (Glasmann et al. 1989; Robinson & Gluyas 1992b), which is about the same length of time it takes for a field to fill with petroleum (Chapter 3). It is also common for the two processes to occur at much the same time (Walderhaug 1990; Nedkvitne et al. 1993). Gluyas et al. (1993) have suggested that in such *a race for space* an early petroleum charge might save a reservoir from the worst ravages of cementation (Fig. 5.37). Here, replacement of most of the water by petroleum would serve to slow cementation by limiting solute access to the reservoir.

Fig.5.35 Diagenetic mineral habits in sandstones. (a) Quartz-cemented quartz arenite, Fontainebleau Sandstone, Paris Basin, France; the quartz cement occurs as syntaxial rims (q) on the grains. Combined back-scattered electron and cathodoluminescence image, scale bar 200 µm. (From Evans et al. 1994). (b) Dolomite with ferroan dolomite rims (d), occurring as a scattered pore-filling cement, Rotliegend Sandstone, Amethyst Field, UK North Sea. This photomicrograph also contains evidence of secondary porosity in the form of partially dissolved feldspar grains (f). Back-scattered electron image, scale bar 200 µm. (From Gluyas et al. 1997a). (c) Pervasive calcite cement occurring as a concretion within an otherwise porous and permeable sandstone, Ula Sandstone, Norwegian North Sea. The top and base of the concretion are curved. The concretion contains secondary pores created on dissolution of bivalve shells (b). Core photograph, scale bar 15 cm. (d) Pore-filling kaolinite (A), Magnus Sandstone, Magnus Field, UK North Sea. Large quantities of microporosity commonly occur between the pseudo-hexagonal clay crystals. Such micropososity may trap water, leading to reservoirs with high water saturations that nonetheless may produce dry oil. Secondary electron image, scale bar 10 µm. (e) Grain-coating chlorite cement, Tuscoloosca Sandstone, Louisiana, USA. The presence of chlorite-coated grains seems to inhibit quartz cementation. Such inhibition may contribute to the high porosity often seen in deeply buried, chlorite-cemented sandstone reservoirs. Secondary electron image, scale bar 100 µm. (Photograph by T. Primmer.) (f) Grain-coating smectite, Tertiary sandstone, UK Western Margin. Smectite is commonly highly sensitive to salinity and may swell to many times its original volume if it comes into contact with fluids that are less saline than the connate water. This can be a serious issue if drill fluids cause swelling and thus permeability reduction in the near-wellbore region. A crystal of zeolite cement occurs in the top right-hand corner of the photomicrograph. Secondary electron image, scale bar 100 µm. (Photograph by T. Primmer.) (g) Fibrous, grain-coating illite, Tertiary sandstone, UK Western Margin. Illite crystals can bridge pore throats (A) and reduce permeability dramatically even in very porous sandstones. Scale bar 250 µm.

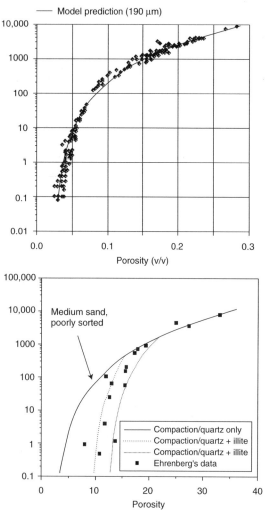

Fig. 5.36 (a) A porosity versus permeability cross-plot for Fontainebleau Sandstone, showing the close correspondence between model predictions based on grain size and cement style, and measured porosity and permeability data. (b) A porosity versus permeability cross-plot for Garn Formation sandstones. The porosity and permeability values are averages for wells with different geographic locations and burial histories. The curves are model derived and match the measured textural and diagenetic characteristics of the Garn Formation sandstones. (From Ehrenberg 1990; Cade et al. 1994.)

5.6.5 Compaction and cementation in limestones

Compaction and cementation in limestones control reservoir quality just as they do in sandstones. However, because carbonate minerals are on the whole more soluble than the sili-

cate minerals in sandstones, limestones are much more prone to complex diagenetic histories. These histories may involve many phases of mineral precipitation, recrystallization, and mineral dissolution. There are no straightforward counterparts to the technologies used to predict the quantitative effects of compaction and cementation on reservoir quality in sandstones. However, this does not mean that we can do nothing.

We offer a simple twofold subdivision of the controls on reservoir quality in carbonates:

- Syn-depositional and near-surface diagenetic processes;
- compaction and burial-linked processes.

This subdivision is neither based on (chemical) processes nor elegant. However, it is a practical classification that can be used when it comes to assessing or predicting the reservoir quality of carbonate reservoirs.

Near-surface processes

By near-surface and syn-depositional processes, we refer to those mineral precipitation and dissolution events that occur near to the sediment/water interface. Sea-level fluctuations, during and soon after deposition of carbonate sediment, will force water flow through the pore system. In many instances the flowing water is likely to vary between fresh, brackish, marine, and possibly hypersaline. Carbonates (± sulfates) will be precipitated and dissolved.

Read and Horbury (1993) developed a model to explain the reservoir quality evolution of carbonates as controlled by near-surface processes. Specifically, they examined the link between sea-level cycles (1–40 Ma tectonic–eustatic and 20–400 ka Milankovich) and climate (humid and arid, hot and cold; Fig. 5.38). The model was wide ranging, allowing explanation of a diverse set of phenomena as observed by carbonate sedimentologists: cathodoluminescence cement stratigraphy, near-surface dolomitization, and karstification. The end members of the sea-level cycle/climate type associations are described below and in Table 5.3.

1 Low-amplitude, high-frequency sea-level changes typical of greenhouse times form thick, meter-scale cycles with tidal-flat tops. The sea-level fluctuations cause little erosion of the tops. In humid climates, cementation is early and aragonite shells are leached. The Swan Hills oilfields of western Canada belong to this category (Prodruski et al. 1988).

2 For sea-level and temperature conditions similar to 1, an arid climate produces cyclic dolomitization and aragonite leaching. Primary intergranular and moldic porosity is preserved, as is intercrystalline and vuggy porosity in the dolomitized portion. The Pukkett Field (Texas, USA; Loucks & Anderson 1985) is an example of a petroleum field devel-

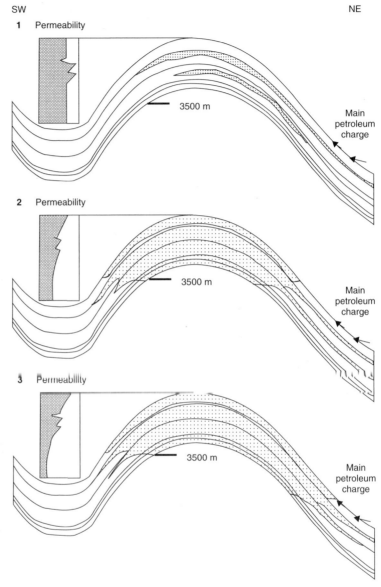

SW   NE

1 Permeability

3500 m

Main
petroleum
charge

2 Permeability

3500 m

Main
petroleum
charge

3 Permeability

3500 m

Main
petroleum
charge

Fig.5.37 A model for the preservation of
porosity and permeability by early oil charge
relative to cementation; faults omitted for
clarity. 1, Initial filling along high-
permeability fairways; cementation starts. 2,
Filling to 3500 m during cementation;
development of a permeability gradient; high
crestal permeabilities permit regular filling;
invasion of high-permeability streaks below
the oil leg in the southwest. 3, Filling to below
3500 m during cementation; further
development of the permeability gradient
below 3500 m, little evolution above; low
flank permeabilities inhibit oil penetration
(except along high-permeability streaks)
below 3500 m in the southwest. (From Oxtoby
et al. 1995.)

oped under such conditions. Ghawar (Saudi Arabia), the
world's largest known oilfield (reserves likely to exceed 100
billion barrels), is another.

3 High-frequency, moderate-amplitude fluctuations in sea
level occur during continental glaciations. Karstic surfaces or
caliche deposits directly overlie the shallow subtidal deposits. Cementation commonly plugs the upper phreatic
zone. Primary, intergranular porosity is preserved in mid-
cycle grainstones and packstones. Much of the Lower
Carboniferous carbonate of northern England was deposited

in a humid climate with moderate sea-level fluctuations
(Horbury 1989; Case history 3.8).

4 In the arid equivalent of 3, there is little mineralogical sta-
bilization and dolomites can form beneath evaporitic tidal
flats. Primary porosity is commonly preserved in unaltered
sediment and intercrystalline pore space dominates in the
dolomitized intervals (Fig. 5.39). The reservoir in the
Whitney Canyon–Carter Field of the US Rocky Mountains
was deposited in arid conditions.

5 Large-amplitude, glacio-eustatic sea-level fluctuations

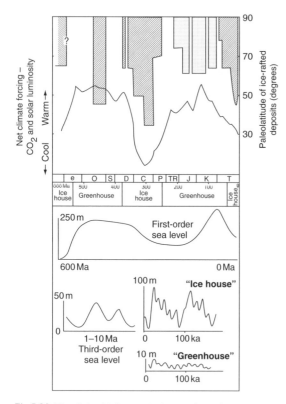

Fig.5.38 The relationship between ice-house and greenhouse conditions, eustasy, and global CO_2. The upper part of the figure shows paleolatitude distributions of ice-rafted glacial deposits in continental regimes (cross-hatched) and those of possible marine origin (stippled) plotted against geologic time. The lower curve on the upper figure shows net forcing of climate due to changes in CO_2, the consequent effect on radiative forcing, and the long-term increase in the solar luminosity. Times of general ice-house versus greenhouse conditions are shown on the time axis. The lower half of the figure shows, at the top, the first-order Vail sea-level curve for the Phanerozoic, for comparison with the above CO_2 – solar luminosity curve. The bottom diagram shows a third-order sea-level curve with 1–10 Ma periods, and schematic fourth- and fifth-order sea-level curves that would be imposed on the longer-term curves. Low-amplitude fourth- and fifth-order curves generally would be typical of times of global greenhouse conditions, but would have been interspersed with short periods when amplitudes were substantially higher. Similarly, during ice-house times, most high-frequency sea-level fluctuations would have been high amplitude, but were probably interspersed with times of lower-amplitude sea-level fluctuations. (From Read & Horbury 1993.)

lead to the development of sinkholes and cave systems. Diagenesis is dominated by large-scale movement of marine, mixed marine, and fresh waters through the sediment. Read and Horbury (1993) record fields with this type of diagenesis in Kansas and Texas.

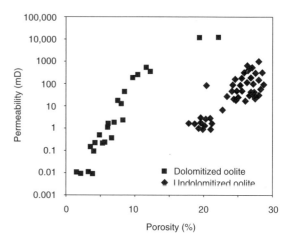

Fig.5.39 A comparison of porosity and permeability for dolomitized and undolomitized oolite (Cretaceous Middle East).

6 Arid-climate, glacio-eustatic effects tend to generate extensive caliche caps and only minor spar precipitation (primary porosity is preserved). The Miocene carbonate reservoirs of Syria, Iraq (Zagros), and the Gulf of Suez have a pore system that has been modified by arid-climate, glacio-eustatic linked diagenesis (Buday 1980). The Permian Zechstein cycles of the North Sea also developed during glacio-eustatic sea-level changes (Taylor 1981).

7 Sea-level falls occurring with a periodicity of one to tens of millions of years (second- or third-order; Vail et al. 1977) often have a major influence on reservoir quality under humid conditions. Porosity changes from intergranular and moldic to vuggy, cavernous, and ultimately to collapse breccias as dissolution progresses. The Golden Lane Trend reservoir of Mexico (Enos 1988) was karstified after tectonic uplift. Similarly, reservoirs developed in the Cretaceous Mishrif Formation of the Middle East owe their vuggy and moldic porosity to large-scale leaching of carbonates during tectonic uplift (Burchette & Britton 1985). The buried hill reservoirs of China (Chapter 4) are also products of deep-penetrating karstification.

8 In arid climates the influence of major sea-level fall may be minor, or at least exceedingly slow. Reflux dolomitization may occur if the surface is covered with hypersaline brines. Read and Horbury (op. cit.) demonstrated that these cycles have a strong influence on the development of many of the world's largest petroleum fields, reservoired in platform top carbonates. They concluded that in greenhouse times giant fields are limited to arid cycles. During the change to high-

Table 5.3 Sea level and climatic controls on near-surface carbonate sediment diagenesis.

	Sea-level change		Climate type	
	Amplitude	Frequency	"Temperature"	Rainfall
1	Low	High	Greenhouse	Humid
2	Low	High	Greenhouse	Arid
3	Moderate	High	Ice house	Humid
4	Moderate	High	Ice house	Arid
5	Large	High	Ice house	Humid
6	Large	High	Ice house	Arid
7	Large	Low	Greenhouse/ice house	Humid
8	Large	Low	Greenhouse/ice house	Arid

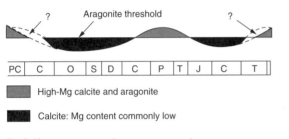

Fig.5.40 The mineralogy of marine inorganic carbonate precipitates through the Phanerozoic. Aragonite is more prone to dissolution than calcite. (From Sandberg 1983; reprinted with permission from *Nature* (305, 19–22); © 1983 Macmillan Magazines Limited.)

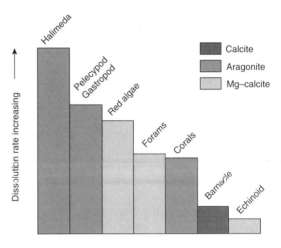

Fig.5.41 The solubility of carbonate sediment grains in sea water as a function of mineralogy and sediment origin. (From Walter 1985.)

amplitude ice-house cycles, giant oilfields occur both in humid and arid cycles. High-amplitude, low-frequency (second- and third-order) sea-level changes in humid climates can also host giant fields.

The degree to which either karstification or selective (moldic) dissolution will increase the quality of a reservoir also depends upon the absolute timing of dissolution (Fig. 5.40) and the sediment composition (Fig. 5.41). Aragonite is more readily soluble than calcite, and some types of bioclastic debris are more soluble than others.

Compaction and burial processes

Burial will lead to compaction and mineral reactions linked to increasing temperature and/or the expulsion of connate waters through the potential reservoir carbonate. The Jurassic Smackover Formation of the southern United States is a carbonate with a dominantly burial-linked control on reservoir quality. It is also well described.

Moore and Haydari (1993) studied the Smackover, and in a

beautifully written, simple, and elegantly argued paper they unravel the diagenesis and reservoir quality history of the Middle Jurassic, petroleum-rich, Smackover Formation. Moore and Heydari (op. cit.) also point out that the importance of burial diagenesis on the quality of many carbonate reservoirs may have been underestimated. They argue that many of the major carbonate reservoir systems around the world — the Upper Jurassic of northern Arabia, west-central Europe, west-central Argentina, and the Silurian–Devonian of the eastern US — have diagenesis, and hence reservoir quality, dominated by reactions associated with burial. None of these systems developed in a cratonic setting. In contrast, they developed during periods of high sea level (highstand systems tracts) in basins undergoing rapid subsidence.

The story for the Smackover was as follows:

1 Deposition of ooid-rich grainstones in a highstand systems tract.

2 Sea-level drop, leading to aragonite dissolution and calcite recrystallization.

3 Dolomitization associated with deposition of the subsequent highstand lagoonal sediments — those areas outside the influence of the lagoon(s) remained as calcite. The dolomitization led also to the reorganization of pores and pore-throat systems, yielding permeable as well as porous rock.

4 Burial and loss of poroperm largely through pressure dissolution and consequent reprecipitation of carbonate.

5 Structuring and emplacement of oil generated from the transgressive systems tract sediment (varved mudstones) in the same sequence.

6 Continued burial and heating leading to pyrobitumen formation and interaction between the carbonate and oil, with metal-rich solutions derived from the underlying clastic Norphlet Sandstone.

For those areas that received an oil charge but escaped the dramatic effects of further burial, diagenesis was arrested with sufficient reservoir quality intact for them to be economically viable reservoirs.

The authors' contentions regarding sources of matter for cementation are supported by isotopic evidence for the carbonate components and some good commonsense reasoning for the other components.

The relationship between reservoir quality and oil emplacement is more marked for carbonates than for sandstones. Wilson (1977) cites many examples of oilfields reservoired with porous carbonates (the Thamama of Abu Dhabi; the Fahud Field, Oman; and the Smackover reservoirs of Mississippi) with tight or low-porosity rock below the oil/water contact. The change from porous to tight rock was ascribed to precipitation of cement in the water leg after oil emplacement. The presence of oil limited further cementation in the oil leg.

In addition to preservation of porosity, oil can also react with carbonates at high temperatures, to produce methane, hydrogen sulfide, and native sulfur. Such reactions are believed to be responsible for many of the sour gasfields of Abu Dhabi (Worden et al. 1995).

5.7 RESERVOIR DESCRIPTION FROM SEISMIC DATA

5.7.1 Introduction

Seismic data are now commonly used within a reservoir des-cription. Gone are the days when seismic information was used only to define the trap geometry. We can now expect the best of 3D seismic data sets to yield information on lithology, net pay, fluid type, porosity, and lateral variations in these properties (reservoir layering, compartmentalization, and net pay distribution). Repeat seismic surveys, so-called 4D seismic, may also allow fluid and rock changes to be monitored during field production (Chapter 6). Such data can be used to aid reservoir management. Seismic data can also be used to complement work done with cores and logs (Chapters 2, 3, & 4).

Seismic information is unique insofar as it is currently the only tool that provides a volumetrically complete, albeit fuzzy, picture of the reservoir. Logged wells and cores provide spots of high detail in what, without seismic, would be volume largely empty of information.

An outline of what "seismic" is and details on the acquisition and processing of seismic data were given in Chapter 2. Here, we will look at the sorts of reservoir information that can be derived from seismic data and the tools used to gather such data. To determine lithology, net pay, fluid type, and porosity, combinations of five different analytic tools are commonly used. These are velocity analysis, amplitude versus offset, absolute acoustic impedance, seismic modeling, and geostatistics.

A brief explanation of these terms is as follows:

• Velocity analysis is part of the seismic processing sequence. It can be used to define the velocity across intervals in the subsurface. This may help to resolve layers with thicknesses greater than about 100 m (300 ft).

• Amplitude versus offset provides an estimate of the change in Poisson ratio across a reflection boundary, the Poisson ratio being a measure of the difference between the P-wave and S-wave properties of the rock.

• The acoustic impedance is a product of the P-wave velocity and density of the layer. Acoustic impedance is aimed at layers that are at or slightly below the level of seismic resolution. If such layers are encased within a homogeneous but different lithology, this type of analysis can be used to measure reservoir lithology, fluid pressure, porosity, and possibly net pay.

• Seismic modeling can be used to test the validity of different interpretations of the subsurface. A seismic model is constructed using geologic information from wells or outcrop and then compared with the seismic section. If the two match, then the criteria on which the model was constructed can be used as a description of the subsurface.

• Geostatistics is used to quantify the detailed character of seismic traces and link them to geologic variations in the reservoir. The use of displays of dip or azimuth on a surface can help to define reservoir structure. Various attributes of

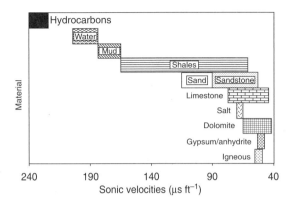

Fig.5.42 P-wave velocities for various lithologies. (Reproduced courtesy of BP.)

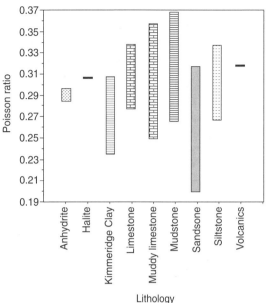

Fig.5.43 The Poisson ratio for various lithologies. (Reproduced courtesy of BP.)

the seismic, such as instantaneous phase or instantaneous frequency, can sometimes be used to map gas/water contacts or stratigraphic boundaries.

5.7.2 Lithology description

An estimate of reservoir lithology can be made in a number of ways. Each method relies on comparing the value of a derived parameter such as the Poisson ratio, absolute acoustic impedance, or P-wave velocity with the value for the same parameter in known lithologies. Although a single parameter can be used to estimate lithology, greater confidence will come from using two or more parameters. The P-wave velocities and Poisson ratios for various lithologies are shown in Figs 5.42 and 5.43.

The estimation of lithology is probably of greatest importance in complex reservoir packages such as delta systems or submarine fans in areas with a fairly extensive petroleum exploitation history. In such systems, there is a need to find the ever more subtle reservoir — the one that the "other company" has missed — but equally there is much information from which to calibrate your search.

5.7.3 Porosity determination

Porosity can be calculated from seismic data for both clastic and carbonate reservoirs. Such a process is commonly used in the appraisal and development stage of a field where there are sufficient well data to allow the seismic to be tied to the wells, but too few wells for a porosity map to be constructed with any confidence from the wells alone.

The most common approach is to invert the seismic to acoustic impedance, simply the product of P-wave velocity of the rock and density. Variations in porosity of a rock cause changes in the rock's bulk density and velocity characteristics. This is because porosity contains low-density, acoustically slow matter such as water, oil, or gas.

The bulk density for a rock is given by:

$$\rho(\text{bulk}) = \rho(\text{matrix})(1 - \Phi) + \rho(\text{fluid})(\Phi)$$

where ρ is the density (g cm^{-3}) and Φ is the porosity (as a fraction).

An example of porosity estimation from seismic is given by Eyles and May (1984). They studied the Natuna D-alpha block, offshore Indonesia, where a discovery of 70 trillion cubic feet of gas (75% CO_2!) had been made in Upper Tertiary carbonates. The reservoir is thick (>5000 ft) and covers an area of about 110 square miles.

The objective of this particular study was to map the average porosity of the gross reservoir and so calculate gas reserves. The database consisted of five wells and 1000 km of seismic data.

An analysis of well data showed that a well-developed negative relationship existed between the porosity and the interval velocity (Fig. 5.44). About 700 velocity analyses were performed for the reservoir over the study area, although only about 33% were found to be useful when compared with check shot data from the wells. The velocity map and its transposition to porosity in Fig. 5.44 could not have been generated from the five wells alone. An estimate of the maximum error on this surface was ±6%.

This particular application was successful because:
• A clear relationship existed between the interval velocities and porosity;
• the reservoir is very thick and is bounded by strong reflectors;
• many interval velocity measurements were made from the seismic data, thus allowing rogue points to be rejected;
• care was taken to remove the velocity effect due to a water leg before the gas reserves were estimated.

5.7.4 Lateral variations and reservoir heterogeneity

The internal characteristics of a reservoir may be visible on seismic data or a derivative of the seismic data. For example, variations in the acoustic impedance of a reservoir interval may be used to infer and thus map lateral variations in porosity. In gas reservoirs, it may even be possible to go one stage further and map the distribution of gas-bearing sandstone/limestone.

One of the more exciting developments during the early 1990s was the use of attribute analysis, in which instantaneous seismic attributes and windowed attributes are calculated from reservoir intervals on the seismic data. Attributes that appear to mimic the required reservoir feature are calibrated to well data and displayed as maps.

One such pair of attributes are dip and azimuth. These displays for a single horizon have produced a wealth of hitherto hidden information. An example of where such displays have helped geoscientists and reservoir engineers to manage a reservoir is the Gyda Field in the Norwegian Sector of the North Sea (Gluyas et al. 1992). The field was brought into production in July 1990. By the end of the same year, a number of wells had seen dramatic pressure declines. These data were interpreted to indicate that three wells in the northeastern part of the field were drawing oil from a relatively small box with little or no natural water inflow (aquifer support). Prior to field start-up, small, apparently discontinuous faults had been mapped in the area of these wells, and a model run to test the sensitivity of production to the then outside possibility that the faults formed a continuous barrier. The reservoir seemed to be performing in accordance with the pessimistic case of the faults being sealing. At about the same time, the first dip and azimuth displays became available for the

Mandal Mudstone that directly overlies the reservoir. This display is illustrated in Fig. 5.45. The continuous feature in the top right is a fault scarp. Clearly, the three wells to the northeast of this feature are in a small reservoir compartment. Had such a display been available at the time at which the development wells were drilled, the operators would not have drilled three of the eight wells into a portion of the field that may have been isolated from the main part.

5.7.5 Reservoir correlation

Correlation of sands between wells in an oilfield is a fundamental part of building a reservoir description (Section 5.5.4). On the basis of these correlations, a forward calculation will be made of how the reservoir will perform during production. Join the wrong sands together and your reservoir model will give the wrong prediction. This may not appear to be an important issue, but it must be remembered that the projected performance of the reservoir will determine how the production facilities are designed.

Seismic information can often help this correlation. The display in Fig. 5.46 shows three wells projected onto a seismic line. The SP and resistivity logs for the wells are also displayed. It is clear from the figure that there is seismic continuity, implying sand continuity, between the upper two wells (A & B), but the sands in these two wells appear to be unconnected (in this 2D view) from similar sands in the downdip well (C). Were we to inject water into the reservoir in the downdip well to maintain reservoir pressure, and hence oil production, we would fail.

This same sort of data on reservoir continuity can be used to help steer high angle (a driller's term for "near horizontal") through prime reservoir. Figure 5.47 shows the trajectory of a high-angle well which was threaded through patches of low acoustic impedance rock. Such rock was known from earlier wells to be highly porous reservoir sandstone.

5.7.6 Identification of fluid type and contacts

Seismic data can deliver a number of real and imaginary "direct hydrocarbon indicators." Flat spots, dim spots, bright spots, and amplitude versus offset (AVO) anomalies all find their place in the annals of exploration and reservoir definition. The use of such phenomena in exploration was examined in Chapter 3. Flat spots are horizontal reflectors that cross-cut the stratigraphy elsewhere on the seismic (Fig. 3.13). They can represent a reflection generated at a gas/oil or gas/water contact. They can also represent a mineralogical change (e.g., from opal to quartz) or a reflection multiple from a shallower

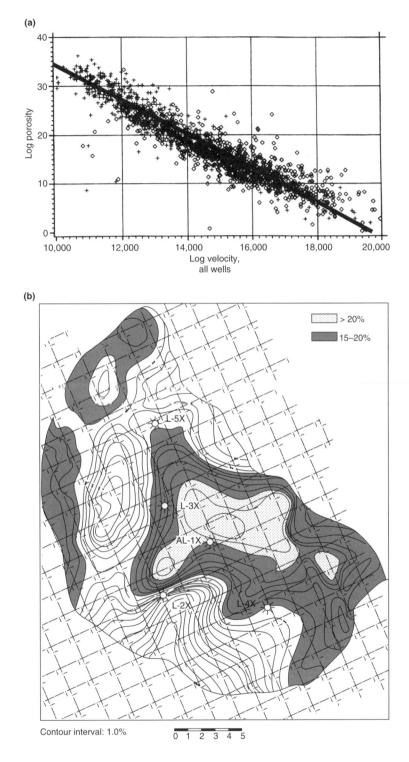

Fig.5.44 (a) The relationship between porosity and interval velocity for five wells from the Natuna License, Indonesia. (b) The porosity map, generated from seismic data. (From Eyles & May 1984.)

Small fault scarp

Fig.5.45 A dip–azimuth display for top Mandal Mudstone (a little above top reservoir) across the Gyda Field, Norwegian continental shelf. The display clearly shows the faults that cut the Mandal Mudstone and the continuity of those faults. Prior to production start-up, three wells were drilled into the small northeastern compartment of the field. This error could have been avoided had such data been available at the time of development well planning.

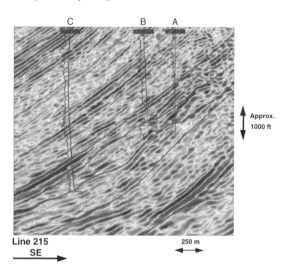

Fig.5.46 A seismic line overlaid with wireline curves from wells, Pedernales Field, eastern Venezuela. (Reproduced courtesy of BP.)

horizon. Interpretation of flat spots should be done after depth migration of the seismic section. After all, there is little point struggling with the interpretation of a flat spot on a time section if, after time-to-depth conversion, it proves not to be flat.

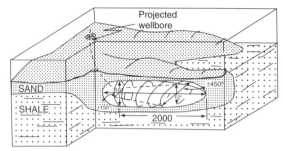

Fig.5.47 A horizontal well trajectory threaded through the low acoustic impedance (high-porosity) parts of a reservoir. (From Soloman et al. 1994.)

Amplitude anomalies (including dim and bright spots) can be generated from large quantities of oil and small quantities of gas, so again great care must be exercised when making interpretations. Where amplitude analysis works best is in highly porous, poorly consolidated sands that contain oil with a high GOR. The Plio-Pleistocene sediments of the Gulf of Mexico (GOM) are a prime example of the technique at its best. The GOM sedimentation was rapid, and in consequence deep sands at 4500 m (15,000 ft) remain highly porous because of burial disequilibrium overpressuring (Chapter 3).

5.8 PETROLEUM IN PLACE, RESERVOIR MODELS, AND RESERVES

5.8.1 Introduction

The purpose of appraisal is, of course, to determine the commercial viability of a petroleum pool. The preceding sections of this chapter have outlined the investigative processes involved in appraisal and the descriptions of trap, reservoir, and fluid distribution produced during appraisal. All the data gathered will be used to generate or qualify two important sets of numbers. These are petroleum (oil or gas) originally in place and that which can be recovered (reserves). The quantity of petroleum in place for a field has a definite and fixed value (although we may not know very precisely what the value is). The reserves figure is the volume of petroleum believed to be recoverable. This figure is less than the petroleum in place but, unlike "petroleum in place," its value is not absolute. The reserves will depend upon the parameters of the reservoir (segmentation, permeability, etc.), the petroleum fluid (phase, viscosity, and pressure), the number of wells, the recovery method (primary, water-flood, etc.), and the location (the proximity to other infrastructure, power supplies, and the market).

Petroleum in place is commonly referred to as STOIIP or GIIP. These terms are acronyms for "STandard barrels of Oil Initially In Place" and "Gas Initially In Place" respectively.

In this section, we will look at the transformation of the data gathered during appraisal into petroleum-in-place figures and reserves calculations, from which a development decision can be made.

5.8.2 Petroleum in place

The methodology for calculation of oil and gas in place was introduced within Chapter 4 and its associated case histories. It is, of course, a simple volume calculation in which the total volume within the trap is multiplied by the reservoir fraction, then progressively by the net pay fraction, porosity, petroleum saturation, and the formation volume factor (or the gas expansion factor). At the exploration stage, the quantity of data available is often limited. As a consequence, the values used for net pay, porosity, petroleum saturation, and the formation volume factor are commonly either single figures or, more sensibly, most likely estimates with attached uncertainty ranges. More data are usually available for trap volume, although even here volume may be calculated from a sparse coverage of 2D seismic data.

The same calculation of petroleum in place during or at the end of appraisal is quite a different process. The appraisal data will have delivered more information. Such information will undoubtedly show that spatial differences exist in reservoir properties across the field and that the geometry of the trap is more complex than could be appreciated during exploration.

There are several ways in which the petroleum volume may be calculated using deterministic and stochastic methods. Here, we will describe the basic deterministic approach, while stochastic methods will be examined in the following subsection.

Using the combined seismic and well data, a number of maps will be produced. These will include structure, annotated with petroleum/fluid contacts, gross rock thickness, and then, using net to gross data, net pay, porosity, and water (or petroleum) saturation. The structural maps are of course contoured elevation data, while the remaining maps are isopleth maps for the other parameters. These can be merged either manually or, more commonly today, by using the appropriate computer software. From such merged data, the volume of petroleum in place can be calculated. A simple, useful check on the petroleum-in-place calculation is to use the well data to produce petroleum column height data at all of the well points. This is simply net pay × porosity × petrole-

um saturation × formation volume factor. This then can be planimetered and the area and petroleum column height data used to calculate the oil in place. The method also has a strong visual impact and utility when it comes to deciding where to place development wells (Fig. 5.48). It is possible to see the distribution of petroleum at a glance. Production wells can be targeted at petroleum column "thicks" and, if appropriate, water injection wells can be targeted at "thins" or beyond the margins of the petroleum accumulations.

In the preceding sections we have covered most of the errors that can arise with depth conversion, porosity, and saturations. One more error needs to be highlighted, that of mapping or averaging net to gross. It is a very common error that can lead to large overestimates or underestimates in the petroleum in place figures. One should map net pay, not net to gross. The error introduced by mapping or averaging net to gross data is best illustrated using a simple averaging calculation. Consider two wells, in which well A has a gross pay thickness of 10 m and a net to gross of 10% and well B has a gross pay of 40 m and a net to gross of 50%. Taking the net to gross figures, the average net to gross would seem to be 30%. However, if we consider the true net pay, well A has 1 m and well B 20 m in a gross reservoir thickness of 10 m + 40 m = 50 m. Thus the true net to gross is 21 m to 50 m, or 42%. Were well B to have not 40 m gross pay but 100 m gross pay and 50 m net pay, the apparent net to gross would still seem to be 30% were the net to gross figures averaged. However, the true net to gross is (50 + 1) m in (100 + 10) m, or 46%. Overestimation can occur when thin high net to gross intervals are averaged using net to gross data rather than net pay data.

5.8.3 Geologic models

Geologic models have already been introduced in the sense of trying to determine the depositional environment of a reservoir interval and then make some estimate of the reservoir architecture. The geologic model can be created deterministically by hand or with the aid of geologic modeling software. Alternatively, a computer-generated geologic model can be built using stochastic (statistical) data. Each of the computer-based methods, be they deterministic or stochastic, generates a model composed of a 3D tessellation. The cells within such tessellations are commonly oblate rectangular prisms. The height of each cell, perpendicular to stratigraphic surfaces, is commonly a few meters, while their widths and lengths are commonly between 10 m and a few tens of meters. Each of the cells is assigned rock and fluid property attributes. Geologic models for large fields may contain more than one million cells. Quite clearly, a subset of the property data contained within each cell can be used to calcu-

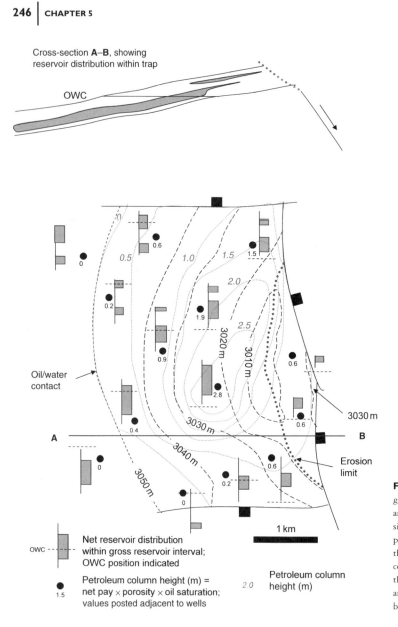

Cross-section **A–B**, showing
reservoir distribution within trap

OWC

Oil/water
contact

A

3030 m

3020 m

3010 m

3030 m

3040 m

3050 m

Erosion
limit

B

1 km

OWC - - - - ▭ Net reservoir distribution
within gross reservoir interval;
OWC position indicated

● Petroleum column height (m) =
1.5 net pay × porosity × oil saturation;
values posted adjacent to wells

2.0 Petroleum column
height (m)

Fig.5.48 A petroleum column height map,
generated from combining a structural map
and a reservoir isopleth map. In order to
simplify the diagram, the oil saturation ×
porosity function is assumed to equal 20%
throughout the area. The resulting petroleum
column height map shows that the petroleum
thick is located slightly off the crest of the
anticlinal structure because the reservoir sand
body thickens off structure.

late the petroleum in place for that cell, and then by addition
of all the cells in the model it is possible to calculate the
petroleum in place.

The scale of the cells in a geologic model tends to be
smaller than that in a reservoir model. This is because the
pressure, and hence the flow, calculations performed within
a reservoir model become too cumbersome if the model has
millions of cells. The process of reducing the number of
cells is called "upscaling," and is something that we will
investigate further in the next subsection.

For a geologic model generated by hand, it is possible to fit
it directly into a reservoir model (by dividing it into layers
and then cells). Upscaling is not performed in such a process.
A common issue with this approach, and indeed with any
deterministic approach, is that it is difficult to preserve the
heterogeneity of the real world in a reservoir model.

In geocellular (computer) modeling, it is possible to use
both some or all of the seismic data and much or all of the well
data. Whatever approach is used, the aim is to capture the
surface data (stratigraphic surfaces and faults) and the pro-

perty data (lithology, porosity, permeability, saturations, etc.). Coarse-scale stratigraphic data are usually generated directly from mapped seismic surfaces. However, in the absence of seismic data it is also possible to interpolate between wells. Finer-scale surface data may also come from seismic, but more commonly such data are interpolations between wells. In the absence of definitive (or suggestive) seismic data, it is often necessary to define the nature of the surface (erosional or depositional) from well and core data. The finest scale of layering is usually interpolated from well data. As with any of the interpolated data, it is necessary to define the relationships between layers. Layers can be defined as parallel, converging/diverging, thickening upward, thinning upward, and so on.

Each layer in the model is divided into cells. The cells are then populated with lithology and then with properties. In the least sophisticated models these lithologies and properties are simply interpolated between wells. Stochastic methods use a variety of different ways of filling in the gaps between the known well points. The most widely used stochastic methods are object-based modeling and grid-based modeling. Each has its merits and each has its shortcomings.

In object modeling, recognizably geologic-shaped bodies (geobodies) are created using guidelines derived from analog studies. Such geobodies are scattered at random throughout the model volume, while being forced to obey conditioning points such as wells and conditioning rules such as abundance or net to gross (derived from well data). Different objects will be successively added to the model to represent the component parts of the reservoir as described. For example, in a meandering fluvial reservoir system, the com-

ponent parts may be channel fills, crevasse splay sands, and mudstones (lagoonal, interdistributary, etc.; Fig. 5.49). For the channel fills, the range in thickness will be derived from the well data, while analog data will be used to determine the ranges in widths, orientation, reach lengths, and reach angles (Fig. 5.50). Fewer data are required to define the crevasse splay sands, but again well data will be used to define thickness. The nonreservoir lithologies are commonly lumped together in low to moderate net to gross reservoirs and used to fill in the background between the porous and permeable geobodies. In high net to gross systems, such as sand-rich submarine fans, the nonreservoir elements may be modeled as discrete objects and the reservoir used to fill in the background (Fig. 5.51).

Object models tend to look like convincing bits of the Earth. Modeled channels look like channels, mouthbars look like mouthbars, and even sandstone dikes look like sandstone dikes! However, one of the shortfalls of object-based modeling is that it is not possible to condition such models to soft data, such as that which might be derived from seismic. For example, it may be possible to define from inverted seismic data a volume in the reservoir that has a 70% probability of being sandstone. Object modeling cannot use such data. The seismically defined object would have to be (or not be) sandstone. The same is true of well data. The geologist will have to define, using logs and cores, what every meter of the penetrated reservoir interval is. There is no opportunity to incorporate uncertainty at the conditioning points. In consequence, many of the soft conditional data are not used in object modeling.

In grid-based modeling, a regularly gridded volume is filled with values of a particular parameter. The grid-filling process is controlled by the frequency of values and the spatial

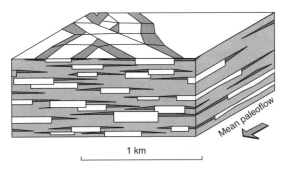

Fig.5.49 An object model of a meandering system. The model comprises three components: channels, channel margins, and background sediments. The background sediments in this particular model include both crevasse splay sandstones and siltstones and intra-channel mudstones.

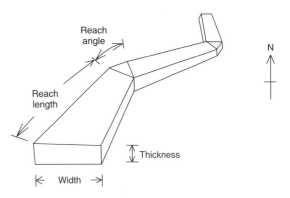

Fig.5.50 A numerical description of a meandering system.

B-18   A-39   A-29   A-42   A-3a   A-8b

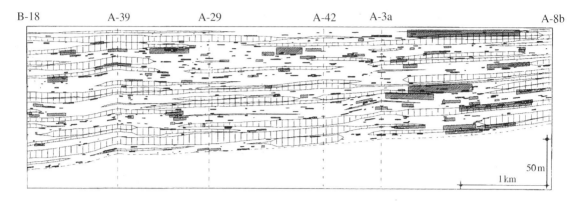

50 m

1 km

Fig.5.51 Modeled nonreservoir objects (stochastic shales) within a reservoir model for the Upper Statfjord Sandstone, Statfjord Field, Norwegian North Sea. Stochastic shales are shaded. (From MacDonald & Halland 1993.)

correlation of values (Fig. 5.52). The frequency data are expressed as histograms and the data are derived from wells. Spatial correlation data are expressed using semivariograms that are taken from a combination of well data and geologic analogs.

A semivariogram is a way of expressing the spatial correlation of data. It is based on a simple (observational) fact that the values for a parameter measured for two points separated in space become less strongly correlated the greater is the distance between the points. At short distances the values between points are correlated, but at larger distances there is no correlation between values. The height of the noncorrelation sill and the distance to the noncorrelation sill (range) are calculated from experimental and analog data and then used in the stochastic modeling (Fig. 5.53). The semivariogram can be used to measure the spatial structure of discrete variables (sequential indicator simulation), such as the lithotype, and continuous variables (sequential Gaussian simulation) such as porosity and permeability. Semivariograms are required for all three dimensions of data. It is commonly possible to derive the vertical semivariogram data from wells, while the other two, orthogonal semivariograms are generated from analog outcrop data.

The down side of grid-based modeling is that the output rarely looks like an adequate representation of geology. On the positive side, grid-based modeling is able to incorporate soft data such as seismic. It is also less likely than object modeling to "fail" when applied to heavily conditioned simulations (fields with many wells).

It is possible to combine object-based and grid-based models in a single reservoir model so long as the methodologies are used at different scales of heterogeneity. For example, in a fluvial reservoir system, object modeling may be used for

the sand bodies, sequential indicator simulation for the lithotypes, and sequential Gaussian simulation for the permeability distribution. When generating a stochastic geologic model, it is usual to perform many realizations. The gross properties of the simulation will vary between simulations. Each realization will produce a different net pore volume, oil in place, average permeability, and so on. The gross properties output can be represented by distributions. From such distribution data, individual simulations can be chosen for input into reservoir models. For example, individual simulations corresponding to P_{90}, P_{50}, and P_{10} cases for oil in place may be selected from the full distribution for upscaling and imported into a reservoir simulator.

5.8.4 Reservoir models

The aim of building a reservoir model is to simulate what will happen in the real reservoir when it is put on production. A reservoir model built during the appraisal stage allows different development options to be tested before major expenditure is committed to drilling production wells and building facilities. During the production life of a field, the reservoir model will be calibrated and recalibrated to the production data. After some time, it may prove necessary to build a new reservoir model to simulate accurately past behavior of the field and so predict future performance. Thus the reservoir model built during the appraisal stage represents just the initial part of field life reservoir management.

Reservoir engineers, not petroleum geoscientists, build "reservoir models." It is not the intention of this book to introduce reservoir engineering. However, it is important that the geoscientist does not simply hand over his or her geologic model to the reservoir engineer. If a robust reservoir model is

(a)

Reservoir (sand)

Nonreservoir (shale)

(b)

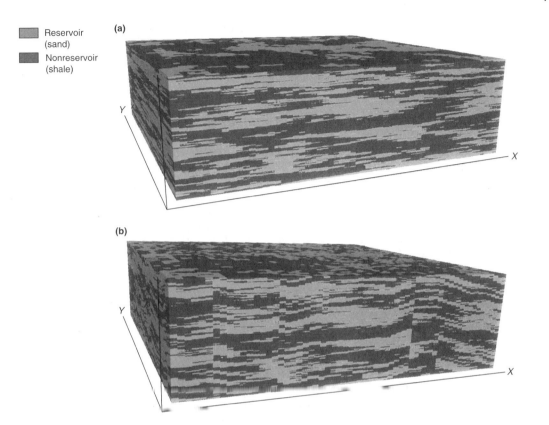

Fig.5.52 (a) A grid-filled model using sequential indicator simulation for reservoir (sand) and nonreservoir (shale) areas; Upper Jurassic submarine-fan sandstone, Magnus Field, UK North Sea. (b) The same model as in (a) with faults simulated. (Reproduced courtesy of BP.)

to be produced, the two disciplines must work together. The geologic model apart, the geoscientist will have a better grasp of the spatial aspects of the reservoir going into development than will the reservoir engineer. In the initial attempts at reservoir simulation, it may be the geologist who has a better feel for the validity of the output, since he or she has a conceptual Earth model with which to compare that output; whereas at the appraisal stage the field has no production history with which the reservoir engineer can numerically compare the output of the reservoir model.

The reservoir model is a numerical representation of the reservoir and its associated fluids. Like the geologic models described above, it comprises a 3D array of cells. However, each cell is significantly larger than its counterpart in the geologic model. Typically, the height of cells will be measured in a few tens of meters, although thin high-permeability streaks or low-permeability baffles and barriers of a meter or

less can be modeled. Cell widths are commonly measured in many tens of meters and lengths commonly in hundreds of meters (Fig. 5.54). If the oil- or gasfield is believed to have an active aquifer, this too will be modeled within the reservoir simulation. However, to reduce the computational complexity, the aquifer may be represented by a few very large (long) cells.

The point about computational complexity is an important one, for it controls the way in which a reservoir model is constructed from a geologic model. To generate a reservoir model from a geologic model, upscaling the properties within each chosen cluster of cells reduces the numbers of cells. Most reservoir models contain tens of thousands of cells rather than the millions of a geologic model. This reduces the time needed to run a simulation to hours. The method used for upscaling is critical, because the aim is to preserve the heterogeneity of the reservoir as captured by the geologic

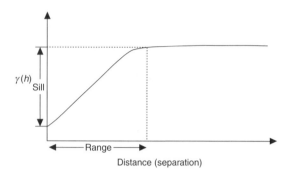

Fig.5.53 The variogram is a measure of the spatial correlation of a variable. It captures data on similarity versus distance. In heterogeneity modeling, the variable is usually a rock type (lithotype) or permeability. The variogram quantifies the fact that as the distance from a given starting point increases, the similarity between the properties at the starting and those at new points becomes less strongly related. A variogram typically has the form shown in the figure. It has two important features. The "sill" is approximately equivalent to the spread of the data, and the "range" marks the point beyond which values are not correlated.

description. Much effort has been put into determining the most appropriate methods.

Like the geologic model, the cells within the reservoir model are posted with a range of rock and fluid property data. In addition to the data used in the geologic model, reservoir model cells commonly include data such as rock compressibility, relative permeability, fluid viscosity, gas to oil ratios, temperature, and pressure. Unlike the geologic model, which is static, it is a prerequisite part of the reservoir model that it is possible to observe the effects of extracting petroleum and injecting water and/or gas. The model is set to run with particular cells designated as production points or injection points. Modifying the pressure in the cells simulates the effects of production or injection. The pressure perturbations at the points of production (or injection) are transmitted through the model as simulated fluid flows toward the pressure lows (and away from the highs).

The key outputs from a reservoir model are the production profiles for all fluids (oil, gas, and water). Such data will include the time to and location of water (or gas) breakthrough

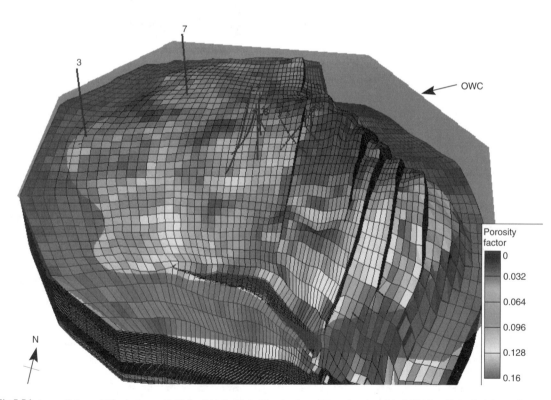

Fig.5.54 A geocellular model for the Lennox Field, East Irish Sea Basin. The view is an oblique plan view of the field. The cells are shaded according to the porosity of the reservoir. The model continues down into the aquifer. (From Yaliz & Chapman 2003; reproduced courtesy of BHP Petroleum Ltd.)

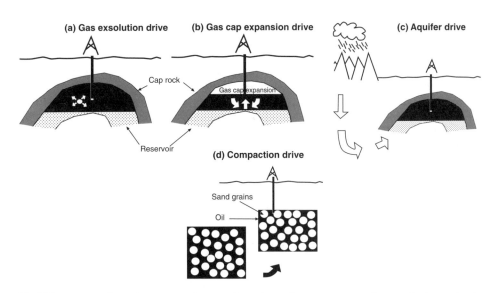

Fig.5.55 The recovery factor: the proportion of oil or gas that can be won from a reservoir. It is a function of both natural conditions and economic criteria. In nature, there are four common drive mechanisms: gas exsolution, gas cap expansion, aquifer drive, and compaction drive. (a) Gas exsolution drive is analogous to the situation that occurs when a champagne cork is popped from a bottle. Gas exsolves and causes a volume increase for the liquid plus gas. Both liquid and gas are expelled from the bottle, but much is left behind. (b) Gas cap expansion has an analog in the production of beer from a pressurized keg; the CO_2 cap expands as the tap is opened, so liberating beer into the glass. Simple gas expansion is used for gas reservoirs. (c) An alcoholic analog for aquifer drive does not readily come to mind. Production is effected through the pressure generated by a hydraulic head. (d) Compaction drive can occur when part of the rock matrix is supported by overpressured fluid (oil and water). On penetrating the seal, the overpressure can be released. The rock begins to compact, pore space is reduced, and the displaced petroleum is produced. These are primary recovery mechanisms. The application by man of similar water or gas injection schemes designed to improve petroleum yield is known as secondary recovery, and the injection of chemicals such as surfactants is known as tertiary recovery. Recovery factors vary greatly from reservoir to reservoir. For example, only 10% of the petroleum might be recovered from a complex, delta top, sandstone reservoir with a viscous oil (<20° API) and a low dissolved gas content using natural depletion, while more than 60% of the oil might be recovered from a homogeneous, highly pressured shallow-marine reservoir with light oil and a favorable oil-to-water mobility ratio.

into production wells, the sustainable production rate per well, the ultimate recovery per well, and the proportion of petroleum recovered from the field (recovery factor). In most instances, the reservoir model will be run a number of times using different development options to optimize recovery and minimize cost. However, there may not be a unique solution to such a problem, and in consequence it is common to incorporate some flexibility in the development, such that as production proceeds the production data can be used to refine the development process.

5.8.5 Reserves

Reserves are the quantity of oil or gas that will be recovered from a field under a specified development plan. Reserves are a portion, commonly a small portion, of the volume of petroleum in place. The reserves volume divided by the oil-in-place volume is called the "recovery factor." This can be expressed as a fraction of one or as a percentage. Recovery factors can vary from near zero in some oilfields to >95% in some gasfields. Control on recovery factor is a function of both intrinsic reservoir and fluid properties, and on the method (ultimately cost) used to win the petroleum from the ground (Fig. 5.55). A field will be abandoned once the cost of extracting the oil or gas is the same as the value of the petroleum being extracted. That is at least the case for multinational and independent petroleum companies, although in some parts of the world government-run companies will continue to extract petroleum long after it has ceased to be economically viable to do so.

Natural factors that help to increase the recovery factor are high-permeability reservoirs, low-heterogeneity reservoirs, unsegemented reservoirs, low-viscosity oil, active aquifers, or—often in the case of gasfields—inactive aquifers. Naturally, the opposites of these factors can yield a reservoir with a very low recovery factor. The natural recovery mechanisms are commonly called "primary recovery." Man-controlled "secondary recovery" is implemented to improve recovery. Such secondary recovery mechanisms include pressure maintenance through injection of water, gas, or both water and gas (WAG). Recovery can also be improved by drilling more wells, or wells at particular angles and orientations, to maximize the sweep efficiency of natural aquifers, water injection and gas-cap expansion, and/or gas-cap development. Decisions on whether to drill more wells to increase recovery often demand that the operating cost of the field be reduced so that the cost of the new wells can be afforded.

Table 5.4 illustrates the range of recovery factors for fields in two basins, and contrasts the natural properties of the reservoir and the man-controlled factors. One basin (Trinidad) is an old onshore province and the other (the North Sea) is a relatively new offshore province.

The oil industry has developed a variety of descriptions of reserves categories. In broad terms, these are proven reserves, probable reserves, and possible reserves. Proven reserves commonly correspond to that oil or gas that can be extracted with the wells in the ground. Probable reserves are the oil and gas that can be extracted from planned and budgeted wells; and possible reserves are those that might exist, but for which no specific recovery plan has been made. The value of a field to a company is based upon the reserves that can be booked for that field. In consequence, reserves calculations and subsequent certification are of great importance to most companies.

5.9 CASE HISTORY: ALNESS SPICULITE, UK CONTINENTAL SHELF

5.9.1 Introduction

This case history examines the appraisal of an unusual, high-porosity but low-permeability reservoir. Oil was discovered in the Oxfordian Alness Spiculite of the Inner Moray Firth, UK North Sea (Fig. 5.56), while drilling for deeper targets. At least five serendipitous oil discoveries were made in the spiculite. A few of the discovery wells were tested, but despite the presence of light, 38° API, oil in the pore space, the flow performance varied between poor and nonexistent (Table 5.5). Scout data available shortly after the wells were drilled suggested that four of the discoveries might contain an aggregate 700 mmstb of oil in place (Table 5.5). However, none of the discoveries were treated to appraisal drilling and by 1990 most of the blocks containing these discoveries had been relinquished.

Given the poor well-test performance of this reservoir and the poor permeability measured during core analysis, it is reasonable to assume that the companies owning the blocks opted for relinquishment on the basis that the reservoirs were incapable of sustaining viable economic performance when compared with contemporaneous projects. In this case history, we reexamine the discoveries to determine whether they might now be brought into production through the application of drilling and completion technologies developed during the 1990s. For example, the performance of this low-permeability reservoir might be improved by using horizontal wells or through fracturing the reservoir around the wellbore. Either process could enhance flow from the reservoir into the wellbore.

5.9.2 The database and work program

The data available for this project were a 1993-vintage 2D, well tie seismic survey, together with well and core data released into the public domain. Wireline logs were available for eight wells. Core was available from three wells. This was logged for sedimentology logging and sampled for petrography and petrophysics.

The structures containing the discovered oil pools were mapped, as were some new prospects. The cores were logged and samples taken for both petrography and mercury porosimetry. Mercury porosimetry was used to measure the distribution of pore sizes and pore-throat sizes as a function of the entry pressure required to force incremental volumes of mercury into the pore system of the rock. The purpose of this work program was to determine the validity of the pool sizes quoted in the scout data reports, to determine whether other undrilled prospects existed, and to evaluate the reservoir quality of the spiculite. This involved calculation of:
- The area of structural closure;
- the net pay thickness;
- porosity;
- oil saturation;
- permeability.

No new data were available for the petroleum formation volume factor, and as such published data for the nearby Beatrice Field were used.

5.9.3 The geologic setting

The Inner Moray Firth Basin is an east–west component of the failed rift system of the North Sea (Stephen et al. 1993). The spiculite-bearing Oxfordian interval lies just beneath the syn-rift megasequence. It was the final expression of pre-rift sedimentation in the area (Fig. 5.57). Deposition of the spiculite occurred in two regressive cycles in an overall

Table 5.4 Contrasting recovery factors: North Sea and Trinidad.

| North Sea (from Abbots 1991) | | Trinidad | |
Field	Recovery factor (%)	Field (Pliocene and Miocene reservoirs throughout)	Recovery factor (%)
Alwyn (Middle Jurassic Brent Group)	50	Balata	6
Alwyn (Lower Jurassic Statfjord Formation)	35	Barrackpore/Penal	11
Auk (Zechstein + Rotliegend)	19	Brighton Marine	11
Brae Central (Upper Jurassic)	40	Cat's Hill	32
Brae South (Upper Jurassic)	33	Coora Quarry	13
Buchan (Devonian, fractured)	16	Forest Reserve	16
Claymore (Lower Cretaceous)	45	Fyzabad	22
Cormorant (Middle Jurassic Brent Group)	39	Guaya Guayare	19
Ekofisk (Upper Cretaceous)	40	North Soldado	16
Forties (Paleocene)	57	Oropuche	9
Fulmar (Upper Jurassic)	57	Palo Seco	12
Glamis (Upper Jurassic)	50	Parrylands	10
Maureen (Paleocene)	54	Rock Dome	17
Piper (Upper Jurassic)	70	Tabaquite	22

Table 5.5 Scout data on pool size and well flow rate for the "spiculite" discoveries.

Well	Operator	Oil or gas	Pool size (mmstb)	Test results
11/25-1	Shell	Oil	100	Untested
12/21-2	Britoil	Oil	150	3 bopd*
12/21-3	Britoil	Oil	215	200 bopd
12/22-2	Arco	Oil	225	280 bopd*
12/27-1	Premier	Gas	Unknown	9.5 mmscf d^{-1}
12/29-1	Kerr McGee	Oil	Unknown	100 bopd

* Only completion fluids recovered, no oil recovered.

transgressive sequence (Fig. 5.58). Specifically, the spiculite accumulated as winnowed deposits on a lower shoreface.

Syn-rift and post-rift sedimentation continued in the area, leading to the maturation of two or possibly three source-prone intervals by the early Tertiary. However, late Tertiary movement on the Great Glen Fault produced uplift and tilting within the western part of the Inner Moray Firth. Residual oil stain in sandstones beneath the Beatrice Oilfield testifies to westward spillage of structures during the late Tertiary.

5.9.4 Spiculite petrography and reservoir quality

The spiculite is composed almost entirely of altered sponge spicules (*Raxella perforata*) or the molds created from their dissolution (Fig. 4.8b). Individual spicules have a grain size

equal to that of fine to very fine sand. They are kidney shaped and highly rounded. At deposition, the spicules would have been opaline silica. Those that remain are now composed of chalcedony.

The spicules occur in what on first examination appears to be a brown clay matrix, but which on closer examination proves to be oil-stained microcrystalline quartz cement. This cement occupies what would have been the pore space between the spiculite grains at the time of deposition. The only other significant minerals in the spiculite are a variety of pore-filling and grain-replacive carbonate cements (calcite, ferroan calcite, and ankerite).

Despite the original pore space being occupied by microcrystalline quartz cement, the rock has up to 35% porosity, with most samples having between 20% and 30% pore space. This is because few of the original spicules remain. Most have

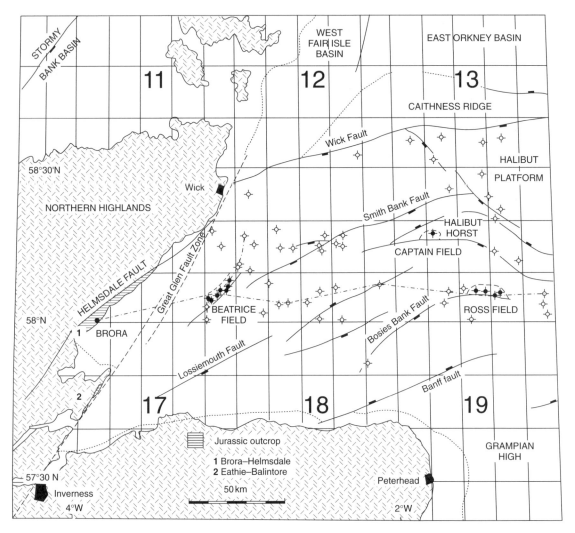

Fig.5.56 A structural map of the Inner Moray Firth area. The wells referred to in the case history are marked on the map. (Adapted from Stephen et al. 1993.)

been dissolved, to leave spicule-shaped pores lined by crystals of the micro-quartz cement. Given that dissolution of the spicules is likely to have been the source of the cement, we are left with the curious situation that the rock has turned inside out, with grains becoming pore space and pore space becoming crystalline matrix. The processes whereby this can occur whilst preserving much of the original fabric of the rock are unclear, but the effect on reservoir quality is readily observable. The spiculite has high porosity but very low permeability (typically less than 1 mD).

5.9.5 Oil in place

In order to determine whether the discoveries in the spiculite reservoir were worth appraising and new prospects worth drilling, it was necessary to determine the oil in place for each of the pools and prospects. It was also necessary to evaluate the reservoir characteristics. To do this, the trap area, net pay thickness, porosity, oil saturation, and formation volume factor for each of the discoveries were evaluated.

The area of closure for each of the prospects was mapped in

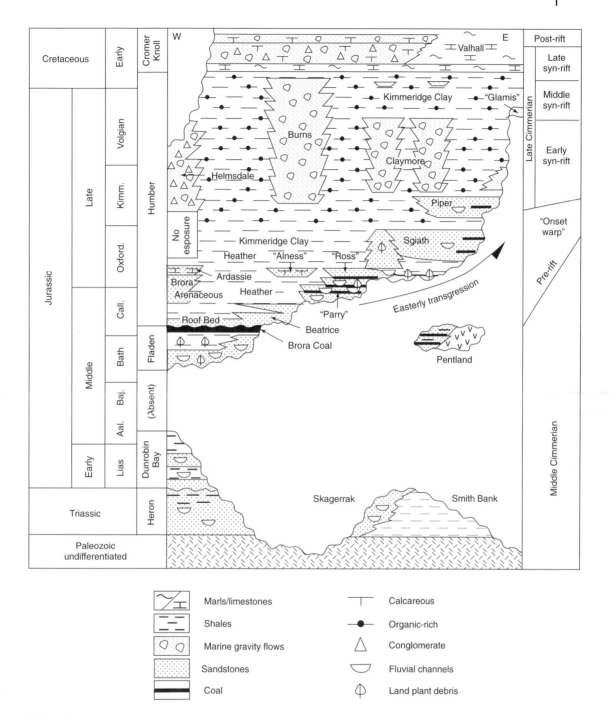

Fig.5.57 The chronostratigraphy of the Jurassic interval within the Inner Moray Firth Basin. (From Harker & Rieuf 1996.)

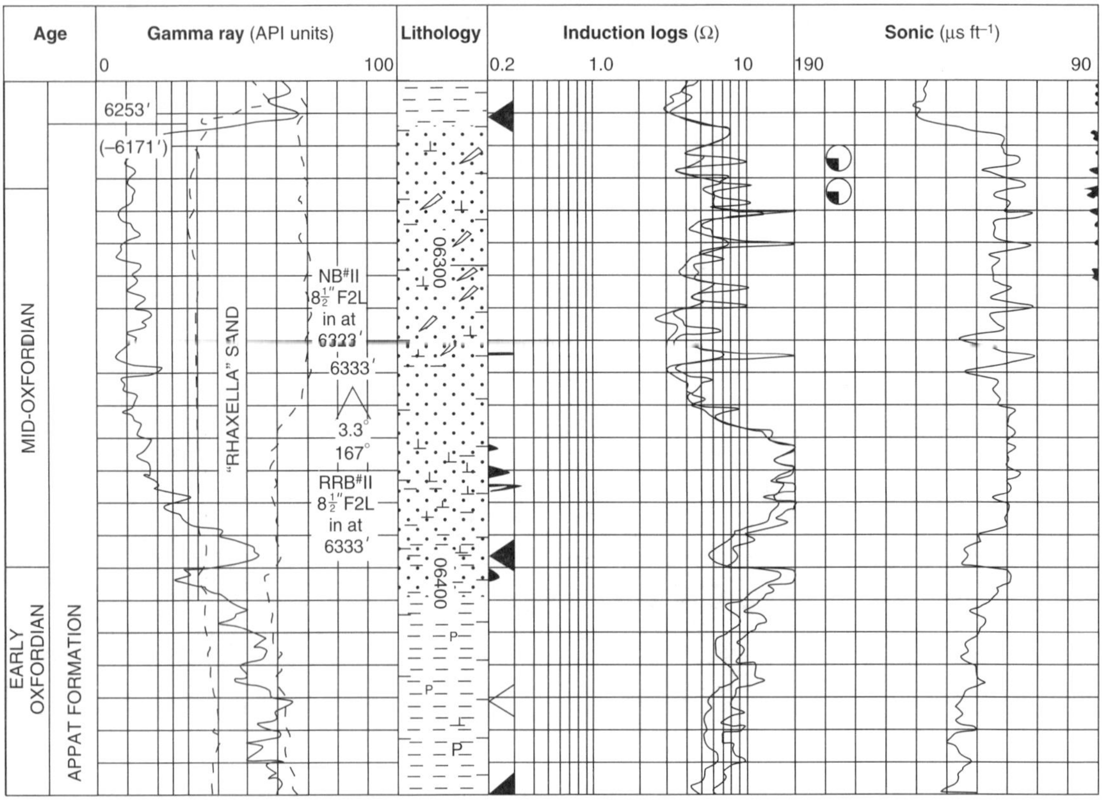

Fig.5.58 A sample of a composite log for well 12/22–3, Inner Moray Firth Basin. The sandstone interval is clearly divisible into two cycles by the gamma, resistivity, and sonic log spike at 6328 ft. The lowermost cycle is upward-cleaning and strongly regressive (progradational). The uppermost cycle is only marginally progradational, with the top 30 ft being essentially aggradational.

two-way time from the seismic data (Fig. 5.59). A time-to-depth conversion was made from the published data on the composite logs and cross-checked with information from the well tie seismic survey. Gross and net pay thicknesses were calculated from the well data and area–depth plots generated using the same methodology as in the Viboral case history (Case history 4.11).

Porosity and oil saturation data were calculated from the wireline log responses and cross-checked against the core data. The results from this piece of work were of particular importance to the appraisal project. We had initially assumed, on the basis of the scout data and published core analysis data, that porosity was high and oil saturation in the range 70–80%. The petrophysics work showed that both statements could not be true. Either oil saturation was low and porosity high or, more likely, oil saturations were normal but only part of the rock was oil saturated. Thus, much of the

porosity was ineffective. If this were true, then the resistivity logs would see a resistive rock plus fluid system comprising large, very poorly interconnected pores. Evidence for this conclusion came from three data sets: grain densities calculated during conventional core analysis, mercury porosimetry data, and the wireline logs themselves.

The details of the petrophysics are important because the estimated volume of oil in place diminishes as the effective (oil-saturated) porosity diminishes. The grain density data provided the clue that part of the porosity was isolated. Density values varied between about 2.48 and 2.65 g cm^{-3}, lower than would be expected for a quartz, chalcedony, and carbonate mixture. This indicates that even helium was unable to access the isolated pores during the routine core analysis procedure (for core analysis methods, see Chapter 2). The mercury porosimetry data indicated that the rock comprised large pores connected by exceedingly

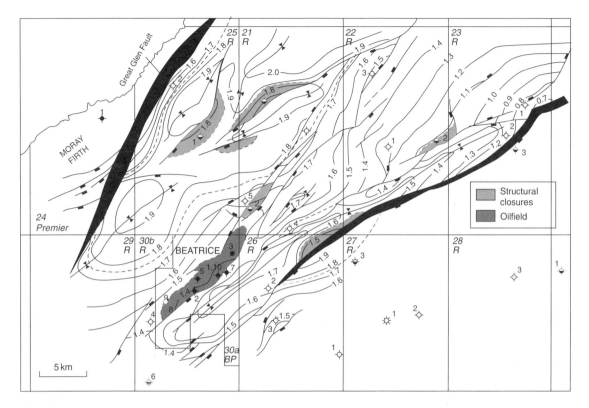

Fig. 5.59 Structural closures at the "top Alness Spiculite" level. Contours are two-way time in seconds.

small pore throats. Finally, the deep resistivity logs were unable to differentiate between petroleum-filled porosity and unconnected porosity, while a comparison of porosity from the sonic density and neutron tools indicated the presence of high porosity. It is probable that only about one-third of the porosity measured during routine core analysis is oil saturated.

Oil-in-place and likely reserves figures were calculated using a Monte Carlo simulation, to enable assessment of the range of uncertainty and main sensitivities. Table 5.6 contains the results for two discoveries (wells 12/21-3 and 12/21-2) and the prospect lying across the boundary between Blocks 21/21 and 12/26 (Fig. 5.59). Quite clearly, the oil in place in each of the discoveries is significantly less than the scout data would have led us to believe.

Details of the assumptions used and forecasts generated for the 21/21-3 pool are listed in Table 5.7. The major factors that influence the forecast reserves are the recovery factor and porosity. The correlation coefficient between the reserves and the recovery factor is 0.71 and that between the reserves and porosity 0.5.

Table 5.6 Spiculite pools, unrisked oil in place and reserves.

Well	Oil in place (mmstb)			Reserves (mmstb)		
	P_{90}	P_{50}	P_{10}	P_{90}	P_{50}	P_{10}
12/21-3	31	50	73	4	8	14
12/21-2	18	37	68	3	6	12
12/21, 12/26 prospect	21	45	86	3	7	16

5.9.6 Conclusions

Scout data led us to believe that the Inner Moray Firth contained a small number of moderate-size oil discoveries in poor-quality, spiculite reservoir. Those companies that made the discoveries assumed that the flow characteristics of the reservoirs were such that development of the pools in the early 1980s was not possible. We reasoned that drilling and completion technology developed since the discoveries

Table 5.7 Alness Spiculite, Inner Moray Firth, UK North Sea: assumptions, forecasts, and sensitivities for oil in place in the well 12/21-3 discovery.

Assumption	Distribution	Minimum	Most likely /mean	Maximum	Standard deviation	Influence on forecast (correlation coefficient)
Area (km^2)	Normal	—	6.5	—	0.65	0.2
Gross pay (m)	Triangular	30.0	38.5	45	—	0.17
Net to gross (ratio)	Triangular	0.5	0.7	0.8	—	0.18
Porosity (fraction)	Triangular	0.04	0.1	0.16	—	0.50
Oil saturation (fraction)	Normal	—	0.60	—	0.05	0.16
Formation volume factor (ratio)	Triangular	1.04	1.09	1.5	—	0.16
Recovery factor (fraction)	Triangular	0.02	0.20	0.30	—	0.71
Forecast	—	P$_{90}$	P$_{50}$	P$_{10}$	—	—
Oil in place (mmstb)	—	31	50	74	—	—
Reserves (mmstb)	—	4	8	14	—	—

might give us the possibility of developing the fields today. However, a new assessment of oil in place revealed that the assumptions in the scout data were wrong. Approximately two-thirds of the porosity in the spiculites is ineffective, isolated, and devoid of petroleum. In consequence, individual pools only have a few tens of millions of barrels in place. Even with the new drilling and completion technologies, and in consequence relatively efficient recovery, there are insufficient reserves to warrant further exploitation today.

5.10 CASE HISTORY: THE FYNE AND DANDY FIELDS, UK CONTINENTAL SHELF

5.10.1 Introduction

In this case history, we examine the reservoir system for the informally named Fyne and Dandy discoveries of the central North Sea. Between 1971 and 1989, six petroleum-bearing wells were drilled on Block 21/28a, yet none of the discoveries were developed. Part of the reason for this long period of appraisal was a function of reservoir continuity and field segmentation. Much uncertainty surrounded the lateral continuity of the reservoir intervals and the likely volume of oil that could be produced by a single completion. Further appraisal was carried out in 1997 and 1998. This involved reservoir studies and correlation work, the acquisition and interpretation of a 3D seismic survey, and eventually an appraisal well with sidetrack. A development was planned,

but suspended indefinitely in late 1998 as the oil price collapsed.

5.10.2 The license background

Block 21/28 was awarded to Mobil in 1965 with 100% equity (Fig. 5.60). Mobil relinquished the southern part of the block in 1971. In the same year, the company drilled a well on the western margin of the block. The primary target for this well was Jurassic sandstone. This target, though present, was dry. However, a 28 ft petroleum column was penetrated at shallow depth in thick (96 ft), high-quality Eocene sandstones (Tay Formation). This discovery has yet to be named. Moreover, because the Eocene section was drilled at high speed, the hole condition was poor. Shortly thereafter, foul weather led to logging difficulties. As a consequence, there remains confusion as to whether the number 1 well discovered oil, gas, or a combination of both. Sixteen years were to elapse before a second well was drilled on the block.

In 1987 Mobil drilled a well in the northeastern corner of the block. This time the target for the well was Eocene sandstone. The prospect was defined by four-way dip closure around a nonpenetrating salt swell. The well was a success. Well 21/28a-2 discovered 25° API oil in the Eocene Tay Formation (Fig. 5.61). Jurassic sandstones were penetrated in the deeper section, but were again found to be dry. Paleocene sandstones, also listed as a secondary target, were absent. The well was tested and flowed 3612 bopd from a 145 ft perforated interval (4489–4644 ft below rotary table) after acidization. The discovery received the informal name Fyne. The discovery was

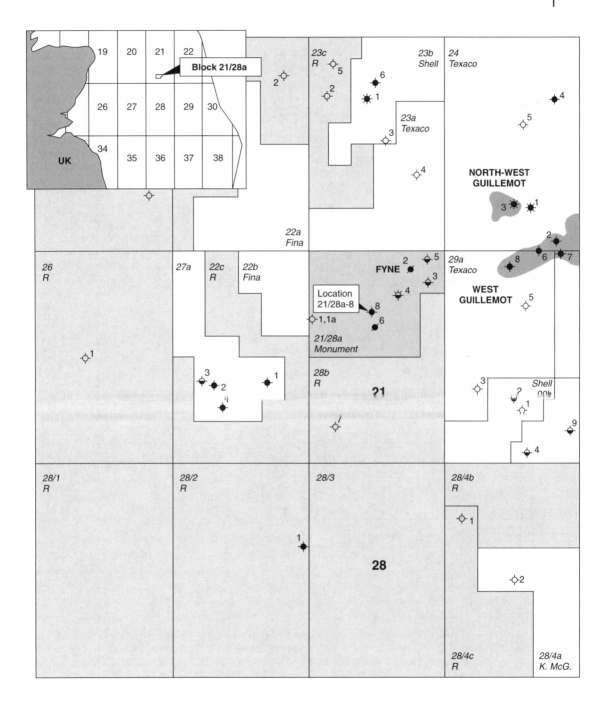

Fig.5.60 UK continental shelf Block 21/28a of the Fyne Field, central North Sea.

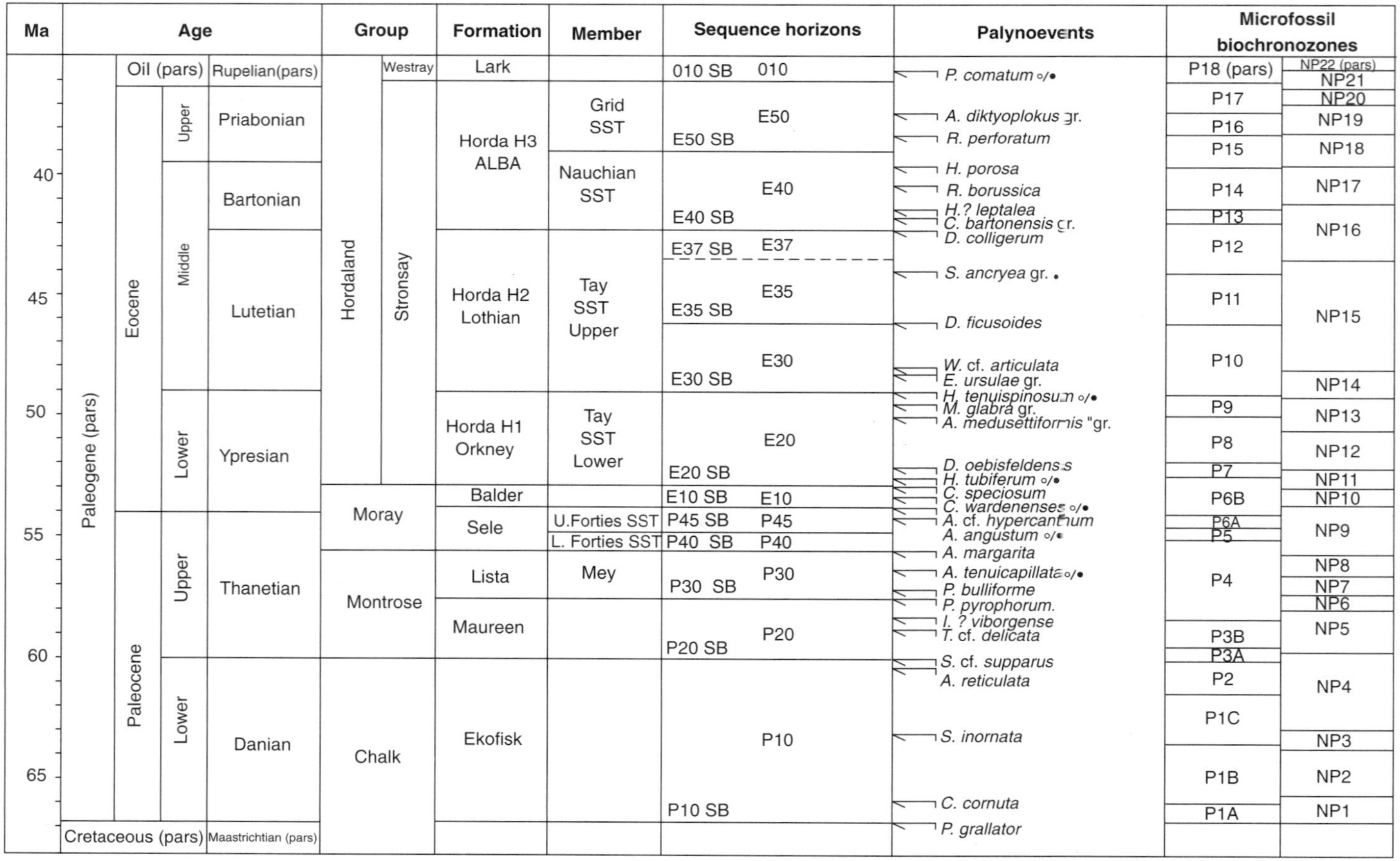

Fig.5.61 The stratigraphy of the Lower Tertiary of the central North Sea.

appraised by wells 21/28a-3 (1987) and 21/28a-5 and 5A (1988). Results from both appraisal wells were disappointing. Although both contained oil in the same Eocene section, the columns were shorter than expected, and RFT-derived pressure data implied that the field was compartmentalized.

Two further wells were drilled on the block at about the same time. Wells 21/28a-4 (1987) and 21/28a-6 (1989) were drilled to test the hypothesis that amplitude anomalies mapped on the sparse 2D seismic data set were produced by petroleum-filled sandstones. A small closure could be mapped around the number 4 location, but no structural closure could (or can) be mapped around the westernmost number 6 location. Both wells struck oil and gas in the same Eocene Tay sandstones. The number 6 well was tested; it flowed about 1000 bopd of 21° API oil. During the short test, gas cut rose rapidly and the test was stopped. The source of this gas was unclear at the time. However, once again RFT data were used to show separation from the Fyne Field (wells 2, 3, and 5A). The well results, combined with analysis of the stratigraphy and attempts at sand-body correlation between wells, clearly showed that in the area of Block 21/28a the Tay Formation has a highly complex architecture. However, given the limited quantity of 2D seismic data and the poor resolution provided by the biostratigraphic analysis, it was unclear as to whether the fields contained many small oil-bearing sandstone reservoirs or whether the stratigraphy was relatively simple and structure (faulting) complex.

5.10.3 Appraisal phase: 1997–8

Monument Oil and Gas acquired the block in 1997. Their aim was to drill a further appraisal well on the block in 1998 and, if appropriate, define a development plan. To this end, much effort was put into understanding the distribution of the Tay sandstone reservoir in the block. An appraisal well location was determined on the basis of this geologic evaluation. At about the same time, Western Geophysical shot a high-quality 3D survey over Block 21/28a and adjacent areas. Data from this survey were largely unavailable at the time of well planning. However, they were used in the post-well block evaluation.

When the block was acquired by Monument, it was clear that the petroleum in the block could be divided into four pools. Most was known about the pool known as Fyne, proven by well 2 and appraised by wells 3 and 5. Fyne is a single structure and the petroleum-bearing reservoir appeared to be the same throughout the three wells. However, pressure data obtained from the three wells indicated the field to be segmented. One thousand feet up dip of Fyne, well 1 proved

the presence of gas, possibly with associated viscous oil. The reservoir was thought to be stratigraphically younger than that found in Fyne. Pressure data (RFT) from the oil column in well 6 (informally called Dandy) indicated that the pool was separate from Fyne and well 21/28-1, but possibly in contact with well 21/28a-4. From the same data the oil column was inferred to extend downward to the elevation of well 4. The main petroleum-bearing sandstone was also thought to be younger than any encountered in previous wells.

Regional data combined with stratigraphic analysis of the existing 2D seismic data indicated that, in Eocene times, the area around Block 21/28a was on the lower part of a submarine slope. Armstrong et al. (1987) had already described contemporaneous, base-of-slope fan deposits from immediately east of Block 21/28a. The presence of thin deposits of Eocene sandstone in Block 21/28a, relative to the thicker shelfal deposits to the west and the thicker basinal deposits to the east, was taken to indicate that the slope was not of uniform gradient. At the time of deposition, the slope probably contained pinches and swells capable of trapping sediment on what was mainly a sediment bypass area.

In an attempt to refine the rather coarse stratigraphic analysis available for the Eocene section in the wells, the original biostratigraphic data were reanalyzed. This led to some improvement in stratigraphic resolution. Ultimately, the cores and cuttings were resampled and their microfossil content analyzed anew. The new correlation obtained is shown in Fig. 5.62. Oil-bearing sandstones from the Tertiary of Block 21/28a occur within both the Lower Eocene and Middle Eocene. Lower Eocene (Ypresian) petroleum-bearing sandstones occur within the Fyne Field (wells 2, 3, and 5A), while both Ypresian and younger, Middle Eocene, Lutetian sandstones occur in wells 4, 6, and 1. However, in these western wells only the Lutetian sandstones contain petroleum. Importantly, the petroleum-bearing intervals in wells 1 and 6 proved to be the same, and the uppermost sandstone in well 4 seemed to correlate with the oil leg in well 6. Reduction in the perceived number of petroleum-bearing horizons to two helped to allay fears that every well would find a separate oil accumulation.

An appraisal well location was chosen to the north and east of well 21/28a-6. The well was placed in the same seismic amplitude anomaly identified prior to drilling well 6 and using the same 2D seismic data. The location of the new well was chosen on the basis that if successful, the Dandy pool was likely to contain in excess of 30 million barrels of reserves. Success was defined as delivering a petroleum column comparable to that of well 21/28a-6 in a reservoir interval that could be shown to be in continuity with well 21/28a-6.

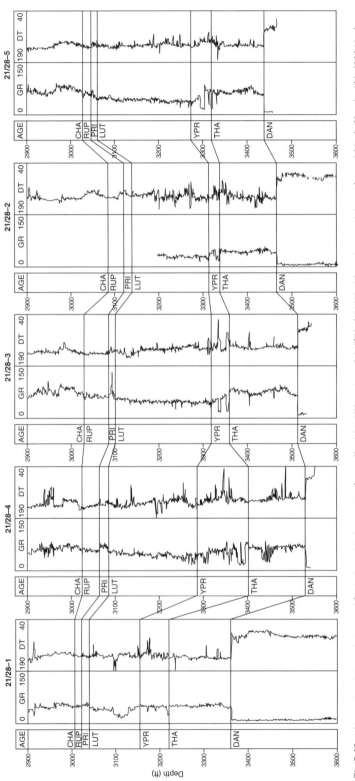

Fig.5.62 A biostratigraphic correlation panel for the Lower Tertiary sections from wells within UK continental shelf Block 21/28a. The main reservoir intervals are within the Ypresian (64–48 Ma) in the east and the Lutetian (48–41.5 Ma) in the west.

Well 21/28a-8 and its sidetrack 21/28a-8z were drilled in the spring of 1998. The prediction of static properties for the well was accurate with the exception of the thickness of the oil column. It was only about half that expected. Moreover, once pressure data were examined, the well was deemed not to be in communication with well 6. There was a possibility that the well had penetrated a small fault at the reservoir level and that this could have caused a reduction in the reservoir thickness. The well was sidetracked about 600 ft toward well 21/28a-6. The sidetrack did find a larger oil column, but it also confirmed the separation from well 21/28a-6. The number 8 well was tested and flowed at over 1000 bopd.

Analysis of the newly acquired 3D seismic data volume at much the same time as the well reached its terminal depth revealed the amplitude anomaly to be much more complex than hitherto thought. Both high- and low-amplitude (bright and dim) anomalies occurred at the reservoir level throughout the block and into the adjacent Block 21/27a. Many of the anomalies had distinctive sedimentary shapes — meandering channels with associated oxbow lakes. Others were more equant patches. The area around wells 6, 8, and 8z could be divided into at least three high-amplitude (bright) patches (Fig. 5.63). Significantly, a dark channel-like feature separates wells 6 and 8. Preparatory work on rock physics and fluid substitution, synthetic seismic modeling had indicated that many of the bright anomalies visible on the seismic data could be produced from small quantities of gas in these highly porous Eocene sandstones. This tallied with well data. However, the combination of thin reservoir and a similarity of rock and fluid physics meant that it was not possible to properly differentiate oil-filled sand and mudstone. Given the pressure data obtained from well 8, it seemed probable that well 8 was separated from well 6 by a mud-filled, abandonment submarine channel.

Following the well, a reassessment was made of the likely oil-in-place figures for the multitude of small pools in the block. Given the likely low recovery factor for the complex reservoir and the prevailing low oil price ($10 per barrel), development plans were postponed.

5.10.4 Conclusions

The North Sea is a mature exploration area. There remain few moderate-sized oil pools to be found. Aside from the few discoveries that are made, most new developments will be of "old" oil. There are many one-well discoveries scattered across the area, which were made in the 1970s and 1980s and have not been developed. There are a multitude of reasons as to why the pools were not developed, some of which may no longer be valid.

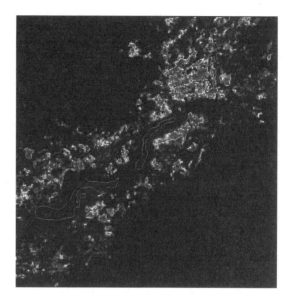

Fig.5.63 Maximum negative amplitude extraction for the Lutetian interval across Dandy. The Dandy area is clearly segmented, north (21/28a-8 and 21/28-8z) being divided from south (21/28a-6) by a dark, low-amplitude, channel-like feature. Well test and pressure data indicate that the dark channel is probably filled with impermeable mudstone.

Fyne (and Dandy) are relatively small oil pools in a reservoir of complex architecture. That the license became available to a new operator in 1996 was reason enough for them to be reappraised. Since the oil was first discovered in the block, the cost of drilling and operating has reduced manifold, drilling and completions technology have improved dramatically, and seismic imaging of reservoirs is so much the better. Had well 21/28a-8 been drilled during a period of higher and stable oil price, it is likely that development would have proceeded.

5.11 CASE HISTORY: THE IRANIAN OILFIELD

5.11.1 Introduction

Toward the end of the 1990s, Iran began to open its petroleum industry to foreign involvement. The schemes

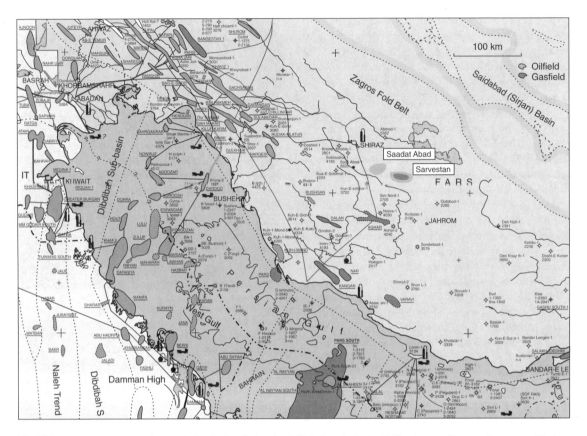

Fig.5.64 A petroconsultant's map showing the distribution of oil- and gasfields in Iran, the Persian Gulf, and surrounding areas. Sarvestan and Saadat Abad lie to the southeast of the main Zagros Trend.

launched by the National Iranian Oil Company (NIOC) were called buy-backs, because of the financial arrangement format used for foreign investors. Many of the projects offered within the first few license rounds involved the rehabilitation of fields that were either old or had suffered damage during the Iran–Iraq war. A few were new fields that had not progressed beyond the early stages of appraisal.

In this case history, we examine a pair of petroleum accumulations that were discovered in the early 1970s and yet still require appraisal. The Sarvestan and Saadat Abad Fields are onshore and within the Zagros Province (Fig. 5.64). However, they are outside the main Zagros production area, the Dezful Embayment, and instead lie southeast of the city of Shiraz. The important issues addressed by this case history were to determine whether the fields merited investment and what would be an optimal appraisal program. Specifically, the quality of the reservoir in the two pools was little understood. The lack of knowledge associated with reservoir quality

generated unacceptable levels of uncertainty concerning petroleum in place, well deliverability, and ultimate well recovery.

Specifically, there were too few data to allow an analysis of the fracture and matrix porosity for the Bangestan reservoirs (Ilam and Sarvak formations; Fig. 5.65) or to understand the distribution of conductive fractures in the subsurface.

To the north, south, and east of the two oilfields are fold mountains within which the Bangestan Group rocks crop out. In consequence, a request to do fieldwork and to examine core was made to NIOC. The plan was to examine the reservoir quality and fracture patterns in both core and outcrop.

5.11.2 The database

Few data were available on the five wells drilled into the two pools. Some geologic and wireline log data were available, as were the test data from the wells. Some secondary

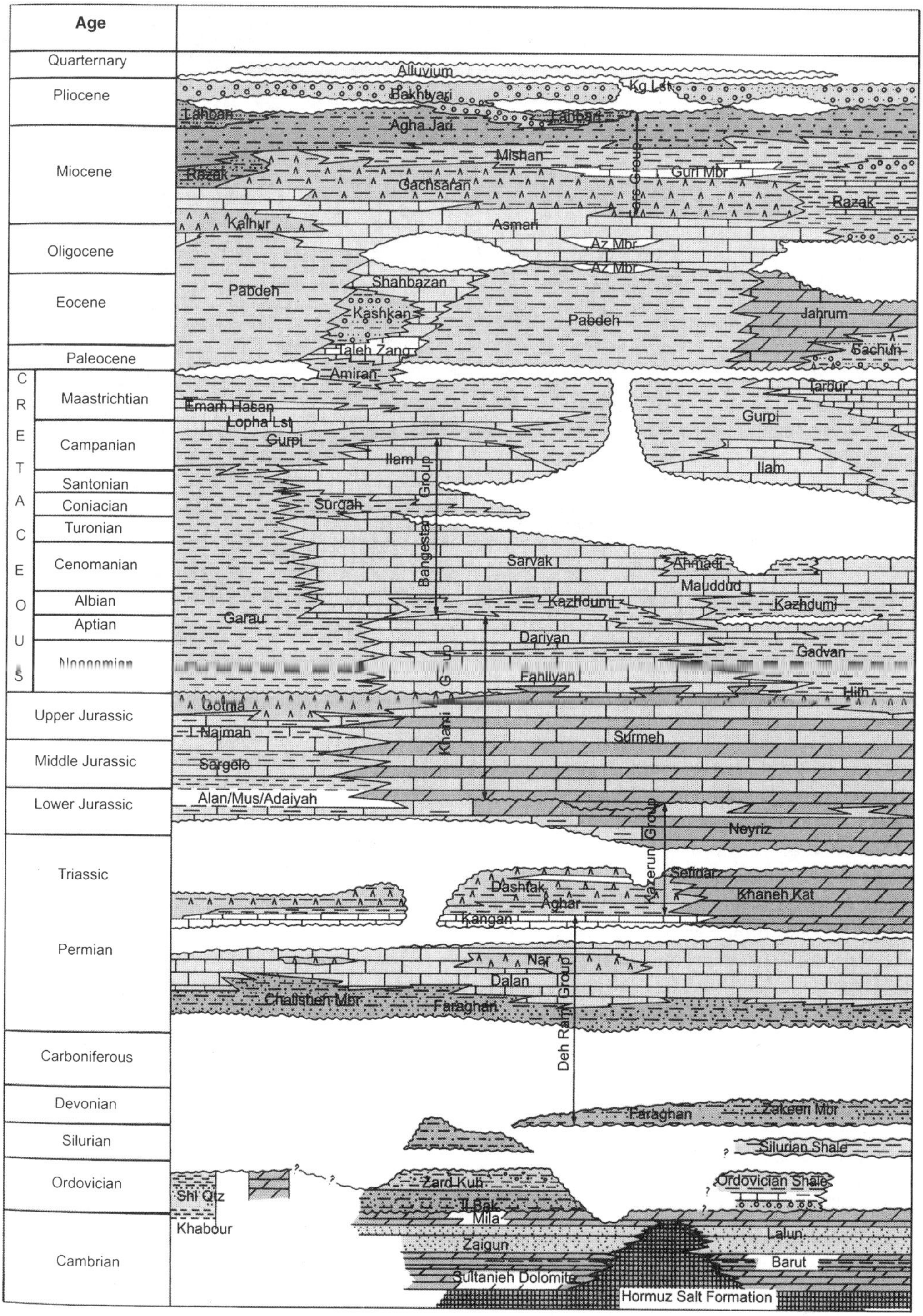

Fig.5.65 A stratigraphic column for southwestern Iran. (Reproduced courtesy of Schlumberger, Gatwick, UK.)

Hanging-wall anticline
of Sarvestan fault
(Nazerabad)

Fig.5.66 A Landsat image of the Sarvestan Valley, with major structures superimposed. The syncline in the northwest corner of the image is also shown on Fig. 4.14 (Image supplied by Nigel Press and Associates.)

data from evaluation reports and subsurface geologic maps were also available. A sparse 2D seismic survey was shot in the area and these data were present for review. It was also possible to obtain published data on the area. BP (then the Anglo-Persian Oil Company) made high-quality geologic maps over much of the Zagros oil province during the 1950s. These maps were revised and updated in the 1970s by the National Iranian Oil Company (NIOC). NIOC have also published SPOT satellite maps with overlain contour detail. Landsat images were bought from commercial vendors.

5.11.3 The regional geology

The Zagros of Iran is a classic fold belt formed by the collision of Asia and Arabia. Typically, the deformation produced almost cylindrical buckle folds with wavelengths of around 10 km, amplitudes of up to 5 km, and strike lengths of 100 km or more. Detachment occurs at the level of the Eo-Cambrian Hormuz salt. Several north–south or NNE–SSW faults cut across the dominant NW–SE strike of the belt. Regionally, these structures appear to be right-handed strike-

slip zones, broadly related to the molding of the structural trends between the rigid indenters of Arabia, the Caspian Sea, and the Lut block. They deflect the fold trends locally and pull-apart basins are developed at releasing bends. The bounding fault of the so-called "Dezful Embayment" is the most obvious of these structures. Convention has it that this structure also bounds the prolific source basin of the Zagros and is, therefore, the limit of its main prospective area. It is also the line along which the southern sector of the Zagros rides forward ahead of the rest of the fold belt. Both of these observations tend to suggest that the Dezful Fault Zone reactivates an older structure.

5.11.4 The geology of the Sarvestan and Saadat Abad Fields

The Sarvestan and Saadat Abad Fields lie beneath a flat-bottomed valley approximately 70 km east of the town of Shiraz (southwestern Iran). The valley is also approximately 70 km long and at a maximum 20 km wide. The fields lie at the southeastern end of the valley (Fig. 5.66). The only relief within the valley is a small set of hills marking the crest of the

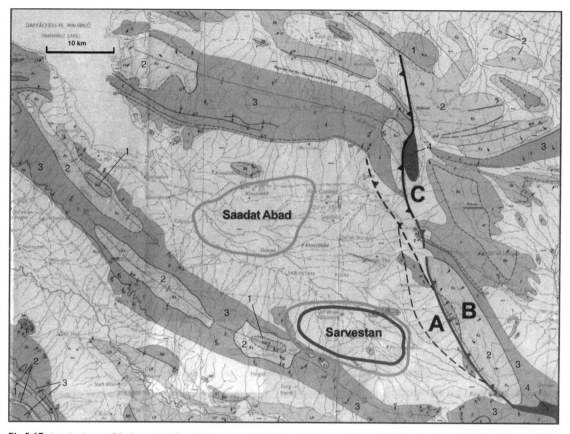

Fig.5.67 A geologic map of the Sarvestan Valley. The important lithologies are Bangestan sediments (1), Gurpi mudstones (2), Eocene limestones (3), and salt, probably of Tertiary age (4).

Sarvestan Field. There is no obvious surface expression linked to the presence of the Saadat Abad structure. A salt lake, Daryacheh-Ye Maharlu, occupies the northwestern end of the valley.

The valley is inclosed on three sides by mountains formed from Eocene and younger limestones. The valley is terminated in the east by mountains formed from both Bangestan and Tertiary sediments, displaced from further north by a right-lateral wrench fault. The mountain-tops are up to 1300 m above the valley floor.

The Sarvestan and Saadat Abad Fields comprise asymmetric anticlines (strictly, periclines) with steeper northern limbs (up to 20°) and less steep southern limbs (up to 14°). Reservoir occurs within the limestones of the Cretaceous Bangestan Group (Ilam and Sarvak Formations). Seal is provided by the Gurpi mudstones (Figs 5.65 & 5.67). The Kazhdumi Formation, which underlies the Sarvak, is believed to be the source of oil in the system.

Both fields contain oil and gas. For Sarvestan, the gas cap has been estimated to be 350 m thick in comparison with the oil leg at 162 m. No gas cap was penetrated in the Saadat Abad wells. However, its presence had been predicted from the tested wells on the field. One of these wells cut large quantities of gas soon after the initiation of a test. This well, it was said, penetrated the very top of the oil leg in the field.

No specific data were available on the fractures within either field, and porosity data were limited to derivative sonic log data in one well and reported average values in field evaluation reports.

5.11.5 Field studies: Sarvestan Fault Zone

The Sarvestan Fault Zone parallels the more important Dezful Fault Zone and, like it, it may be a long-lived structure. The Sarvestan fault lies at the eastern edge of the valley containing the fields. West of the fault is low ground

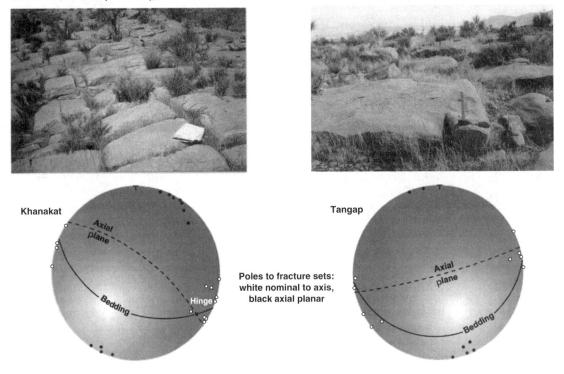

Fig.5.68 Fracture distribution within the Khanakat and Tangap Bangestan exposures.

that probably represents a pull-apart basin. The two fields lie within this basin.

The surface manifestation of the fault zone is a fold trend deflection from NW–SE to east–west. At the eastern end of the Sarvestan basin (east of Nazarabad), the fault zone appears as a significant westward-directed thrust that puts Bangestan (Ilam) over Eocene. The periclinal geometry in the hanging wall of this thrust can be seen as the interference between a hanging-wall anticline and the eastward continuation of the broad, east–west-trending Sarvestan Anticline (Fig. 5.67). Northeast of Sarvestan, the fault zone appears as an easterly-directed thrust putting Eocene over Miocene (Fars) shales and evaporites. The precise geometric relationship of these elements is unclear. The simplest possibility is a positive flower structure with associated thrusts. However, the thrusts lie along a deflected sector of the fault zone, and they would have acted as a releasing bend during pure dextral strike-slip along the dominant fault trend. It follows that the thrusting and the formation of the Sarvestan pull-apart must have occurred at different times and in response to different regional stress states.

5.11.6 Field studies: the folds

The outcropping folds of the Sarvestan area are typical mountain-scale Zagros anticlines, with limb dips of up to 60°. The tightness of the folds that form the fields is significantly less than that of the exposed folds. Core from the Sarvestan Field also seems to show a lower degree of recrystallization in equivalent stratigraphic units than that visible in outcrop. It would seem that the regional folds outside the basin continued to develop after amplification of the Sarvestan and Saadat Abad folds had ceased.

Significant erosion of the post-tectonic section is indicated by the residual synclinal outlier of Pliocene channel sands at Kamalabad, west of Sarvestan (Fig. 4.14). A hundred meters or more of Neogene sediment must have been removed before the deposition of the deposits that make up the present land surface of the basin.

5.11.7 Fractures

Two inliers in the main Kuh e Garh anticline south of the

(a)   **(b)**

Fig.5.69 Core from Sarvestan, containing (a) argillaceous, bioturbated, nonporous Sarvak Formation limestones and (b) bioclastic Sarvak Formation limestones with vuggy, biomoldic porosity.

Sarvestan Basin expose Bangestan beneath the Gurpi shales. Both areas present fine continuous exposure of the Ilam Formation. These structures, together with periclinal hanging wall of the Sarvestan Fault Zone east of Nazerabad, made up the main field study locations. All of them show the same fundamental fracture pattern, with one set normal to the fold hinges and the other parallel to the fold hinges and fanning around the fold. All of the fractures are sub-orthogonal to bedding (Fig. 5.68).

At Tangap, the two strike directions are about 170° and 080°, indicating a fold hinge trending around 080°. These orientations are compatible with the trend of the anticline at reservoir level visible on the Landsat image. They are incompatible with the overall trend of the Kuh-e Garh fold in the younger stratigraphy, which is NE–SW. The mismatch indicates detachment within the Gurpi shales between competent limestones, the fundamental origin of which remains obscure.

In contrast, the outcropping reservoir in the Kuh e Ghar fold at Khanakat trends parallel with the hinge of the structure at a higher stratigraphic level, and the fractures trend at 120° (axial plane parallel) and 030° (axial plane normal). In both outcrops, the fracture density is low–half metric spacing at best and locally (the hinge zone at Khanakat) very low (no more than metric). Subsidiary fractures and veins in the rock between the main weathered-out joints are nowhere densely distributed. The hinge of the Tangap fold has the densest fracture distribution, with axial plane parallel fractures dominant.

In the hanging wall of the Sarvestan Fault east of Nazerabad, the fractures are more densely spaced than elsewhere—locally at 5–10 cm. There is considerable irregularity, and total recrystallization in the most important fracture zones. Calcite-filled fracture sets at 080° (hinge parallel) and 155° (hinge normal) can be identified. Unfilled (later?) fractures running 030° to 040° cross them. This set of structures has an orientation that can be interpreted either as hinge normal to the hanging-wall anticline of the Sarvestan fault, or at right angles to the extension direction of the Sarvestan pull-apart basin. In either case, the intensity of these fractures should be expected to be greatest around the Sarvestan Fault Zone.

Throughout the fieldwork, all the Bangestan (Ilam) outcrops contained nonporous, impermeable limestones. Texturally, the limestones ranged from argillaceous wackestones to bioclastic grainstones. Recrystallization has been extensive, particularly in the coarser lithologies.

5.11.8 Core examination

Cores were cut within the Bangestan Group sediments of all three wells in the Sarvestan Field. In SV-1, core was only cut in the Sarvak Formation from beneath the tested interval, while in SV-2 part of the tested interval was cored. The relationship of the cored and tested intervals in SV-3 was unknown.

The overwhelming appearance of the bulk of the core in all three wells is of nonporous, variably muddy, heavily cemented limestone (Fig. 5.69). The Ilam Formation was not cored. Much of the cored Sarvak Formation comprises pale gray to dark gray micritic wackestones. Trace fossils are abundant and shell debris locally common. Near-complete bioturbation commonly makes specific ichnogenera difficult to identify, although *Chondrites* and ?*Ophiomorpha* have been recorded. A small-scale shaling-up and cleaning-up sequence can be observed. Bioclastic debris forms the only coarse material in the sediment.

Only one porous lithotype was identified throughout the three sets of cores. Cores 8, 9, and 10 in SV-2 contain white, coarse-grained (grainstone) recrystallized limestone, with abundant vuggy biomoldic porosity (Fig. 5.69). Individual pores are equant to rectangular in cross-section and up to 5–6 mm long. A visual estimation of porosity in this system would be about 5–6%. All of the vuggy pores are heavily stained with bitumen. However, the matrix appears to be completely unstained by petroleum. Diagenesis has reduced the porosity of the matrix to near zero.

The net to gross within the generally porous interval of cores 8, 9, and 10 in SV-2 was calculated. A simple method was devised whereby the obviously porous intervals were measured and this "net pay" was compared with the whole of the Sarvak interval. In SV-1, the uncored porous interval was identified from sonic-log derivative data supplied by NIOC. Three, or possibly four, porous intervals occur beneath the gas/oil contact. The composite thickness of the net pay within the four intervals amounts to about 62 m in SV-1. In SV-2, the maximum possible porous pay can be calculated by comparing the cored and tested intervals. Here, it amounts to about 70 m. There were insufficient data to correlate the porous intervals between wells.

High-angle and low-angle fractures occur throughout the cored intervals. Many appear to be partially cemented and partially open. Bituminous petroleum can be observed on many of the fracture surfaces. Stylolites too are common. Unusually, the stylolites have many different orientations, which indicates that tectonic stress as well as gravity-induced compaction is responsible for their formation. Like the fractures and the vuggy pores, the stylolite surfaces are commonly stained with bitumen. Given that such stylolites are a product of either gravitational or tectonic loading (i.e., they are in compression), there must have been a pressure release at some stage for them to open and admit petroleum.

5.11.9 Discussion: oil in place and productivity

The combined field and core work indicated that while both the Ilam and Sarvak formations appear permeable, because they are fractured, the porosity of the system is generally low.

The exception to this generality occurs within the Sarvak Formation, where 60–70 m of vuggy porous limestone is present. Work on the cores indicated where in the Bangestan Group such limestone occurs and its likely abundance. In order to generate oil-in-place figures for the fields, it would be necessary to conduct new core analytic work on both the porous and nonporous core, to determine its precise porosity and the matrix permeability.

The likely well productivity from the fields has been analyzed by analogy with the intensity of fracturing in the surrounding outcrops. Fractures are most intense at the eastern end of the valley that contains the two fields; that is, adjacent to the Sarvestan Fault. Fracturing is least evident in outcrops at the western end of the valley. If the fields contain fracture patterns that mimic those in the outcrop, then we might expect to see a degradation of well performance from east to west. This was observed for instantaneous rate in respect of the exploration and appraisal wells. The same could also be true of ultimate volume delivery from wells.

5.11.10 Conclusions

Geologic fieldwork still has a place in exploration geoscience in the 21st century. In this case history, analysis of reservoir rocks in the area surrounding a discovered but undeveloped field was able to provide important information on the quality of the reservoir rocks in the subsurface. This led to a reduction in uncertainty in the oil in place and the likely productivity of development wells.

FURTHER READING

Archer, J.S. & Wall, P.G. (1986) *Petroleum Engineering, Principles and Practice.* Graham & Trotman, London.

Cubitt, J.M. & England, W.A. (1995) *The Geochemistry of Reservoirs.* Special Publication No. 86, Geological Society, London, 321.

Kupecz, J., Gluyas, J.G., & Bloch, S. (1997) *Reservoir Quality Prediction in Sandstones and Carbonates.* American Association of Petroleum Geologists, Memoir 69.

Tearpock, D.J. & Bischke, R.E. (1991) *Applied Subsurface Geological Mapping.* Prentice-Hall, Englewood Cliffs, New Jersey.

CHAPTER 6

DEVELOPMENT AND PRODUCTION

6.1 INTRODUCTION

Field appraisal will either close with a decision to abandon a project or a mandate to take the field into production. In this final chapter, we examine the role of the geoscientist in the processes of field development, production, abandonment, and reactivation. In each of the preceding chapters, the emphasis has been placed upon a description of the Earth using static data. This chapter is different. Static data are generated during production, but many of the new data available to the geoscientist and reservoir engineer are dynamic (production) data. These dynamic data are commonly available in time series. They include production profiles for wells, fluid pressure data through time, and fluid composition data collected during production. Production geoscience and reservoir engineering are intimately linked; as a consequence, field production teams commonly comprise a mixture of geoscientists and reservoir engineers. Another profound difference that separates the way in which geoscience is used in production from that in exploration is the timescale involved. It can take years for acreage to be acquired, seismic surveys shot, prospects identified, and exploration wells drilled. In such situations, the mistakes that geoscientists make may never catch up with them. They will be long gone onto the next exploration job before their poorly considered prospect is drilled and proved dry. Curiously, the same seems not to be true of success. For one of us, as a young geologist joining BP in the early 1980s, it seemed that all of the older generation in the company had a personal hand in discovering the giant Forties Field ten years previously! However, we digress; production geoscience allows no such luxuries with time. In the most intense of operating situations, it may be necessary to plan a new well every two weeks and have the wells drilled within a few weeks of planning. Such rapid testing of one's understanding of field geometry, fault distributions, and reservoir architecture and quality can be very exciting

(or demoralizing). Moreover, the expectation at the production stage is that the wells will be successful. It is not so easy to hide behind a failure when the pre-drill probability of success was 80%, compared with an exploration prospect at 20%.

Field development is covered in Section 6.2, production in Section 6.3, and changes to reserves—be that through revisions, additions, or field reactivation—in Section 6.4. It will become clear as this chapter unfolds that the discovery of an oilfield can be ascribed to a specific point in time, but no such temporal definition can be made for the point of final depletion of a field. Abandonment of a field is driven by economic criteria. At the point of abandonment the field still contains petroleum; it is just uneconomic to extract it. This fact leads to the possibility that changing fiscal regimes, increases in the oil price, the development of new technology, and reductions in operating costs can cause old fields to gain new leases of life and abandoned fields to be reactivated. The ultimate extension of this is in the so-called "stripper wells" of North America. These wells produce minuscule quantities of petroleum, measured in barrels per day. During periods of high oil price the wells become economic and production is restarted, only to be shut-in again as the oil price falls. Many of these wells are exceptionally old, and because the fields repressurize during periods of low oil price, recovery factors can often be very high indeed (more than 60%), something that can rarely be achieved in expensive offshore situations.

We use six case histories to illustrate this chapter (Table 6.1).

6.2 WELL PLANNING AND EXECUTION

6.2.1 Facilities location and well numbers

The process that links appraisal and production is called

Table 6.1 Case histories in development, production, reserves revisions/additions, and reactivation.

Case history	Location	Illustrating
Wytch Farm Field	UK onshore	Extended-reach drilling
Amethyst Field	UK continental shelf	The resolution of differences between the estimated gas in place from static and dynamic data
Fahud Field	Onshore Oman	Stages in the development of a giant field over a 40-year period
Heather Field	UK continental shelf	Management of field decline
Crawford Field	UK continental shelf	A possible reactivation project
The Trinidadian oilfields	Onshore Trinidad	The possibilities for countrywide rehabilitation of old oilfields

development. For petroleum accumulations that are offshore, development is a distinct phase that accompanies construction of the offshore facilities. For an onshore field, appraisal, development, and production can form an overlapping spectrum of processes that take a field from discovery to peak production. This is because, in many land locations, it is possible to produce and sell oil soon after discovery. In such situations, development might reasonably be taken to cover the major period of well drilling between discovery and plateau production.

Appraisal of an oil- or gasfield has delivered a measure of petroleum in place. At the onset of development, extensive modeling of the reservoir will have been completed to give an estimate of how much oil or gas is likely to be recovered and how many wells will be needed. However, such modeling of the reservoir will have used little or perhaps no dynamic data. For example, many of the recent field developments in the deep-water Gulf of Mexico were not production-tested during the exploration or appraisal phases. Companies are reluctant to test such wells because of the high expense. Instead, analog data are used to predict the flow characteristics of the field. This is a high-risk strategy.

The field-life production forecast will commonly include a prediction for water production as well as oil and gas (Fig. 6.1). All oilfields will produce gas as well as oil (Fig. 6.2). This first production forecast is of particular importance for the design criteria for the production facilities. The estimated plateau oil production rate will naturally define the size of the export facility for the oil. The estimate of gas production in an oilfield will be used to determine whether a separate gas export facility is required, whether compression is required for gas reinjection, or if there is just sufficient gas for power generation at the facility. For oilfields where the production of associated gas is low, it may be necessary to import gas to the facility. The timing and quantity of water production will control how and when water handling will be installed. For

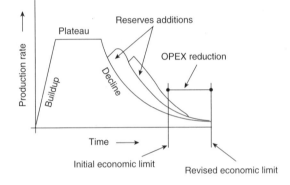

Fig.6.1 An idealized production profile for an oilfield. Production buildup is commonly planned to be rapid, plateau stable, and decline rapid to maximize the use of the facilities. Reserves additions from satellite pools will be used to maintain the plateau. Abandonment of the field will occur before production has declined to zero, the point being determined by the cross-over between the value of the produced oil and the cost of producing it. A reduced cost of operation (OPEX) will commonly allow production to continue at economic rates beyond the initial planned abandonment date.

example, preliminary modeling may indicate that water will not be produced for a few years after field start-up. In consequence, it may not be necessary to install the full water-handling capacity for the first day of production. The total production profile thus controls the whole design of the facilities. During the period of design and construction, much work is done on the reservoir. Indeed, it is common nowadays for production wells to be drilled from a template during construction of the production facility. Such wells clearly provide a wealth of information about the reservoir. They also allow a rapid buildup of production, so as to enable an early start to paying-off the cost of development.

The position of production and injection wells will be

optimized using the reservoir model. The factors that will be taken into consideration are the position and strength of aquifers, the distribution of gas caps, the dip of the reservoir, its quality and the degree of layering and lateral heterogeneity, and the location and number of potential baffles and barriers to fluid flow (faults, cemented horizons, and sealing lithologies). The well positions will also depend upon the chosen method for production, which itself may in part be influenced by the distribution of the natural factors listed above.

For example, in the super giant oilfields of the Zagros fold belt of Iran and Iraq, production wells are often places on the crest of the fields. Many of the fields are without primary gas caps, oil columns can be measured in kilometers, reservoirs are fractured, and aquifers are poor. But for the thick oil column one would expect rapid production of water as it floods up fractures into the production wells. However, with such large fields and such thick oil columns it is possible to produce oil wells at a high rate, and it will still take many tens of years (perhaps more than 100 years) for water cut to become appreciable. In smaller fields, it may be necessary to be less cavalier about well placement. Fields deemed to have very active aquifers might need to have production wells placed not just at their crests but in lower-quality reservoir. In such situations, high-permeability conduits could cause much oil to be bypassed, such that production wells quickly run to water. This situation is a progressively greater problem as oil viscosity increases.

Oil reservoirs with gas caps require a different well placement strategy. A reservoir containing a large gas cap, a poor aquifer, and an oil rim may be exploited by placing production wells low in the oil column or even at the top of the underlying water interval. However, if the aquifer is large and active, wells tend to be drilled in the lower part of the oil leg. In this instance, it is commonly easier to manage a downward migration of the gas/oil contact by injecting gas into the gas cap.

If the field is to be produced using secondary recovery techniques, the position of wells for water or gas injection must also be planned. Water injection is commonly planned either as line drives or pattern floods. These terms are reasonably self-explanatory (Fig. 6.3). Line drives are commonly used in dipping reservoirs, while pattern floods may be used in com-

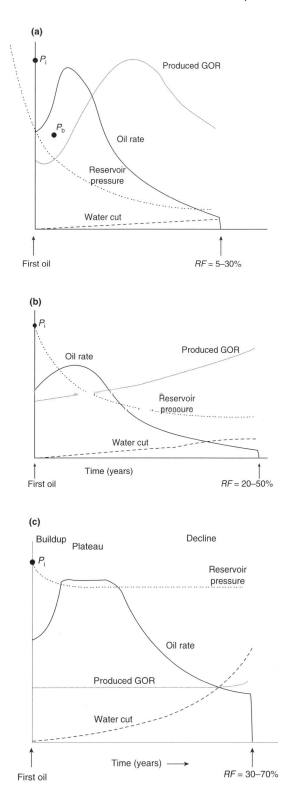

Fig. 6.2 Typical, time-dependent, fluid changes in an oilfield being produced under (a) primary gas exsolution drive, (b) gas cap drive, and (c) water drive. P_i = initial pressure, P_b = bubble point pressure, RF = recovery factor.

plex reservoirs or near flat-lying reservoirs. If a reservoir is heavily segmented, it may be necessary to treat each segment as a separate accumulation and use paired production and injection wells for each segment. In recent years, some less than obvious water-flood techniques have been tried with considerable success. In the giant Prudhoe Field (Alaskan North Slope), water has been injected into the gas cap. Because the oil and water legs are laterally separated in this low-dipping reservoir, this method has compromised neither the oil nor the gas production, but has helped to sustain reservoir pressure and thus oil production. The injected water sinks to the base of the reservoir section and then flows down dip, so displacing oil toward the production wells.

Fields with natural gas caps are often considered for gas cap drive (Figs 6.2 & 5.55). Associated gas, that exsolved from the oil during production, can be injected into the gas cap to maintain pressure. Even in reservoirs that lack natural gas caps, gas may still be put into field crests because it is an extremely efficient displacement process, enabling production of attic oil that may otherwise be inaccessible. This technique has been used in the western panel of the Douglas Field (East Irish Sea Basin; Yaliz & McKim 2003). In the Lennox oil- and gasfield close to Douglas, wet gas (methane plus higher homologs) has been injected into the gas cap to maintain pressure (Yaliz & Chapman 2003). Preliminary analysis of the data for Lennox indicate that gravity segregation of the "heavier" components within this gas may limit gas breakthrough in oil production wells by reducing the viscosity contrast at the gas/oil contact. In some fields, a combination of water and alternating gas injection (WAG) may be chosen.

The development team uses the reservoir model to determine the number of wells required. The reservoir model will also yield the expected volume of petroleum that is recoverable from each well. Although every effort will be taken to ensure that the reservoir model mimics the behavior of the field when in production, it is of course not conditioned to the actual performance of the field for which it has been built. However, after the field has been on production for some time, it will become clear just how much each well is capable of producing and the degree to which the model was able to forecast the behavior of the field. Naturally, the new production data can be used to modify or rebuild the reservoir model and so refine the forecasts for production from both individual wells and for the field as a whole.

The quantity of petroleum that can be recovered from a well depends upon a range of criteria. These include the permeability of the reservoir, the viscosity of the oil, the degree of reservoir heterogeneity, well management (pressure drawdown history), the presence or absence of aquifers or gas caps,

and well longevity. The most prolific wells in the world's giant fields can deliver over 50 million barrels during their lifetimes, while the productive history of most wells can be measured in hundreds of thousands of barrels. For example, wells in the Magnus Field (UK North Sea) need to deliver on average 50 mmbbl; while Helena-1, the discovery well for the Forest Reserve/Fyzabad complex (Trinidad), celebrated its 75th birthday in 1995 having delivered about 750,000 bbl. In reaching this quantity of oil, the well has been worked over about six times in its 75-year history. The ultimate life-of-field productivity for gas wells is commonly measured in tens to hundreds of billions of standard cubic feet.

6.2.2 Well geometries

Most exploration and appraisal wells are vertical. Few development or production wells are vertical. A vertical or near-vertical well is the natural choice during exploration. The aim of the well will be to penetrate a primary, and possibly several secondary, target horizons. A vertical well cuts across the stratigraphy. It is easier to plan on the sparse seismic data commonly available during most exploration programs than would be a well with more exotic geometry. A vertical well will also allow collection of data such as pressure gradients and it will possibly penetrate fluid contacts (gas/oil, gas/water, and oil/water).

Quite clearly, a development/production well has quite different requirements. The aim for such wells is to produce as much petroleum as possible, and to produce that petroleum at economic rates. This can be achieved in a variety of ways, although ultimately the effect is much the same; optimization of permeability × height (measured in millidarcy-feet, mD ft, or millidarcy-meters, mD m). Such optimization may mean that the field has many vertical wells or a few high-angle or horizontal wells (Fig. 6.4). The decision on the type of well to be drilled will be a function of the location, the petroleum distribution, cost, and the reservoir properties. In an onshore location, vertical wells may be easy to locate, shallow, and cheap to drill. In such situations, it would be perverse to try to drill expensive and complex geometry wells. Offshore, the situation is different. It is rarely possible to have lots of wellheads scattered over a large area, except in shallow lakes and seas subject to mild weather conditions (e.g., Lake Maracaibo, Venezuela; the Gulf of Paria, Venezuela/Trinidad). In offshore settings such as the Gulf of Mexico, offshore West Africa, and the North Sea, wellheads will be clustered at the production facility (Fig. 6.5). Well tracks will be deviated away from the facility location to cover the productive area of the field. The angle of the wells as they pass through the reservoir can vary

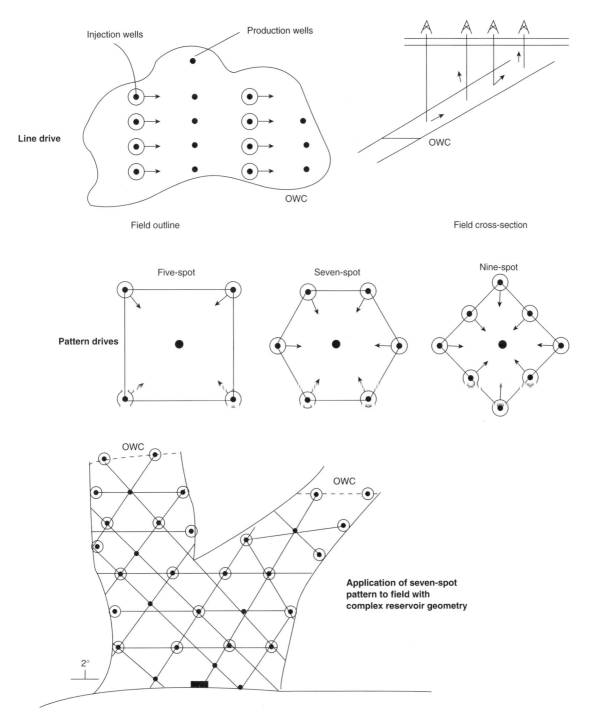

Fig.6.3 Well patterns used for water drive, secondary recovery.

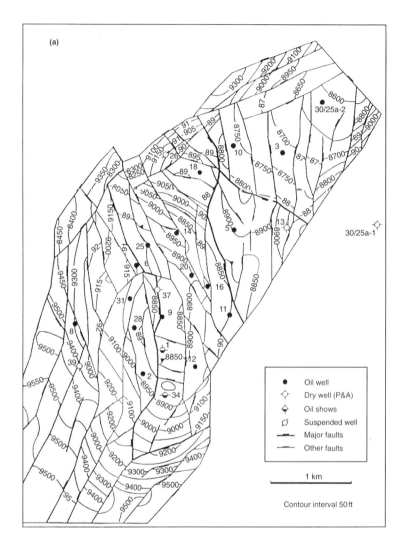

Fig.6.4 Vertical versus horizontal wells (contrasted development methods for the Argyll and Foinaven fields, UK continental shelf. (a) The Argyll Field was developed with vertical and near-vertical wells from a floating facility during the 1970s and 1980s. Each well penetrated all reservoir horizons, while commonly the wells were completed in just one horizon. (From Robson 1991.) (b) The panel map and development well locations for the Foinaven Field, UK continental shelf. The wells are drilled from two centers (DC1 & DC2), individual, high-angle wells targeting specific reservoir, sandstone lobes. Development wells were drilled during the late 1990s and early 2000s. (From Carruth 2003.)

from vertical to horizontal. As discussed earlier, high-angle and horizontal wells will tend to be more productive than vertical wells insofar as they have longer sections in reservoir than do equivalent vertical wells. The choice of high-angle versus horizontal is commonly made depending upon the vertical heterogeneity of the reservoir. Strongly layered reservoirs in which there are barriers to vertical fluid flow may not benefit from horizontal wells — or, more properly, those that are parallel to the stratigraphy — because the barriers will prevent petroleum from being produced from horizons not specifically penetrated by the wellbore. In such instances, it is commonly the practice to drill wells that cross-cut the reservoir stratigraphy.

High-angle and horizontal well technology was developed during the 1980s. The product of the 1990s is the multilateral well. In such wells there are two or more branches in the reservoir section. Clearly, although such wells are complex to drill, much money is saved in drilling the top-hole section. It is drilled only once for each multilateral cluster. An example of multilateral wells being used in field development comes from the Lennox Field in the East Irish Sea Basin (Fig. 6.6; Yaliz & Chapman 2003). The field is a broad domal structure, with a thick gas cap of about 700 ft and a thinner oil leg of about 150 ft. The unmanned platform lies approximately above the center of the field, from which wells are drilled vertically downward toward the field, before being deviated

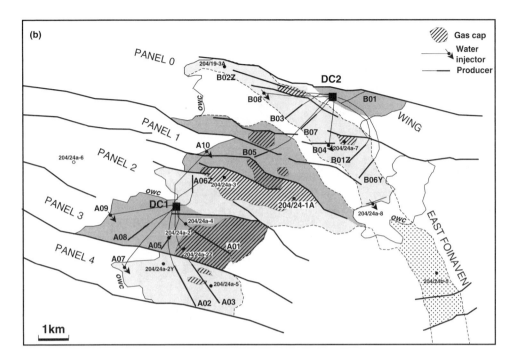

Fig. 6.4 *continued*

to horizontal within the oil leg. The field contains one trilateral well drilled in the east and south and several bilateral wells.

6.2.3 Well types

Wells can be divided into three broad categories; production wells, injection wells, and utility wells. We have already covered production wells. Injection wells may be required for either water or gas, depending upon the recovery method used or the field. Utility wells can include those drilled as a source of water for injection, those drilled such that produced water and cuttings can be disposed of, and those used for observation.

The primary purpose of injection wells, be they for water or for gas, is to maintain the pressure in the reservoir as petroleum is extracted. The efficiency with which the water or gas will sweep oil from the reservoir is also important. Water injection will be of little value in a field if the injected water runs from injection well to production well and bypasses much of the oil in the reservoir. Most oils are more viscous than water and the problem is exacerbated as the viscosity contrast increases. Thus attempts are made to place injection wells where pressure support can be maximized and the potential for water breakthrough minimized. Whether or not this can be achieved will be determined by the segment size of the field and the permeability heterogeneity. Clearly, it is best to avoid high-permeability intervals with injection wells, as these can often be conduits direct to production wells (Fig. 6.7).

The injection of cold water into hot rock can lead to thermal fracturing as the rock shrinks around the wellbore area. The direction of fracture propagation can be determined by measuring the local stress field in the subsurface. Clearly, fractured rock will allow greater injectivity for a given injection pressure. This natural phenomenon is commonly used to help injectivity in hot and deep reservoirs that have low matrix permeability.

The design of injection wells, like that of production wells, has become highly sophisticated. For example, paired groups of multilateral production and injection wells are being used to produce oil from poor-quality carbonate reservoirs in the Middle East. In this particular example, productivity was improved by two orders of magnitude compared with the production rate in the original, vertical wells. This has allowed development of hitherto uneconomic oil. All of these injection wells need to be designed by a team that includes drilling engineers, reservoir engineers, and petroleum geoscientists.

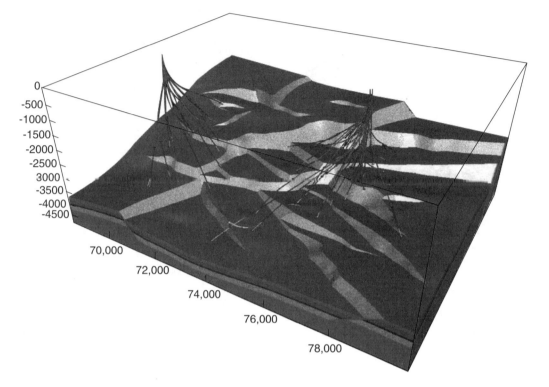

Fig.6.5 Clustered wellheads and radiating wells beneath the two production platform locations. (Dynamic Graphics Earth Vision model, published with permission.)

Utility wells, be they for injection of produced fluid waste or cuttings or for source water, tend to be simple and are likely to be near vertical. Although such wells will be simple in geometric terms, the geologist will be involved in determining their position in much the same way as would be required for production and injection wells. For example, the distribution of Paleocene submarine-fan sandstones overlying the Jurassic-age reservoir of the Gyda Field (Norwegian North Sea) was studied from the point of view of cuttings injection.

Observation wells are rarely drilled in their own right, except to monitor flood fronts in enhanced oil recovery (EOR) pilot projects. More often, pressure gages are inserted in an old wellbore.

6.2.4 Drilling hazards

In addition to determining the best locations for production and injection wells, it commonly falls to the geoscientist to identify potential drilling hazards for a well. It is most important for the geologist while planning a well to identify

possible sources of danger to human health. In many instances, the singularly most important and abundant shallow hazard is gas (methane). It is most usual to conduct high-resolution (seismic) site surveys to enable identification of such gas.

Mobile mudstones and those that are chemically unstable with respect to drilling fluids are also commonly mapped. Similarly, hard horizons such as flint layers tend to be identified and their abundance calculated. Once the potential hazards have been identified, the drilling engineer can plan to minimize their effect. This may be manifest as a decision to change the drill bit before entering hard lithologies, to manage the mud weight in sensitive formations, or to modify casing schemes (the process of stabilizing an open hole by cementing a pipe in place).

6.2.5 Well completion and stimulation

The process that takes place between the drilling of a well and the production of petroleum from it is called completion. The type of completion can vary from the most simple —

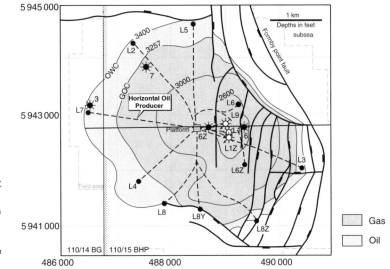

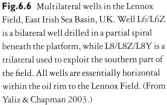

Fig.6.6 Multilateral wells in the Lennox Field, East Irish Sea Basin, UK. Well L6/L6Z is a bilateral well drilled in a partial spiral beneath the platform, while L8/L8Z/L8Y is a trilateral used to exploit the southern part of the field. All wells are essentially horizontal within the oil rim to the Lennox Field. (From Yaliz & Chapman 2003.)

open hole—to rather more sophisticated methods using slotted liners, wire mesh screens, gravel packs, and fracture stimulation methods. The main aim of completion is to optimize petroleum production while maintaining the integrity of the wellbore. In many instances, the technology required to stabilize the wellbore has a detrimental effect upon the productivity. For example, in poorly consolidated formations it may be necessary to place a gravel pack (a sheath of clean, monodisperse sand or glass beads) across the producing interval, so as to reduce or eliminate the tendency for the well to produce sand as well as petroleum. Clearly, in placing a gravel pack across the production interval, the permeability in the wellbore has been compromised. The role of the geoscientist is to describe the rocks penetrated by the well in terms of their physical properties, such as permeability, the degree of consolidation or cementation, and the natural stress regime.

Well stimulation technology is designed to improve the near-wellbore permeability. Such stimulation may be either through physical or chemical methods. Acid washes can be used to dissolve minerals in the near-wellbore region. Hydrochloric acid is used for carbonates, while hydrofluoric acid may be used for low-permeability sandstone formations. Apart from the intrinsic health hazards of using such acids, work is also required on the reservoir formations prior to acid treatment, to determine their suitability for such treatment. For example, direct treatment of the Rotliegend Sandstone of the North Sea's Southern Gas Basin with hydrofluoric acid can result in reduction of permeability, as the hydrofluoric

acid reacts with calcite and dolomite to produce insoluble calcium fluoride (fluorite). In consequence, such sandstones are commonly pretreated with hydrochloric acid to remove calcium carbonates before the hydrofluoric treatment. Problems may also be encountered with the production of iron hydroxide gels following acid treatment. It is therefore important to have a mineralogical analysis of the formation prior to any chemical treatment.

Physical stimulation of reservoirs commonly involves fracturing. In this instance, fractures are created hydraulically, using a suspension of tough beads in a proppant gel. The mixture is injected into the formation until the rock fractures. The mix of proppant gel plus beads migrates along the fractures. At the end of the fracture process, the pressure is reduced but the fractures are kept open by the injected beads. The proppant gel is then broken down. This may require further chemical treatment. Alternatively, the proppant may be designed to degrade naturally.

Quite clearly, the important geoscience work to perform prior to any "frac pack" treatment is that of analyzing the local stress regime. The ideal fracture geometry is one that is parallel to the wellbore, rather like a pair of "Mickey Mouse" ears (Fig. 6.8). This can normally be achieved when the principal stress direction is vertical (gravity). However, in neotectonic areas the principal stress direction may not be vertical, and in consequence the fracture may open up perpendicular to the wellbore. This leads to minimal improvement in near-wellbore permeability (Fig. 6.8). Indeed, this is believed to have happened when attempts were first made to reactivate the Peder-

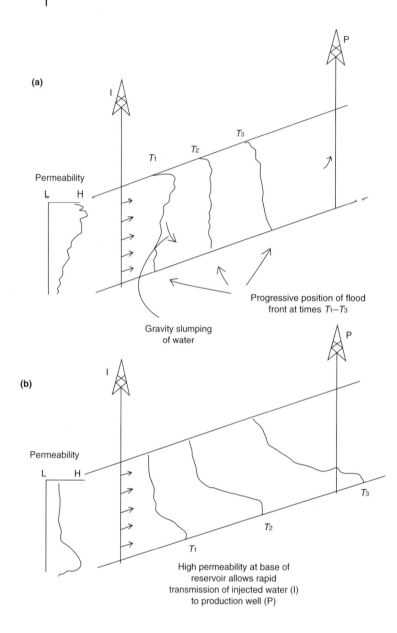

(a)

Permeability

L   H

T_1 T_2 T_3

Progressive position of flood
front at times T_1–T_3

Gravity slumping
of water

(b)

Permeability

L   H

T_1 T_2 T_3

High permeability at base of
reservoir allows rapid
transmission of injected water (I)
to production well (P)

Fig.6.7 A sketch showing the control of
permeability profile in a well on the injected
water-flood front. (a) A downward decrease in
permeability results in an even flood front,
with the denser water slumping to the base of
the sandstone, so displacing oil from the
lower-quality rock. (b) The high-permeability
conduit at the base of the sandstone promotes
water under-run and early breakthrough of
water into the production well.

nales Field in eastern Venezuela (Jones & Stewart 1997): a
wrongly designed frac-pack is thought to have led to complete
loss of productivity in the first of the reactivation wells drilled.

6.2.6 Formation damage

Formation damage is the inadvertent and opposite effect to
well stimulation. Permeability is reduced and the flow rate of
the well diminishes or is stopped altogether. Two categories
of mechanisms can be responsible for formation damage;
reduction of absolute permeability, or reduction of relative
permeability in the near-wellbore region. Although the
overall effects of formation damage can be categorized simply
into one of these two mechanisms, the individual causes are
multifarious and a whole array of individual terms has been
developed to describe various situations. Just a few of these

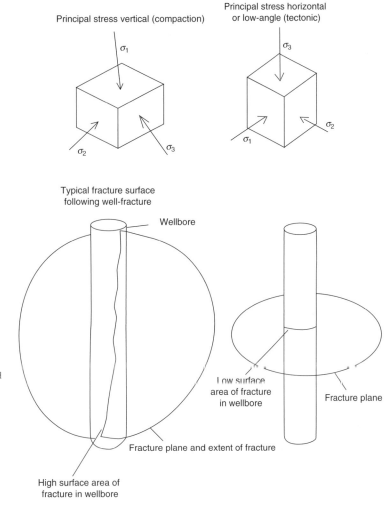

Fig.6.8 Artificial fracturing of wells can stimulate production by increasing the effective contact area between the wellbore and the formation. Under normal hydrostatic or even normal compactional overpressure, the fracture surface radiates from the wellbore as two "ear-shaped" surfaces. However, in situations in which the principal stress is horizontal and the minimum stress vertical, the fracture is horizontal, and in consequence the surface area of fracture in the wellbore is minimal.

are water block, condensate block, fines migration, asphaltene drop-out, and scaling (mineral precipitation). Some effects may be cured or reduced by remedial action, but the best option is to reduce the likelihood of the problem in the first place. As will be clear from some of the terms described above, formation damage results from the physical effects on the reservoir and fluids caused by drilling, or by the interaction between the drilling fluids and formation fluids or reservoir minerals.

The degree of damage to a well is referred to as the "skin factor." This skin effect can be considered as a rate-proportional steady state pressure drop, given as a product of a flow rate function and a dimensionless skin factor (Archer &

Wall 1986). The numerical value of the factor is only a semi-quantitative measure. Nonetheless, such numbers do help to describe formation damage. For example, for a vertical, unfractured, and undamaged well, the skin factor would be zero. As the degree of damage increases, so does the skin factor. Skin factors of less than about five indicate a little damaged well. Skin factors reported as many tens or even hundreds are obtained from severely damaged wells. It is possible to calculate a negative skin factor. This can be real, insofar as it is a product of a naturally or artificially fractured well, or simply a product of the calculation, as can be the case in a high-angle or horizontal well. In both instances, it implies that the flow into the wellbore will be greater for a given pressure drop than would be the case in an undamaged, equivalent vertical well.

Absolute permeability reduction occurs when pore throats in the near-wellbore area are blocked. Such blocking can occur when fines such as clay minerals become detached from their host grains and migrate toward the wellbore. This may result from both chemical interaction between the drilling fluids and the rock and then from the physical effects of fluid flowing from formation into the wellbore. The same absolute reduction can occur when drilling fluids and formation fluids react to produce insoluble precipitates. A particularly dramatic example of such formation damage occurred when the discovery well of the Miller Field (UK North Sea) was drilled in December 1982. One of the authors was involved in the detective story associated with tracking down the problem of a nonproductive well test on the Miller discovery well.

The Miller reservoir is almost pure quartzite and is one of the cleanest sandstone reservoirs known (Gluyas et al. 2000). However, on test the discovery well failed to flow at an appreciable rate. The initial reaction of the drilling, completions, and testing teams was that the rock was of poor quality (permeability). This was not compatible with either sedimentological or core analysis data. The porosity was known to range between about 13% and 25%, and the absolute permeability was measured in hundreds of millidarcies and darcies. Routine analysis of the rock using a scanning electron microscope showed that although it comprised quartz-cemented quartz grains, many of the pores were partially filled with patches of a microcrystalline solid. Coincident with the work on the Miller discovery well, an EDAX (X-ray elemental analysis) facility had been fitted to the microscope. The microcrystalline mass was found to contain barium and sulfur. At first, it was thought that this barium sulfate (barite) was a contaminant from the drilling mud. However, this proved not to be so. The formation waters in Miller were found to contain between 1000 and 2000 ppm of dissolved barium, and the exploration well had been drilled with a seawater-based (sulfate-rich) mud system. Barium and sulfate had reacted to yield a near-impermeable membrane around the wellbore. But for this work on formation damage, a cooperative venture between geoscientists and completion engineers, the barite problem might never have been properly appreciated, the research work on scale prevention not initiated, and the full quality of the Miller reservoir never recognized.

Reduction of absolute permeability can also occur when drilling fluid filters into the formations and solids are deposited around the wellbore. Such formation damage is commonly very difficult to remove or reduce during the well-completion process.

Reduction of absolute permeability can also occur as a result of precipitation of asphaltene from the reservoired oil. Such precipitation can occur in some oils when the pressure is reduced. This is a very common problem in production strings. Here it can be removed, albeit with difficulty, by scraping the production string or giving it a chemical wash with a nonpolar solvent such as toluene or xylene. However, if precipitation occurs in the reservoir close to the wellbore, amelioration of the problem can be near impossible. The well may have to be abandoned. The propensity for asphaltene precipitation needs to be studied by the geochemist, using fresh oil samples taken, if possible, at reservoir conditions. Asphaltene precipitation, both natural and induced, is a problem in parts of the El Furrial Province in eastern Venezuela.

Relative permeability reduction is a product of having several phases of fluid (gas, oil, and water) in the near-wellbore region (Fig. 5.32). The relative permeability to the desired phase, usually oil, can be impaired or even reduced to zero by the presence of another phase. The use of water-based drilling fluids can in some reservoirs lead to so-called "water block," in which only water can be produced back into the wellbore. A similar situation can be induced if the pressure drawdown during testing and production causes gas to break out in the reservoir.

Formation damage can occur in injection wells just as easily as in production wells. The former leads to a loss of injectivity and the latter to a loss of production. Although here we have concentrated on near-instantaneous effects caused by drilling, formation damage can progressively reduce the productivity/injectivity of a well. Particularly acute problems are commonly encountered once injection water starts to break through into production wells. Many formation damage problems can be avoided or minimized by appropriate geoscience work on formation mineralogy, formation fluids, and the physics of the reservoir and fluid system.

6.2.7 Well logging and testing

In the preceding chapters, a variety of well logs and the data obtained from them have been described. These data have in all instances been static data, rock properties and fluid properties. However, it is also possible to take measurements in flowing production and injection wells. The aim of such measurements is to obtain information on the rate of fluid flow, the type and locations of different fluids as they enter the wellbore and where fluids are flowing (inside the well casing or behind pipe). It is also possible to measure pressure data from wells during flow periods or pressure changes when a well is not flowing. In this section, we will concentrate on the geologic information that can be obtained from such logging and well testing. Such data can help the geoscientist to improve his or her description of the near-wellbore region, as well as yielding data on the reservoir architecture and the proximity to barriers such as faults.

Production logging methods are highly varied. They include mechanical devices (spinner surveys) and radioactive methods to detect fluid flow in the wellbore. The fluid type can be measured using density tools or fluid capacitance instruments, in which the dielectric properties of the fluid are used to differentiate between petroleum and water. Temperature data are also very useful in allowing the identification of flowing and nonflowing units within producing wells, injecting wells, and even shut-in wells.

The spinner type of survey uses an instrument that contains an impeller. As the instrument is pulled up the well, across the flowing reservoir interval, the impeller spins ever faster in response to the flow into the wellbore. If particular intervals are not contributing to flow, then the rate of spin does not increase as the tool passes across them. In an undamaged well, with all zones contributing, there will be a close match between a composite permeability height plot for the interval and spinner results. The radioactive devices depend upon the injection of a slug of radioactive material into the wellbore. The radioactivity is then measured across the reservoir interval and the results may be interpreted in terms of flowing and nonflowing intervals. Given the obvious environmental and health issues of such a process, the application of such technology is commonly limited to injection wells.

The capacitance and density methods allow identification not just of petroleum versus water but of where the different fluids are entering the wellbore. In a geoscience context, such information will enable improved reservoir description, the application of which may help to avoid water inflow in subsequent wells, as problem intervals need not be perforated.

In much the same fashion that mechanical spinner tools can be used to measure fluid flow, so too can temperature response tools (Fig. 6.9). For example, in an injection well, cold injection water is introduced into a hot formation. If a particular horizon is accepting most of the injected water, then it will be cold relative to those horizons in which injection is limited. In much the same way, temperature data in a shut-in well can help to identify cross-flow between formations.

Pressure data from wells are used to define local and field-average reservoir pressures. Such data, when combined with production information on petroleum and water, and with fluid and rock property data, can be used to calculate the petroleum in place and the expected recovery factor. Unlike measurements based upon core or well logs, which only yield direct data from within or close to the wellbore, the depth of investigation of time series pressure data can encompass an entire field or field segment.

Repeat formation tests (RFT^{TM}s) can be made in flowing wells in addition to static wells (Chapter 2). Data from such tests can be used to indicate the boundaries to individual flow units, barriers to flow, and compartments in the reservoir. The interpreted data from the RFT^{TM}s can then be used to aid field development design.

Pressure data from wells, rather than individual formations, may be obtained in a variety of situations. Data can be collected during constant production from a well (pressure drawdown), while a well is shut in (pressure buildup), during multiple-rate flow periods, in one or more shut-in wells while producing from another, in injection wells, and in drill stem tests (DST). The essential features of all of these different situations in which pressure is measured is that a time series of data may be obtained. The temporal response of the field to a pressure disturbance is being measured. We will concern ourselves here with the description of such test data. The mathematical background is given in Matthews and Russell (1967).

Pressure buildup tests tend to be performed more often than drawdown tests. This is for several reasons; under buildup the well is responding only to the natural restoration of pressure and is not under the influence of "man-induced" flow, and therefore the more subtle pressure changes are likely to be captured. Secondly, a combination of legislation and much greater awareness of environmental issues means that waste emission during testing of exploration and appraisal wells (flaring) is minimized. The consequence of this is that wells tend to be flow-tested for short periods and then the pressure is allowed to build up over a few days.

Whether during drawdown or buildup, the pressure change through time can be divided into three periods, characterized by different pressure change behavior. In the period immediately following the pressure disturbance (test), the wellbore pressure is unaffected by the drainage boundaries of the well and the pressure buildup behaves as if it were occurring in an infinite system. This interval of time is often referred to as the "transient period." At some time later, the influence of the nearest boundary is seen in the wellbore pressure response. Other boundaries may then be observed during this so-called "late transient period." Finally, the pressure buildup (or drawdown) reaches steady state (the rate of change of reservoir pressure is constant with time) once stabilized flow conditions have been established.

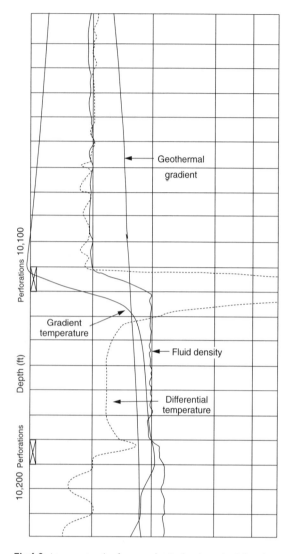

Fig.6.9 A temperature log from a production logging suite. Below the perforations ($\underline{X}$) at 10,115 ft there is a column of water in the well, extending to 10,500 ft. Except for some minor water production at 10,115 ft, the water column is inactive; therefore the portion of the temperature log below 10,300 ft has a slope like that of the geothermal gradient. Both the temperature and fluid density decrease significantly at 10,115 ft, which indicates substantial gas production at that depth. (From Western Atlas 1982.)

The change from transient to steady state depends in particular upon reservoir geometry, porosity, and permeability. Ideal behavior (Fig. 6.10) is rarely achieved, and a number of techniques have been developed for analysis of the data in the absence of constraining geologic information (Fig. 6.11). Naturally, the corollary is that when combined with geologic data (fault distributions, reservoir pinchout maps, and fluid composition data), the pressure data provides a very powerful tool for reservoir description.

Most tests are performed at a constant production rate. However, it is possible to derive much of the same data by performing tests on a well that is subject to multiple-rate flow periods. This is particularly common on gas wells, which may require such testing to satisfy regulatory bodies. This method is also used on wells where a shut-in prior to a drawdown test or for a buildup test may not be economically desirable.

One further category of well test, the interference test, deserves particular mention. In a single well test, although the pressure buildup information may be highly accurate, it is commonly difficult to interpret the data in geologic terms because it is scalar: the distance to boundaries may be observed, but the direction to those boundaries is not known. Even when high-quality geologic data are available, it may not be easy to identify the likely geologic features responsible for particular effects on the pressure buildup. The value of time series pressure data can be maximized in interference tests. In such tests, a well is put on production or subject to injection while one or more surrounding wells are shut in, with pressure gages installed. The effects of the production or injection can be recorded in the surrounding wells. Clearly, the response in these wells will be delayed relative to initiation of flow in the production/injection well. Factors that will influence the response time and response effect are distance, permeability, connected porosity, and directional reservoir flow patterns. That is, when combined with the geologic data, it should be possible to describe the location and nature of baffles and barriers to fluid flow and the permeability anisotropy of the reservoir of the system.

Where pressure data are not available, it may be possible to use the production history of wells to effect a similar analysis of reservoir performance behavior. The technique is simply one of observing the effects on existing wells when a new well is brought on stream. The method is particularly useful when applied to old fields with many closely spaced wells. Often in such situations, the production data and well histories are the only recorded data for the field. Such production data are a valuable tool for the geoscientist.

The petroleum production rate obtained during a well test is an important indication of how the well, and indeed the field, will behave when on production. An important derivative of a production test is the productivity index (PI). This is because the actual production rate in a test varies as a function

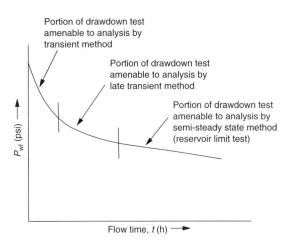

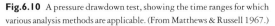

Fig.6.10 A pressure drawdown test, showing the time ranges for which various analysis methods are applicable. (From Matthews & Russell 1967.)

of the pressure drawdown imposed upon the reservoir as well as the intrinsic properties of the reservoir and its oil. The PI is thus a measure of the performance of the reservoir, normalized for the pressure difference between the reservoir and the well-head (drawdown):

$$PI = \frac{\text{rate of production}}{\text{drawdown}} = \frac{q}{\overline{P} - P_{wf}}$$

where q is the rate of production, in $m^2\,d^{-1}$ or $b\,d^{-1}$; $\overline{P}$ is the average reservoir pressure; and P_{wf} is the flowing bottom-hole pressure at the rate q.

The estimation of PI from static data is possible using permeability, reservoir thickness, and oil viscosity data (Case history 6.6). However, this tends to be only an approximation, since the PI is commonly rate dependent.

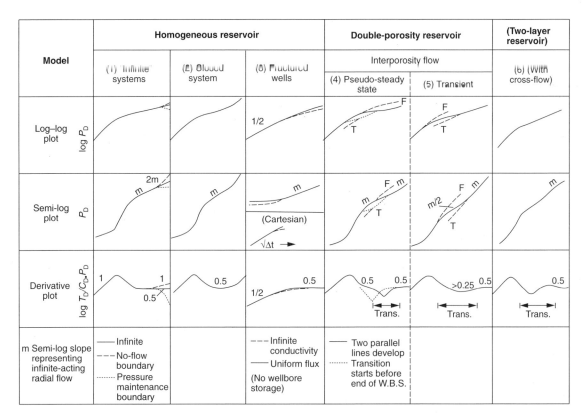

Fig.6.11 A summary of well/reservoir responses to testing in different reservoir systems. (From Archer & Wall 1986.)

6.3 RESERVOIR MANAGEMENT

6.3.1 Reservoir description from production data

Reservoir description from production data is an essential part of maintaining production from a field on plateau and extending the life of a field in decline. There are few things about which there can be certainty when producing a field, and one of those certainties is that the field will not perform as the development plan suggested it would! The main factors of interest to the production geoscientist are establishing what petroleum has been produced from where, and how both petroleum and water reach the production wellbore. These two pieces of knowledge will allow the geoscientist, working with the reservoir engineer, production engineer, and drilling engineer, to target unswept oil and possibly reduce the unwanted water flowing into the production wells. Moreover, during production, information arrives very quickly and in large quantities. Most fields have many tens or even hundreds of wells and there may also be many layers within the reservoir. The production geoscientist needs to be able to capture and assimilate the data and then use it to guide further production of the field. In consequence, graphic and pictorial methods are often used to synthesize the mass of data into something understandable. In fields with many tens, hundreds, or even thousands of wells, it will not be possible to understand the full details of field performance without a lot of work, which takes a lot of time. In such circumstances, time is often short and as a consequence it is commonly necessary to study just the anomalous wells, those that either over-perform or under-perform. In doing so, it may be possible to understand about 80% of the field performance characteristics by working on about 20% of the wells.

A very simple analysis of production data is to plot produced volumes on a map of the field. Bubble plots (Fig. 6.12) can be constructed simply, using the proportionality between the area or radius of the bubble and the produced volume. A slightly more sophisticated method would be to overlay the bubble plot on a contoured petroleum pore-volume map for the field. In a highly layered reservoir in which individual layers are completed sequentially, it may be appropriate to produce a map for each layer. Derivatives of the same basic method would be to plot cumulative water volume, time to water breakthrough, or changing gas to oil ratio. From any and all of these maps, it will be possible to visualize how the field is behaving through time.

The spatially mapped production data can be used to identify areas of poor recovery. By comparison of the production data with the static reservoir description, it could be concluded that such areas are poor because the intrinsic oil volume is low (a thin reservoir or low porosity) or because of poor production characteristics. The course of subsequent action could then be either to do no more if there is insufficient petroleum to merit further production, or to attempt to stimulate the poor production. Another possibility is to examine the recovery factor of areas with high cumulative production. These areas could form targets for new wells. This approach was used in the reactivation of the Pedernales Field in eastern Venezuela (Gluyas et al. 1996). The best-producing wells in Pedernales had delivered 5–8 million barrels of oil, and yet a simple calculation on well spacing and total reservoired volume indicated that even in the field segments penetrated by the high-productivity wells, recovery was still below 10%. The old wells were not usable, but new infill wells in the same field segments gave excellent performance.

Information on the performance of specific horizons in a well can be obtained from production logging (Section 6.2.7) or alternatively on drilling new wells. Wells drilled after start-up of a field may show modified petroleum and pressure distributions (Fig. 6.13). If the well cuts a petroleum/water contact, the contact might have moved (upward) compared with the original position in the virgin reservoir. There may also be intervals above the petroleum/water contact that have become water bearing, with only residual petroleum saturation as a result of petroleum production. Pressure profiles in the new wells may indicate horizons that are more or less depleted than the condition of the field before production began. These data can all be used to help construct a more robust description of the reservoir architecture than is possible from static (geologic) data alone. The utility of pressure test analysis in wells as a reservoir description tool has already been covered in the previous section.

Analysis of produced waters can also provide insight into the way in which the field's aquifer behaves during production and the permeability anisotropy of the reservoir. Bulk fluid chemistry can be of some use, although methods using either injection of tracer chemicals or naturally occurring stable isotopes of hydrogen and oxygen tend to produce less equivocal data, since they are unaffected by water–rock interaction. Coleman (1998) reported on water production data from the giant Forties Field (North Sea). Unfortunately, no isotope measurements were made before field start-up. Nonetheless, Coleman (op. cit.) was able to demonstrate that there were two different naturally occurring waters in the oilfield, produced in different places from different layers and at different times. In addition to these natural waters, there was

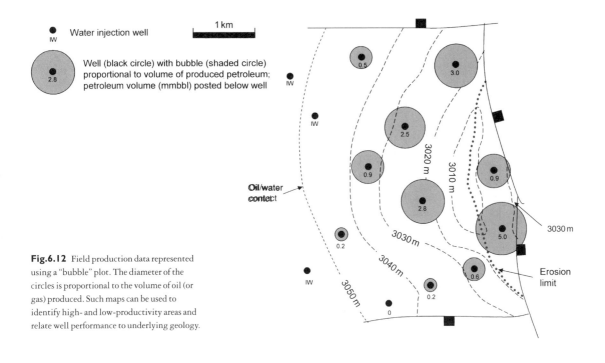

Fig.6.12 Field production data represented using a "bubble" plot. The diameter of the circles is proportional to the volume of oil (or gas) produced. Such maps can be used to identify high- and low-productivity areas and relate well performance to underlying geology.

also the injection water that eventually broke through to the production wells. The use of all of this data helped to improve the reservoir description, and in so doing helped to extend production from the field.

6.3.2 Reservoir visualization

It is important that the geoscientists and those working in other disciplines share a common understanding of the petroleum accumulation under development and production. Most fields are complex, be that complexity in structure, reservoir architecture, or fluid flow behavior. Reservoir visualization techniques provide an opportunity for all of the technical and managerial disciplines to share a common understanding of the field. Most of the visualization techniques that we report in the following paragraphs have already been mentioned earlier in the book, usually in the context of the way in which they are used as part of the geoscience component of exploration, appraisal, and development. Here, we will concentrate on the visualization characteristics.

For technical and managerial staff without specific geo-

science training, it can be very difficult to create a mental picture of a reservoir from scattered seismic lines and a couple of well logs. Yet, clearly, such interpretations are part of the geoscientist's skill base. Three-dimensional seismic displays, geocellular models, virtual reality simulations, and outcrop analog data can all help to share information between disciplines because of their highly visual nature.

Readily available software now allows 3D seismic data to be displayed as a data cube (Fig. 6.14). Such data cubes can be rotated in any desired direction, sliced in any desired direction, and surfaces or individual seismic lines peeled away. Short movie sequences can be constructed to view the cube of data as surfaces (vertical or horizontal) are stripped away. Data cubes can be made to be semitransparent, with perhaps only peaks and troughs appearing as opaque voxels. When displayed in this way, it is possible to look inside the data volume. All of these types of seismic data display allow trap configurations and structural elements to be appreciated.

The same sort of image manipulation methods as applied to seismic data may be applied to geocellular models (Section 5.8.3). Geocellular models can in themselves be particularly

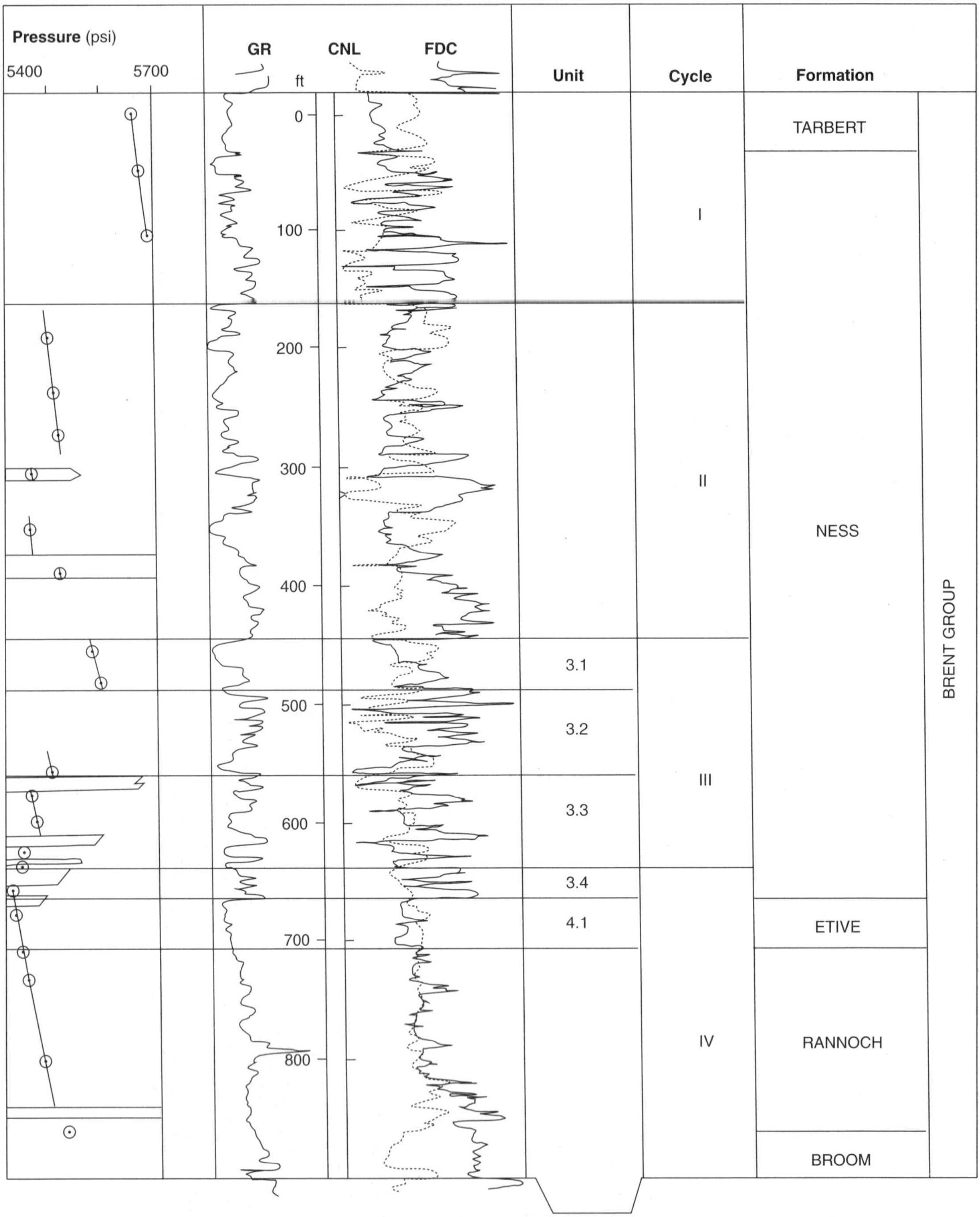

Fig.6.13 Well logs and pressure data from a well in the Brent Field, UK continental shelf, to show the subdivision of the Brent reservoir into geologic formations and petroleum engineers' "cycles" and "units." The RFT[TM] pressure measurements show pressure differences between and within cycles, which have developed during production as a consequence of vertical pressure barriers of baffles within the sequence. (From Bryant & Livera 1991.)

powerful images of a reservoir. Object modeling is an especially strong tool for conveying a 3D image of a reservoir (Fig. 6.15). With such tools, it is possible to visualize pay distribution within a reservoir or connected volumes of reservoirs (geobodies). On these same models it is often possible to display well paths, and possibly well-log data. Indeed, 3D seismic visualizations or geocellular models can form a common basis on which geoscientists and drilling engineers can plan well locations and well paths.

Seismic work and geocellular models are commonly displayed on workstations that are, typically, about the size of a television. However, it is possible to display any of this visual data on a much larger "person-sized" scale. This can be through simple projection or, alternatively, virtual reality technology can be used. It is of major benefit to be able to walk around inside a virtual reservoir, displayed as either its seismic response or as some sort of geocellular model (Fig. 6.16). The full aspect of the data can be explored from different angles; moreover, supplementary data can be displayed using audio signals or even tactile responses. Such technology is currently and mostly to enable sharing of information between geoscientists and engineers, rather than to be used as an interpretation medium by geoscientists.

Curiously, perhaps the most powerful reservoir visualization remains the outcrop. Suitable analog-outcrop visits form an important part of creating a shared vision of the Earth between the various disciplines (Fig. 6.17). They can be examined on a variety of scales, from cliff face (seismic scale) to sand grain (microscopic scale).

6.3.3 Time-lapse seismic

Repeat seismic surveys, sometimes called 4D or time-lapse seismic, are currently being developed as a methodology to examine changes in the reservoir and fluids during production. The premise is that production of petroleum and possibly injection of cold water will alter the acoustic properties of the rock plus fluid. An examination of the difference in seis-

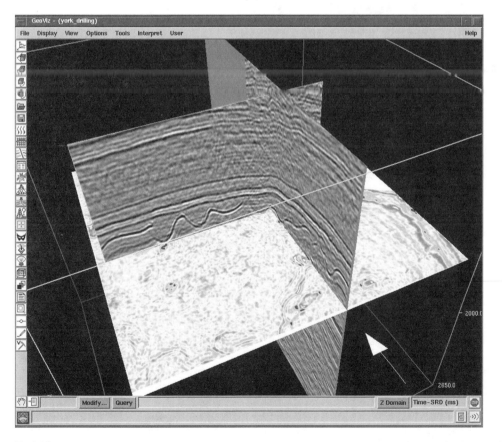

Fig.6.14 A 3D seismic data cube, showing two orthogonal cross-sections and a time slice. (Image supplied by Schlumberger, Gatwick, UK.)

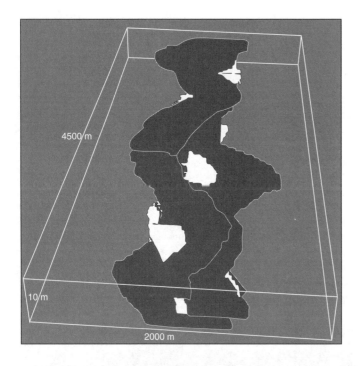

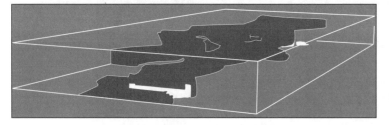

Fig.6.15 An object model of a channel belt consisting of four channels. (From Tyler et al. 1994.)

mic response between surveys conducted at different times can then highlight these changes. In particular, the properties that might be expected to change during petroleum extraction include pore pressure, pore fluids (saturation, viscosity, compressibility, and fluid type), and temperature. There is also some evidence to suggest that mineralogical changes (precipitation) can also take place during petroleum production.

Reservoir pore pressure can drop during petroleum extraction, particularly around the wellbore. It can also increase around injection wells. Barriers to fluid flow can cause pore-pressure discontinuities either laterally at the same depth or vertically at the same locality.

Naturally, reservoir fluids change during production and injection. The fluid properties of light oil and gas can be particularly sensitive to pressure changes. If the sonic properties of a reservoir are changed sufficiently by injection or by pressure loss (such that the petroleum fluid drops below the bubble point), it might be possible to follow water-flood fronts and identify areas of bypassed oil (Jack 1998; Koster et al. 2000; Fig. 6.18).

Temperature changes induced by injection of cold water, hot steam, or through *in situ* combustion can lead to changes in both the acoustic properties of the fluids (viscosity changes) and the rock (possible fracturing).

Secondary effects of production include changes in the state of compaction of the rock, porosity reduction, and hence increase in bulk density and changes in the overburden stress.

Fig.6.16 Seismic data can be viewed and interpreted within a visionarium. The images may be projected flat or in 3D (using suitable headgear or glasses) at a "human" scale. Where 3D imaging is used, it is possible to walk around inside the seismic cube. (Image supplied by Schlumberger, Gatwick, UK.)

Fig.6.17 A virtual walk in an outcrop, Miocene sandstones, Gulf of Suez (Egypt). The sandstones are exposed in an arid desiccated landscape cut by numerous small wadis. Erosion has been such that it is possible to examine many sections of different orientations with individual outcrops at scales similar to those of individual cells within a reservoir model. (Photograph by J.G. Gluyas.)

These changes may also be of sufficient magnitude to be detected through repeat seismic surveys. Thus, in the optimum case, it may be possible to detect directly spatial changes in the reservoir that can be used to monitor field exploitation. Even with the most sophisticated of reservoir models and well-test data, it is only possible to infer the internal characteristics of reservoir performance.

Although the promise of 4D seismic is great, the reality is much less promising. Seismic data obtained today is often quite different from data acquired in years gone by. Not only has seismic quality improved dramatically in recent years, but it is also difficult to ensure that both the acquisition and processing parameters are the same for the various surveys. The undesirable factors that can change with time include ambient noise, environmental changes (buildings, rigs, new road cuts, and major developments such as airports), and natural cycles of deposition and erosion such as those seen in modern fluviodeltaic systems.

6.3.4 Managing decline and abandonment

At plateau, the oil production rate of a field is controlled by design. The limiting factor will be some aspect of the facilities above ground, and not the reservoir or its fluids. The plateau production of a field may be measured in months, years, or tens of years, but sooner or later oil production will

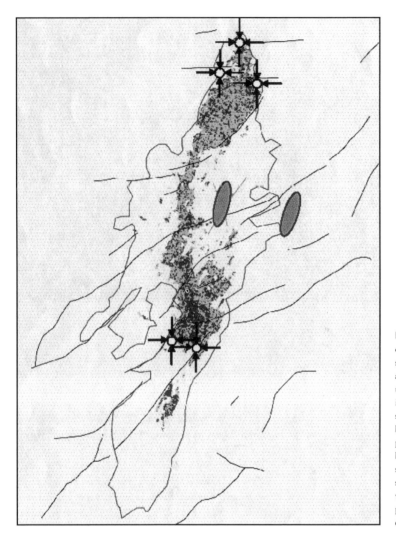

Fig.6.18 A map of the seismic amplitude difference between the initial 3D seismic survey and the second similar survey acquired after field start-up and after water injection in the northern and southern parts of the reservoir in the Draugen Field, Norwegian continental shelf. The difference between the surveys can be explained in terms of changes in the acoustic properties of the reservoir interval following both injection of water at the northern and southern ends of the field (shown by darker shade) and the natural aquifer influx from the west. The ovals are areas of no data beneath the platform and production facility. (From Koster et al. 2000.)

naturally drop below the plateau level. At this point, the field goes into decline. There are a variety of reasons as to why oil production may decline, not all of which are indicative of the failing potential of the reservoir to deliver oil. Nonetheless, whatever the reason for the declining oil production, one important aspect of the field remains the same. Without intervention, the operating cost of a facility will remain much the same whether it is producing oil at plateau rate or at some lesser rate. Indeed, if a declining oil flow rate is accompanied by a large increase in the gas and/or water flow rates, then the OPEX can increase. When the income from producing oil drops below the cost of operating the facility, the oilfield

ceases to make a profit. It will then be abandoned. Of course, in practice things are not quite so simple, as there will be a cost associated with abandonment (removing the facility and cleaning the site). However, the cost of abandonment will have been factored into the overall plan for the field, and the operating partnership will know what oil price/oil rate combination will make the field uneconomic. The reservoir model (Chapter 5) and history matches performed during the plateau period will also have given the operating company a reasonable idea of when the field is likely to be abandoned. The key point from the above discussion, as far as the geo-scientist and the reservoir engineer are concerned, is that at abandonment the field is still capable of producing oil; it is just not economic to do so. Only in exceptional circumstances will production be continued until oil production drops to zero.

There are two important aspects to managing a field in decline. The first is cost reduction. If the operating cost for the facility can be reduced, then it follows that the point at which it becomes uneconomic to produce the field will be shifted to ever lower oil rates. In an efficient operation, the facilities and production engineers will devise ever more ingenious ways of reducing the OPEX while maintaining a safe production facility.

At the same time as the OPEX is being reduced, the job of the geoscientist, petrophysicist, and reservoir engineer is to arrest or slow the decline in petroleum production. There are a number of ways in which this can be done, and most have been mentioned in earlier chapters or earlier sections of this chapter. We will address some of these possibilities shortly, but before doing so we need to consider the cost of maintaining production. We have already said in the previous paragraph that the objective during decline is the reduction of the OPEX. However, if the geologist or other subsurface technologists are to deliver more oil faster, then there will be a cost associated with that. For example, new wells may be needed to target unswept oil. The cost of the well can be regarded as capital expenditure (CAPEX) rather than OPEX but, nonetheless, the cost of drilling the well has to be rewarded by oil production that is sufficient to pay for the well and make a profit. The same rigorous economic criteria will be applied to the cost of collecting new data such as cores, or performing studies. The case history on the Heather Field (Case history 6.8) at the end of this chapter is an example of managing a field in decline.

6.4 RESERVES REVISIONS, ADDITIONS, AND FIELD REACTIVATION

6.4.1 Introduction

Reserves revision is an increase or decrease of reserves without change in the perceived oil in place. In other words, the reserves are won as a result of improving the recovery factor. Reserves additions result from the finding of new oil; that is, oil not considered as part of the original development. Field rehabilitation is a process whereby an oil- or gasfield is revamped, leading to a significant increase in production and possibly reserves. Considerable investment is made to improve the production rate and possibly increase the reserves. The reserves growth can be achieved through either revisions and or additions. We choose to limit the term "reactivation" to describe the situation in which a field is rehabilitated following its former abandonment.

6.4.2 Reserves revisions

Upward revision of reserves can be achieved in a number of ways. In the Heather Field, much of the increase was achieved through a combination of OPEX reduction and identification of areas within the field from which oil had not been swept. The effect of OPEX reduction is to allow the field to produce economically at a diminished oil production rate. In consequence, the field lasts longer and the recovery factor increases.

The integration of production data and the reservoir description information derived from either the geologic model or the reservoir model allows analysis of the relatively productive and relatively nonproductive parts of the field. This should allow identification of areas from which the oil has not been swept. Such areas may then be developed into targets for infill wells. For example, much of the late field life exploitation of the giant Forties Field (North Sea) has been concerned with the identification of unswept oil. In this instance, much of the remaining oil in the field occurs in inter-channel areas within the Paleocene submarine-fan complex that forms the reservoir.

6.4.3 Reserves additions

Reserves additions come from pools that were not originally considered to be part of the field development. Such reserves additions may come from small satellite fields close to the major development. Alternatively, secondary horizons, either above or below the main reservoir, may be produced. The quest for reserves additions commonly leads to a small phase of exploration in the areas surrounding the field. The key aspects of such exploration are that the new pools must either be within tie-back distance or, alternatively, it must be possible to reach them by extended-reach drilling from existing facilities.

In introducing this section, it was emphasized how the cost of any new work would come under rigorous scrutiny. In order to sell an idea for a new well or a new set of perforations in an existing well to management, it will be necessary to demonstrate the value of that work. A number of measures, or "metrics" as they are commonly called, can be used. Typical measures can be as follows:

- The time to payback;
- the production cost per unit of petroleum (barrel of oil or million standard cubic feet of gas);
- the income per day.

In order to work out these figures, it will be necessary to calculate the following parameters:

- The cost of the work;
- the likely production rate from the new interval;
- the likely production volume from the new interval.

The drilling, production, and facilities engineers can supply the cost components for these sums. The petroleum rate data and volume data come, of course, from the geoscientists and reservoir engineers. Late in the life of a field, it should be possible to make fairly accurate estimates of the production rate and volume for a new well or perforated interval on the basis of experience in the field to date.

Table 6.2 is an example of new well and new interval perforation opportunities for a gasfield in Asia. The field is mature and close to designed plateau production. The owners wished to extend the life of the field, and in consequence a small team of geoscientists, reservoir engineers, and a petrophysiscist investigated the possibilities for new wells, sidetrack wells, and perforation of hitherto unproduced sandstones in existing wellbores. The background information is that the field produces principally from the E and D sandstones. The stratigraphically higher F and G sandstones are commonly thin and often of poor reservoir quality. Although most of these intervals are gas bearing, they had not been included in the original development plan for the field. The stratigraphically deeper C, B, and A sandstones, though of moderate to high quality, were also not included in the development plan because in many instances they are below the gas/water contact. A combination of core data, log data, and experience from existing wells allowed estimates to be made of likely production rates for the various sands across the field, whether the estimate was for new wells or new perforations in existing wells. Similarly, the production history to date allowed calculation of the likely volume to be delivered from a well over its history. Thus it was possible to rank the opportunities in terms of cost, ultimate reserves, or time to pay back the cost of recompletion.

In existing wells, the choice of sands that could be perforated rested on a straightforward analysis to identify which of the sands had not been perforated already. These were listed along with the estimates of production rate and volume. The cost involved in recompleting the wells could generally be divided into two categories, depending upon the arrangement of the "production jewelry" in the wells. For some wells, it would be a simple matter of perforating through the casing string (cost c. \$0.1 million), while in others the tubing would require resetting (cost c. \$1.5 million). New wells were estimated to be capable of delivering 50 bcf on the basis of analog data from the existing wells. The cost of such wells, including tie-back of the well into the production facility, was estimated to be about \$5 million. The total of all the available reserves additions for the field was in excess of 100 bcf at negligible risk.

6.4.4 Field rehabilitation and reactivation

Rehabilitation of an oilfield is the process whereby production is improved in late field life. Reactivation involves much the same processes, but applied to a field that has been shut in or abandoned. Oilfield rehabilitation and reactivation became important parts of many companies' business during the latter part of the 1990s. There are several reasons why companies have chosen to reactivate and rehabilitate fields. These include paucity of high-quality exploration acreage and expectations of early production, and hence early cash flow, from such activities. However, such drivers are not new. What is new is that political changes have created opportunities in countries that were previously unavailable to Western companies. Areas with long production histories, such as those in the former Soviet Union, South America, and parts of the Middle East (Iran and Kuwait) are now open for investment. Many of the fields in these areas are old, have been developed without the benefit of modern technology, and have lacked reinvestment as they approached senility. However, despite the appearance of old age, recovery in many of these fields has been particularly low. Perhaps only 10% of the original oil in place has been produced. A similar field developed today might reasonably be expected to deliver 30–40% of its oil in place, perhaps more. So, while it may not be possible to replicate a new field development, it may be possible to dramatically improve upon the initial recovery of the field.

Field rehabilitation and reactivation usually involve a change in operatorship. The new operator is likely to want to redescribe the field as well as recomplete existing wells and perhaps drill new wells. There is often a tendency for the new operator to believe that reactivation will simply involve the application of new technology. Old data from the field are commonly difficult to find and are often ignored. This is invariably a mistake. The old reservoir data may not be familiar to the modern geoscientist, but they are rarely without value. Moreover, the development practices developed on a 50-year-old field may appear quaint, but more often than not they are derived from a substantial quantity of empirical experiences. We, the authors, have been involved in field reactivations

Table 6.2 New well, sidetrack, and perforation options for an Asian gasfield.

Well/sand	Activity	Activity cost ($ million)	Expected flow rate (mcf per day)	Gas volume (bcf)	Gas cost ($ per mmscf)	Cash flow ($ thousand per day)	Time to payout (days)
7/G	Perforate	0.5	5–10	5	100	10–20	80
9/G	Perforate	0.1	5	<5	20	10	10
9/F	Perforate	1.5	20	10+	150	40	40
4/E	Sidetrack	5–8	30	15–40	125–530	60	80–130
4/C	Perforate	0.1	10	1	100	20	5
5/E	Perforate	3–4	20–30	1–6	500–4000	40–60	50–100
5/F	Perforate	0.1	10	5–20	20–100	20	5
8/G,D,E	Perforate	5	5–20	5–20	250–1000	10–40	125–500
6/D	Perforate	0.1	5–10	2–5	10–20	10–20	5–10
6/G	Perforate	0.5	5	5	100	10	50
1/F,G,C,D	Perforate	1.0	30	10–20	50–100	60	16
3/G	Perforate	0.5	5	5	100	10	50
3/all	Sidetrack	5–8	25	5–15	300–1600	50	100–160
10/D	Perforate	0.1	5	1–3	33–100	10	10
11/E	New well	8.0	30	30	2.3	60	150

across three continents. In each instance, the old geologic and production data have proved to be important for the renewed development of the field. The potential for reactivation of a North Sea field is discussed in the Crawford case history, and the opportunity for rehabilitation of a whole country's onshore petroleum production is examined in the final case history in the book, on Trinidad.

6.5 CASE HISTORY: WYTCH FARM FIELD, UK ONSHORE

6.5.1 Introduction

The Wytch Farm oilfield lies beneath the coast of Dorset in southern England, extending offshore under Poole Bay (Fig. 6.19). By 1996 the reserves estimate was revised to 365 mm bbl, making it Europe's largest onshore oilfield (Hogg et al. 1996). The first pay to be discovered was the Middle Jurassic Bridport Sandstone at a depth of about 900 m, in 1974. The discovery well was drilled onshore. Subsequent appraisal led to the development of an onshore terminal capable of exporting 6000 bopd by rail (Stage I). The ultimate reserves for this Bridport pay are believed to be 35 mmbbl. The pool of oil in the Bridport Sandstone does

not extend offshore and therefore all of this production will come from onshore wells.

A second, deeper pay was discovered in Triassic Sherwood Sandstone in 1977 (Fig. 6.20). Although the first well to penetrate the Sherwood Sandstone was drilled onshore, the field was subsequently proven to have a significant offshore extension. Moreover, the initial oil/water contact at about 1500 m was common to both the onshore and offshore portions of the field. Development of the onshore portion of the Sherwood Sandstone reservoir (Stage II) began in 1988 during appraisal of the offshore portion. Thirty-two production wells and 12 water injection wells were drilled from nine onshore sites. The gathering station was built to handle 60,000 bopd. Thus, by 1990, production facilities were in place to recover the onshore reserves; that is, about two-thirds of the estimated total. This case history tells the story behind the development of the offshore portion of the field (Stage III).

6.5.2 Offshore development: the environmental issue

Appraisal of the offshore portion of the Sherwood oil pool proved that the field extended up to 10 km offshore and to the east of the landward portion of the field (Fig. 6.19). In order to develop Wytch Farm Stage III, the original conceptual design was to build an artificial island in Poole Bay. At the time,

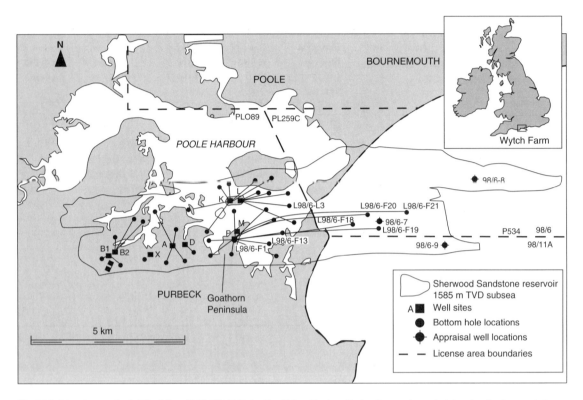

Fig.6.19 A location map for the Wytch Farm Field, UK. Drill sites F and M, on Goathorn Peninsula, were the top-hole locations for the extended-reach wells used to develop the offshore portion of the Sherwood reservoir oil pool. (From Hogg et al. 1996.)

it was realized that such a decision was going to be controversial, both from the ecological and the leisure amenity perspective. The areas around Poole Harbour and Poole Bay:

• Are one of Britain's most popular vacation spots;
• are home to all six of Britain's native reptile species, internationally protected migrating birds, and oyster beds;
• contain some of England's most valued and environmentally sensitive habitats, including heathlands, forests, salt marshes, and tidal mud flats.

By 1990 two things had happened which, when combined, were sufficient to cause the cancellation of any proposal to build an artificial island. The operators decided that the environmental cost was likely to be greater than the value of oil within the offshore portion of the field. In addition, drilling technology had advanced to the point at which a 5 km, sub-horizontal, extended-reach well was considered feasible. If drilled onshore from Goathorn Peninsula (Fig. 6.19), such wells could win 90% of the estimated offshore reserves without detriment to the commerciality of the project. The decision was taken; Wytch Farm Stage III would be developed by

extended-reach wells drilled from onshore to offshore (Fig. 6.21). All such wells would require substantial horizontal or near-horizontal sections.

6.5.3 Geologic constraints on Stage III development

The decision to produce the Wytch Farm offshore reserves using extended well technology may have meant that the drilling engineers were gleeful at the prospect of displaying their skills, but that decision worried the geoscientists and reservoir engineers working the project. Their concerns were threefold:

• Was it possible to guarantee hitting the Sherwood target with the wells?
• Would the high-angle wells deliver the high petroleum flow rate required?
• Would the high-angle wells deliver the volume of petroleum required?

In this case history, we will investigate the well-site operational issues; that is, the way in which the development team

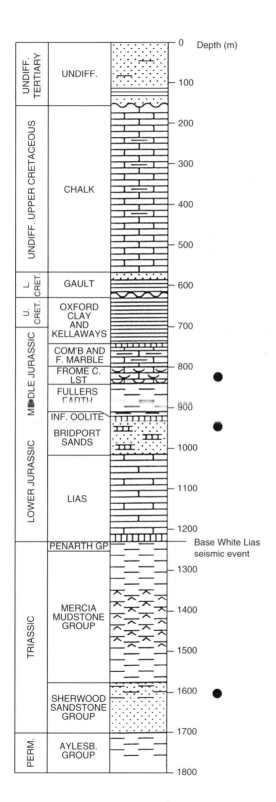

delivered the wells to target and to the desired production rate. The uncertainty associated with whether the wells could deliver large volumes of oil throughout their lives is reported by McClure et al. (1995).

6.5.4 Hitting the target

The first offshore, extended-reach well was planned to intersect the Sherwood reservoir at about 3.5 km east of the drill site on Goathorn Peninsula. At this point, the gross pay interval for the Sherwood Sandstone was estimated to be 50 m. In order to minimize coning of water into the wellbore from the underlying aquifer, the well had to be at least 20 m above the oil/water contact along the entire completed interval. Indeed, this "stand-off" of the well from the oil/water contact would need to be maintained for a distance of at least 300 m. To put these target figures into context, it needs to be appreciated that the estimated error on mapping the top reservoir was ±45 m. This large error was a composite of several factors:

- Only 2D seismic data were available over much of the area;
- the near-shore area had no seismic coverage (because acquisition of seismic data in the area would have caused environmental damage);
- the top reservoir was without distinct seismic response, but had been calculated using an isopach method, simply by adding a predetermined thickness to a distinct seismic marker in the overlying Mercia Mudstone (Fig. 6.20).

To add to these problems, there were drilling issues. The vertical location at the target site (3.5 km from the top-hole location) could only be estimated to ±11 m, by the drilling engineers while actually drilling the well. Moreover, in order to deliver a substantial horizontal section in the target Sherwood Sandstone, the borehole had to be steered through the overburden. The build angle for the well had to be clearly monitored and executed. The consequences of a wrong trajectory were clear. If the build angle was too low, the well would penetrate the Sherwood Sandstone and oil/water contact without delivering sufficient horizontal section. If the build angle was too high, a substantial, wasted horizontal section would be in the cap rock, the Mercia mudstones.

Fig.6.20 Stratigraphy in the Dorset area, UK. Oil has been found at three stratigraphic levels in the Wytch Farm Field. The Sherwood pool is the major accumulation and Bridport Sandstone contains the secondary accumulation. Small quantities of oil are also produced from the fractured Frome Clay Limestone. The key seismic marker, base White Lias, is also highlighted. (From McClure et al. 1995.)

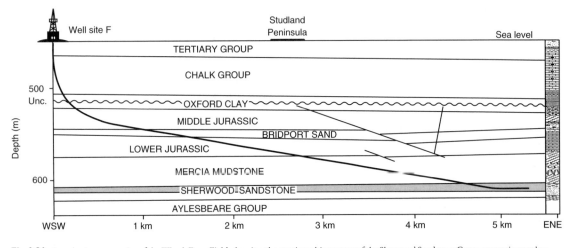

Fig.6.21 A geologic cross-section of the Wytch Farm Field, showing the stratigraphic context of the Sherwood Sandstone Group reservoir together with a genetic design for Wytch Farm extended-reach wells. (From Payne et al. 1994.)

The solution to both the geologic and the drilling issues was to steer the well through the overburden stratigraphy and into the reservoir. This process is commonly referred to as geosteering. The Mercia Mudstone was examined in detail to determine if lithological and grain-size variations had local (across the field) stratigraphic significance. The intention was to check the predicted depth against the actual depth each time a marker horizon was penetrated by the drill bit. In this way, the trajectory of the well could be adjusted continuously as the target was approached (Fig. 6.22). A variety of mudstone/anhydrite couplets and subtle grain-size variations could be recognized on gamma and sonic logs in cuttings and related to the rate of penetration (of the drill bit). All of these data types would be available while drilling the horizontal wells. Moreover, there were sufficient marker horizons at the level where the well angle was to be built from 70° to 85°. The utility of these markers proved themselves on the first well, where stratigraphic control was better than ±2 m, a considerable improvement on the ±45 m that might have been expected from use of the top Sherwood map alone. Geosteering of the wells through the Mercia Mudstone has continued to deliver the horizontal wells on target, even in the case of one well where top reservoir was encountered at a depth 40 m shallower than the prediction based on the top Sherwood map.

6.5.5 Delivering the productivity

The decision to drill extended-reach wells with horizontal sections in the reservoir also impacted the number of wells to be drilled in the Stage III development of Wytch Farm. Far fewer horizontal wells would be needed than there would have been vertical wells, because the oil production rate from the horizontal wells was expected to be much higher. In order to deliver the target volume of oil per day from the field, each horizontal well was set to deliver 10,000 bopd. This figure was increased to 15,000 bopd just before the first well was to be spudded. To fulfill this new objective, it was necessary to develop a method to quantify the accumulated millidarcy-feet (mD ft; that is, the flow potential) as each well was drilled.

There were few data available while drilling from which permeability could be calculated. These data were the logging-while-drilling (LWD) information and the cuttings returns. In order to determine whether such data could be used to calculate permeability while drilling, the petrography and petrophysics of the extant onshore wells was studied in detail. Sedimentological studies on core cut from these onshore wells showed substantial variations in reservoir properties (porosity 5–30%, permeability 0.01–1000 mD, and grain size very fine to coarse). It was also known that one of the LWD logs could supply a measure of sandstone porosity. In addition, there was the possibility that the grain size might be measured from the cuttings returns. It was considered that such data might allow calculation of permeability from cuttings, and that such a calculation could take place while the well was being drilled.

Before permeability could be measured during drilling, it was important to verify that:

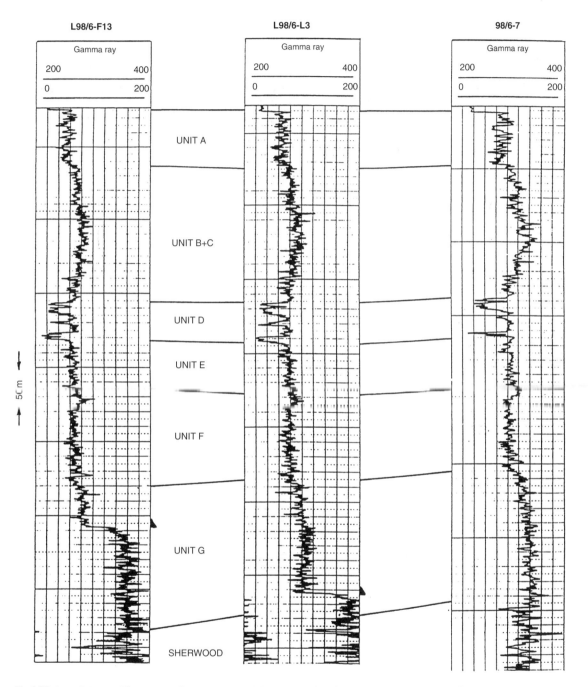

Fig.6.22 A correlation panel for the gamma log response through the Mercia Mudstone, Wytch Farm Field. The details of the correlation were used in conjunction with cuttings returns, during drilling, to determine the position of the drill bit within the overburden. (From McClure et al. 1995.)

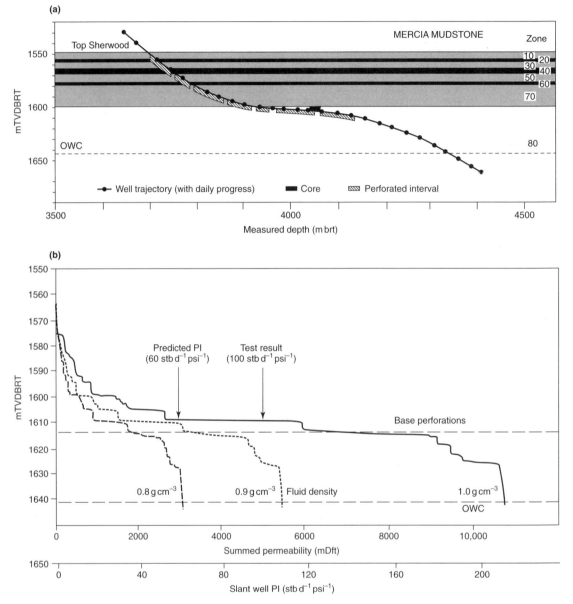

Fig.6.23 A well trajectory (a) and cumulative permeability prediction curves (b) for Wytch Farm well L98/6-F18. The tested productivity indexes are also shown.

- There was a strong correlation between grain size, porosity, and permeability in the Sherwood sandstones;
- the grain size of the sandstone could be measured accurately from the disaggregated cuttings returns;
- the porosity log could be calibrated with sufficient accuracy.

It was relatively easy, although somewhat time-consuming, to answer the first point. The Sherwood reservoir had been extensively cored during both the onshore and offshore appraisal programs. Porosity and permeability data were already available from these cores. These cores were then logged for sedimentology and grain size. There are distinct

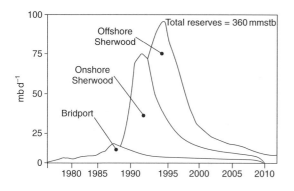

Fig.6.24 Production profiles for the three stages of Wytch Farm development. In 1996, one year after these curves were published, production broke the 100,000 bopd barrier. (From Harrison & Mitchell 1995.)

differences between the porosity/permeability relationships for different grain sizes, and similarities for similar grain sizes in different wells. The Sherwood sandstones contain little mineral cement and as such the coarser sands, with their larger pores and pore throats, were more permeable than the finer lithologies.

The relationship between grain size measured from core logging and from sieve analysis of cuttings returns was determined by examining both sets of data from equivalent depths in three calibration wells. The results showed that for most samples the mean particle size for the cuttings was comparable with the grain-size determinations made on core. There was no significant difference in the particle size distributions for washed and unwashed cuttings. In the few instances where there were discrepancies, the cuttings returns were always finer than those measured in core. Such differences were between one and two grain-size classes. Clearly, a bias toward calculation of a slightly low permeability compared with reality was not going to be an issue.

The determination of porosity while drilling was achieved using a density tool attached to the drillstring. There is clearly a very simple linear relationship between bulk density and porosity for a simple sand. However, the density tool measures the conditions down hole, where the situation is a complex function of both the fluid and the matrix (rock) densities. The average core grain density for the calibration wells was $2.64\,\mathrm{g\,cm^{-3}}$. This was taken as the matrix density. The fluid density was more difficult to establish. The drilling mud system to be used on the horizontal wells was different from that used on the older calibration wells. Calculations based upon

one of the calibration wells showed that the permeability value and productivity index (PI) derived through the transforms from the density log data were highly sensitive to the fluid density used. A variation in density of $\pm 0.1\,\mathrm{g\,cm^{-3}}$ could change the productivity index by a factor of two. The grain-size and porosity/permeability transforms were tested using data from the cored Sherwood reservoir appraisal wells before being used actively within the Stage III development drilling.

In order to make the permeability calculations while the well was drilling, well-site grain-size measurements were made from the cuttings returns throughout drilling of the reservoir sections. The data derived from sieve analysis were combined with those from the LWD tools. Data were presented as summation graphs of permeability × height and productivity index against true vertical depth (of the well) below rotary table. The results from the first well are shown in Fig. 6.23. In this well, together with the second and third wells, the productivity indexes as measured from the oil production were within the range predicted using the algorithms derived from the grain size and LWD data. For the second and third wells, the two estimates of PI were within 10% of that measured once the wells were brought into production (Hogg et al. 1996).

Geosteering of wells and calculation of PI while drilling were continued throughout the drilling campaign. These techniques, combined with continued success with delivering long-reach, high-rate wells, allowed daily production to exceed the 100,000 bopd day figure in 1996 and so improve on the original Stage III production estimates (Fig. 6.24).

6.5.6 Conclusions

The Stage III development of Wytch Farm has succeeded because drilling technology has improved enormously over the past decade. However, such technology would have been useless without the detailed and concise work of petroleum geoscientists and reservoir engineers, who developed methodologies for guiding the wells to target and then through the oil-bearing reservoir.

6.6 CASE HISTORY: AMETHYST FIELD, UK CONTINENTAL SHELF

6.6.1 Introduction

Gas production from the Amethyst Field (UK North Sea; Fig. 6.25) began in October 1990. About two years later, parts of the field began to show pressure declines, which could have indicated that the volume of gas in the trap was significantly smaller than had been calculated from the static (geologic) data before field start-up. Clearly, the subsurface team responsible for managing the field were eager to understand the origin of the discrepancy between the gas-in-place figure derived from mapping the structure and that calculated from the dynamic production data (Gluyas et al. 1997a).

In an effort to reduce the uncertainty associated with gas-in-place estimates, the subsurface team examined old appraisal and development well reports. It soon became clear that some of the wells had penetrated reservoir of very low quality, described as being cemented by sulfate minerals. These wells either failed to produce any gas or had very poor production rates. Two hypotheses were suggested in which the distribution of such cements could be used to explain the

two sets of gas-in-place figures. The first possibility was that the sulfate minerals wholly cemented large parts of the reservoir as originally mapped. The second possibility was that such sulfate cements were of much more limited distribution, but that they occurred as sheets that were barriers to gas flow. If the first hypothesis was correct, then the gas in place was indeed smaller than originally mapped. If the second possibility proved correct, then the gas in place was as mapped, but the wells drilled to that date had simply failed to access all the gas-bearing segments of the field. A project was initiated to examine the abundance, distribution, and origin of the sulfate cements.

6.6.2 The geologic background

The reservoir in the Amethyst Field is the Lower Permian Rotliegend Sandstone (Garland 1991). The Rotliegend

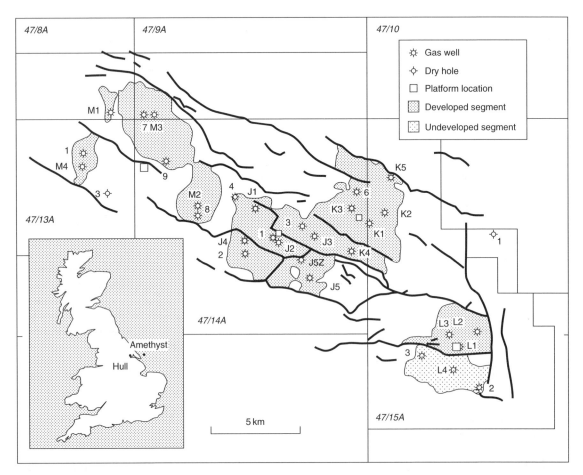

Fig.6.25 A location map for the Amethyst Field, UK continental shelf. (From Gluyas et al. 1997a.)

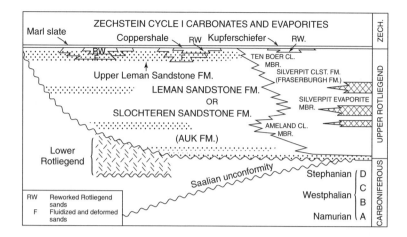

Fig.6.26 The summary stratigraphy of the southern North Sea. (After Glennie 1986.)

Sandstone together with associated mudstone (Fig. 6.26) accumulated in a large nonmarine basin, which occupied much of what is now northern Europe. Sandstone was deposited in aeolian and fluvial environments. The finergrained lithologies were deposited in sabkhas and lakes. In the Amethyst Field, the reservoir is between 15 m and 40 m thick (45–120 ft).

Catastrophic flooding that created the Zechstein Sea terminated Rotliegend sand deposition. The effects of the flooding can be seen at the top of the Rotliegend Sandstone in some of the Amethyst wells; the sands have been reworked and redeposited by mass-flow processes.

Evaporites precipitated from the Zechstein Sea during Late Permian times (Taylor 1986). The evaporite sequences comprise limestone and dolomite, followed by anhydrite, then halite, and finally polyhalite and sylvite. The main seal to the Rotliegend Sandstone reservoir in Amethyst is a combination of anhydrite and dolomite of the Zechsteinkalc and Werranhydrit formations. The Kupferschiefer Mudstone occurs locally, and where present is thin (Fig. 6.26).

The Amethyst trap is structural and low relief (Garland 1991). There are three main closures and a number of smaller ones, spread over an area of 32 km by 10 km (20 miles by 6.25 miles). Fault seal supplies a substantial part of the field closure. The elevation of the gas/water contact varies across the field. The gas/water contact (GWC) is about 21 m (70 ft) shallower in the eastern segments compared with the central and western segments.

The reservoir sandstones unconformably overlie Westphalian A and Namurian (Carboniferous) fluviodeltaic sedimentary rocks. Thermally mature coals within the Carboniferous interval delivered hydrocarbon gas to the Rotliegend reservoir (Garland 1991).

A representative burial history for the Rotliegend reservoir is shown in Fig. 6.27. Periods of rapid burial occurred in the Late Permian, Lias (Lower Jurassic), and Late Cretaceous. A minor phase of uplift probably occurred at the Jurassic/Cretaceous and Cretaceous/Tertiary boundaries. Most of the Tertiary stratigraphy is missing from over the field. It has been estimated, by comparison of the Tertiary stratigraphy above Amethyst with that preserved off structure, that about 500 m of Tertiary sediment was eroded across much of the field area (pers. comm. D. Richards, December 1992). In the past few million years there has been a resumption of burial.

6.6.3 Cement mineralogy and timing of diagenesis

Large quantities of anhydrite and gypsum were described in the reservoir sandstones of two old appraisal wells (47/15a-3 and 47/14a-5). Core from these two wells was sampled for reanalysis of the mineralogy and geochemistry of the cements. The reservoir in a further, new, production well, 47/15a-L02, was also sampled. Petrography was performed using three different, but complementary, methods; transmitted light optical microscopy, back-scattered electron microscopy, and cathodoluminescence microscopy. A scanning electron microscope was used for both the back-scattered and cathodoluminescence imaging. The electron microscope was also fitted with an energy-dispersive system to enable elemental analysis.

The petrographic analyses revealed three, not two, abundant and locally pervasive cements; anhydrite, barite, and ferroan dolomite. In the original work, the barite had been misidentified as gypsum. It was also clear from the petrographic analysis that these cements precipitated after all the

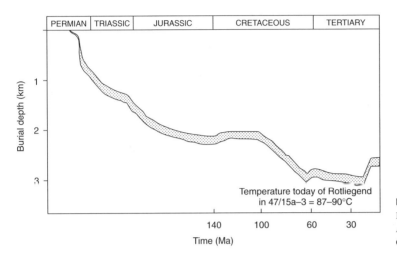

Fig.6.27 The burial history for the Rotliegend Sandstone of well 47/15a-3, Amethyst Field, UK continental shelf. (From Gluyas et al. 1997a.)

other diagenetic phases identified in the sandstone (microrhombic dolomite, kaolinite, illite, and quartz).

Quantitative data on the time or temperature of anhydrite, barite, or ferroan dolomite precipitation were not available. However, additional studies on the illite and quartz cements provided some constraint on when and where the three cements precipitated. The time of illite precipitation was dated directly using the potassium–argon method (Emery & Robinson 1993). The range of ages obtained was from 140 Ma to 100 Ma. The timing of quartz cementation was not measured directly, but indirect evidence pointed toward a Late Cretaceous (100–65 Ma) date for precipitation. Given that the anhydrite, barite, and ferroan dolomite cements precipitated after the illite and quartz, it seemed most probable that all three cements precipitated during the Tertiary; a time of inversion and uplift (Fig. 6.27).

6.6.4 The source of the cements

The three cements responsible for the poor reservoir performance contain calcium, magnesium, iron, barium, sulfate, and carbonate. The budget and time constraints of the project meant that it was not possible to try to determine the sources of all these components. Moreover, for some of the components (calcium and magnesium), there was no readily available technology to allow sources to be typed. However, it is possible to use the stable $^{34}S/^{32}S$ isotope ratio in sulfates to determine the origin of the cement. This is because the isotope ratio in primary sulfate deposits (evaporites) has varied through geologic time. The same situation exists for the oxygen isotope ratio in sulfates; but not in most minerals, where

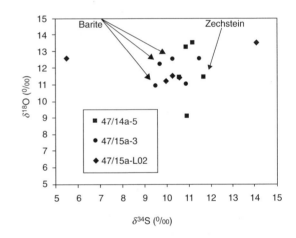

Fig.6.28 Sulfur and oxygen isotope compositions for the anhydrite and barite cements of the Rotliegend reservoir in the Amethyst Field. (From Gluyas et al. 1997a.)

it reflects the ambient water composition and temperature. This means that a combined $\delta^{34}S$, $\delta^{18}O$ analysis of the sulfate in both anhydrite and barite can give data that helps constrain the cement source.

The sulfur and oxygen isotope ratios for samples of both anhydrite and barite from the Amethyst Field reservoir are plotted in Fig. 6.28. On the same figure, the isotope ratios for a sample of Zechstein anhydrite (from well 47/14a-5) are also plotted. Most of the data form a tight cluster, with a 3‰ range in both $\delta^{18}O$ and $\delta^{34}S$. The ranges of both the sulfur and the oxygen isotope values are identical to those expected for

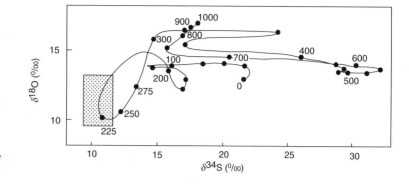

Fig.6.29 A secular curve for $\delta^{34}S$ and $\delta^{18}O$ variations in the world's oceans over the past one billion years (Claypool et al. 1980). The annotated points on the graph are ages in millions of years. The shaded area marks the range of data for the sulfate cements from the Amethyst Field. (From Gluyas et al. 1997a.)

Upper Permian evaporites as determined by Claypool et al. (1980; see also Fig. 6.29). The clear implication of these data is that the sulfate in the Lower Permian Rotliegend Sandstone was derived from the Upper Permian (Zechstein) evaporites.

Direct evidence for the source of barium in barite was not available. However, circumstantial evidence was taken to indicate that the Carboniferous sandstones and shales that underlie the Rotliegend are the most likely source. Barium sulfate (barite) is extremely insoluble. Any primary barium in the Zechstein is likely to be present as immobile barite. However, barium is liberated on dissolution of alkali feldspars and micas. Feldspar is uncommon in the Rotliegend Sandstone locally, but partly dissolved feldspars are abundant in the Carboniferous sandstones (Ramsbottom et al. 1974).

In summary, the mineralogical and geochemical evidence indicates a Zechstein-age source for the sulfate and possibly a Carboniferous-age source for the barium. Moreover, the anhydrite, barite, and ferroan dolomite all precipitated at some time in the Tertiary during basin inversion. No data are available on the timing of the dissolution events, in either the Zechstein or the Carboniferous, that yielded the solutes. Yet the combination of sources and the timing (syn-basin inversion) of the mineral precipitation event point toward a catastrophic mixing of different formation fluids.

6.6.5 The distribution of the cements

The distribution of the three cements was examined on two scales: within individual wells and across the field. The abundance and distribution of anhydrite, barite, and dolomite within individual wells could not be measured directly because too few petrographic thin sections were available. However, it was possible to make use of the high density of all three minerals to map the distribution in individual wells. Average

Table 6.3 Density and approximate mineral abundance data for the main mineral phases within the Rotliegend sandstones of the Amethyst Field. Density data are from Deer et al. (1966).

Mineral	Density (g cm^{-3})	Abundance range
Quartz	2.65	90%+ of detritus
Feldspar	2.55–2.63	<5% of detritus
Illite	2.6–2.9	A few percent; Garland (1991)
Kaolinite	2.61–2.68	A few percent; Garland (1991)
Dolomite/ferroan dolomite	2.86	Up to 20%
Anhydrite	2.9–3.0	Up to 20%
Barite	4.5	Up to 20%

grain density data are collected as a matter of routine during core analysis. The reservoir in Amethyst is extensively cored, and as such about 2000 sets of density and poroperm data were available. The density data for the three cements plus the common detrital minerals in the Rotliegend Sandstone are shown in Table 6.3.

The core analysis grain density data can be used to identify the location of the abundant barite, dolomite, and anhydrite within the stratigraphy of each well. However, there appears to be no consistent pattern of cement development as a function of stratigraphy. For example, the cements are not all located close to the overlying Zechstein unconformity or at the base of the reservoir, near the Carboniferous.

On the larger scale, the distribution of cements across the field is highly heterogeneous. Some wells contain up to 15% of late-diagenetic, pore-filling anhydrite, barite, or dolomite, while others contain none. As a broad generalization, the more southerly wells in the field tend to contain more of these

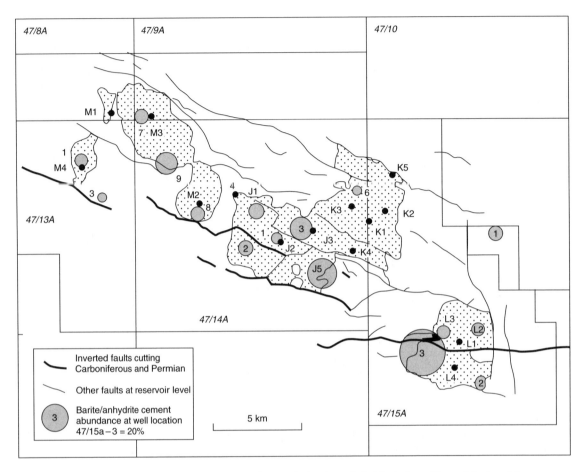

Fig.6.30 A map of the Amethyst Field, showing the relationship between the abundance of anhydrite, barite, and ferroan dolomite and the occurrence of inverted east–west faults that cut both the Permian (Zechstein) and Carboniferous intervals. (From Gluyas et al. 1997a.)

cements than other wells, but the pattern is not especially distinctive. Indeed, this pattern made little sense to the team investigating the problem until the results of the isotope and petrographic studies became available. The wells with thoroughly cemented Rotliegend reservoir sandstone share one thing in common. All wells either cut, or are adjacent to, an east–west fault array (Fig. 6.30). These faults extend down into the Carboniferous and they commonly juxtapose Permian and Carboniferous sediments. These normal faults underwent inversion during the Tertiary. Thus the evidence afforded by the distribution of the mineral cements complements that obtained from analysis of mineralogy and stable isotope ratios. That is, the mineral cements destroy porosity only in limited areas (volumes) adjacent to east–west normal faults that cut both the Permian and Carboniferous intervals.

Precipitation of the minerals occurred during mixing of formation fluids derived from both the Zechstein and the Carboniferous.

6.6.6 Conclusions

The volume of gas in place in the Amethyst Field was calculated by two methods using static (geologic) data and dynamic (pressure decline) data. The static data gave a much larger figure than the dynamic data. Either there is much less reservoir than calculated or the wells from which the dynamic data were derived were not in contact with the whole of the gas pool.

Mineralogical, isotopic, and spatial analysis of anhydrite, barite, and ferroan dolomite in some of the Amethyst wells indicates that the field is more segmented, rather than smaller, than hitherto appreciated. Extensive and pervasive cementation by these minerals appears to have occurred late in the history of the sandstone, and been caused by mixing along and across faults of formation fluids derived from the overlying Zechstein and the underlying Carboniferous.

The distribution of the cements in the field, and in particular their association with the east–west faults that cut down into the Carboniferous, provided the development team with a methodology for assessing field segmentation and identifying portions of the field not yet accessed by production wells.

6.7 CASE HISTORY: THE FAHUD FIELD, OMAN

6.7.1 Introduction

The Fahud Field is a giant oil accumulation in northern Oman (Fig. 6.31). The field was discovered in 1964. Production began three years later. The initial phase of production was through natural depletion. One year later, in 1968, injection of gas into the original gas cap of the field was used to aid production.

Fahud contains about 6.3 billion barrels of oil originally in place. However, by 1969 the high offtake rates (220,000 bopd) had resulted in a rapid drop in reservoir pressure, a 224 m lowering of the gas/oil contact, and the gassing out of some of the oil production wells (O'Neill 1988). In order to combat the pressure decline and production problems, water injection was begun in 1972. Within a few years it was clear that pressure decline was not being arrested and the production problems continued. Water breakthrough in production wells was added to the list of problems. All was not well; the field had produced but a tiny portion of its original oil in place.

In this case history, we examine the development phases of the Fahud oilfield and the relationship between development of the field and improved understanding of the reservoir character. Despite the early problems with pressure decline, gas, and water breakthrough, the current estimate for ultimate recovery from the field is 910 million barrels, about 15–20% of the original oil in place (Terhen 1999). Thus there is a substantial quantity of remaining oil to be won from the field.

6.7.2 The long search for Fahud

The search for oil in Oman began in the winter of 1924–5. Dr George Lees and K. Washington Gray of D'Arcy Exploration made the first geologic survey of the sultanate. They found little evidence of oil, and exploration activity ceased in Oman. Interest in the potential for finding oil in Oman was only reawakened once oil had been discovered in Saudi Arabia. In 1937, a 75-year exploration concession agreement was signed between the Sultan of Oman and Petroleum Concessions Limited, a subsidiary of the Iraq Petroleum Group. A new company, Petroleum Development (Oman and Dhofar) (PD(O)) was formed to operate the license. World War II interrupted the search for petroleum, and it was not until 1954 that the Fahud structure was identified during an aerial reconnaissance flight. By October of the same year, geologists were on the ground surveying the structure, escorted by the Sultan's soldiers.

On that first geologic survey, the Fahud structure was mapped as an elongate (NW–SE) anticlinal dome. The decision was made to drill, and on January 18, 1956, the first oil exploration well was spudded in Oman; it was Fahud well No. 1. Eighteen months later, the well had penetrated 3000 m of rock. It was declared a dry hole and abandoned. Strictly speaking, the well was not dry. It encountered a few meters of oil-bearing Wasia (Cretaceous) Limestone, although this was not recognized at the time. Interest in the Fahud structure evaporated and, indeed, by 1962 interest in Oman in general was low. A total of $12 million had been spent on four dry exploration wells. Some of the partners in the original company withdrew, leaving only Shell and Partex in PD(O). Their persistence paid off when the giant Yibal Field was discovered in 1962. PD(O) returned to Fahud in 1963. A second well was spudded only 1.5 km from Fahud No. 1. At only 2000 ft below the surface, the Wasia Limestone was encountered and it was oil bearing. A further 1400 ft of oil-bearing strata were found. The giant Fahud Field had been discovered. A reevaluation of the Fahud structure revealed that it was not a simple anticline. A large normal fault occurs along the crest of the field (Fig. 6.32). Fahud No. 1 had been drilled on the hanging wall. It penetrated the fault and then a sliver of Wasia Limestone on the footwall. Production wells have since been drilled only a few hundred meters away from the original dry hole.

6.7.3 An initial description of the structure and reservoir architecture

The Fahud structure is currently described as a northeasterly-dipping monocline. Dip averages about 15° and the length of closure is about 16 km (Fig. 6.33). The structure is dip closed to the northwest and southeast. A large low-angle (45°), normal fault, with a throw in excess of 1000 m, defines closure to

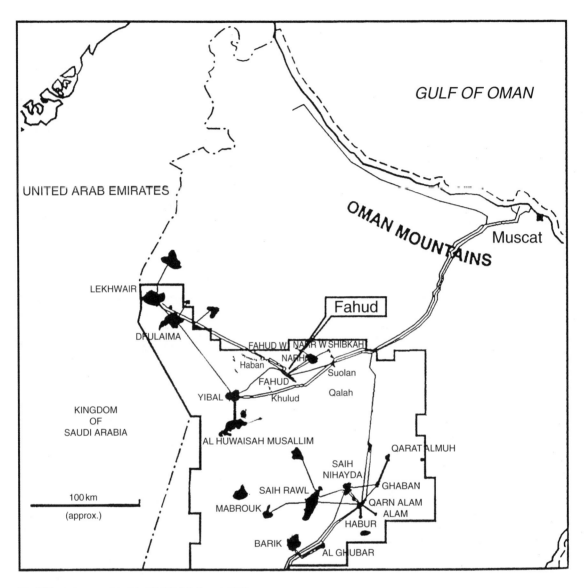

Fig.6.31 The location of the Fahud Field (Nicholls et al. 1999).

the southwest. This fault scarp surface was modified by erosion before being onlapped and draped by the sealing Fiqa Shale.

Fahud contains both 31° API oil and gas. The field crest is at only 50 m subsea and the deepest oil/water contact is at 530 m subsea, giving a maximum petroleum column of 480 m. Under initial conditions, the field contained a small gas cap (a 50 m column). Before production began, pressure data were interpreted to indicate that the whole field had a common free water level and that there was pressure communication throughout the field (Tschopp 1967).

The field contains three main and four minor oil-producing reservoirs in late Albian–early Turonian limestones of the Natih Formation (Wasia Group). It is up to 440 m thick and is composed of seven shallowing-upward cycles. A fully developed cycle has mudstones at its base and a lime grain-

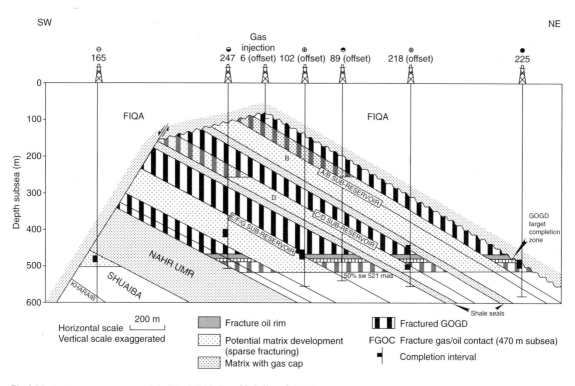

Fig.6.32 A schematic cross-section of the Fahud Field. (From Nicholls et al. 1999.)

stone at its top, and each cycle improves in reservoir quality from base to top. The cycles are labeled "a" at the top to "g" at the base. The most productive reservoir occurs in cycles "a," "c/d," and "e." Collectively, the cycles comprise packstones, wackestones, and lime mudstones. Bioclastic material is the most common grain type (rudist and other bivalve fragments, benthic foraminifera, and gastropod debris). Cycle "b" contains bituminous wackestones, while a tight dolomite is present at the base of cycle "c." Chert commonly occurs at the base of "e." Porosity is high throughout the limestones (25–32%), while the permeability of the main productive intervals was estimated to be measurable in darcies to tens of darcies (Table 6.4). The poorer-quality intervals have a permeability of only a few millidarcies. Initial well potentials were estimated to be 5000–7000 bopd. In the 1967 publication, no mention was made of the reservoir being fractured. Despite the presence of basal mudstones to the cycles, the reservoir is in good fluid communication—as indicated by reservoir geochemistry—and in good pressure communication.

6.7.4 The history of field development

Production began from Fahud in 1967, using only natural depletion (gas cap expansion and solution gas drive). Within a year, production had risen to about 100,000 bopd ($16 \times 10^3 \, \text{m}^3 \, \text{d}^{-1}$). In 1968 gas injection, at the field crest, was begun in response to a rapid decline in reservoir pressure. Production built up to more than 220,000 bopd ($35 \times 10^3 \, \text{m}^3 \, \text{d}^{-1}$) in 1969. The high offtake rates led to what was believed to be a rapid lowering of the gas/oil contact and the resultant gassing out of a number of downdip production wells. Production fell rapidly and the gas to oil ratio of the produced fluid rose rapidly (O'Neill 1988).

Simultaneous with the first phase of production, geologic and petrophysical studies had shown the reservoir to be much more heterogeneous than had hitherto been believed. Specifically, there were thought to be high-permeability streaks in the reservoir in vuggy and/or rudist-rich layers, intervals from which core recovery had been particularly poor during field appraisal. A conclusion was drawn that these high-per-

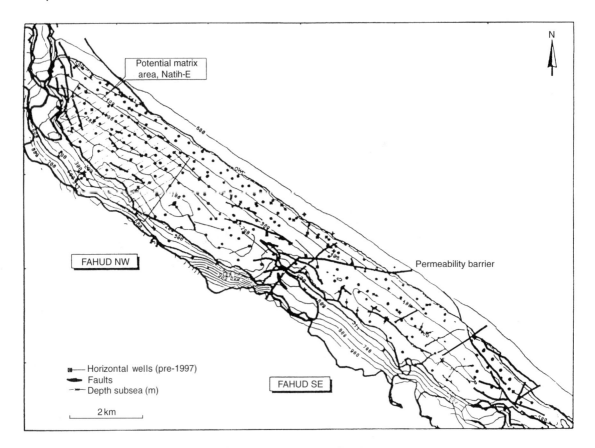

Fig.6.33 A structure map for the top Natih reservoir of the Fahud Field. (From Nicholls et al. 1999.)

meability layers had been quickly drained and gas produced without significant production from the bulk of the reservoir.

In order to stop early gas breakthrough into the two main rows of production wells, two strategies were considered. The first was to limit production to levels that matched the sustainable drainage rates of each reservoir. The second possibility was to maintain pressure by injection of water into the high-permeability layers. Reservoir simulation was performed for both methods. The work demonstrated that gas injection (gravity drainage) could be expected to deliver an ultimate recovery of 15–20%, while water injection was between 50% and 100% more efficient. The main difference turned out to be the sweep efficiency of the lower-permeability layers. However, water-flooding of the lower layers could only occur through gravity-induced cross-flow, as capillary effects were ignored.

Water injection field trials were conducted in 1971. They were not a success. Water was injected into an updip well. It broke through into one downdip well, some 900 m deeper, in

only 8 days. Water breakthrough occurred in a second well, 450 m down dip from the injection well, in 10 days. The rapid water breakthrough could only be explained in terms of transmission through a fracture system or a high-permeability layer. Following the rapid lowering of the gas/oil contact, it had been suspected that a dual-porosity system existed, although at that time the favored cause was a high-permeability layer. This hypothesis was disproven when pressure buildup tests from an array of wells in the northwest of the field showed average permeability values of 30 D over the "a" unit, irrespective of where in the "a" unit the completions were made. The pervasive nature of the high permeability in the reservoir could only be explained by the presence of an extensive, conductive fracture system. Further evidence of a fracture system was provided by the nature of the water breakthrough in the two production wells. The deeper, more distant, production well cut water before the one nearer to the injection well. Further injection tests were made close to the

Table 6.4 The reservoir quality of limestones within the Fahud Field. (Adapted from O'Neill 1988; Nichols et al. 1999.)

Sub-unit	Thickness (m)	Porosity (%)	Matrix permeability (10^{-3} D)	Fracture permeability (10^{-3} D)
A_1	5	33	1–20	10,000
A_2	12	33	1–20	20,000
A_3	10	32	1000	30,000
A_4	6	28	20–100	6000
A_5	11	32	5–20	7000
A_6	18	32	1–5	1000
A_7	20	20–30	1	100
B_{1-3}	62	17–20	0.1–5	Not fractured
B_4	15	Mudstone	Baffle	Baffle
C_1	12	29	200–1000	1500
C_2	29	32	1–20	500
C_{3-4}	9	—	0–1	Not fractured
D_1	18	29	1–20	Not fractured
D_{2-3}	12	—	1	Not fractured
D_4	11	Mudstone	Baffle	Baffle
E_1	18–22	16–24	1–5	300
E_2	42–52	24–27	15	300
E_3	52–62	25–31	5–13	300
E_{4A}	19–34	25–31	4–7	300
E_{4B}	9–18	18–24	2–4	300
E_5	1	Mudstone	Baffle	Baffle
F	31	28	10	Not fractured
G	17	26	3	Not fractured

original oil/water contact. The results here were a little better, with water breakthrough occurring at 1.5% and 3% sweep efficiency. Despite this poor result, the sweep efficiency was still calculated to be twice that of the recovery efficiency induced by gas injection.

By 1972, the natural water influx was calculated to be only 23% of the fluids produced. Pressure maintenance was clearly needed. A decision was taken to install water injection, and to accept that large volumes of water would need to be recycled in order to maintain pressure and have a high recovery factor for the oil. The pilot water injection schemes were extended and water cuts rose rapidly. In 1976, water was produced in equal volume to that injected, at around 45,000 bwpd (7×10^3 m³ d⁻¹). Despite the water injection, the pressure continued to decline and the oil production rate to fall (95,000 bopd; 15×10^3 m³ d⁻¹).

The poor performance of the water-flood led to a period of reassessment of the field during the period 1981–3. Many production logs were run and tracer studies employed to determine the flow paths between the injection wells and

the production wells. These data further demonstrated the fractured nature of the reservoir and that it was oil wet. The decision was made that full-scale, gas/oil gravity drainage was the appropriate method for future production. From 1984, water injection was phased out and gas injection reestablished. Downdip production wells were drilled to produce from the oil rims in the various reservoirs, whilst oil drained out of reservoir matrix areas to slowly feed the oil rims via the fractures. Reservoir simulation on dual-porosity systems was conducted to enable injection sites to be optimized. In the early 1990s, horizontal wells began to be employed to continue to develop the thinning oil rims. Existing vertical wells were sidetracked to horizontal drainage points. In 1994, a 3D seismic survey was acquired over the field to help optimize development drilling and to help characterize the fault and fracture patterns. 3D seismic had not been acquired earlier, due to the difficulty and cost of acquiring seismic over the rocky and topographically rugged Fahud surface jebel. By 1999, it was calculated that gas–oil gravity drainage could deliver 17% of the oil in place (Nicholls et al. 1999).

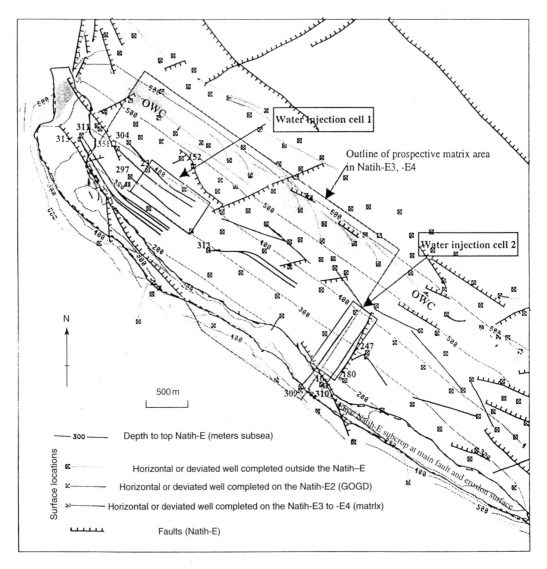

Fig.6.34 A map of the northwest part of the Fahud Field, showing prospective "matrix" development area and the two pilot injection cell plots. (From Nicholls et al. 1999.)

Gravity drainage was not, however, the end of the story, for in the same paper Nicholls et al. (op. cit.) concluded that more detailed reservoir characterization was the key to successful further development of Fahud. They further went on to predict ultimate recoveries in the order of 40%. In particular, Nicholls et al. (op. cit.) studied the Natih-E reservoir in the northwest part of the field. Sub-units Natih-E_3 and Natih-E_4 contain very few, widely spaced fractures. Wells completed in these intervals could be characterized in terms

of production from fractures or production from matrix using their production performance. The data from these wells was interpreted to indicate that the sparsely fractured area covers about 5 km^2 and is 100 m thick. The same well data were also used to aid the design of two pilot-scheme water-flood projects. Both pilot projects used high-angle/horizontal production and injection wells. The major difference was orientation. One pilot scheme was strike-parallel and the other dip-parallel (Fig. 6.34). In the first pilot scheme, in

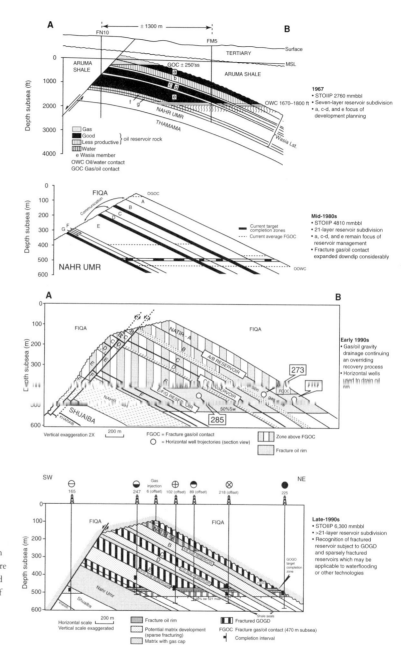

Fig.6.35 The development of the Fahud Field. The four cross-sections illustrate the changing perception of structure and stratigraphy, fluid distribution, reservoir character, and development schemes. Note in particular the recognition of the erosive nature of the fault scarp between the early 1980s and the early 1990s. Note also the appreciation of the fractured and nonfractured layers in the early 1990s. (Modified from Tschopp 1967; O'Neill 1988; Kharusi 1991; Nicholls et al. 1999.)

which the wells were parallel to strike, the results were impressive. Water breakthrough took 18 months and there was no evidence of increasing gas production, something to be expected should a fracture have been encountered. The second pilot scheme, using a dual-bore production well, was less successful due to the well crossing the main fault to the field.

Future plans for the field are now to continue the gas/oil gravity drainage, but with selective unfractured areas undergoing water-flood. The wells are planned as alternating injector and producer pairs, drilled parallel to strike and with 150 m spacing. Dual lateral wells will be used, one lateral completed in the E_7 and the other in the $\bar{E}_4$.

Today, the Fahud Field is producing 50,000 bopd. It could still be producing oil in 100 years' time, should that oil be needed and provided that it remains economic.

6.7.5 Conclusions

The main purpose of choosing the Fahud Field as one of the case histories in this chapter is to illustrate the point that the development of a field is a dynamic and long-term process. Fahud in particular was chosen because it is an onshore, giant field, producing from both fractures and matrix limestone. Proper integration of static geologic data and dynamic production data can lead to a significant improvement in the way in which a field is managed. The case history also illustrates that large fields can undergo a variety of development technologies, both sequentially and simultaneously. For the Fahud Field, the temporal sequence of drilling technology used was vertical wells, horizontal wells, multilateral wells, and laterals drilled from existing wells. That a variety of technologies were used to develop the field does not mean that the early phases of development were incorrect; indeed, they were both innovative and appropriate for their time.

This case history also illustrates another aspect of the point made in Chapter 1, where we discussed the progression from frontier exploration to production. In that progression, the quantity and quality of information and knowledge increases as the scale of observation decreases from basin to play fairway to prospect to field. A similar progression of knowledge has taken place during the development of Fahud. The initial description of the field was based on a coarse grid of data and the reservoir was interpreted to be spatially homogeneous. Today, the characterization of the reservoir is highly detailed and heterogeneity of the reservoir is recognized (Fig. 6.35).

Another important point to draw from this case history is that big fields get bigger, or at least there is a tendency for oil in place, and particularly reserves, to be increased throughout the life of the field. Oil in place in the Fahud Field was initially calculated to be about 2.76 billion barrels; today, the STOIIP is calculated as 6.3 billion barrels. For small fields, particularly those offshore, the luxury of size and hence time is absent, the data are fewer, and the chances of failure are higher.

6.8 CASE HISTORY: THE HEATHER FIELD, UK CONTINENTAL SHELF

6.8.1 Introduction

Well 2/5-1 was drilled early in the exploration history of what was to become the North Sea's Brent Province. The well, drilled in 1973, proved to be the discovery well for the Heather Field (Fig. 6.36). In 1982, oil production from the field peaked at an annual average of about 33,500 bopd. This production rate was short-lived and by the early 1990s the oil rate had fallen to less than 10,000 bopd. The decline in oil rate occurred in parallel with falling oil prices. Heather production looked uneconomic and, indeed, by 1993 the then operator prepared a "cessation of production" document. However, one of the original, albeit minor, partners (Den Norske Oljeselskap, DNO) believed that it was possible to extend the field life beyond the original abandonment date and maintain economic production levels. In this case history we examine the process of managing declining production from an aging offshore oilfield. Unlike old onshore fields that can be produced when the oil price is high and shut in when the oil price is low, an offshore field must be maintained in production, since there will always be an operating cost (OPEX) even if no oil is being produced. The story is taken from three papers, Gray and Barnes (1981), Penny (1991), and Kay (2002), that traced the history first through the years of peak production and then into the years of decline.

6.8.2 The geologic background

The Heather Field lies at the western edge of the Brent Province. The tilted fault block that contains the field and its satellite accumulations is substantially larger than the fields either to the north (the Hudson Field) or the south (the Emerald Field). Oil is reservoired in the five Brent sandstones (Broom, Rannoch, Etive, Ness, and Tarbert). Oil has also been encountered in younger Callovian sandstone (Emerald Sandstone) and in the older Triassic, Cormorant interval in and around the Heather Field. Petroleum occurs between about 9000 ft (2750 m) and 12,500 ft (3800 m). The field and its satellites are both heavily segmented (Fig. 6.36), and the reservoir sandstones are locally of poor quality. Early diagenetic precipitation of calcite cement reduced net pay. Moreover, in the deeper parts of the field, precipitation of authigenic illite caused a dramatic reduction in permeability. The initial oil in place was about 464 mmstb. In 1991 the ultimate (economically recoverable) reserve was thought to be about 100 mmbbl (a recovery factor of 21.5%). However, by mid-1999 the field had produced 120 mmstb. Moreover, a further 57 mmstb is now believed to be economically recoverable from the Heather Field and potential satellites. The

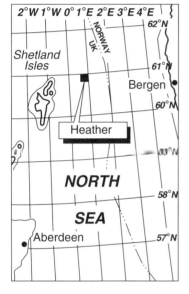

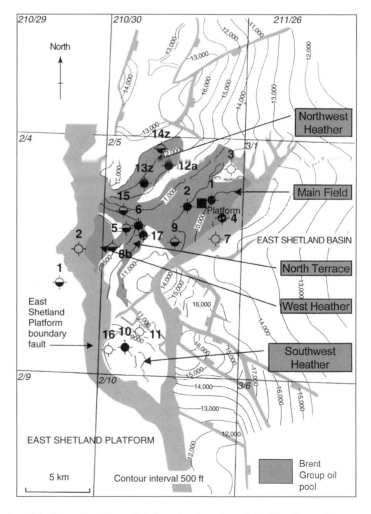

Fig.6.36 A depth map of the Heather Field, UK continental shelf, at top Brent Group. Only the exploration and appraisal well locations are shown. (From Kay 2003.)

additional potential is highly dependent upon the oil price and the operating cost. The figure of 57 mmbbl is based upon operating costs similar to those achieved in 1999 and an average oil price of $15 per barrel.

6.8.3 Arresting production decline

DNO took full operatorship in July 1997, although they were already driving the development of Heather. The company conceived a phased plan to arrest the decline in oil production and maintain the economic life of the field. The three components were:

• To revise and refine the reservoir model through acquisition and interpretation of new 3D seismic data, mapping of reservoir parameters, and recalculation of oil in place. The calculated petroleum volume was then to be compared with the produced volume.

• To map in detail the distribution of zones in the downflank area that have higher-than-average reservoir properties. Such zones could provide new well targets.

• To appraise satellite prospects and identify drilling locations, both in the main field and in the satellites.

The decision to acquire new 3D seismic data is an interesting one, because although the main reason for doing so was that the original survey was of insufficient quality, this was not the only consideration. When DNO adopted the field, they had access to only paper copies of the older 3D seismic data. No electronic versions could be found. Although these paper copies were scanned for use on a workstation, the data were substantially degraded when compared with the original, and they were not of sufficient quality to be able to be used for remapping the field in the detail required.

By mid-1999, the seismic remapping was complete. The results of the mapping were used to refine the reservoir simulation model; for example, to identify bypassed oil in fault-bounded compartments in the field. These compartments would form new well targets. A decision was taken not to rebuild the full-field simulation model but, rather, to use the new seismic mapping data with the preexisting data on calcite and illite distribution to build local reservoir models for the potential target locations.

The combined effect of this work was to enable identification of additional reserves. It was calculated that 13 mmbbl could be won from the Brent sandstones in the main part of the field through infill drilling, and up to 32 mmbbl of reserves in the satellite accumulations in the North Terrace, Southwest Heather, and West Heather, to be won from both Brent and Emerald sandstones. Only part of these potential

reserves is included in the official figure for the field (146 mmbbl). Kay (2003) predicts that ultimate recovery could pass 170 mmbbl even in a "$15 oil" world.

6.8.4 Conclusions

The degree of complexity of the Heather Field was woefully underestimated at the exploration and appraisal stage. Soon after production from the field began, it was necessary to downgrade reserves dramatically. Since that point in the early 1980s, the reserves of the field have been steadily revised upward to a point today at which they are slightly above what was originally estimated during the appraisal phase. This restoration of reserves has been achieved despite the high degree of segmentation in the field and patchy distribution of adequate-quality reservoir.

Three factors have contributed to the steady improvement of reserves:

• Improvements in technology, and in particular the use of 3D seismic data to allow identification of new targets;

• reduction of the cost base such that it has been possible to maintain production during a period of falling oil price (Fig. 6.37);

• integration of the production data with the geologic and seismic information, so allowing an understanding of the spatial distribution of production within the geologic framework.

Indeed, it is clear from Fig. 6.38 that DNO and the operators that preceded them (Unocal) were improving their understanding of the field throughout its life, and so enabling the upward revision of reserves over a period of almost two decades.

There are also some less specific conclusions that can be drawn from the Heather story that are appropriate to other case histories in this chapter. Throughout the life of a field, the people working it will change. Such changes may occur in a regulated fashion, with "handover" periods. The changes may also occur abruptly when operatorship changes. Inevitably, data and experience are lost. In this example, it was necessary to acquire new 3D seismic data partly as a result of data loss. In older fields, particularly those subject to reactivation, it can be a major task just to find well locations, let alone specific geologic and production data. Yet all such data can be critical in allowing an understanding of the field to be developed quickly. This can make the difference between managing an efficient economic project and one that is doomed to financial disaster.

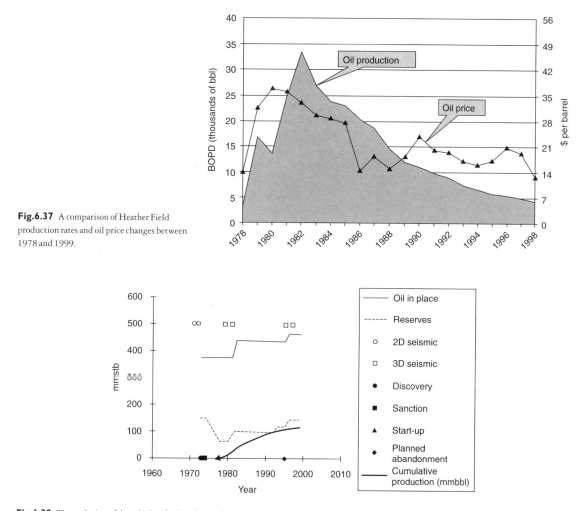

Fig.6.37 A comparison of Heather Field production rates and oil price changes between 1978 and 1999.

Fig.6.38 The evolution of the calculated oil in place and reserves figures for the Heather Field. The field was discovered using a 2 km spaced grid of 2D seismic. Four wells were used to appraise the field. Shortly after the field began production in 1978, reserves were cut dramatically and the first 3D seismic survey shot. Continued production coupled with interpretation of the 3D seismic data allowed upward revision of both STOIIP and reserves. Abandonment was initially planned for 1995, but not executed. Instead, DNO, one of the minor partners, increased their holding, took operatorship, shot modern 3D seismic, increased the reserves, and continued to produce oil. The current reserves estimate is much the same as that at sanction and there remains upside potential.

6.9 CASE HISTORY: THE CRAWFORD FIELD, UK CONTINENTAL SHELF

6.9.1 Introduction

Hamilton Brothers drilled the second well in UK Block 9/28 during the spring of 1975 and discovered oil in what was later to be called the Crawford Field. The first well on the block, although not a commercial oil discovery, contained about

208 ft of Tertiary sandstone with residual oil stain. Oil shows were also encountered in the Cretaceous, Jurassic, and Triassic sections. Discovery well 21/28a-2 contained oil in Triassic sandstones. The well tested at 2610 bopd. As the appraisal program progressed, oil was found in the Zechstein of 21/28a-3, the Middle Jurassic of 21/28a-4, and the Paleocene of 9/28a-10A. Gas and condensate were tested from Cretaceous limestones in 9/28a-4. One appraisal well, No. 7, was dry despite its structurally elevated position. Elsewhere in

the block, exploration wells proved the presence of oil in non-commercial quantities in Upper Jurassic sandstones.

It was clear from these well results that the local structure and petroleum-charging history for the area was very complicated (Fig. 6.39). However, on the basis of the sparse 2D seismic data available at the time, the degree of structural complexity was not appreciated. During the appraisal phase (in 1983), at least one of the participating companies believed that Crawford might contain about 1 billion barrels of oil in

place. One of the authors (J.G.) was privy to this information. In consequence, it came as somewhat of a shock to learn some years later that the oil in place was estimated to be 130 million barrels (Yaliz 1991) and the recoverable reserves only 9 million barrels (Yaliz 1991). Indeed, the field was abandoned in 1991 after producing only 3.7 million barrels. Clearly, in the years since 1983, the oil-in-place and reserves estimates for the field have been reduced massively. Such a dramatic change in perception of a field's worth demands

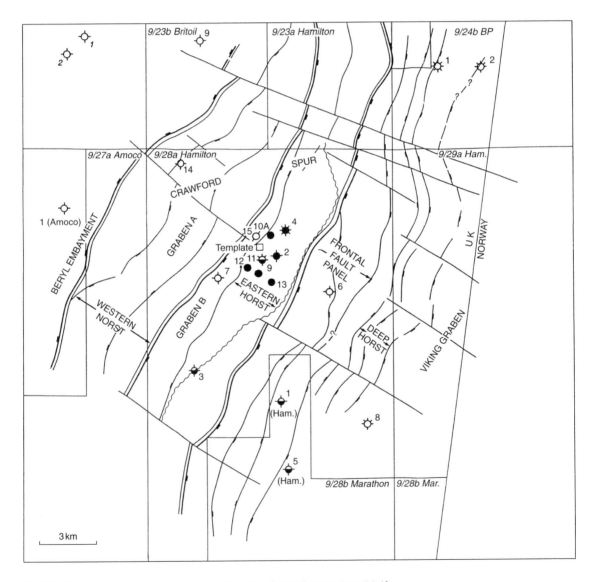

Fig.6.39 A location map and local structural elements for the Crawford Field, UK continental shelf.

further investigation. Crawford Field may be a candidate for reactivation.

In this case history, we examine the field history up to the point of field abandonment and then evaluate oil in place and possible reserves using the data available at the time of abandonment. Finally, we try to determine whether the field is suitable for reactivation.

6.9.2 The field history

The Crawford Field was discovered with well 9/28a-2 in 1975 (Fig. 6.39). The field lies within a major structural high known as the Crawford Ridge (or Spur). This ridge comprises tilted fault blocks, some terraces of which contain sections of preserved Triassic and Middle Jurassic strata (Fig. 6.40). Immediately to the east of the field is a major, down to the east, fault system that marks the western edge of the Central Graben. Within the graben there is a substantial thickness of Upper Jurassic syn-rift sediment. The tilted Triassic and Middle Jurassic sediments are extensively eroded beneath the base Cretaceous unconformity. Successively younger Cretaceous sediments onlap the ridge from east (graben) to west

(graben margin; Fig. 6.41). Tertiary sediments drape the structure.

The trap for oil reservoired in the Middle Jurassic, Triassic, and older intervals is a combination of dip and fault closed. Seal is provided by intra-Triassic and Jurassic mudstone, together with seal beneath the base Cretaceous unconformity. The trap configuration for the Cretaceous limestones has not been adequately explained, while that for the Tertiary almost certainly comprises a combination of sedimentological pinchout of channelized submarine-fan sandstones and compactional drape over the Crawford Ridge. The field is approximately elliptical in overall geometry, being about 3.5 miles long in a NNE–SSW direction and about 1.5 miles wide.

The Triassic reservoir section contains fine- to medium-grained sandstones, mudstones, and intraformational, mud-pebble conglomerates. Two types of sequence have been recognized, a high net to gross interval (80–90% sandstone) and a lower net to gross interval (<30% sandstone). Rarely are both intervals present in the same well, and where they do exist the high net to gross interval is above the low net to gross interval. The sandstones have been interpreted as products of

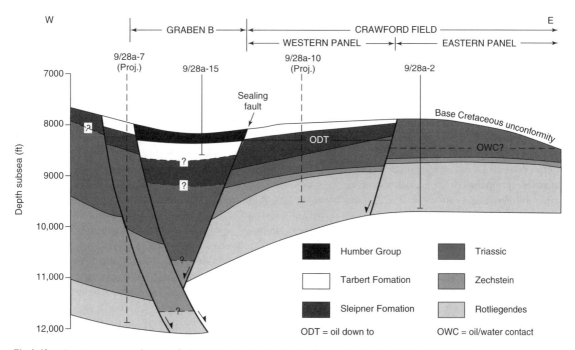

Fig.6.40 A sketch cross-section of the Crawford Field, showing the distribution of reservoir and seal units. (From Yaliz 1991.)

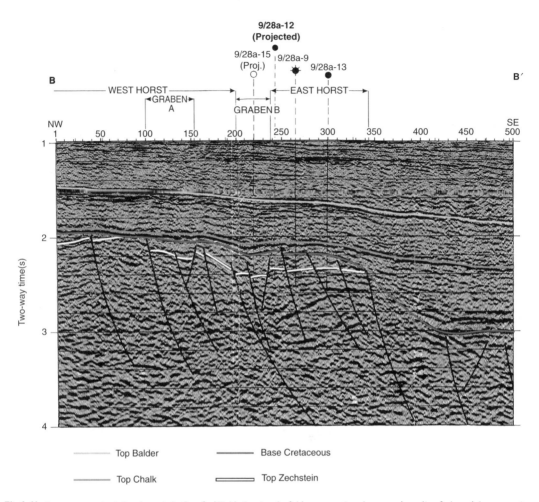

Fig.6.41 An east–west seismic line through the Crawford Field, showing the field segmentation, the eastern bounding fault, and the progressive onlap of Cretaceous sediments onto the Crawford Ridge.

deposition within a braided fluvial system. The sediment transport direction was axial along the incipient graben. The siltstones and mudstones were deposited as sheet-floods, within ephemeral lakes or interfluves. Calcrete horizons testify to soil development in a predominantly arid setting (Yaliz 1991).

The quality of the Triassic sandstones varies widely. However, many of the cleaner sandstones have a porosity in the range 20–25% and a permeability between 10 mD and 2500 mD (most are in the range of a few 100 mD).

The Middle Jurassic reservoir (Sleipner and Tarbert Formations) similarly contains interbedded sandstones, siltstones, and mudstones. The deposition of these sediments occurred on a lower delta plain and in shallow-marine environments. The prime reservoir sandstones were deposited as fluvial channel fills and as shoreface sandstones. Grain size is commonly fine to medium, although some intervals are pebbly. Net to gross is commonly in the range of 30–50%.

Porosities of around 20% are typical for the prime reservoir sandstones; and permeability is measured in hundreds of millidarcies, with a range from 0.01 mD to 6000 mD.

The Paleocene Sele and Andrew Formations (which are petroleum bearing in 9/28a-10A and 9/28a-12) were deposited in a basin-floor submarine fan. The reservoir qualities of this medium- to coarse-grained lightly consolidated sand are excellent.

The oil gravity varies across the Crawford Field. Most of the produced oil had a gravity between about 25° API and 31° API.

Segmentation of the field was recognized to be an issue

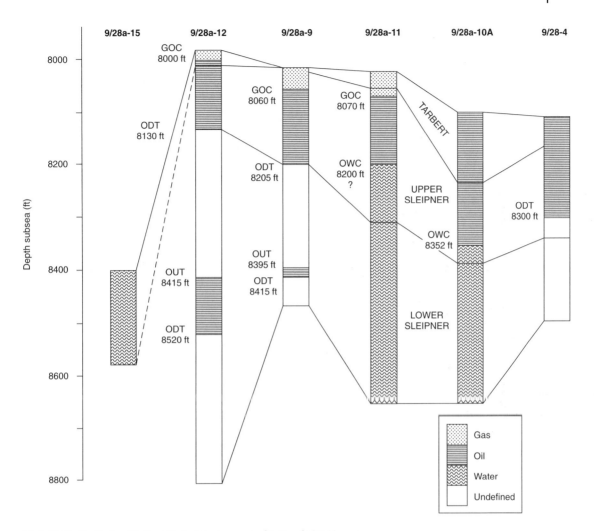

Fig.6.42 The distribution of fluids within the Triassic reservoir of the Crawford Field.

from the early parts of the appraisal program. The distribution of oil, gas, and water in the Triassic and Jurassic levels could not be adequately explained, given the paucity of seismic data from which to recognize faults (Fig. 6.42). However, well-test data indicated that at least some wells had very small drainage volumes caused by sealed local faults (Table 6.5).

Production from the field began in April 1989, using a floating production facility. The first well to be brought on stream was 9/28a-9, the Middle Jurassic Sleipner Formation being perforated. It began to flow at over 7000 bopd and with little associated gas, but within 50 days the oil rate had declined to 2000 bopd and the gas rate had risen to 2000 scf bbl^{-1} (Fig. 6.43a). After 200 days, the well was producing only a few hundred barrels of oil per day and the gas to oil ratio was in excess of 6000 scf bbl^{-1}. Thereafter, the oil rate stayed steady but the gas to oil ratio declined. This behavior was taken to be typical of exsolution gas drive in a limited-volume reservoir. In total, the well produced 351 mstb of oil, 1.32 bcf of gas, and no water. In July 1990, the well was shut in.

The second well on stream was 9/28a-10A. It too was completed in the Middle Jurassic Sleipner Formation. The Middle Jurassic Tarbert Formation was also perforated. The Triassic at this location was water wet. Initial tests showed that the well was capable of producing in excess of 10,000 bopd, although this high rate was not sustainable. The well was brought on at 5000 bopd (Fig. 6.43b). At 1980 scf bbl, the gas to oil ratio was higher than anticipated and analyses

Table 6.5 Well-test data for Crawford Field wells.

Well DST	Reservoir	Perforations (ft below rotary table)	Flow rate (bopd)	PI*	Permeability (mD)	Skin	Remarks
9/28a-2							
1	Triassic	8146–8156	2450	4.15	163	−1.2	32-acre drainage area
2	Rotliegend	9337–9484	—	—	—	—	Dry test
3	Triassic	8593–8633	—	—	—	—	Dry test
4	Triassic	8403–8445	707	0.43	6	−2.2	No depletion seen
5	Triassic	8343–8498	1150	0.55	5	−2.4	No depletion seen
6	Triassic	8223–8268	1850	1.78	43	−3.4	No depletion seen
7	Triassic	8145–8202	2100	2.20	—	—	
7A	Triassic	8146–8202	2167	2.25	60	+1.0	32-acre drainage area
9/28a-4							
1	Triassic	8812–8890	75	—	—	—	
2	Triassic	8602–8794	495	0.25	5	−1.2	No depletion seen
3	Jurassic	8224–8416	1749	1.00	30	−0.2	Depletion suggested
3A	Jurassic	8224–8416	1045	1.40	30	−0.2	Depletion suggested
9/28a-9							
1	Jurassic	8162–8568	3600	30.00	3.01	−1.4	92-acre drainage area
2	Triassic	8642–8868	12	0.02	0.3	−1.7	
9/28a-10A							
1	Jurassic	8660–8914	7648	18.40	300	+3.2	Skin attributed to restricted flow in tubing
9/28a-12							
1	Andrew (Tertiary)	7495–7508 7536–7564	5931–1431	27	650	−2.6	
2	Sele (Tertiary)	6012–6030	1131	2.7	484	−3.0	Completed buildup
3	Jurassic	9603–9653 9682–9693	No flow	—	—	—	
4	Jurassic	9074–9155	2026–1126	0.50	100	−2.0	Rapid decline
9/28a-13							
1	Triassic	10,150–10,294 10,310–10,416 10,426–10,552	4542	3.9	30	−3.0	

* PI = productivity index, measured in barrels per day per 1 lb per 1 psi of pressure drawdown on the flowing formation.

indicated that a small initial gas cap probably existed updip of the well. A pressure buildup test on the well indicated that the reservoir was laterally extensive, with at least 10 million barrels of oil in the segment. After the initial decline from 10,000 bopd, the oil rate remained steady until the end of the first year of production (c.5000 bopd). During the same period, the gas to oil ratio declined, indicating depletion of the small primary gas cap. The oil rate then declined to about 2500 bopd as the gas rate increased. In total, the well produced 2.20 mmstb and 7.07 bcf of gas. No water was produced.

The third of the first batch of development wells was 9/28a-12 (Fig. 6.43c). The Triassic target proved to be only poor-quality reservoir from which no petroleum flow was obtained. However, two intervals in the Paleocene were oil bearing and the thicker one of these (the Andrew Formation) was put into production. In the short time for which the well was in production, it delivered 474 mstb of oil and over 2 million barrels of water and associated solids. Little gas was produced. The production history was complex, with several dramatic increases in wellhead pressure being experienced. These increases in pressure were followed by increases in the oil rate.

Well 9/28a-13 was the last well to be produced in the initial four-well program. The Middle Jurassic was absent at the well location, but 380 ft of pay was encountered in the Upper Triassic section. Well 13 behaved differently from those completed in the Middle Jurassic. The initial rate was 6000 bopd. This declined rapidly to 1000 bopd within about 60 days (Fig. 6.43d). Thereafter, the rate remained constant for over

200 days. The rate then declined further to 600 bopd until the cessation of production. Cumulative oil production was 439 mstb, with no water.

Finally, 9/28a-16 was drilled as a horizontal production well within the Triassic section (Fig. 6.44). The well

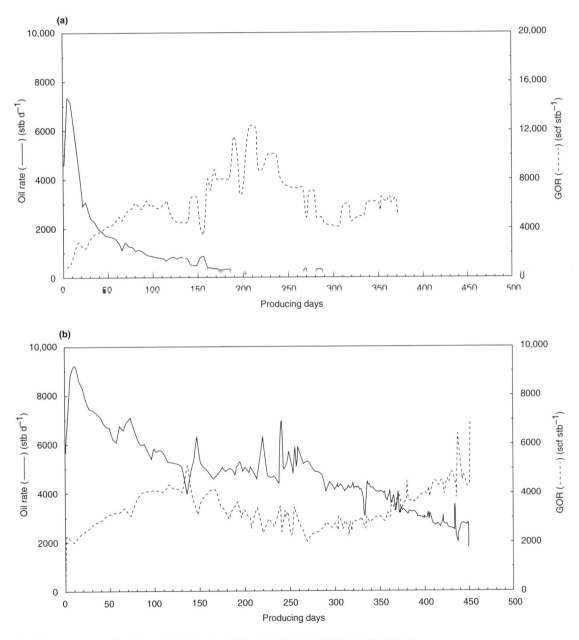

Fig.6.43 Production profiles for Crawford Field wells: (a) 9/28a-9; (b) 9/28a-10A; (c) 9/28a-12; (d) 9/28a-13.

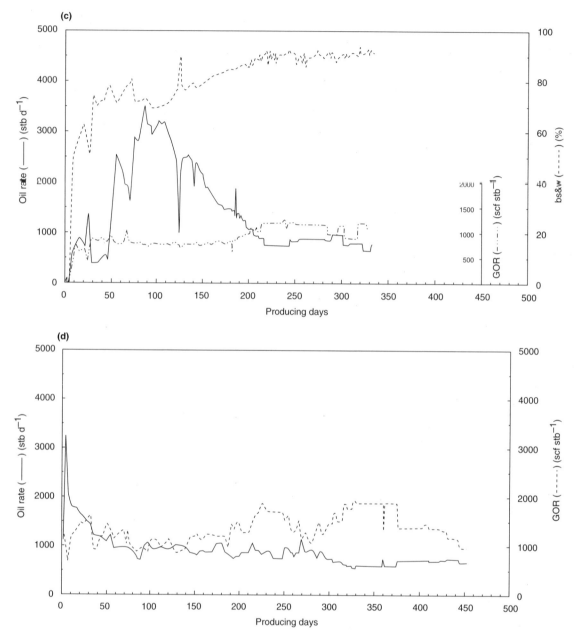

Fig. 6.43 *continued*

performed at high rate initially but, as with well 13, the rate declined rapidly to a steady 1100–1200 bopd. The ultimate recovery from the well was 291 mstb.

Abandonment of the field was justified on the basis of the low total oil rate and the projected continuing decline in production. In an evocative metaphor, the field was compared to

shattered glass, in which each shard was likened to a fault-bounded field segment.

6.9.3 Oil in place and reserves

There have been many calculations of oil in place for

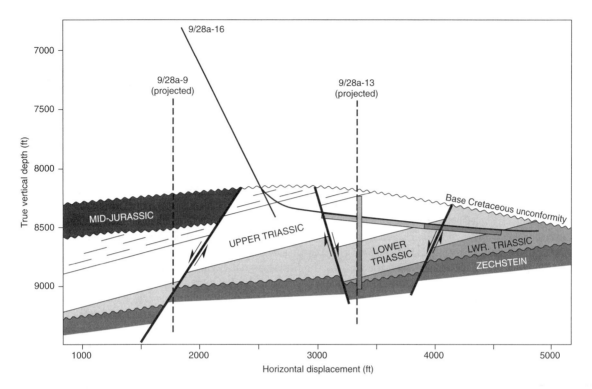

Fig.6.44 The trajectory of the high-angle well 9/28a-16 through the Triassic reservoir of the Crawford Field. This assumes a structural dip in the Triassic of 12° in the plane of the section.

Table 6.6 Oil-in-place and reserves changes through time for the Crawford Field.

Interval	Unattributed source, 1983 (mmstb)	Production consent, 1988 (mmstb)	Yaliz (1991) (mmstb)	Cessation of production, 1990 (mmstb)	Internal review, 1997 (mmstb)	This case history (mmstb)
Tertiary	Not recognized	Not recognized	Not included	4	Not included	Not included
Jurassic	Not quoted	83	Not quoted	48	40	Not calculated
Triassic	Not quoted	70	Not quoted	205	198 or 206*	Not calculated
Total STOIIP	1000	153	130	253	238	320–490
Reserves	Unknown	3.5†	9	3.7	3.7‡	Not calculated

* Geocellular model 206 mmstb, history match 198 mmstb.
† The minimum reserve was given at 2.5 mmbbl.
‡ Volume produced at abandonment, rehabilitation deemed uneconomic.

Crawford. We have access to some of these data, but not to all of them. No formal note of the hearsay figure of one billion barrels in place has been found. However, it is possible to speculate that at the time this figure was postulated (1983), the southern limit of the field may have been considered to be 9/28a-3 (approximately 7 km SSW of 9/28a-2). Moreover,

the structurally high but dry well, 9/28a-7, had not then been drilled. Production consent was given in 1988 on the basis of estimated oil in place of 153 mmstb (Table 6.6). The earliest published reference to reserves, rather than oil in place, is in the UK government annual publication for the oil and gas industry for 1991, the so-called "brown book." It gives reserves for what is described as a small and isolated field as 0.5 tonnes. This approximates to 3.5 million barrels.

Yaliz (1991) quotes the oil in place for the Triassic and Jurassic combined as 130 mmstb. The same document contains a figure of 9 mmstb as the probable reserves. The mapping methodology from which these figures were derived was not quoted. However, contemporaneous internal company documentation suggests that the basic mapped horizon was the base Cretaceous, from which deeper horizons were generated using isochores derived from well data. Although this document, which is in the public domain, is dated 1991, the information that it contains is older than that available in the cessation-of-production documentation.

The document requesting cessation of production from the Crawford Field contains an increase in the estimated oil in place relative to Yaliz (1991), but a downgrade in reserves to 3.7 mmstb. The oil-in-place estimate is based on specific mapping of the Triassic and Jurassic reservoir sequences in the eight fault blocks recognized at the time.

The oil in place was reviewed again in 1997. On this occasion, two methods were used. A geocellular model was built from the static data for the Triassic reservoir data, using the same petrophysical and mapping parameters as used for the cessation-of-production document. Perhaps not surprisingly, the oil in place calculated using this method was the same as generated in 1990 (Table 6.6). An alternative approach based on reservoir engineering data (a history match) delivered oil-in-place figures of 198 mmbbl for the Triassic reservoir and 40 mmbbl for the Jurassic reservoir.

We have attempted to estimate the oil in place for Crawford on the basis of the published field area and petrophysical data. The field is approximately 3.5 miles (5.6 km) long by 1.5 miles (2.4 km) wide. Net pay, porosity, and saturation data are given in Table 6.7. The formation volume factor for the Crawford Field crude has been measured at between 1.3 and 1.4. The volume of the small gas cap has been ignored. It is clear from Table 6.7 that net pay distribution, irrespective of stratigraphic interval, is highly heterogeneous. Within the field boundaries, calculated oil in place and net pay are highly correlated. Average (mean) net pay is 60 m. These figures transpose to between 320 mmstb (P_{50}, Monte Carlo simulation) and 490 mmstb (deterministic mean) for a field area of 5.6 km × 2.4 km. Given the inadequacy of the seismic data upon which the older figures for oil in place were calculated, we see no reason to suppose that those oil-in-place values generated from mapped horizons and fault segments are any more accurate than our statistically derived estimate. Given the highly fragmented nature of the field and the small

volume of oil produced, the history match figure is also likely to be prone to large errors.

The main point to draw from the above discussion is that the oil in place is poorly constrained but nonetheless large, and probably in excess of 250 mmstb for all the horizons in the Crawford Field. This makes Crawford the largest known undeveloped (or partially developed) pool of medium to light crude in the North Sea.

6.9.4 Can the Crawford Field be reactivated?

The clear answer to the above question is "Yes." However, we need to determine whether this is economically feasible. This in turn requires that the field segments be mapped and the quantity of oil that could be won from each segment calculated.

The 2D seismic data coverage available at the time of the initial development and appraisal of the Crawford Field was inadequate for the purposes of determining trap geometry and reservoir architecture. As part of the evaluation process for reactivation of the field, a 3D seismic survey was shot in 1997. Additional studies were also undertaken in an attempt to elucidate the structure and reservoir architecture. Considerable use was made of the dipmeter logs and cores from the original wells, to determine the nature and orientation of faulting and reservoir architecture. The Mesozoic reservoir sequences were reexamined in order to understand the stratigraphy. In the absence of biostratigraphic information from the red-bed Triassic sequences, heavy mineral analysis was used to divide the stratigraphy.

In parallel with the geologic work, the production data and well data were reexamined in order to determine whether new wells could produce at sustainable rates were reactivation to take place.

Analysis of the dipmeter data indicated that the Jurassic and Triassic intervals dip steeply in most of the penetrations. Typically, the Mesozoic reservoir intervals dip at between 20° and 30°. However, in 9/28a-12 the Triassic and Jurassic dip at between 40° and 70°.

Analysis of the fractures in cores showed an abundance that ranged from one fracture every 10 m up to one fracture every 0.5 m. Fractures were observed in clean sandstones and in mixed sandstones and mudstones. In the clean sandstones, most fractures could be identified by the presence of a cataclastic granulation seam, while those in shaley sandstone often contained a clay smear along the fracture plane. In both instances, the fractures are likely to have low permeability and therefore act as baffles and barriers. Only a few percent of

Table 6.7 Petrophysical parameters for Crawford Field wells.

Well	Reservoir	Net pay (m)	Porosity (%)	Oil saturation (%)
9/28a-2	Tertiary	3	30.1	34
	Jurassic Tarbert	0	—	—
	Jurassic Sleipner	0	—	—
	Triassic	128	20.5	65
9/28a-4	Tertiary	0	—	—
	Jurassic Tarbert	2	21.1	54
	Jurassic Sleipner	8	23.8	64
	Triassic	17	19.8	52
9/28a-9	Tertiary	0	—	—
	Jurassic Tarbert	0	—	—
	Jurassic Sleipner	23	19.6	63
	Triassic	7	18.5	53
9/28a-10A	Tertiary	4	32.2	75
	Jurassic Tarbert	25	19.8	78
	Jurassic Sleipner	0	—	—
	Triassic	0	—	—
9/28a-11	Tertiary	2	29.4	38
	Jurassic Tarbert	0	—	—
	Jurassic Sleipner	7	18.9	58
	Triassic	0	—	—
9/28a-12	Tertiary	13	25.0	70
	Jurassic Tarbert	1	19.4	73
	Jurassic Sleipner	36	19.0	65
	Triassic	0	—	—
9/28a-13	Tertiary	0	—	—
	Jurassic Tarbert	0	—	—
	Jurassic Sleipner	0	—	—
	Triassic	181	20.6	65
9/28a-15	Tertiary	8	30.0	21
	Jurassic Tarbert	0	—	—
	Jurassic Sleipner	0	—	—
	Triassic	0	—	—

the fractures seen in core were classified as open and likely to contribute to improved fluid flow.

Data from heavy mineral analysis were used to revolutionize the perception of the Triassic stratigraphy in the Crawford Field. Hitherto, a lithostratigraphic approach was used to construct a stratigraphy in which an upper high net to gross interval overlies a lower low net to gross interval. The five new units are given in Table 6.8.

These data indicate that the stratigraphy of the Triassic is quite different from that conveyed by the simple lithostratigraphic subdivision. The Triassic from the central portion of the field is younger than that in the east, completely opposite to the conclusion reached using lithostratigraphic data. This conclusion clearly changes the inter-well correlation and the structural interpretation.

Table 6.8 Triassic stratigraphy within the Crawford Field.

Youngest	Unit 5	Only seen in wells −4 and −9, this unit has an increased garnet/zircon (G/Z) compared with Unit 4
	Unit 4	Lower G/Z compared with Unit 5, lower apatite/tourmaline (A/T) than Unit 3
	Unit 3	Marked change in garnet geochemistry (pyrope/almandine rich) relative to Unit 2
	Unit 2	Transitional between Units 1 and 3, lower chrome spinel/zircon (C/Z) than Unit 1 but similar garnet chemistry
Oldest	Unit 1	High A/T, high G/Z, and stable ratio trends, grossular rich garnets observed in wells −13, −2, and −7

From the 3D seismic data, new maps were produced on the top Balder (lowermost Eocene), the top Ekofisk (top Cretaceous), the base Cretaceous unconformity, and the top Zechstein. Using these same surfaces, amplitude extractions were made and instantaneous dip maps created. These illustrated the degree of faulting at all levels. Moreover, it was clear that fault patterns in the Tertiary and base Cretaceous are dissimilar to those in the Zechstein, indicating sole-out of some faults in the pre-Cretaceous to Zechstein interval.

From a field reactivation perspective, the main point to emerge from analysis of the seismic data was that the whole of the area of the field is heavily faulted. Reservoir imaging on the 3D seismic volume is not wonderful, but the resolution is sufficient to show that the wells drilled and completed during the brief production period for the field were not drilled in unrepresentative parts of the field where faulting was worse than in other parts. This information, when combined with the results of the fracture analysis performed on core, confirmed that the field is heavily segmented and that individual wells, whatever their position and orientation, are unlikely to yield large quantities of oil.

The reservoir engineering studies conducted in parallel with the geologic and geophysical work produced similar results. The southeastern (Triassic) panel of the field was investigated. It contains between about 50 mmbbl and 60 mmbbl of oil in place. Simulation modeling indicated that under natural depletion a high-angle well drilled into this segment would recover only 1–2 million barrels. Reserves greater than 10 million barrels could be won from the area. However, this would require cross-segment communication to be high and a secondary, water-flood recovery scheme to be successful. Clearly, given the relatively poor imaging of this complex reservoir on the 3D seismic, the implementation of a successful water-flood scheme carries substantial risk.

The combination of reservoir complexity, inadequate reservoir imaging, the likelihood of poor-recovery wells, and a prevailing low oil price all combined at the end of 1998 to the decision to postpone any plans to reactivate the Crawford Field.

6.10 CASE HISTORY: THE TRINIDADIAN OILFIELDS

6.10.1 Introduction

This, our final case history, examines the possibilities for oilfield rehabilitation on two scales: a countrywide evaluation and a specific field project. The island of Trinidad has a very old oil industry and an even longer history of petroleum exploitation. Sir Walter Raleigh caulked his boats with bitumen that flowed from La Brae (a tar lake) in southwest Trinidad in March 1595. About three hundred years later, Captain W. Darwen spudded Trinidad's first oil well in September 1866. By January 1867, the well had encountered oil-bearing strata at 36 m. The remains of this well exist today, and although nothing was made of the discovery at the time, the well will still flow light oil! Commercial production was eventually established in 1908, and today petroleum remains one of the key industries for the island.

6.10.2 Remaining oil potential: southern Trinidad

For the first part of this case history, we will examine onshore oilfields in southern Trinidad, discovered between 1911 and 1952 (Fig. 6.45). Most of the data that we present come from Ablewhite and Higgins (1965), who reviewed the status of oil production up to the end of 1964. Two dogmas regarding the Trinidad petroleum industry are buried deep within the corporate psyche of many oil companies. The first is that pool size is small (a few tens of millions of barrels in place being the norm) and the second is that well flow rates are very low (from tens to hundreds of barrels per day per well). Ablewhite and Higgins (op. cit.) demonstrated that the dogma associated with pool size is simply a function of the way in which the exploration licenses were awarded in the first half of the 20th century. Some individual pools of oil may have up to six field names, each for a different license and different operator. Thus in Table 6.9 the fields are grouped into pools as defined by Ablewhite and Higgins (op. cit.). Similarly, production rates achieved using modern practices may be up to ten times better that those achieved 50–70 years ago.

Most of the traps of southern Trinidad are structural. Many are rollover anticlines associated with thrusting and/or diapiric movement of mud. Trap generation has happened within the past few million years. Trap depths are commonly from a few hundred to a few thousand feet below the surface. The Trinidadian reservoirs are young (Miocene to Pliocene). Miocene reservoirs are commonly turbidite sands, while those of the latest Miocene and Pliocene were deposited in deltaic environments created by the proto-Orinoco river. Porosity and permeability are both high, since the sands are consolidated but uncemented. The Cretaceous Naparima Hill Formation is a prolific source rock. It has sourced most of the Trinidadian oil, although there may be a component of waxy oil produced from a Tertiary source. The Naparima Hill oil ranges from heavy to light with minor to high sulfur and metal contents (dependent upon source facies and maturity).

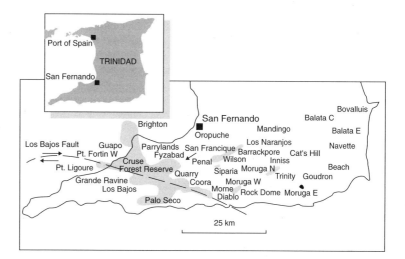

Fig.6.45 The oilfields of southern Trinidad.

Table 6.9 Areas and thicknesses of pay of the Trinidadian oilfields. (From Ablewhite & Higgins 1965.)

Pool group	Individual fields	Thickness of pay (ft)	Area (acres)
Guaya Guayare	Goudron, Beach, Navette	250	1270
Balata	Balata East, Balata Central, Bovallius	150	435
Cat's Hill	Cat's Hill	75	850
Moruga	Moruga East	180	220
Rock Dome East	Inniss, Trinity, Moruga North	130	850
Rock Dome West	Moruga West, Rock Dome	60	860
Barrackpore/Penal	Penal, Barrackpore, Wilson/Siparia, Mandingo, Los Naranjos	155	5375
Oropuche	Oropuche	65	250
Coora Quarry	Coora Quarry, Morne, Diablo/Quinam	270	2460
Fyzabad	Fyzabad, Forest Reserve, Apex, San Francique, Pt. Fortin (East), Los Bajos	240	8140
Palo Seco	Palo Seco, Erin/Los Bajos, Grand Ravine	260	5535
Parrylands	Guapo, Parrylands, Pt. Fortin, Cruse	230	6949
Brighton Pt. Reserve	Brighton Land & Marine, Pt. Ligoure Land & Marine	120	4730

The first steps in this problem are to determine how much oil was originally in place and how much was produced up to the end of 1964. To do this, the data in Ablewhite and Higgins (1965) are used. Table 6.9 lists the oil pools (and fields), their estimated areal extents, and their net pay thicknesses. Production data to the end of 1964 are also given. All of the data presented by Ablewhite and Higgins (1965) were in imperial units. Our preference is to work largely in SI units. The conversion factors are as follows:

acres to m^2 multiply by 4047
feet to meters   multiply by 0.3048
m^3 to barrels   multiply by 6.29

For bulk rock volume we have simply multiplied area by net pay thickness. This is clearly an approximation, but for this case history, which is a screening exercise, the short-cut is worth taking. The sands are clean and largely uncemented and, as such, porosity can be calculated from burial depth. Burial depth today is minimal, but was greater in the past before folding and uplift. Given the absence of hard data, we have estimated porosity to be about 25%, which would equate to a maximum burial under hydrostatic conditions of about 2.5 km. We have chosen an initial oil saturation of about 75% (fairly typical for a sandstone; see Table 4.3). A formation volume factor of 1.2 has been used. Such a figure is

Table 6.10 Estimated oil in place and reserves for Trinidadian oilfields. (Production data from Ablewhite & Higgins 1965.)

Pool group	STOOIP (mmbbl)	Production by December 1964 (mmbbl)	Percent recovery	Reserves at 25% RF (mmbbl)	Left for recovery (mmbbl)
Guaya Guayare	373	28	8	93	65
Balata	76	2	3	19	17
Cat's Hill	75	15	19	19	4
Moruga	47	1	3	12	11
Rock Dome East	130	12	9	32	20
Rock Dome West	61	7	11	15	8
Barrackpore/Penal	**979**	78	8	245	167
Oropuche	19	2	12	5	3
Coora Quarry	781	82	10	194	114
Fyzabad	2296	353	15	574	221
Palo Seco	1691	89	5	423	334
Parrylands	1879	119	6	470	351
Brighton Pt. Reserve	667	47	7	167	120

probably high for the 20–30° API oil expected, but will tend to deliver a slightly conservative answer for oil in place. Thus:

$$\frac{\text{area} \times \text{net pay} \times \text{porosity} \times \text{oil saturation} \times 6.29}{\text{formation volume factor}}$$
$$= \text{oil in place (in barrels)},$$

where area is in m^2, net pay is in meters, porosity and oil saturation are fractional, and the formation volume factor is dimensionless (oilfield volume/stock tank volume).

The recovery factor has been set at 25%, low by today's standards but possibly reasonable given that our intention is to rehabilitate already depleted oilfields. Clearly, it is possible to test the sensitivity of the outcome to uncertainties in the estimates of porosity, oil saturation, area, net pay, and formation volume factor.

The results are given in Table 6.10. The total oil remaining for production (as of January 1965) in 13 pools was 1.4 billion barrels. A profit margin of about $4 per barrel in 1995 currency could therefore deliver about $5 billion. You might think it worthwhile farming into the industry of Trinidad. However, before you do, it would be wise to check on production to date (mid-1995). For example, the Fyzabad/Forest Reserve complex has produced about 425 mmbbl (pers. comm. T. Rambarran, July 1995) from our estimated 574 mmbbl (25% recovery). This leaves about 150 mmbbl remaining to be produced. That is about two-thirds of the 1965 figure. If this pattern is repeated for all 13 pools, then there remains 1 billion barrels to be produced from these shallow, easy-to-

drill onshore Trinidadian oilfields. It may be worth checking the facts for today!

6.10.3 Rehabilitation: Palo Seco Field, Trinidad

The collective remaining potentials of the onshore oilfields of Trinidad were used for the first part of this case history. In the second part of the case history, we examine the rehabilitation potential for a single field. The Palo Seco Field (Fig. 6.46) has been examined in greater detail to determine whether field rehabilitation is commercially viable. The Palo Seco Field was selected because it looked to have the greatest remaining potential. The choice of Palo Seco in March 1996, for this case study, was somewhat fortuitous, because six months later the government of Trinidad and Tobago invited offers for partial rehabilitation of this field. The ministry invited tenders to work over about 80 wells in the Petrotrin (Trinidad's state oil company) concession.

Any field development plan for a petroleum pool is based upon how much oil (or gas) is in place and how much might be recovered. From Table 6.10, we calculated the original oil in place for Palo Seco to be about 1.7 billion barrels. The field covers about 5500 acres (22.3 km^2) and has 260 ft (79 m) of pay. On the basis of data published for 1989 (Persad & Persad 1993) for both the produced volume and the decline rate, we can project the estimated volume recovered to date as 256 million barrels. This is equivalent to a recovery factor to date of 12%. Clearly, if we could improve the recovery factor to 25%, the reserves could be doubled. Specifically, the amount

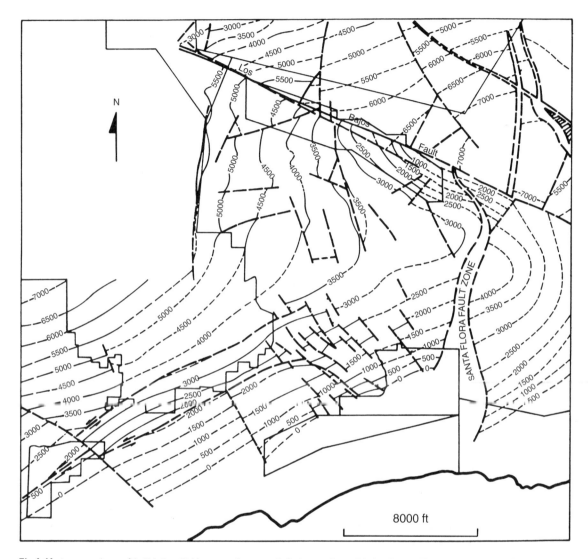

Fig.6.46 A structural map of the Palo Seco Field: structural contours (in feet) on top Cruse, Palo Seco/Erin Field. (From Sealy & Ramlackhansingh 1985.)

of oil remaining to be produced for an ultimate recovery factor of 20% would be 162 mmstb, and for 25% it would be 266 mmstb. Such a large volume of oil should be attractive to most oil exploration and production companies. We now need to examine the field, the current equity holdings, and the fiscal regime in more detail, to determine whether such a project is indeed viable.

6.10.4 The history of the Palo Seco Field

The original discovery of what is now the Palo Seco Field was

made on a concession granted in 1910. Drilling sites were chosen to be at the location of oil seeps. Six production wells were drilled on the south flank of the Erin Syncline, between 1912 and 1916. The wells only produced between one and five barrels each per day, and the field was shut in during 1921. Further activity from several companies between 1921 and 1926 on the south flank of the Erin Syncline and along the Los Bajos Fault delivered more discoveries, none of which were commercial.

Like all good oil industry stories, the commercially successful Palo Seco 1 well was to be the last well drilled by the

Trinidad Petroleum Development Company. The company was near bankruptcy. Profit from the well saved the company. Discoveries continued to be made in the 1920s and the 1930s, extending the limits of Palo Seco. In 1963, the Erin extension was discovered. By the end of 1995, the field had produced about 256 mmstb and some thermal recovery projects had been initiated.

6.10.5 Palo Seco geology

The Palo Seco Field has a synclinal trap geometry (Fig. 6.46). The official published trapping mechanism is said to be faulting (Sealy & Ramlackhansingh 1985). Northeast closure is against the Los Bajos Fault, while south and southeast closure occurs against the faulted anticline of the Southern Ranges Uplift. Down dip to the west, the Skinner Fault completes the triangular-shaped trap. This latter fault may be superfluous to the trap, since it is in a downdip position.

The reservoir comprises lenticular delta-top sandstones of the Forest Formation (Pliocene), overlying more extensive delta-front turbidites that belong to the Cruse Formation (Fig. 6.47). Given the geometry of the sandstones, it is highly likely that there is a large stratigraphic element to the trap. Oil within the Palo Seco Field is 20° API gravity crude, derived from the Naparima Hill, Cretaceous source mudstones.

6.10.6 Palo Seco production characteristics

The production profile for the Palo Seco Field since 1962 is shown in Fig. 6.48. The historical production performance before that date is not available. Over the past 40 years, annual peak production was about 11 mmbbl in the late 1960s. By the mid-1980s, annual production had fallen to about 4 mmbbl, and it has varied between 4 mmbbl and 6 mmbbl since then. The projected decline rate is 1.5% per annum. A number of features about Fig. 6.48 warrant further investigation. At 1.5% per annum, the decline rate is extremely low, and typical of either completely depleted fields or inadequately exploited fields. Since we know that Palo Seco contains much more oil than has been produced, it is reasonable to conclude that the current well suite and production practices are not exploiting the field to the full.

The second feature of the graph is that there are a couple of small production peaks in the middle and late 1980s. On the data available to us at the time of writing, we do not know what caused these peaks. However, we could speculate that these peaks in production correspond to periods of activity in the field when new wells were drilled or old ones worked over. If this is the case, then given that there are about 3000 wells

on the field, the significant rise in production can only have been associated with a small portion of the well inventory, because it takes time to drill or work over a well. Perhaps 12 new drills or workovers could be managed in a year using a single rig. The rise and fall in production in the early 1980s lasted about 4 years. Given that the period of rising production (2 years) is likely to have accompanied any drilling and recompletion of wells, then a single rig could have worked on about 24 wells. It is easy to imagine what an extensive well program might achieve.

6.10.7 Opportunities for improved recovery and production rate

The simple calculation on the remaining potential of Palo Seco belies just how difficult it could be to target unswept oil. The key to understanding the structural and stratigraphic complexities of this reservoir could be provided by a 3D seismic survey. The spatial information derived from such a survey could be integrated with the historical production data to enable identification of poorly exploited oil within the trap volume. These data could also be used for the construction of a reservoir model in which opportunities for secondary recovery could be investigated. Such secondary projects could include the steam soak and flood already being used in part of the field. However, cheaper and more easily sustainable possibilities, such as injection of water, gas, or water and alternating gas (WAG), could also be modeled.

Correct targeting of wells for either production or injection purposes would enable more oil to be won for each well drilled. However, if the wells are to deliver the expected volume and deliver the oil at an acceptable (commercial) rate, such positioning must be applied alongside improved drilling and completion practices. The sorts of processes that are currently being used by drilling and completions engineers to improve well performance include the use of low-invasion (drilling) muds and deep-penetrating perforation charges with low shot density. The former reduces the amount of damage to the permeability of the rock around the wellbore. The latter is designed to penetrate any damage without destroying the fabric of the rock and so reduce sand production from the well. Other measures, such as prepacked screens to inhibit sand production, could also be used. The geometry of the wellbore in the reservoir is also important. Higher production rates can often be achieved by using high-angle wells, so that more formation contacts the wellbore.

The sorts of well rate improvement that one could hope to produce with all of this careful drilling and completion can be estimated from analog data in easternmost Venezuela. A few

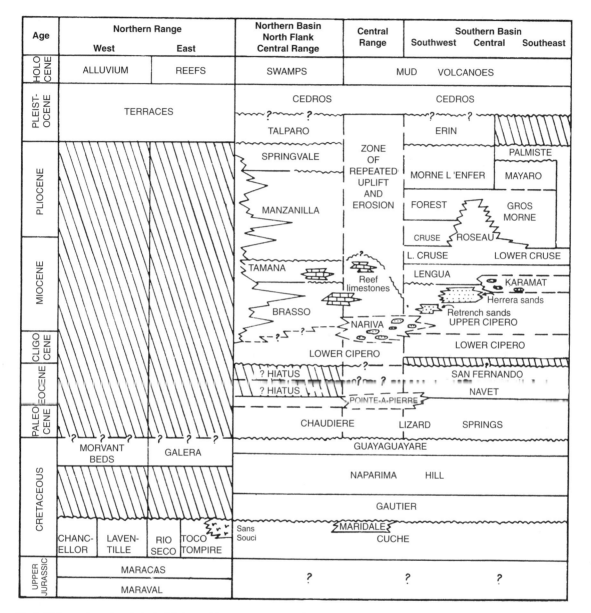

Fig.6.47 Reservoir stratigraphy in the Palo Seco Field. (From Sealy & Ramlackhansingh 1985.)

tens of kilometers due west of Palo Seco lies the Pedernales Field. It too is old and has been produced for over 60 years. A recent reactivation by BP Venezuela has delivered well rates twice that achieved at field start-up in the 1930s, despite the reservoir pressure being 1000 psi lower than at the time of discovery (Gluyas et al. 1996). This improvement has been achieved using the well technology described above.

With few data available in the open literature, it is not possible to make a meaningful estimate of how much oil each new well might produce. At its most simple, the flow potential of a well can be calculated using a radial flow equation (Archer & Wall 1986). Input data for this equation include the viscosity of the oil, the temperature of the oil in the trap, the height of the net hydrocarbon column, the reservoir pres-

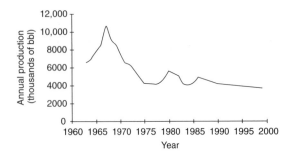

Fig.6.48 The production history for the Palo Seco Field. (Modified from Persad & Persad 1993.)

Table 6.11 Flow potential for new Palo Seco wells at 300 psi drawdown.

Input	
Stock tank oil gravity	22° API
Formation pressure	2000 psi
Formation temperature	45°C
Permeability	125 mD
Net pay thickness	75 m
Fracture radius	10 m
Calculated	
Live oil viscosity	6.115 cp
Gas to oil ratio	297.19 scf stb^{-1}
Formation volume factor (oil)	1.1561
Oil rate (drawdown = 300 psi)	
Barrels per day	1200 stb d^{-1}
Barrels per day (fractured reservoir)	2800 stb d^{-1}

sure, and the pressure drawdown. The equation may be further adapted to accommodate artificially induced fractures in the near-wellbore region. The appropriate data for Palo Seco are listed in Table 6.11. The output data take no account of any formation damage, which can generate a so-called skin and reduce the flow of oil into the wellbore. Nonetheless, the flow rate estimated for a 300 psi drawdown is 1200 bpd (barrels per day), and for a well with a 10 m radial fracture the figure is 2800 bpd.

6.10.8 A redevelopment plan for the Palo Seco Field

When it comes to the rehabilitation of an old oilfield, there are no set rules and few guidelines. There are many variables and therefore many ways in which a development could be

planned. However, there are a few things about which we can be certain, or at least fairly sure:
- Almost 3000 wells have been drilled on the 7000 acre (c.28 km²) oilfield;
- the reservoir pressure is much lower now than it was at discovery;
- secondary gas caps may have formed in the reservoir;
- there are likely to be sizable areas of the field that have been poorly swept.

We need to assume at this stage that a new 3D seismic survey, when combined with the production data for the field, will be able to deliver sufficient data for the selection of well sites. In order to calculate how many wells will be required to drain the field, we are going to make the assumption that new wells can deliver, on average, 2 million barrels throughout their lifetime. Such volumes are the best that has been achieved by the current well suite. However, they are not unreasonable when compared with analogs offshore western Trinidad. From the estimated remaining reserves, the requirement becomes about 100 wells at an average spacing of 64 acres. Since the earlier production will have depleted reservoir pressure, it is likely that pressure support (water injection) will be needed from day one. A cautious approach would be to plan to drill production well/injection well pairs. To this end, the cost of water injection wells has been built into the well costs below. To make an economic evaluation for a part-produced field, we have readjusted the initial conditions of S_o (oil saturation) to today's estimated value, and we have calculated a recovery factor that will give a final (end of field life) recovery factor of 25%.

6.10.9 Commercial evaluation

This book is not the place to present a detailed commercial analysis of a petroleum exploitation opportunity and, more importantly, the authors do not have the skills to make such an analysis. Instead, we have chosen to present a relatively simple Monte Carlo simulation of reserves and net present value (NPV). The input data includes that presented above together with a number of fixed (by us) costs and equipment efficiencies. These include the cost of a facility. We have combined with this the cost of obtaining a seismic survey. There will also be costs associated with health, safety, and the environment (HSE). The total royalty, tax, and tariff bill for operating in Trinidad is variable and depends upon production rate and oil quality, but we have chosen to run the simulation with a fixed 75% government take. The discount rate has been set at 7%. There are also costs associated with operating (OPEX) and intervening in the operation.

Table 6.12 Assumptions used for the Palo Seco NPV calculation.

Assumption	Distribution	Minimum	Likeliest or mean	Maximum	Standard deviation
Well rate ($\times 10^3$ bopd)	Normal	—	1.50	—	0.33
Oil price ($)	Triangular	12	15	19	—
Well cost ($ million)	Triangular	1.0	1.5	3.0	—
Area (acres)	Triangular	6100	6835	7400	—
Thickness (ft)	Uniform	235	—	280	—
Porosity (fraction)	Triangular	0.23	0.25	0.28	—
Oil saturation (fraction)	Triangular	0.55	0.61	0.65	—
Net to gross (fraction)	Uniform	0.9	—	1.0	—
Formation volume factor (ratio)	Normal	—	1.2	—	0.03
Recovery factor (fraction)	Uniform	0.08	—	—	0.14
Well availability (%)	Triangular	81	90	99	—
Facilities efficiency (%)	Triangular	91	95	99	—
Export efficiency (%)	Triangular	91	95	99	—

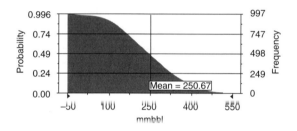

Fig.6.49 An exceedance curve for the net present value of Palo Seco rehabilitation.

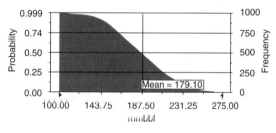

Fig.6.50 A cumulative probability curve for the reserves of rehabilitation of the Palo Seco field.

In respect of the field performance, we have assumed that half of the reserves can be produced on plateau and that the reserves to production ratio is about 5. We have also set an economic minimum well rate of 100 bopd. The assumptions (variables) used in the calculation are in Table 6.12. Of these, the only assumption that needs further comment is the oil price. After the giddy $35+ per barrel price of the early 1980s, the late 1980s and the 1990s saw a return to relatively low oil prices. As such, the mid-price of $15 per barrel used may seem too cautious. However, many companies now plan their activity on a commercially defensible oil price and then consider the extra revenue generated by a high oil price as a windfall.

We have chosen to generate two forecasts, net present value (Fig. 6.49 & Table 6.13) together with reserves (Fig. 6.50 & Table 6.14). The main controlling factors on the net present value are given in the sensitivity chart (Fig. 6.51). The important NPV forecast indicates that there is a finite possibility that the project will lose money (up to $16 million), but

there is a 90% probability that it will deliver $116 million or more. The 50% probability is $243 million. The reserves forecast indicates that there is a 90% probability that the project will deliver 140 mmbbl or more and a 50% probability of producing about 175 mmbbl. The net present value of the project is most sensitive to recovery factor (Fig. 6.51), while oil price and drilling costs are also important. The oil price cannot be controlled, but action can be taken to minimize the negative impacts of the drilling cost by ensuring that wells are drilled to time and to budget.

6.10.10 Conclusions

Even the most pessimistic of assumptions would appear to indicate that the rehabilitation of Palo Seco is a commercially viable, indeed highly profitable, project. However, to achieve the rehabilitation and thus realize the net present value of $243 million ($P_{50}$), a number of conditions have to be met. Most importantly, access to the whole field is required. New

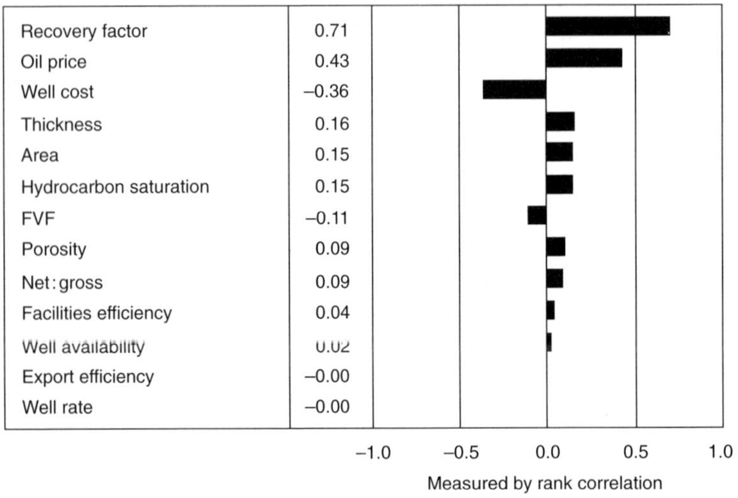

Recovery factor	0.71	
Oil price	0.43	
Well cost	−0.36	
Thickness	0.16	
Area	0.15	
Hydrocarbon saturation	0.15	
FVF	−0.11	
Porosity	0.09	
Net : gross	0.09	
Facilities efficiency	0.04	
Well availability	0.02	
Export efficiency	−0.00	
Well rate	−0.00	

−1.0   −0.5   0.0   0.5   1.0

Measured by rank correlation

Fig.6.51 A sensitivity graph for the net present value (NPV) of rehabilitation of the Palo Seco field.

Table 6.13 Net present value probability exceedance data for Palo Seco.

Percentile (exceedance)	Value ($ million)
100%	−49.04
90%	98.20
80%	146.92
70%	182.73
60%	206.80
50%	231.54
40%	258.53
30%	293.71
20%	333.14
10%	384.51
0%	605.38

Table 6.14 Possible reserves available on rehabilitation of the Palo Seco field.

Percentile (exceedance)	Reserves (mmbbl)
100%	118.62
90%	149.62
80%	159.39
70%	166.14
60%	172.13
50%	177.57
40%	183.66
30%	189.75
20%	197.66
10%	208.38
0%	257.42

wells are required, and these wells need to target areas with large quantities of unswept oil. In order to achieve the well targets, a 3D seismic survey is required. All of these requirements lie outside the offer made by the ministry for a workover of 80 wells. Thus negotiation would be required to deliver a project of sufficient size to be attractive to most oil companies.

In addition to obtaining a large enough project, the production rate from individual wells has to be substantially greater than has been achieved previously on Palo Seco.

FURTHER READING

Abbots, I.L. (1991) *United Kingdom Oil and Gas Fields, 25 Years Commemorative Volume.* Geological Society Memoir No. 14, London.

Archer, J.S. & Wall, P.G. (1986) *Petroleum Engineering, Principles and Practice.* Graham & Trotman, London.

Beaumont, E.A. & Foster, N.H. (1990–2) *Structural Traps I to VIII and Stratigraphic Traps I to III* (11 vols). Treatise of Petroleum Geology Atlas of Oil and Gas Fields, American Association of Petroleum Geologists, Tulsa, Oklahoma.

Gluyas, J.G. & Hichens, H. (2003) *United Kingdom Oil and Gas Fields Millennium Commemorative Volume.* Geological Society Memoir No. 20, London (in press).

REFERENCES

Abbots, I.L. (ed.) (1991) *United Kingdom Oil and Gas Fields, 25 Years Commemorative Volume*. Geological Society Memoir No. 14.

Abbott, W.O. (1990) Maui Field. In Beaumont, E.A. & Foster, N.H. (eds.) *Structural Traps I: Tectonic Fold Traps*. Treatise of Petroleum Geology Atlas of Oil and Gas Fields, American Association of Petroleum Geologists, Tulsa, Oklahoma, 1–25.

Ablewhite, K. & Higgins, G.E. (1965) A review of Trinidad, West Indies, oil development and the accumulations at Soldado, Brighton Marine, Grande Ravine, Barackpore-Penal and Guayagayare. In *Proceedings of the Fourth Caribbean Geological Conference, Trinidad, 1965*, 41–100. Trinidad and Tobago Geological Society, Port-of-Spain, Trinidad.

Adams, J.E. (1936) Oil pool of open reservoir type. *American Association of Petroleum Geologists Bulletin* 20, 780–96.

Ajdukiewicz, J.M. (1995) A model for quartz cementation in the Norphlet Formation, Mobile Bay, offshore Alabama. American Association of Petroleum Geologists, Annual Convention, Houston, 1.

Allen, P.A. & Allen, J.R. (1990) *Basin Analysis, Principles and Applications*. Blackwell Scientific, Oxford.

Alsharhan, A.S. (1990) Geology and reservoir characteristics of Lower Cretaceous Kharaib Formation in Zakum Field, Abu Dhabi, United Arab Emirates. In Brooks, J. (ed.) *Classic Petroleum Provinces*. Special Publication No. 50, Geological Society, London, 299–316.

Alsharhan, A.S. & Nairn, A.E.M. (1997) *Sedimentary Basins and Petroleum Geology of the Middle East*. Elsevier, Amsterdam.

Aplin, A.C., Macleod, G., Larter, S.R., Pederson, K.S., Sorensen, H., & Booth, T. (1999) Combined use of confocal laser scanning microscopy and *PVT* simulation for estimating the composition and physical properties of petroleum in fluid inclusions. *Marine and Petroleum Geology* 16, 97–110.

Archer, J.S. & Wall, P.G. (1986) *Petroleum Engineering, Principles and Practice*. Graham & Trotman, London.

Armstrong, L.A., Ten Have, A., & Johnson, H.D. (1987) The geology of the Gannet Fields, Central North Sea, UK sector. In Brooks, J. & Glennie, K. (eds.) *Petroleum Geology of North West Europe*. Graham & Trotman, London, 533–48.

Atkinson, J.H. & Bransby, P.L. (1978) *The Mechanics of Soils: An Introduction to Critical State Soil Mechanics*. McGraw-Hill, London.

Aymard, R., Pimentel, L., Eitz, P., et al. (1990) Geological integration and evaluation of Northern Monogas, Eastern Venezuela Basin. In Brooks, J. (ed.) *Classic Petroleum Provinces*. Special Publication No. 50, Geological Society, London, 37–54.

Baars, D.L. & Stevenson, G.M. (1982) Subtle stratigraphic traps in Palaeozoic rocks of the Paradox Basin. In Halbouty, M.T. (ed.) *The Deliberate Search for the Subtle Trap*. American Association of Petroleum Geologists Memoir 32, 131–58.

Bain, J.S. (1993) Historical overview of exploration of Tertiary plays in the UK North Sea. In Parker, J.R. (ed.) *Petroleum Geology of Northwest Europe: Proceedings of the 4th Conference*. Geological Society, London, 5–13.

Baldwin, B. & Butler, C.O. (1985) Compaction curves. *American Association of Petroleum Geologists Bulletin* 69, 622–6.

Barss, D.L., Copland, A.B., & Ritchie, W.D. (1970) Geology of the Middle Devonian reefs, Rainbow area, Alberta, Canada. In Halbouty, M.T. (ed.) *Geology of Giant Petroleum Fields*. American Association of Petroleum Geologists Memoir 14, 19–49.

Bathurst, R.G.C. (1975) *Carbonate Sediments and their Diagenesis*. Developments in Sedimentology 12. Elsevier, Amsterdam.

Baudouy, S. & LeGorjus, C. (1991) Sendji Field — People's Republic of Congo, Congo Basin. In Beaumont, E.A. & Foster, N.H. (eds.) *Structural Traps V*. Treatise of Petroleum Geology Atlas of Oil and Gas Fields, American Association of Petroleum Geologists, Tulsa, Oklahoma, 121–49.

Beard, D. & Weyl, P. (1973) Influences of texture on porosity of unconsolidated sand. *American Association of Petroleum Geologists Bulletin* 57, 349–69.

Beaumont, E.A. & Foster, N.H. (1990) *Structural Traps VI: Tectonic and Nontectonic Fold Traps*. Treatise of Petroleum Geology Atlas of Oil and Gas Fields, American Association of Petroleum Geologists, Tulsa, Oklahoma.

Beaumont, E.A. & Foster, N.H. (1990–2) *Structural Traps I to VIII and Stratigraphic Traps I to III* (11 vols). Treatise of Petroleum Geology Atlas of Oil and Gas Fields, American Association of Petroleum Geologists, Tulsa, Oklahoma.

Bebbington, C. (1996) The Orinoco challenge. *BP Shield*, issue 2, 18–22.

Beckley, A., Dodd, C., & Loss, A. (1993) The Bruce Field. In Parker, J.R. (ed.) *Petroleum Geology of Northwest Europe: Proceedings of the 4th Conference.* Geological Society, London, 1453–64.

Belfield, W.C. (1988) Characterisation of a naturally fractured carbonate reservoir: Lisburne Field, Prudhoe Bay, Alaska. SPE 18174, Society of Petroleum Engineers.

Betancourt, R. (1978) *Venezuela's Oil.* George Allen & Unwin, London.

Beydoun, Z.R. (1991) *Arabian Plate Hydrocarbon Geology and Potential.* American Association of Petroleum Geologists Studies in Geology #33.

Beydoun, Z.R., Cheng, W.B., Toksoez, C.H., & Nafi, M. (1985) Detection of open fractures with vertical seismic profiling. *Journal of Geophysical Research* **90**, 4557–66.

Bischke, R.E. (1994) Interpreting sedimentary growth structures from well log and seismic data (with examples). *American Association of Petroleum Geologists Bulletin* **78**, 873–92.

Bjørnseth, H.M. & Gluyas, J.G. (1995) Petroleum exploration in the Ula Trend. In Hanslien, S. (ed.) *Petroleum Exploration and Exploitation in Norway.* Norwegian Petroleum Society (NPF), Special Publication No. 4, 85–96.

Blackwell, D.D. & Steele, J.L. (1989) Thermal conductivity of sedimentary rocks: measurement and significance. In Naeser, N.D. & McCulloh, T.H. (eds.) *Thermal History of Sedimentary Basins: Methods and Case Histories.* Springer-Verlag, New York, 13–36.

Blakey, S.E. (1985) *Oil on their Shoes, Petroleum Geology to 1918.* American Association of Petroleum Geologists, Tulsa, Oklahoma.

Bond, J. (1997) Late Triassic stratigraphy of the Beryl Field. In Oakman, C.D., Martin, J.H., & Corbett, P.W.M. (eds.) *Cores from the Northwest European Hydrocarbon Province, An Illustration of Geological Applications from Exploration to Development.* Geological Society, London, 79–95.

Bowen, J.M. (1991) Introduction. In Abbots, I.L. (ed.) *United Kingdom Oil and Gas Fields, 25 Years Commemorative Volume.* Geological Society Memoir No. 14, 9–20.

Bowman, M.B.J., McClure, N.M., & Wilkinson, D.W. (1993) Wytch Farm oilfield: deterministic reservoir description of the Triassic Sherwood Sandstone. In Parker, J.R. (ed.) *Petroleum Geology of Northwest Europe: Proceedings of the 4th Conference.* Geological Society, London, 1513–17.

Boyles, J.M. & Scott, J.A. (1982) A model for migrating shelf-bar sandstones in Upper Mancos Shale (Campanian), NW Colorado. *American Association of Petroleum Geologists Bulletin* **66**, 491–508.

Braithwaite, P., Armentrout, J.M., Beeman, C.E., & Mslecek, S.J. (1988) East Breaks Block 160 Field, Offshore Texas: a model for deepwater deposition of sand. Offshore Technology Conference, Houston, 145–56.

Breckels, I.M. & van Eekelen, H.A.M. (1982) Relationship between horizontal stress and depth in sedimentary basins. *Journal of Petroleum Technology* **34**, 165–78.

Brennan, P. (1990) Greater Burgan Field. In Beaumont, E.A. & Foster, N.H. (eds.) *Structural Traps I: Tectonic Fold Traps.* Treatise of Petroleum Geology Atlas of Oil and Gas Fields, American Association of Petroleum Geologists, Tulsa, Oklahoma, 103–28.

Brewster, J. (1991) The Frigg Field, Block 10/1 UN North Sea and 25/1 Norwegian North Sea. In Abbots, I.L. (ed.) *United Kingdom Oil and Gas Fields, 25 Years Commemorative Volume.* Geological Society Memoir No. 14, 117–26.

Brognon, G.P. & Verrier, G.R. (1966) Oil and geology in Cuanza basin of Angola. *American Association of Petroleum Geologists Bulletin* **50**, 108–58.

Brooks, J., Cornford, C., & Archer, R. (1987) The role of hydrocarbon source rocks in petroleum exploration. In Brooks, J. & Fleet, A.J. (eds.) *Marine Petroleum Source Rocks.* Special Publication No. 26, Geological Society, London, 17–46.

Brown, A.M., Milne, A.D., & Kay, A. (2003) The Thistle Field, Blocks 211/18a and 211/19a, UK North Sea. In Gluyas, J.G. & Hichens, H. (eds.) *United Kingdom Oil and Gas Fields, Millennium Commemorative Volume.* Geological Society Memoir No. 20 (in press).

Brown, A., Mitchel, A.W., Nilssen, I.R., Stewart, I.J., & Svela, P.T. (1992) Ula Field: relationship between structure and hydrocarbon distribution. In Larsen, B.T. & Larsen, R.M. (eds.) *Structural and Tectonic Modelling and its Application to Petroleum Geology.* Norsk Petroleumsforening/NFP, Special Publication 1, Elsevier, Amsterdam.

Browne, E.J.P. (1995) The international oil industry: challenges and opportunities. A speech given to the Oxford Energy Seminar, September 7, 1995.

Bruno, L., Roy, D.L., Grinsfelder, G.S., & Lomando, A.J. (1991) Alabama Ferry Field—USA East Texas Basin, Texas. In Beaumont, E.A. & Foster, N.H. (eds.) *Stratigraphic Traps II.* Treatise of Petroleum Geology Atlas of Oil and Gas Fields, American Association of Petroleum Geologists, Tulsa, Oklahoma, 1–28.

Bryant, S., Cade, C., & Mellor, D. (1993) Permeability prediction from geologic models. *American Association of Petroleum Geologists Bulletin* **77**, 1338–50.

Bryant, I.D. & Livera, S.E. (1991) Identification of unswept oil volumes in a mature field by using integrated data analysis: Ness Formation, Brent Field, UK North Sea. In Spencer, A.M. (ed.) *Generation, Accumulation and Production of Europe's Hydrocarbons.* Special Publication of the European Association of Petroleum Geoscientists No. 1. Oxford University Press, Oxford, 75–88.

Buday, T. (1980) *The Regional Geology of Iraq, Volume 1, Stratigraphy and Palaeogeography.* State Organisation for Minerals, Baghdad, Iraq.

Burchette, T.P. & Britton, S.R. (1985) Carbonate facies analysis in the exploration for hydrocarbons: a case study from the Cretaceous of the Middle East. In Brenchley, P.J. & Williams, B.P.J. (eds.) *Sedimentology, Recent Developments and Applied Aspects.* Blackwell Scientific, Oxford, 311–38.

Burchette, T.P. & Wright, V.P. (1992) Carbonate ramp depositional systems. *Sedimentary Geology* **79**, 3–57.

Burke, K. & Dewey, J.F. (1973) Plume generated triple junctions: key indicators in applying plate tectonics to old rocks. *Journal of Geology–Chicago* 81, 406–33.

Byerlee, J. (1993) Model for episodic flow of high-pressure water in fault zones before earthquakes. *Geology* 21, 303–6.

Cade, C.A., Evans, I.J., & Bryant, S.L. (1994) Analysis of permeability controls–a new approach. *Clay Minerals* 29, 491–501.

Cai, D.Q. (1987) A preliminary analysis of hydrocarbon exploration of Dongsha Massif. In Collection of papers from the International Petroleum Geological Convention, northern South Sea continental shelf, China. *China Oil Magazine* (Hong Kong), 12–23.

Campbell, C.J. (1997) *The Coming Oil Crisis.* Multiscience, Brentwood, UK.

Carnevali, J.O. (1992) Monagas thrust-fold belt in the Eastern Venezuela Basin: anatomy of a giant discovery of the 1980's. In *Thirteenth World Petroleum Congress, Buenos Aires, Argentina, 1991.* John Wiley, Chichester, 47–58.

Carruth, A.G. (2003) Foinaven Field, Blocks 204/19 & 204/24a. In Gluyas, J.G. & Hichens, H. (eds.) *United Kingdom Oil and Gas Fields, Millennium Commemorative Volume.* Geological Society Memoir No. 20 (in press).

Cazier, E.C., Hayward, A.B., Espinosa, G., Velandia, J., Mugnoit, J.-F., & Leel, W.G. Jr. (1995) Petroleum geology of the Cusiana Field, Llanos Basin foothills, Colombia. *American Association of Petroleum Geologists Bulletin* 79, 1444–63.

Chadwick, R.A., Holliday, D.W., Holloway, S., & Hulbert, A.G. (1993) The evolution and hydrocarbon potential of the Northumberland–Solway Basin. In Parker, J.R. (ed.) *Petroleum Geology of Northwest Europe: Proceedings of the 4th Conference.* Geological Society, London, 717–26.

Chalmers, L.S. (1943) Colinga East extension area of the Colinga oil fields, geologic formations and economic development of oil and gas fields of California. *California Department of Natural Resources Bulletin* 118, 486–90.

Cherry, S.T.J. (1993) The interaction of structure and sedimentary processes controlling deposition of the Upper Jurassic Brae Formation conglomerate, Block 16/7 North Sea. In Parker, J.R. (ed.) *Petroleum Geology of Northwest Europe: Proceedings of the 4th Conference.* Geological Society, London, 387–400.

Chiang, K. (1990) Hoadley Gas Field. In Beaumont, E.A. & Foster, N.H. (eds.) *Stratigraphic Traps I.* Treatise of Petroleum Geology Atlas of Oil and Gas Fields, American Association of Petroleum Geologists, Tulsa, Oklahoma, 123–36.

Choquette, P.W. & Steinen, R.P. (1980) Mississippian non-supratidal dolomite, Ste. Genevieve limestone, Illinois basin: evidence for mixed water dolomitization. In Zenger, D.H., Dunham, J.B., & Ethington (eds.) *Concepts and Models of Dolomitization.* Special Publication No. 28, Society for Economic Paleontologists and Mineralogists, Tulsa, Oklahoma, 163–96.

Clark, D.N. (1980) The diagenesis of Zechstein carbonate sediments. In Füchtbauer, H. & Peryt, T.M. (eds.) *The Zechstein Basin with Emphasis on Carbonate Sequences.* Contributions to Sedimentology 9. E. Schweizerbart'sche, Stuttgart, 167–203.

Claypool, G.E., Holser, W.T., Kaplan, I.R., Sakai, H., & Zak, I. (1980) The age curves of sulfur and oxygen in marine sulfate and their mutual interpretation. *Chemical Geology* 28, 199–260.

Cochran, J.R. (1983) Effects of the finite extension times on the development of sedimentary basins. *Earth and Planetary Science Letters* 75, 157–66.

Coleman, M.L. (1998) Novel methods for measuring chemical compositions of oil-zone waters: implications for appraisal and production. *PETEX 98 Conference Proceedings and Published Abstracts* M3.

Collinson, J.D. (1972) The Røde Ø Conglomerate of Inner Scoresby Sund and the Carboniferous (?) and Permian rocks west of the Schuchert Flod. *Meddelelser om Grønland* Bd 192, No. 6, 1–48.

Coney, D., Fyfe, T.B., Retail, P., & Smith, P.J. (1993) Clair appraisal: the benefits of a co-operative approach. In Parker, J.R. (ed.) *Petroleum Geology of Northwest Europe: Proceedings of the 4th Conference.* Geological Society, London, 1409–20.

Cook, H.E., Field, M.E., & Gardner, J.V. (1982) Continental slopes. In Scholle, P.A. & Spearing, D. (eds.) *Sandstone Depositional Environments.* American Association of Petroleum Geologists, Tulsa, Oklahoma, 329–54.

Cooles, G.P., Mackenzie, A.S., & Quigley, T.M. (1986) Calculation of masses of petroleum generated and expelled from source rocks. In Leythaeuser, D. & Rüllkotter, J. (eds.) *Advances in Organic Geochemistry 1985.* Pergamon Press, Oxford, 235–45.

Cornford, C. (1990) Source rocks and hydrocarbons of the North Sea. In Glennie, K.W. (ed.) *Petroleum Geology of the North Sea, Basic Concepts and Recent Advances.* Blackwell Science, Oxford, 376–462.

Craddock, C., Addicott, W.O., & Richards, P.W. (1983) *Plate Tectonic Map of the Circum-Pacific Region: Antarctica Sheet.* American Association of Petroleum Geologists, Tulsa, Oklahoma.

Craig, B.H. (1990) Yates and other Guadalupian (Kazanian) oil fields, U.S. Permian Basin. In Brooks, J. (ed.) *Classic Petroleum Provinces.* Special Publication No. 50, Geological Society, London, 249–63.

Creek, J.L. & Schraeder, M.L. (1985) East Painter reservoir: an example of a compositional gradient. SPE 14411, Society of Petroleum Engineers, Las Vegas.

Cromer, J.B. & Littlejohn, R. (1976) Content, composition and thermal history of organic matter in Mesozoic sediments, Falkland Plateau. In Barker, P.F. & Dalziel, I.W.D. (eds.) *Initial Reports DSDP,* vol. 36. US Government Printing Office, Washington, DC, 941–4.

Cubitt, J.M. & England, W.A. (1995) *The Geochemistry of Reservoirs.* Special Publication No. 86, Geological Society, London, 321.

Cuevas Gozalo, M.C. & Martinius, A.W. (1993) Outcrop data-base for the geological characterisation of fluvial reservoirs: an example from distal fluvial fan deposits in the Loranca Basin, Spain. In North, C.P. & Prosser, D.J. (eds.) *Characterisation of Fluvial and Aeolian Reservoirs.* Special Publication No. 73, Geological Society, London, 74–94.

Curiale, J.A., Larter, S.R., Sweeney, R.E., & Bromley, B.W. (1989) Molecular thermal maturity indicators in oil and gas source rocks.

In Naeser, N.D. & McCulloh, T.H. (eds.) *Thermal History of Sedimentary Basins: Methods and Case Histories*. Springer-Verlag, New York, 53–72.

Dale, C.T., Lopes, J.R., & Abilio, S. (1992) Takula Oil Field and the Greater Takula area, Cabinda, Angola. In Halbouty, M.T. (ed.) *Giant Oil and Gas Fields of the Decade, 1978–1988*. American Association of Petroleum Geologists Memoir 54, 197–216.

Davis, J.C. (1986) *Statistics and Data Analysis in Geology*, 2nd edn. John Wiley, New York.

Deer, W.A., Howie, R.A., & Zussman, J. (1966) *An Introduction to the Rock Forming Minerals*. Longman, London.

de la Beche, H.T. (1839) *Report on the Geology of Cornwall, Devon and West Somerset*. Longman, London.

Demaison, J.G. & Moore, G.T. (1980) Anoxic environments of oil source bed genesis. *American Association of Petroleum Geologist Bulletin* 64, 1179–209.

Denison, C.N. & Anthony, J.S. (1990) New Late Jurassic subsurface lithostratigraphic units, PPL100 Papua New Guinea. In *Proceedings of the First PNG Petroleum Convention, Port Moresby*, 153–8. Geological Society of Papua New Guinea, Port Moresby.

Deroo, G., Herbin, J.P., & Roucache, J. (1983) Organic geochemistry of Upper Jurassic – Cretaceous sediments from Site 511, Leg 71, Western South Atlantic. In Ludwig, W.J. & Krasheninnikov, V.A. (eds.) *Initial Reports DSDP*, vol. 71. US Government Printing Office, Washington, DC, 1001–13.

Dewey, J.F. (1972) Plate tectonics, In *Planet Earth (1974), Readings from Scientific American*. W.H. Freeman, San Francisco, 124–35.

D'Heur, M. (1984) Porosity and hydrocarbon distribution in the North Sea chalk reservoirs. *Marine and Petroleum Geology* 1, 211–39.

Dickinson, W.R. & Seely, D.R. (1979) Structure and stratigraphy of forearc regions. *American Association of Petroleum Geologists Bulletin* 63, 2–31.

Dickson, J.A.D. & Saller, A.H. (1995) Identification of subaerial exposure surfaces and porosity preservation in Pennsylvanian and Lower Permian shelf limestones, eastern central Basin Platform, Texas. In Budd, D.A., Saller, A.H., & Harris, P.M. (eds.) *Unconformities and Porosity in Carbonate Strata*. American Association of Petroleum Geologists Memoir 63, 239–58.

Dingle, R.V. (1980) Large allochthonous sediment masses and their role in the construction of the continental slope and rise of southwestern Africa. *Marine Geology* 37, 333–54.

Dromgoole, P. & Spears, R. (1997) Geoscore; a method for quantifying uncertainty in field reserves estimates. *Petroleum Geoscience* 3, 1–12.

Duval, B.C., Choppin de Janvry, G., & Loiret, B. (1992) The Mahakam delta province: an ever-changing picture and a bright future. 24th Offshore Technology Conference, Houston, Texas, May 4–7, 1992, 393–404.

Duval, R., Cramez, C., & Vail, P.R. (1998) Stratigraphic cycles and major marine source rocks. In de Graciansky, P.C., Hardenbol, J., Jacquin, T., & Vail, P.R. (eds.) *Mesozoic and Cenozoic Sequence Stratig-*

raphy of European Basins. Special Publication No. 60, Society for Economic Paleontologists and Mineralogists, Tulsa, Oklahoma, 43–51.

Earnshaw, J.P., Hogg, A.J.C., Oxtoby, N.H., & Cawley, S.J. (1993) Petrographic and fluid inclusion evidence for the timing of diagenesis and petroleum entrapment in the Papuan Basin. In Carmen, G.J. & Carmen, Z. (eds.) *Petroleum Exploration and Development in Papua New Guinea: Proceedings of the Second PNG Petroleum Convention, Port Moresby*, May 31 – June 2, 1993.

Edwards, C.W. (1991) The Buchan Field, Blocks 20/5a and 21/1a, UK North Sea. In Abbots, I.L. (ed.) *United Kingdom Oil and Gas Fields, 25 Years Commemorative Volume*. Geological Society Memoir No. 14, 253–9.

Ehrenberg, S.N. (1990) Relationship between diagenesis and reservoir quality in sandstones of the Garn Formation, Haltenbanken, Mid-Norwegian Continental Shelf. *American Association of Petroleum Geologists Bulletin* 75, 1579–92.

Elliott, T. (1978) Deltas. In Reading, H.G. (ed.) *Sedimentary Environments and Facies*. Blackwell Scientific, Oxford, 97–142.

Emery, D. & Myers, K.J. (1990) Ancient subaerial exposure in sandstones. *Geology* 18, 1178–81.

Emery, D. & Myers, K.J. (1996) *Sequence Stratigraphy*. Blackwell Science, Oxford.

Emery, D. & Robinson, A.G. (1993) *Inorganic Chemistry: Applications to Petroleum Geology*. Blackwell Scientific, Oxford.

Emery, D., Smalley, P.C., & Oxtoby, N.H. (1993) Synchronous oil migration and cementation in sandstones reservoirs demonstrated by quantitative description of diagenesis. *Philosophical Transactions of the Royal Society of London* A315, 187–202.

England, W.A. & Fleet, A.J. (1991) *Petroleum Migration*. Special Publication No. 59, Geological Society, London.

England, W.A., Mann, A.L., & Mann, D.M. (1991) Migration from source to trap. In Merrill, R.K. (ed.) *Source and Migration Processes and Evaluation Techniques*. Treatise of Petroleum Geology Handbook of Petroleum Geology, American Association of Petroleum Geologists, Tulsa, Oklahoma, Chapter 3, 23–46.

Enos, P. (1988) Evolution of pore space in the Poza Rica trend (Mid-Cretaceous), Mexico. *Sedimentology* 35, 287–325.

Estaban, M. (1991) Palaeokarst: practical applications. In Wright, V.P. (ed.) *Palaeokarsts and Palaeokarstic Reservoirs*. University of Reading Postgraduate Research Institute for Sedimentology Contribution No. 152, 89–119.

Eubank, R.T. & Makki, A.C. (1981) Structural geology of the Central Sumatra back-arc basin. In *Indonesian Petroleum Association, 10th Annual Convention, Proceedings*, 1–53. Pertamina, Jakarta.

Eva, A.N., Burke, K., Mann, P., & Wadge, G. (1989) Four-phase tectonostratigraphic development of the southern Caribbean. *Marine and Petroleum Geology* 6, 9–21.

Evamy, B.D. Harembourne, J., Kamerling, P., Knapp, W.A., Molloy, F.A., & Rowlands, P.H. (1978) Hydrocarbon habitat of Tertiary Niger Delta. *American Association of Petroleum Geologists Bulletin* 62, 1–39.

Evans, I.J., Hogg, A.J.C., Hopkins, M.S., & Howarth, R.J. (1994) Quantification of quartz cements using combined SEM, CL, and image analysis. *Journal of Sedimentary Research* **A64**, 334–8.

Eyles, D.R. & May, J.A. (1984) Porosity mapping using seismic interval velocities, Natuna L-structure. In *Proceedings of the Indonesian Petroleum Association, 13th Annual Convention*, 301–15. Pertamina, Jakarta.

Fielding, C.R. (1984) A coal depositional model for the Durham Coal Measures of NE England. *Journal of the Geological Society of London* **141**(5), 919–31.

Finney, J. (1968) Random packings and the structure of the liquid state. Unpublished Ph.D. thesis, University of London.

Foster, P.T. & Rattey, P.R. (1993) The evolution of a fractured chalk reservoir: Machar Oilfield, UK North Sea. In Parker, J.R. (ed.) *Petroleum Geology of Northwest Europe: Proceedings of the 4th Conference.* Geological Society, London, 1445–52.

Frank, J.R., Cluff, S., & Bauman, J.E. (1982) Painter reservoir, East Painter reservoir and Clear Creek Fields, Unita County, Wyoming. In Powers, R.B. (ed.) *Geologic Studies of the Cordilleran Thrust Belt.* Rocky Mountain Association of Geologists, Denver, Colorado, 601–18.

Fraser, A.J. & Gawthorpe, R.L. (1990) Tectonostratigraphic development and hydrocarbon habit of the Carboniferous in northern England. In Brooks, J. (ed.) *Classic Petroleum Provinces.* Special Publication No. 50, Geological Society, London, 49–86.

Fraser, A.J., Nash, D.F., Steele, R.P., & Ebdon, C.C. (1990) A regional assessment of the intra-Carboniferous play of northern England. In Brooks, J. (ed.) *Classic Petroleum Provinces.* Special Publication No. 50, Geological Society, London, 417–40.

Galloway, W.E. (1975) Process framework for describing the morphologic and stratigraphic evolution of the deltaic depositional systems. In Broussard, M.L. (ed.) *Deltas, Models for Exploration.* Houston Geological Society, Houston, Texas, 87–98.

Galloway, W.E. (1989) Genetic stratigraphic sequences in basin analysis: architecture and genesis of flooding surface bounded depositional units. *American Association of Petroleum Geologists Bulletin* **73**, 125–42.

Galloway, W.E., Hobday, D.K., & Magara, K. (1982) *Frio Formation of the Texas Gulf Coast Basin–Depositional Systems, Structural Framework and Hydrocarbon Origin, Migration, Distribution and Exploration Potential.* University of Texas, Bureau of Economic Geology, Report of Investigations No. 122.

Garland, C.R. (1991) The Amethyst Field, Blocks 47/8a, 47/9a, 47/13a, 47/14a, 47/15a, UK North Sea. In Abbots, I.L. (ed.) *United Kingdom Oil and Gas Fields, 25 Years Commemorative Volume.* Geological Society Memoir No. 14, 387–93.

Garland, C.R. (1993) Miller Field: reservoir stratigraphy and its impact on development. In Parker, J.R. (ed.) *Petroleum Geology of Northwest Europe: Proceedings of the 4th Conference.* Geological Society, London, 401–14.

Gérard, J., Wheatley, T.J., Richie, J.S., Sullivan, M., & Bassett, M.G. (1993) Permo-Carboniferous and older plays, their historical development and future potential. In Parker, J.R. (ed.) *Petroleum Geology of Northwest Europe: Proceedings of the 4th Conference.* Geological Society, London, 641–50.

Giles, B.F. (1968) Pneumatic acoustic energy source. *Geophysical Prospecting* **16**, 21–53.

Glasmann, J.R., Clark, R.A., Larter, S., Briedis, N.A., & Lundegard, P.D. (1989) Diagenesis and hydrocarbon accumulation, Brent Sandstone (Jurassic), Bergen Area, North Sea. *American Association of Petroleum Geologists Bulletin* **73**, 1341–60.

Glennie, K.W. (1986) Early Permian—Rotliegend. In Glennie, K.W. (ed.) *Introduction to the Petroleum Geology of the North Sea*, 2nd edn. Blackwell Scientific, Oxford, 63–86.

Glennie, K.W. (ed.) (1998) *Petroleum Geology of the North Sea, Basic Concepts and Recent Advances*, 4th edn. Blackwell Science, Oxford.

Gluyas, J.G. (1985) Reduction and prediction of sandstone reservoir potential, Jurassic North Sea. *Philosophical Transactions of the Royal Society* **A315**, 187–202.

Gluyas, J.G. & Bowman, M.B.J. (1997) Edale No. 1 Oilwell, Derbyshire, England, 1938. *Marine and Petroleum Geology* **14**, 191–9.

Gluyas, J.G. & Cade, C.A. (1997) Prediction of porosity in compacted sands. In Kupecz, J., Gluyas, J.G., & Bloch, S. (eds.) *Reservoir Quality Prediction in Sandstones and Carbonates.* American Association of Petroleum Geologists Memoir **69**, 19–28.

Gluyas, J.G. & Coleman, M.L. (1992) Material flux and porosity changes during sediment diagenesis. *Nature* **356**, 52–3.

Gluyas, J.G. & Hichens, H. (eds.) (2003) *United Kingdom Oil and Gas Fields, Millennium Commemorative Volume.* Geological Society Memoir No. 20 (in press).

Gluyas, J.G. & Leonard, A.J. (1995) Diagenesis of the Rotliegend sandstone: The answer ain't blowin' in the wind. *Marine and Petroleum Geology* **12**, 491–7.

Gluyas, J.G. & Oxtoby, N.H. (1995) Diagenesis a short (2 million year) story—Miocene sandstones of Central Sumatra, Indonesia. *Journal of Sedimentary Research* **A65**, 513–21.

Gluyas, J.G., Byskov, K., & Rothwell, N. (1992) A year in the life of Gyda production. In *Advances in Reservoir Technology.* IBC, London.

Gluyas, J.G., Jolley, J.E., & Primmer, T.J. (1997a) Element mobility during diagenesis: sulphate cementation of Rotliegend sandstones, Southern North Sea. *Marine and Petroleum Geology* **14**, 1001–11.

Gluyas, J.G., Leonard, A.J., & Oxtoby, N.H. (1990) Diagenesis and petroleum emplacement: the race for space—Ula Trend North Sea. In *13th International Sedimentological Congress Abstracts Volume.* International Association of Sedimentologists, Utrecht, extended abstract, p. 193.

Gluyas, J.G., Robinson, A.G., & Primmer, T.P. (1997b) Rotliegend Sandstone diagenesis: a tale of two waters. In Hendry, J., Carey, P., Parnell, J., Ruffel, A., & Worden, R. (eds.) *Geofluids II '97, Belfast, March 1997*, 291–4. Geological Society, Belfast.

Gluyas, J.G., Garland, C.R., Oxtoby, N.H., & Hogg, A.J.C. (2000) Quartz cement; the Miller's tale. In Worden, R.H. & Morad, S. (eds.) *Quartz Sedimentation in Sandstones.* Special Publication No.

29, International Association of Sedimentologists, Blackwell Science, Oxford, 199–218.

Gluyas, J.G., Oliver, J.S., Wilson, W.W., & Tineo, M. (1996) Pedernales Field, eastern Venezuela: the first 100 years. *American Association of Petroleum Geologists Bulletin* 80, 1294.

Gluyas, J.G., Robinson, A.G., Emery, D., Grant, S.M., & Oxtoby, N.H. (1993) The link between petroleum emplacement and sandstone cementation. In Parker, J.R. (ed.) *Petroleum Geology of Northwest Europe: Proceedings of the 4th Conference*. Geological Society, London, 1395–402.

Gray, W.T.D. & Barnes, G. (1981) The Heather Oilfield. In Illing, L.V. & Hobson, G.D. (eds.) *Petroleum Geology of the Continental Shelf of NW Europe*. Heyden, on behalf of the Institute of Petroleum, London, 335–41.

Gressly, A. (1838) Observations géologiques sur le Jura Soleurois. *Neue Denkschr. allg. schweiz. Ges. Naturw.* 2, 1–112.

Gutteridge, P. (1991) Aspects of Dinantian sedimentation in the Edale Basin, North Derbyshire. *Geological Journal* 26, 245–69.

Gussow, W.C. (1954) Differential entrapment of oil and gas: a fundamental principle. *American Association of Petroleum Geologists Bulletin* 38, 816–53.

Guzmán, E.J. (1967) Reef type stratigraphic traps in Mexico. *Proceedings of the 7th World Petroleum Congress* 2, 461–70.

Haig, D.B. (1991) The Hutton Field, Blocks 211/28, 211/27, UK North Sea. In Abbots, I.L. (ed.) *United Kingdom Oil and Gas Fields, 25 Years Commemorative Volume*. Geological Society Memoir No. 14, 135–44.

Halbouty, M.T. (1976) Application of Landsat imagery to petroleum and mineral exploration. *American Association of Petroleum Geologists Bulletin* 60, 745–93.

Halbouty, M.T. (1979) *Salt Domes: Gulf Region, United States and Mexico*, 2nd edn. Gulf Publishing, Houston, Texas.

Halbouty, M.T. (ed.) (1982) *The Deliberate Search for the Subtle Trap*. American Association of Petroleum Geologists Memoir 32, 57–75.

Halbouty, M.T. (1986) Basins and new frontiers: an overview. In Halbouty, M.T. (ed.) *Future Petroleum Provinces of the World*. American Association of Petroleum Geologists Memoir 40, 1–10.

Halbouty, M.T. (1991) East Texas Field USA, East Texas Basin, Texas. In Beaumont, E.A. & Foster, N.H. (eds.) *Stratigraphic Traps II*. Treatise of Petroleum Geology Atlas of Oil and Gas Fields, American Association of Petroleum Geologists, Tulsa, Oklahoma, 189–206.

Hancock, J.M. (1984) Cretaceous. In Glennie, K.W. (ed.) *Introduction to the Petroleum Geology of the North Sea*, 2nd edn. Blackwell Scientific, Oxford, 133–50.

Haq, B.U., Hardenbol, J., & Vail, P.R. (1988) Mesozoic and Cenozoic chronostratigraphy and cycles of sea-level change. In Wilgus, C.K., Hastings, B.S., St. Kendall, C.G., Posamienter, H.W., Ross, C.A., & Van Wagoner, J.C. (eds.) *Sea-Level Changes: An Integrated Approach*. Special Publication No. 42, Society of Economic Palaeontologists and Mineralogists, 71–108.

Harding, T.P. (1985) Seismic characteristics and identification of negative flower structures, positive flower structures, and positive structural inversion. *American Association of Petroleum Geologists Bulletin* 69, 582–600.

Harker, S.D. & Rieuf, M. (1996) Genetic stratigraphy and sandstone distribution of the Moray Firth Humber Group (Upper Jurassic). In Hurst, A., Johnson, H.D., Burley, S.D., Canham, A.C., & Mackertich, D.S. (eds.) *Geology of the Humber Group: Central Graben and Moray Firth, UKCS*. Special Publication No. 114, Geological Society, London, 109–30.

Harland, W.B., Armstrong, R.L., Cox, A.V., Craig, L.E., Smith, A.G., & Smith, D.G. (1990) *A Geologic Time Scale 1989*. Cambridge University Press, Cambridge.

Harrison, P.A. & Mitchell, A.W. (1995) Continuous improvement in well design optimises development. SPE 030536, presented at SPE 70th Annual Technical Conference and Exhibition, Dallas, Texas, October 22–25.

Heavyside, J., Langley, J.O., & Pallett, N. (1983) Permeability characteristics of Magnus reservoir rock. In *8th European Formation Evaluation Symposium*, London, March 1983, 1–29. Society of Exploration Well Log Analysts, London.

Hedberg, H.D., Sass, L.C., & Funkhauser, H.J. (1947) Oilfields of the Greater Oficina area. *American Association of Petroleum Geologists Bulletin* 31, 2089–169.

Hein, F.J., Dean, M.E., DeLure, A.M., Grant, S.K., Robb, G.A., & Longstaffe, F.J. (1986) The Viking Formation in the Caroline, Garrington and Harmatten East fields, western south-central Alberta: sedimentology and palaeogeography. *Bulletin of Canadian Petroleum Geology* 34, 91–110.

Hellem, T., Kjemperud, A., & Øvrebø, O.K. (1986) The Troll Field: a geological/geophysical model established by the PL085 Group, In Spencer, A. M. et al. (eds.) *Habitat of Hydrocarbons on the Norwegian Continental Shelf*. Norwegian Petroleum Society, Graham & Trotman, 217–38.

Helmy, H. (1990) Southern Gulf of Suez, Egypt: structural geology of the B-trend oil fields. In Brooks, J. (ed.) *Classic Petroleum Provinces*. Special Publication No. 50, Geological Society, London, 353–64.

Héritier, F.E., Conort, A., & Mure, E. (1990) Frigg Field—UK and Norway Viking Graben, North Sea. In Beaumont, E.A. & Foster, N.H. (eds.) *Stratigraphic Traps I*. Treatise of Petroleum Geology Atlas of Oil and Gas Fields, American Association of Petroleum Geologists Bulletin, Tulsa, Oklahoma, 69–90.

Higgins, G.E. (1996) *A History of Trinidad Oil*. Trinidad Express Newspapers Limited, Port-of-Spain, Trinidad.

Hillier, A.P. & Williams, B.P.J. (1991) The Leman Field, Blocks 49/26, 49/27, 49/28, 53/1, 53/2 UK North Sea. In Abbots, I.L. (ed.) *United Kingdom Oil and Gas Fields, 25 Years Commemorative Volume*. Geological Society Memoir No. 14, 451–8.

Hinze, W.J. & Merritt, D.W. (1969) Basement rocks of the Michigan Basin. In Stonehouse, H.B. (ed.) *Annual Field Excursion*. Michigan Basin Geological Society, East Lansing, Michigan, 28–59.

Hinze, W.J., Kellog, R.L., & O'Hara, N.W. (1975) Geophysical studies of basement geology of the southern peninsula of Michigan. *American Association of Petroleum Geologists Bulletin* **59**, 1562–84.

Hogg, A.J.C., Mitchell, A.W., & Young, S. (1996) Predicting well productivity from grain size analysis and logging while drilling. *Petroleum Geoscience* **2**, 1–15.

Horbury, A.D. (1989) The relative roles of tectonism and eustasy in the deposition of the Unswick Limestone Formation in south Cumbria and north Lancashire. In Arthurton, R.A., Gutteridge, P., & Nolan, S.C. (eds.) *The Role of Tectonics in Devonian and Carboniferous Sedimentation in the British Isles*. Occasional Publication No. 6, Yorkshire Geological Society, 153–69.

Horstad, I. & Larter, S.R. (1997) Petroleum migration, alteration, and remigration within Troll Field, Norwegian North Sea. *American Association of Petroleum Geologists Bulletin* **81**, 222–48.

Hovland, M.T. & Judd, J.G. (1988) *Seabed Pockmarks and Seepages: Impact on Geology, Biology and the Marine Environment*. Graham & Trotman, London.

Hubbard, R.J., Edrich, S.P., & Rattey, R.P. (1990) Geological evolution and hydrocarbon habit of the "Arctic Alaska microplate." In Brooks, J. (ed.) *Classic Petroleum Provinces*. Special Publication No. 50, Geological Society, London, 143–88.

Hubbert, M.K. (1953) Entrapment of petroleum under hydrodynamic conditions. *American Association of Petroleum Geologists Bulletin* **37**, 1954–2026.

Hubbert, J.F. (1975) *Inorganic Chemistry: Principles of Structure and Reactivity*. Harper and Row, New York.

Hunt, J.M. (1979) *Petroleum Geochemistry and Geology*. W.H. Freeman, San Francisco.

Hunt, J.M. (1990) Generation and migration of petroleum from abnormally pressured fluid compartments. *American Association of Petroleum Geologists Bulletin* **74**, 1–12.

Hurley, N.F. & Budros, R. (1990) Albion-Scipio and Stoney Point Fields. In Beaumont, E.A. & Foster, N.H. (eds.) *Stratigraphic Traps I*. Treatise of Petroleum Geology Atlas of Oil and Gas Fields, American Association of Petroleum Geologists, Tulsa, Oklahoma, 1–38.

Hutton, J. (1795) *Theory of the Earth with Proofs and Illustrations*. Edinburgh: William Creech.

Jack, I. (1998) *Time Lapse Seismic in Reservoir Management*. Distinguished Instructor Short Course (notes). Society of Exploration Geophysics, Tulsa, Oklahoma.

Jackson, M.P.A. (1995) Retrospective salt tectonics. In Jackson, M.P.A., Roberts, D.G., & Snelson, S. (eds.) *Salt Tectonics, a Global Perspective*. American Association of Petroleum Geologists Memoir **65**, 1–28.

James, K.H. (1990) The Venezuelan hydrocarbon habit. In Brooks, J. (ed.) *Classic Petroleum Provinces*. Special Publication No. 50, Geological Society, London, 9–36.

Johnson, H.D. (1998) Jurassic. In Glennie, K.W. (ed.) *Petroleum Geology of the North Sea: Basic Concepts and Recent Advances*, 4th edn. Blackwell Science, Oxford, 245–93.

Johnson, H.D. & Baldwin, C.T. (1986) Shallow siliciclastic seas. In Reading, H.G. (ed.) *Sedimentary Environments and Facies*, 2nd edn. Blackwell Scientific, Oxford, 229–82.

Jones, N.E. & Stewart, R.C.S. (1997) Reactivation of the Pedernales Field, Venezuela: a sleeping giant awakes (abstract). In Gluyas, J.G. & Daines, S. (eds.) *Geological Society Petroleum Group, Biannual Meeting, Field Reactivation for the 21st Century*, Bath, England, April 1997. Geological Society, Bath.

Karasek, R.M., Vaughn, R.L., & Masuda, T.T. (2003) The Beryl Field, Block 9/13, UK North Sea. In Gluyas, J.G. & Hichens, H. (eds.) *United Kingdom Oil and Gas Fields, Millennium Commemorative Volume*. Geological Society Memoir No. 20 (in press).

Katsube, T.J. & Coyner, K. (1994) Determination of permeability–compaction relationship from interpretation of permeability–stress data in shales from eastern and northern Canada. In *Current Research 1994-D*. Geological Survey of Canada, 169–77.

Kay, S. (2003) Heather Field, Block 2/5, UK North Sea. In Gluyas, J.G. & Hichens, H. (eds.) *United Kingdom Oil and Gas Fields, Millennium Commemorative Volume*. Geological Society Memoir No. 20 (in press).

Kent, P.E. (1985) UK onshore oil exploration, 1930–1964. *Marine and Petroleum Geology* **2**, 56–64.

Ketter, F.J. (1991a) The Esmond, Forbes and Gordon Fields, Blocks 43/8a, 43/13a, 43/15a, 43/20a, UK North Sea. In Abbots, I.L. (ed.) *United Kingdom Oil and Gas Fields, 25 Years Commemorative Volume*. Geological Society Memoir No. 14, 425–32.

Ketter, F.J. (1991b) The Ravenspurn North Field, Blocks 42/30, 43/26a, UK North Sea. In Abbots, I.L. (ed.) *United Kingdom Oil and Gas Fields, 25 Years Commemorative Volume*. Geological Society Memoir No. 14, 459–67.

Kharusi, M.S. (1991) Evaluating the opportunities for horizontal wells in Oman. SPE 023539, Society of Petroleum Engineers.

Knipe, R.J. (1997) Juxtaposition and seal diagrams to help analyze fault seals in hydrocarbon reservoirs. *American Association of Petroleum Geologists Bulletin* **81**, 187–95.

Koster, K., Gabriels, P., Hartung, M., Verbeek, J., Deinum, G., & Staples, R. (2000) Time-lapse seismic surveys in the North Sea and their business impact. *The Leading Edge*, March, 286–93.

Kupecz, J., Gluyas, J.G., & Bloch, S. (1997) *Reservoir Quality Prediction in Sandstones and Carbonates*. American Association of Petroleum Geologists Memoir 69.

Kurkjy, K.A. (1988) Experimental compaction studies of lithic sands, M.Sc. Thesis, Comparative Sedimentary Laboratory Division of Marine Geology and Geophysics, Rosenstiel School of Marine and Atmospheric Sciences, University of Miami, Florida.

Lagazzi, R., Gallagher, B.C., & Villaba, L. (1994) Impacto de la sismica 3D en el modelo estructural del Campo Boqueron. In Repiso, M.E. (ed.) *VII Congreso Venezalano de Geofisica 1994 Memorias*, 409–13.

Larter, S.R. & Aplin, A.C. (1995) Reservoir geochemistry: methods, applications and opportunities. In Cubitt, J.M. & England, W.A. (eds.) *The Geochemistry of Reservoirs*. Special Publication No. 86, Geological Society, London, 5–32.

Leeder, M.R. (1992) Dinantian. In Duff, P.McL.D. & Smith, A.J. (eds.) *Geology of England and Wales*. Geological Society, London, 207–38.

Leeder, M.R. (1999) *Sedimentology and Sedimentary Basins from Turbulence to Tectonics*. Blackwell Science, Oxford.

Lees, G.M. & Cox, P.E. (1937) The geological basis of the present search for oil in Great Britain by the D'Arcy Exploration Company Ltd. *Quarterly Journal of the Geological Society of London* **93**, 156–94.

Leythauser, D. & Poelchau, H.S. (1991) Expulsion of petroleum from type III kerogen source rocks in gaseous solution: modelling of solubility fraction. In England, W.A. & Fleet, A.J. (eds.) *Petroleum Migration*. Special Publication No. 59, Geological Society, London, 33–46.

Lopez, J.A. (1990) Structural styles of growth faults in the US Gulf Coast Basin. In Brooks, J. (ed.) *Classic Petroleum Provinces*. Special Publication No. 50, Geological Society, London, 203–19.

Loucks, R.G. & Anderson, J.H. (1985) Depositional facies, diagenetic terrains, and porosity development in the Lower Ordovician Ellenberger dolomite, Pukkett Field, West Texas. In Roehl, P.O. & Choquette, P.W. (eds.) *Carbonate Petroleum Reservoirs*. Springer-Verlag, New York, 19–37.

Lowell, J.D. (1985) *Structural Styles in Petroleum Exploration*. Oil and Gas Consultants International, Tulsa, Oklahoma.

Ludwig, W.J. & Krasheninnikov, V.A. (1983) *Initial Reports DSDP*, vol. 71. US Government Printing Office, Washington, DC.

Ma, L., Ge, T., Zhao, X., Zie, T., Ge, R., & Dang, Z. (1982) Oil basins and subtle traps in the eastern part of China. In Halbouty, M.T. (ed.) *The Deliberate Search for the Subtle Trap*. American Association of Petroleum Geologists Memoir **32**, 287–315.

McBride, B.C., Weimer, P., & Rowan, M.G. (1998) The effect of allochthonous salt on the petroleum systems of the northern Green Canyon and Ewing Bank (offshore Louisiana), northern Gulf of Mexico. *American Association of Petroleum Geologists Bulletin* **82**, 1083–112.

McClure, N.M. & Brown, A.A. (1992) Miller Field; a subtle Upper Jurassic submarine fan trap in the South Viking Graben, UK Sector North Sea. In Halbouty, M.T. (ed.) *Giant Oil and Gas Fields of the Decade, 1978–1988*. American Association of Petroleum Geologists Memoir **54**, 307–22.

McClure, N.M., Wilkinson, D.W., Frost, D.P., & Geehan, G.W. (1995) Planning extended reach wells in Wytch Farm Field, UK. *Petroleum Geoscience* **1**, 115–27.

McCollough, C.N. & Carver, J.A. (1992) The giant Cañon Limon Field, Llanos Basin, Colombia. In Halbouty, M.T. (ed.) *Giant Oil and Gas Fields of the Decade, 1978–1988*. American Association of Petroleum Geologists Memoir **54**, 175–96.

MacDonald, A.C. & Halland, E.K. (1993) Sedimentology and shale modelling of a sandstone-rich, fluvial reservoir, Statfjord Formation, Statfjord Field, northern North Sea. *American Association of Petroleum Geologists Bulletin* **77**, 1016–40.

McGann, G.J., Green, S.C.H., Harker, S.D., & Romani, R.S. (1991) The Scapa Field, Block 14/19, UK North Sea. In Abbots, I.L. (ed.) *United Kingdom Oil and Gas Fields, 25 Years Commemorative Volume*. Geological Society Memoir No. 14, 369–76.

McHardy, W.J., Wilson, M.J., & Tait, J.M. (1982) Electron microscope and X-ray diffraction studies of filamentous illitic clay from sandstones of the Magnus Field. *Clay Minerals* **17**, 23–9.

Machour, L., Masse, J.-P., Oudin, J.-L., Lambert, B., & Lapointe, P. (1998) Petroleum potential of dysaerobic carbonate source rocks in the intra shelf basin: Lower Cretaceous of Provence, France. *Petroleum Geoscience* **4**, 139–46.

Mackay, A.H. & Tankard, A.J. (1990) Hibernia Oil Field – Canada, Jeanne d'Arc Basin, Grand Banks, offshore Newfoundland. In Beaumont, E.A. & Foster, N.H. (eds.) *Tectonic Fold and Fault Traps 1990*. American Association of Petroleum Geologists Special Publication, 145–75.

Mackenzie, A.S., Price, I., Leythaeuser, D., Müller, P., Radke, M., & Schaefer, R.G. (1987) The expulsion of petroleum from Kimmeridge Clay source-rocks in the area of the Brae Oilfield, UK continental shelf. In Brooks, J. & Glennie, K. (eds.) *Petroleum Geology of North West Europe*. Graham & Trotman, London, 865–77.

McKenzie, D.P. (1978) Some remarks on the development of sedimentary basins. *Earth and Planetary Science Letters* **40**, 25–32.

McQuillan, H. (1973) Small-scale fracture density in Asmari Formation of southwestern Iran and its relation to bed thickness and structural setting. *American Association of Petroleum Geologists Bulletin* **57**, 2367–85.

McQuillin, R., Bacon, M., & Barclay, W. (1979) *An Introduction to Seismic Interpretation*. Graham & Trotman, London.

Mancini, E.A., Mink, R.M., Brearden, B.L., & Wilkerson, R.P. (1985) Norphlet Formation (Upper Jurassic) of southwestern and offshore Alabama: environments of deposition and petroleum geology. *American Association of Petroleum Geologists Bulletin* **69**, 881–98.

Marshall, J.E.A. (1994) The Falkland Islands: a key element in Gondwana palaeogeography. *Tectonics* **13**, 499–514.

Matthews, C.S. & Russell, D.G. (1967) *Pressure Buildup and Flowtests in Wells*. Monograph of the Society of Petroleum Engineers, American Institute of Mining, Metallurgical and Petroleum Engineers, Dallas, Texas.

Mearns, E.W. & McBride, J.J. (1999) Hydrocarbon filling history and reservoir continuity of oil fields evaluated using $^{87}Sr/^{86}Sr$ isotope ratio variations in formation water, with examples from the North Sea. *Petroleum Geoscience* **5**, 17–27.

Melvin, J. & Knight, A.S. (1984) Lithofacies, diagenesis and porosity of the Ivishak Formation, Prudhoe Bay Area, Alaska: part 3. Applications in exploration and production. In Surdam, R. & MacDonald, D.A. (eds.) *Clastic Diagenesis*. American Association of Petroleum Geologists Memoir **37**, 347–65.

Mencher, E., Fichter, J.H., Renz, H.H., et al. (1953) Geology of Venezuela and its oil fields. *American Association of Petroleum Geologists Bulletin* **37**, 690–777.

Mero, W.E. (1991) Point Arguello Field. In Beaumont, E.A. & Foster, N.H. (eds.) *Structural Traps V*. Treatise of Petroleum

Geology Atlas of Oil and Gas Fields, American Association of Petroleum Geologists, Tulsa, Oklahoma, 27–58.

Michelsen, O., Thomsen, E., Danielsen, M., Heilmann-Clausen, C., Jordt, H., & Laursen, G.V. (1998) Cenozoic sequence stratigraphy in the eastern North Sea. In de Graciansky, P.C., Hardenbol, J., Jacquin, T., & Vail, P.R. (eds.) *Mesozoic and Cenozoic Sequence Stratigraphy of European Basins*. Special Publication No. 60, Society for Economic Paleontologists and Mineralogists, Tulsa, Oklahoma, 91–118.

Miller, D.J. & Eriksson, K.A. (2000) Sequence stratigraphy of Upper Mississippian strata in the central Appalachians: a record of glacioeustasy and tectonoeustasy in a foreland basin setting. *American Association of Petroleum Geologists Bulletin* 84, 210–33.

Min, Y. (1980) Hydrocarbon accumulations in sedimentary basins of non-marine facies in China. In Mason, J.F. (ed.) *Petroleum Geology in China*. PennWell, Tulsa, Oklahoma, 1–4.

Mitchener, B.C., Lawrence, D.A., Partington, M.A., Bowman, M.B.J., & Gluyas, J.G. (1992) Brent Group: sequence stratigraphy and regional implications. In Morton, A.C., Haszeldine, R.S., Giles, M.R., & Brown, S. (eds.) *Geology of the Brent Group*. Special Publication No. 61, Geological Society, London, 45–80.

Mitchum, R.M. Jr., Vail, P.E., & Thompson, S. III (1977) The depositional sequence as a basic unit for stratigraphic analysis. In Payton, C.E. (ed.) *Seismic Stratigraphy—Applications to Hydrocarbon Exploration*. American Association of Petroleum Geologists Memoir 26, 53–62.

Moore, C.H. & Heydari, D. (1991) Burial diagenesis and hydrocarbon migration in platform limestones: a conceptual model based on the Upper Jurassic of the Gulf Coast of the USA. In Horbury, A.D. & Robinson, A.G. (eds.) *Diagenesis and Basin Development*. Studies in Geology No. 36, American Association of Petroleum Geologists, Tulsa, Oklahoma, 213–30.

Moreton, R. (1995) *Tales from Early UK Oil Exploration 1960–1979*. 30th Anniversary Book, Petroleum Exploration Society of Great Britain.

Morgan, W.A. (1985) Silurian reservoirs in upward-shoaling cycles of the Hunton Group, Mt. Everette and southwest Reeding fields, Kingfisher County, Oklahoma. In Roehl, P.O. & Choquette, P.W. (eds.) *Carbonate Petroleum Reservoirs*. Springer-Verlag, New York, 107–20.

Morton, A.C. (1992) Provenance of Brent Group sediments: heavy mineral constraints. In Morton, A.C., Haszeldine, R.S., Giles, M.R., & Brown, S. (eds.) *Geology of the Brent Group*. Special Publication No. 61, Geological Society, London, 227–44.

Morton, A.C., Haszeldine, R.S., Giles, M.R., & Brown, S. (1992) *Geology of the Brent Group*. Special Publication No. 61, Geological Society, London.

Naeser, N.D. & McCulloh, T.H. (eds.) (1989) *Thermal History of Sedimentary Basins: Methods and Case Histories*. Springer-Verlag, New York.

Neal, J.E. (1996) A summary of Palaeogene sequence stratigraphy in northwest Europe and the North Sea. In Know, R.W.O'B., Corfield, R.M., & Dunay, R.E. (eds.) *Correlation of the Early Palaeogene in Northwest Europe*. Special Publication No. 101, Geological Society, London, 15–42.

Nedkvitne, T., Karlsen, D.A., Bjørlykke, K., & Larter, S.R. (1993) Relationship between reservoir diagenetic evolution and petroleum emplacement in the Ula Field, North Sea. *Marine and Petroleum Geology* 10, 185–96.

Newman, M.St.J., Reeder, M.L., Woodruff, A.H.W., & Hatton, R. (1993) The geology of the Gryphon Oilfield. In Parker, J.R. (ed.) *Petroleum Geology of Northwest Europe: Proceedings of the 4th Conference*. Geological Society, London, 123–34.

Nicholls, C.A., Boom, W., Geel, J., Al Khodori, S., & Al Lawati, M. (1999) Fracture modeling as a key to waterflood development of the Fahud Field Natih-E reservoir. SPE 53211, Society of Petroleum Engineers.

North, F.K. (1985) *Petroleum Geology*. Unwin Hyman, Boston.

Oakman, C.D., Martin, J.H., & Corbett, P.W.M. (eds.) (1997) *Cores from the Northwest European Hydrocarbon Province: An Illustration of Geological Applications from Exploration to Development*. Geological Society, London.

O'Conner, R.B. Jr., Castle, R.A., & Nelson, D.A. (1993) Future oil and gas potential in southern Caspian basin. *Oil and Gas Journal*, May, 117–25.

O'Neill, N. (1988) Fahud Field review: a switch from water to gas injection. *Journal of Petroleum Technology*, May, 609–18.

Osborne, M.J. & Swarbrick, R.E. (1997) Mechanisms for generating overpressure in sedimentary basins: a re-evaluation. *American Association of Petroleum Geologists Bulletin* 81, 1023–41.

Oxtoby, N.H., Mitchell, A.W., & Gluyas, J.G. (1995) The filling and emptying of the Ula oilfield (Norwegian North Sea). In Cubitt, J.M. & England, W.A. (eds.) *The Geochemistry of Reservoirs*. Special Publication No. 86, Geological Society, London, 141–58.

Palmer, A.N. (1995) Geochemical models for the origin of macroscopic solution porosity in carbonate rocks. In Budd, D.A., Saller, A.H., & Harris, P.M. (eds.) *Unconformities and Porosity in Carbonate Strata*. American Association of Petroleum Geologists Memoir 63, 77–101.

Payne, M.L., Cocking, D.A., & Hatch, A. (1994) Critical techniques for success in extended reach drilling. SPE 28293, presented at Society of Petroleum Engineers 69th Annual Technical Conference and Exhibition, New Orleans, Louisiana, September 25–8.

Payton, C.A. (1977) *Seismic Stratigraphy—Applications to Hydrocarbon Exploration*. American Association of Petroleum Geologists Memoir 26.

Penny, B. (1991) The Heather Field, Block 2/5, UK North Sea. In Abbots, I.L. (ed.) *United Kingdom Oil and Gas Fields, 25 Years Commemorative Volume*. Geological Society Memoir No. 14, 127–34.

Pepper, A.S. (1991) Estimating the petroleum expulsion behavior of source rocks: a novel quantitative approach. In England, W.A. & Fleet, A.J. (eds.) *Petroleum Migration*. Special Publication No. 59, Geological Society, London, 9–32.

Peridon, P. (1983) Dynamics of oil and gas accumulations. *Bulletin des Centres de Recherches Exploration–Production, Elf Aquitaine* 5, Pau, France, 53–78.

Persad, K. & Persad, M. (1993) *The Petroleum Encyclopedia of Trinidad and Tobago*, 1993 edn. Krishna Persad and Associates, Trinidad.

Phipps, G.G. (1989) Exploring for dolomitized Slave Point carbonates in northeastern British Columbia. *Geophysics* 54, 806–14.

Plumley, W.J. (1980) Abnormally high fluid pressure: survey of some basic principles. *American Association of Petroleum Geologists Bulletin* 64, 414–30.

Powers, M.C. (1967) Fluid release mechanisms in compacting marine mudrocks and their importance in oil exploration. *American Association of Petroleum Geologists Bulletin* 51, 1240–54.

Price, L.C. & Wenger, L.M. (1992) The influence of pressure on petroleum generation and maturation as suggested by aqueous pyrolysis. *Organic Geochemistry* 19, 141–59.

Primmer, T.J., Cade, C.A., Evans, I.J., et al. (1997) Reservoir quality prediction in sandstones during exploration: establishing cementation styles. In Kupecz, J., Gluyas, J.G., & Bloch, S. (eds.) *Reservoir Quality Prediction in Sandstones and Carbonates*. American Association of Petroleum Geologists Memoir 69.

Prodruski, J.A., Barclay, J.E., Hamblin, A.P., et al. (1988) Conventional oil resources of western Canada (light and medium). Geological Survey of Canada, Paper 87–26; see "Devonian system," 20–55.

Pulham, A.J. (1994) The Cusiana Field, Llanos Basin, eastern Colombia: high resolution sequence stratigraphy applied to late Palaeocene – Early Oligocene, estuarine, coastal plain and alluvial clastic reservoirs. In Johnson, S.D. (ed.) *High Resolution Sequence Stratigraphy: Innovations and Applications*. University of Liverpool, Department of Earth Sciences, 63–8.

Purdy, E.G. (1963) Recent calcium carbonate facies of the Great Bahama Bank II. Sedimentary facies. *Journal of Geology* 71, 472–9.

Rahmanian, V.D., Moore, P.S., Mudge, W.J., & Spring, D.E. (1990) Sequence stratigraphy and habitat of hydrocarbons, Gippsland Basin, Australia. In Brooks, J. (ed.) *Classic Petroleum Provinces*. Special Publication No. 50, Geological Society, London, 525–41.

Ramsbottom, W.H.C., Goosens, R.F., Smith, E.G., & Calver, M.A. (1974) Carboniferous. In Rayner, D.H. & Hemmingway, J.E. (eds.) *The Geology and Mineral Resources of Yorkshire*. W.S. Maney and Sons, Leeds, 45–114.

Rattey, R.P. & Hayward, A.B. (1993) Sequence stratigraphy of a failed rift system: the Middle Jurassic to Early Cretaceous basin evolution of the Central and Northern North Sea. In Parker, J.R. (ed.) *Petroleum Geology of Northwest Europe: Proceedings of the 4th Conference*. Geological Society, London, 215–50.

Read, J.F. & Horbury, A.D. (1993) Eustasy and tectonic controls on porosity evolution beneath sequence-bounding unconformities and parasequence disconformities on carbonate platforms. In Horbury, A.D. & Robinson, A.G. (eds.) *Diagenesis and Basin Development*. Studies in Geology No. 36, American Association of Petroleum Geologists, Tulsa, Oklahoma, 155–98.

Reading, H.G. (1986) *Sedimentary Environments and Facies*, 2nd edn. Blackwell Scientific, Oxford.

Reynolds, A.D. (1994) Sequence stratigraphy and the dimensions of paralic sandstone bodies. In Johnson, S.D. (ed.) *High Resolution Sequence Stratigraphy: Innovations and Applications*. University of Liverpool, Department of Earth Sciences, 69–72.

Richards, P.C. (1995) *Oil and the Falkland Islands: An Introduction to the October 1995 Offshore Licensing Round*. British Geological Survey, Edinburgh, for the Falkland Islands Government.

Richards, P.C. & Fannin, N.G.T. (1997) Geology of the North Falkland Basin. *Journal of Petroleum Geology* 20, 165–83.

Richards, P.C., Gatliff, R.W., Quinn, M.F., & Fannin, N.G.T. (1996) Petroleum potential of the Falkland Islands offshore area. *Journal of Petroleum Geology* 19, 161–82.

Rider, M.H. (1986) *The Geological Interpretation of Well Logs*. Blackie/Halsted Press, Glasgow.

Rittenhouse, G. (1972) Stratigraphic-trap classification. In King, R.E. (ed.) *Stratigraphic Oil and Gas Fields: Classification, Exploration Methods, and Case Histories*. American Association of Petroleum Geologists Memoir 16, 14–28.

Roberts, D.G., Thompson, M., Mitchener, B., Hossack, J., Carmichael, S., & Bjørnseth, H.M. (1999) Palaeozoic to Tertiary rift and basin dynamics: mid-Norway to the Bay of Biscay – a new context for hydrocarbon prospectivity in the deep water frontier. In Fleet, A.J. & Boldy, S.A.R. (eds.) *Petroleum Geology of Northwest Europe: Proceedings of the 5th Conference*. Bath Geological Society, Bath, 7–40.

Robertson, G. (1997) Beryl Field: Late Triassic to Cretaceous stratigraphy. In Oakman, C.D., Martin, J.H., & Corbett, P.W.M. (eds.) *Cores from the Northwest European Hydrocarbon Province, an Illustration of Geological Applications from Exploration to Development*. Geological Society, London, 97–108.

Robinson, A.G. & Gluyas, J.G. (1992a) Model calculations of sandstone porosity loss due to compaction and quartz cementation. *Marine and Petroleum Geology* 9, 319–23.

Robinson, A.G. & Gluyas, J.G. (1992b) Duration of quartz cementation in sandstones, North Sea and Haltenbanken. *Marine and Petroleum Geology* 9, 324–7.

Robson, D. (1991) The Argyll, Duncan and Innes Fields, Blocks 30/24 and 30/25a, UK North Sea. In Abbots, I.L. (ed.) *United Kingdom Oil and Gas Fields, 25 Years Commemorative Volume*. Geological Society Memoir No. 14, 219–25.

Rogers, S.J. (1990) Monterey, San Joaquin Basin, onshore California. In Dromgoole, P. & Pepper, A. (eds.) *Petroleum Geoscience Handbook*. Internal BP publication.

Rønnevik, H., Eggen, S., & Vollset, J. (1983) Exploration of the Norwegian Shelf. In Brooks, J. (ed.) *Petroleum Geochemistry and Exploration of Europe*. Special Publication No. 12, Geological Society, London, 71–94.

Rose, P.R. (1992) Expected value and chance of success and its use in petroleum exploration. In *Development Geology Reference Manual*. Special Publication ME 10, American Association of Petroleum Geologists, Tulsa, Oklahoma, 30–4.

Royden, L.H. & Keen, C.E. (1980) Rifting processes and thermal evolution of the continental margin of eastern Canada determined from subsidence curves. *Earth and Planetary Science Letters* 53, 343–61.

Rudwick, M.J.S. (1985) *The Great Devonian Controversy*. The University of Chicago Press, Chicago.

Russo, A.J. (1993) Tecnologia aplicada para el control de arena en yacimientos profundos – campo Boquerón SVIP 041, *X Jornadas Tecnicas de Petroleo*, Puerto la Cruz.

Safarudin, X. & Manulang, M.H. (1989) Trapping mechanism in Mutiara Field, Kutei Basin, East Kalimantan. In *Proceedings of the Indonesian Petroleum Association 8th Annual Convention* 1, 31–54.

Sales, J.K. (1993) Closure versus seal capacity – a fundamental control on the distribution of oil and gas. In Dore, A.G. (ed.) *Basin Modelling: Advances and Applications*. NPF Special Publication 3, 399–414.

Salvador, A. & Leon, H.J. (1992) Quiriquire Field – Venezuela. Eastern Venezuela (Maturin) Basin. In Beaumont, E.A. & Foster, N.H. (eds.) *Stratigraphic Traps III*. Treatise of Petroleum Geology Atlas of Oil and Gas Fields, American Association of Petroleum Geologists, Tulsa, Oklahoma, 313–32.

Sandberg, P.A. (1983) An oscillating trend in Phanerozoic non-skeletal carbonate mineralogy. *Nature* 305, 19–22.

Saunders, J.B. & Bolli, H.M. (1985) Trinidad's contribution to world biostratigraphy. In Carr-Brown, B. & Christian, J.T. (eds.) *Transactions of the Fourth Latin American Geological Conference*, July 7–15, 1979, Port-of-Spain, Trinidad and Tobago, 781–95. Geological Society of Trinidad and Tobago, Port-of-Spain.

Schmidt, V. & McDonald, D.A. (1979) The role of secondary porosity in the course of sandstone diagenesis. In Scholle, P.A. & Schluger, P.R. (eds.) *Aspects of Diagenesis*. Special Publication No. 26, Society for Economic Paleontologists and Mineralogists, Tulsa, Oklahoma, 175–207.

Schneeflock, R. (1978) Permeability traps in Gatchell (Eocene) Sand of California. *American Association of Petroleum Geologists Bulletin* 62, 848–53.

Sealy, E.C. & Ramlackhansingh, A. (1985) The geology of Trinidad – Tesoro's Palo Seco Field including South Erin and Central Los Bajos. In Carr-Brown, B. & Christian, J.T. (eds.) *Transactions of the Fourth Latin American Geological Conference*, July 7–15, 1979, Port-of-Spain, Trinidad and Tobago, 796–802. Geological Society of Trinidad and Tobago, Port-of-Spain.

Secord, J.A. (1986) *Controversy in Victorian Geology: the Cambrian–Silurian Dispute*. Princeton University Press, Princeton, New Jersey.

Selley, R.C. (1978) Porosity gradients in North Sea oil-bearing sandstones. *Journal of the Geological Society of London* 135, 119–31.

Shepherd, M. (1991) The Magnus Field, Blocks 211/7, 211/12a, UK North Sea. In Abbots, I.L. (ed.) *United Kingdom Oil and Gas Fields, 25 Years Commemorative Volume*. Geological Society Memoir No. 14, 153–8.

Sibson, R.H. (1996) Structural permeability of fluid-driven fault-fracture meshes. *Journal of Structural Geology* 18, 1031–42.

Sleep, N.H. & Sloss, L.L. (1978) A deep borehole in the Michigan Basin. *Journal of Geophysical Research* 83, 5815–19.

Sleep, N.H. & Snell, N.S. (1976) Thermal contraction and flexure of midcontinent and Atlantic marginal basins. *Geophysical Journal of the Royal Astronomical Society* 45, 125–54.

Sloss, L.L. (1963) Sequences in the cratonic interior of North America. *Bulletin of the Geological Society of America* 74, 93–114.

Small, J.S., Hamilton, D.L., & Habesch, S. (1992) Experimental simulation of clay precipitation within reservoir sandstone 1: techniques and examples. *Journal of Sedimentary Petrology* 62, 508–19.

Smalley, P.C., Dodd, T.A., Stockden, I.L., Råheim, A., & Mearns, E.W. (1995) Compositional heterogeneities in oilfield formation waters: identifying them, using them. In Cubitt, J.M. & England, W.A. (eds.) *The Geochemistry of Reservoirs*. Special Publication No. 86, Geological Society, London, 59–70.

Soloman, S.T., Ross, K.C., Burton, R.C., & Wellhorn, J.E. (1994) A multidisciplined approach to designing targets for horizontal wells. *Journal of Petroleum Technology*, February, 143–9.

Spencer, A.M. & Eldholm, O. (1993) Atlantic margin exploration: Cretaceous–Tertiary evolution, basin development and petroleum geology, introduction and review. In Parker, J.R. (ed.) *Petroleum Geology of Northwest Europe: Proceedings of the 4th Conference*. Geological Society, London, 899.

Spencer, A.M., et al. (1987) *Geology of the Norwegian Oil and Gas Fields*. Graham & Trotman, London.

Stamp, L.D. (1923) *An Introduction to Stratigraphy, British Isles*. Thomas Murby, London.

Stanistreet, I.G. & McCarthy, T.S. (1993) The Okavango Fan and the classification of subaerial fan systems. *Sedimentary Geology* 85, 115–33.

Steckler, M.S. (1985) Uplift and extension at the Gulf of Suez: indications of induced mantle convection. *Nature* 317, 135–9.

Steele, R.J. & Aasheim, S.M. (1978) Alluvian sand deposition in a rapidly subsiding basin (Devonian, Norway). In Maill, A.D. (ed.) *Fluvial Sedimentology*. Memoirs of the Canadian Society of Petroleum Geologists 5, 385–412.

Stephen, K.J., Underhill, J.R., Partington, M.A., & Hedley, R.J. (1993) The genetic sequence stratigraphy of the Hettangian to Oxfordian succession, Inner Moray Firth. In Parker, J.R. (ed.) *Petroleum Geology of Northwest Europe: Proceedings of the 4th Conference*. Geological Society, London, 485–505.

Stevenson, I.P. & Gaunt, G.D. (1971) *Geology of the Country Around Chapel en le Frith*. Institute of Geological Sciences, Memoirs of the Geological Survey of Great Britain, Her Majesty's Stationary Office, London.

Stoneley, R. (1990) The Middle East Basin: a summary overview. In Brooks, J. (ed.) *Classic Petroleum Provinces*. Special Publication No. 50, Geological Society, London, 293–8.

Stow, D.A.V. (1986) Deep clastic seas. In Reading, H.G. (ed.) *Sedimentary Environments and Facies*, 2nd edn. Blackwell Scientific, Oxford, 399–444.

Struijk, A.P. & Green, R.T. (1991) The Brent Field, Block 211/29, UK North Sea. In Abbots, I.L. (ed.) *United Kingdom Oil and Gas Fields, 25 Years Commemorative Volume*. Geological Society Memoir No. 14, 63–72.

Sturrock, S.J. (1996) Biostratigraphy. In Emery, D. & Myers, K.J. (eds.) *Sequence Stratigraphy*. Blackwell Science, Oxford, 89–110.

Sun, S.Q. (1995) Dolomite reservoirs: porosity evolution and reservoir characteristics. *American Association of Petroleum Geologists Bulletin* 79, 186–204.

Suppe, J. (1985) *Principles of Structural Geology*. Prentice-Hall, Englewood Cliffs, New Jersey.

Swarbrick, R.E. & Osborne, M.J. (1996) The nature and diversity of pressure transition zones. *Petroleum Geoscience* 2, 111–16.

Swarbrick, R.E. & Osborne, M.J. (1998) Mechanisms that generate abnormal pressures: an overview. In Law, B.E., Ulmishek, G.F., & Slavin, V.I. (eds.), *Abnormal Pressures in Hydrocarbon Environments*. American association of Petroleum Geologists Memoir 70, 13–14.

Swarbrick, R.E., & Osbourne, M.J., & Yardley, G.S. (2002) Comparison of overpressure magnitude resulting from the main generating mechanisms. In Huffman, A.R. & Bowers, G.L. (eds.), *Pressure Regimes in Sedimentary Basins and their Prediction*. American Association of Petroleum Geologists Memoir 76, 1–12.

Szabo, D.J. & Meyers, K.O. (1993) Prudhoe Bay: Development history and future potential. SPE 26053, Society of Petroleum Engineers.

Taylor, G.K. & Shaw, J. (1989) The Falkland Islands: new palaeomagnetic data and their origin as a displaced terrane from southern Africa. In Hillhouse, J.W. (ed.) *Deep Structure and Past Kinematics of Accreted Terranes*. Geophysical Monograph 50, IUGG Volume 5, 59–72.

Taylor, J.C.M. (1981) Zechstein facies and petroleum prospects in the central and northern North Sea. In Illing, L.V. & Hobson, G.D. (eds.) *Petroleum Geology of Northwest Europe: Proceedings of the 2nd Conference*. Heyden, London, on behalf of the Institute of Petroleum, 176–85.

Taylor, J.C.M. (1986) Late Permian – Zechstein. In Glennie, K.W. (ed.) *Introduction to the Petroleum Geology of the North Sea*, 2nd edn. Blackwell Scientific, Oxford, 87–111.

Tearpock, D.J. & Bischke, R.E. (1991) *Applied Subsurface Geological Mapping*. Prentice-Hall, Englewood Cliffs, New Jersey.

Telford, W.M., Geldart, L.P., Sheriff, R.E., & Keys, D.A. (1976) *Applied Geophysics*. Cambridge University Press, Cambridge.

Terhen, J.M. (1999) The Natih petroleum system in North Oman. *Geoarabia* 4(2), 157–80.

Thompson, K.F.M. (1988) Gas condensate migration and oil fractionation in deltaic systems. *Marine and Petroleum Geology* 5, 237–46.

Tiratsoo, E.N. (1984) *Oilfields of the World*. Scientific Press, Beaconsfield, UK.

Tissot, B.P., Pelet, R., & Ungerer, Ph. (1987) Thermal history of sedimentary basins, maturation indices, and kinetics of oil and gas generation. *American Association of Petroleum Geologists Bulletin* 71, 1445–66.

Todd, S.P., Dunn, M.E., & Barwise, A.J.G. (1997) Characterising petroleum charge in the Tertiary of SE Asia. In Fraser, A.J., Mathews, S.J., & Murphy, R.W. (eds.) *Petroleum Geology of SE Asia*. Special Publication No. 126, Geological Society, London, 25–47.

Tollas, J.M. & McKinney, A. (1988) Brent field 3-D reservoir simulation. SPE 18306, Society of Petroleum Engineers.

Tong, X.G. & Huang, Z. (1991) Buried hill discoveries of the Damintun Depression in North China. *American Association of Petroleum Geologists Bulletin* 75, 780–94.

Tonkin, P.C. & Fraser, A.R. (1991) The Balmoral Field, Block 16/21, UK North Sea. In Abbots, I.L. (ed.) *United Kingdom Oil and Gas Fields, 25 Years Commemorative Volume*. Geological Society Memoir No. 14, 237–44.

Trewin, N.H. & Bramwell, G.M. (1991) The Auk Field, Block 30/16, UK North Sea. In Abbots, I.L. (ed.) *United Kingdom Oil and Gas Fields, 25 Years Commemorative Volume*. Geological Society Memoir No. 14, 227–36.

Tschopp, R.H. (1967) Development of the Fahud field. In *Proceedings of the 7th World Petroleum Congress* 2, 243–50.

Tucker, M.E. & Wright, V.P. (1990) *Carbonate Sedimentology*. Blackwell Scientific, Oxford.

Turner, P.J. (1993) Clyde: reappraisal of a producing field. In Parker, J.R. (ed.) *Petroleum Geology of Northwest Europe: Proceedings of the 4th Conference*. Geological Society, London, 1503–12.

Tyler, K., Henriquez, A., & Svanes, T. (1994) Modeling heterogeneities in fluvial domains: a review of the influence on production profiles. In Yarus, J.M. & Chambers, R.L. (eds.) *Stochastic Modeling and Geostatistics*. American Association of Petroleum Geologists, Tulsa, Oklahoma, 77–90.

Tyrrell, W.W. & Christian, H.E. Jr. (1992) Exploration history of the Liuhua 11–1 Field, Pearl River Mouth Basin, China. *American Association of Petroleum Geologists Bulletin* 76, 1209–23.

Underhill, J.R. & Partington, M.A. (1993) Jurassic thermal doming and deflation in the North Sea: implications of the sequence stratigraphic evidence. In Parker, J.R. (ed.) *Petroleum Geology of Northwest Europe: Proceedings of the 4th Conference*. Geological Society, London, 377–46.

Vail, P.R., Mitchum, R.M., & Thompson III, S. (1977) Seismic stratigraphy and global changes of sea level, part 4: global cycles of relative changes of sea level. In Payton, C.A. (ed.) *Seismic Stratigraphy – Applications to Hydrocarbon Exploration*. American Association of Petroleum Geologists Memoir 26, 83–97.

Van Ditzhuijzen, P.J.D. & Sandor, R.K.J. (1995) *NMR Logging, the New Measurement*. Shell International Petroleum, Maatschappu.

Van Krevelen, D.W. (1961) *Coal*. Elsevier, New York.

van Vessem, E.J. & Gan, T.L. (1991) The Ninian Field, Blocks 3/3 and 3/8, UK North Sea. In Abbots, I.L. (ed.) *United Kingdom Oil and Gas Fields, 25 Years Commemorative Volume*. Geological Society Memoir No. 14, 175–82.

Van Wagoner, J.C., Mitchum, R.M., Campion, K.M., & Rahmanian, V.D. (1990) *Siliciclastic Sequence Stratigraphy in Well Logs, Cores, and Outcrops: Correlation of Time and Facies*. American Association of Petroleum Geologists Methods in Exploration Series, No. 7.

Vesic, A.S. & Clough, G.W. (1968) Behaviour of granular material under high stresses. *Journal of the Soil Mechanics Foundation Division* 94, 661–88.

Walderhaug, O. (1990) A fluid inclusion study of quartz-cemented sandstones from offshore mid-Norway – possible evidence for continued quartz cementation during oil emplacement. *Journal of Sedimentary Petrology* 60(2), 203–10.

Walter, L.M. (1985) Relative reactivity of skeletal carbonates during dissolution: implications for diagenesis. In Schneiderman, N. & Harris, P.M. (eds.) *Carbonate Cements*. Special Publication No. 36, Society for Economic Paleontologists and Mineralogists, Tulsa, Oklahoma, 3–16.

Walther, J. (1894) *Einleitung in die Geologie als Historische Wissenschaft*, Bd. 3. *Lithogenesis der Gegenwart*. Fischer-Verlag, Jena, 535–1055.

Waples, D.W. (1998) Basin modelling: how well are we doing? In Duppenbecker, S.J. & Iliffe, J.E. (eds.) *Basin Modelling: Practice and Progress*. Special Publication No. 141, Geological Society, London, 1–14.

Warren, E.A. & Smalley, P.C. (1994) *North Sea Formation Water Atlas*. Geological Society Memoir No. 15.

Watson, H.J. (1982) Casablanca Field offshore Spain, a palaeogeomorphic trap. In Halbouty, M.T. (ed.) *The Deliberate Search for the Subtle Trap*. American Association of Petroleum Geologists Memoir 32, 237–50.

Weber, K.J. & Daukoru, E. (1988) Petroleum geology of the Niger Delta. In Foster, N.H. & Beaumont, E.A. (eds.) *Traps and Seals I, Structural/Fault-Seal and Hydrodynamic Traps*. Treatise Of Petroleum Geology Reprint Series No. 6, American Association of Petroleum Geologists, Tulsa, Oklahoma, 61–80.

Weimer, P. (1990) Sequence stratigraphy, facies geometry and depositional history of the Mississippi Fan. *American Association of Petroleum Geologists Bulletin* 74, 425–53.

Western Atlas (1982) *Interpretive Methods for Production Well Logs*.

Whitaker, M.F., Giles, M.R., & Cannon, S.J.C. (1992) Palynological review of the Brent Group, UK sector, North Sea. In Morton, A.C., Haszeldine, R.S., Giles, M.R., & Brown, S. (eds.) *Geology of the Brent Group*. Special Publication No. 61, Geological Society, London, 169–202.

Whyatt, M., Bowen, J.M., & Rhodes, D.N. (1991) Nelson – successful application of a developemnt geoseismic model in North Sea exploration. *First Break* 9, 265–80.

Williams, G. (1992) Palynology as a palaeoenvironmental indicator in the Brent Group, northern North Sea. In Morton, A.C., Haszeldine, R.S., Giles, M.R., & Brown, S. (eds.) *Geology of the Brent Group*. Special Publication No. 61, Geological Society, London, 203–12.

Williams, G.D. & Stelck, C.R. (1975) Speculations on the Cretaceous palaeogeography of North America. *Special Papers of the Geological Association of Canada* 13, 1–20.

Willis, D.G. (1988) Entrapment of petroleum. In Foster, N.H. & Beaumont, E.A. (eds.) *Traps and Seals I, Structural/Fault-Seal and Hydrodynamic Traps*. Treatise Of Petroleum Geology Reprint Series No. 6, American Association of Petroleum Geologists, Tulsa, Oklahoma, 419–86.

Wills, J.M. (1991) The Forties Field, Blocks 21/10, 22/6a, UK North Sea. In Abbots, I.L. (ed.) *United Kingdom Oil and Gas Fields, 25 Years Commemorative Volume*. Geological Society Memoir No. 14, 301–8.

Wilson, A.O. (1985) Depositional and diagenetic facies in the Jurassic Arab-C and -D reservoirs, Qatif Field, Saudi Arabia. In Roehl, P.O. & Choquette, P.W. (eds.) *Carbonate Petroleum Reservoirs*. New York, Springer-Verlag, 319–40.

Wilson, H.H. (1977) "Frozen-in" hydrocarbon accumulations or diagenetic traps – exploration targets. *American Association of Petroleum Geologists Bulletin* 61, 313–21.

Wilson, I.G. (1973) Ergs. *Sedimentary Geology* 10, 77–106.

Wilson, J.T. (1966) Did the Atlantic close then re-open? *Nature* 211, 676–81.

Winstanley, A.M. (1993) A review of the Triassic play in the Roer Valley Graben, SE onshore Netherlands. In Parker, J.R. (ed.) *Petroleum Geology of Northwest Europe: Proceedings of the 4th Conference*. Geological Society, London, 595–607.

Worden, R.H., Smalley, P.C., & Oxtoby, N.H. (1995) Gas souring by thermochemical sulphate reduction at 140°C. *American Association of Petroleum Geologists Bulletin* 79, 854–63.

Wyllie, B.K.N. (1926) Anglo-Persian Oil Company report on Kuwait. In Chisholm, A.H.T. (ed.) *The First Kuwait Oil Concession*. Frank Cass, London, Note 31, 111–18.

Yaliz, A. (1991) The Crawford Field, Block 9/28a, UK North Sea. In Abbots, I.L. (ed.) *United Kingdom Oil and Gas Fields, 25 Years Commemorative Volume*. Geological Society Memoir No. 14, 287–94.

Yaliz, A.M. (1997) The Douglas Oil Field. In Meadows, N.S., Trueblood, S.P., Hardman, M., & Cowan, G. (eds.) *Petroleum Geology of the Irish Sea and Adjacent Areas*. Special Publication No. 124, Geological Society, London, 399–416.

Yaliz, A. & Chapman, T. (2003) The Lennox Field, Block 110/15, UK Irish Sea. In Gluyas, J.G. & Hichens, H. (eds.) *United Kingdom Oil and Gas Fields, Millennium Commemorative Volume*. Geological Society Memoir No. 20 (in press).

Yaliz, A. & McKim, N. (2003) The Douglas Field, Block 110/13b, UK Irish Sea. In Gluyas, J.G. & Hichens, H. (eds.) *United Kingdom Oil and Gas Fields, Millennium Commemorative Volume*. Geological Society Memoir No. 20 (in press).

Yergin, D. (1991) *The Prize: The Epic Quest for Oil, Money, and Power*. Simon & Schuster, London.

Zenger, D.H., Dunham, J.B., & Ethington, H. (eds.) (1980) *Concepts and Models of Dolomitization*. Special Publication No. 28, Society for Economic Paleontologists and Mineralogists, Tulsa, Oklamhoma.

Zhai, G. & Zha, Q. (1982) Buried-hill oil and gas pools in the North China Basin. In Halbouty, M.T. (ed.) *The Deliberate Search for the Subtle Trap*. American Association of Petroleum Geologists Memoir 32, 317–36.

Zubkov, M. Yu. & Mormyshev, V.V. (1987) Correlation and formation conditions in the Bazhenov Suite in the Salym deposit. *Lithology and Mineral Resources* 22, 1167–74.

INDEX

Page numbers in *italics* refer to figures; those in **bold** to tables.